KV-201-156

WITHDRAWN
FROM
UNIVERSITIES
AT
MEDWAY
LIBRARY

Bioprocess Technology

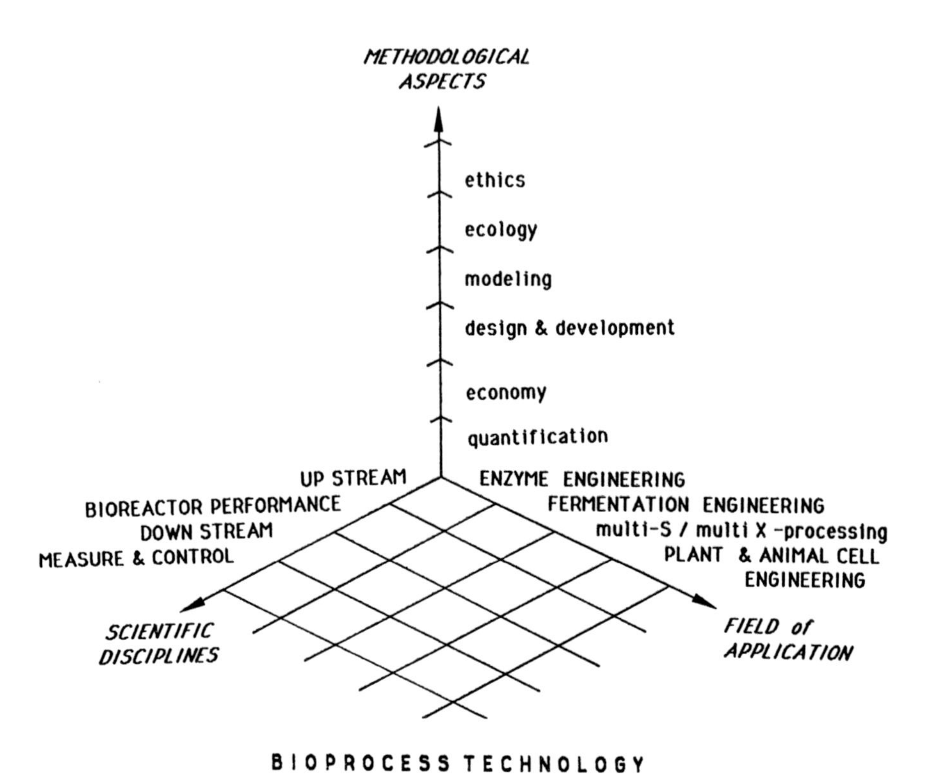

Engineering Disciplines in Biotechnologies
Health care (Pharma)
Agro & Food
Environmental
Industrial

Anton Moser

Bioprocess Technology

Kinetics and Reactors

Revised and Expanded Translation

Translated by Philip Manor

With 279 Illustrations

Springer-Verlag
New York Wien

Professor Dr. ANTON MOSER
Institut für Biotechnologie, Mikrobiologie und Abfalltechnologie,
Technische Universität Graz, A-8010 Graz, Austria

Translator
Dr. PHILIP MANOR

Library of Congress Cataloging-in-Publication Data
Moser, Anton, Dipl.-Ing. Dr. techn.
Bioprocess technology.
Rev. and translated from the German ed.:
Bioprozesstechnik: Berechnungsgrundlagen der Reaktionstechnik biokatalytischer Prozesse.
Bibliography: p.
Includes index.
1. Biochemical engineering. I. Title.
TP248.3.M6713 1988 660'.63 87-26590

Printed on acid-free paper.

Revised and expanded translation from the German edition
Bioprozesstechnik: Berechnungsgrundlagen der Reaktionstechnik biokatalytischer Prozesse.

ISBN-13:978-1-4613-8750-3

Softcover reprint of the hardcover 1st edition 1988

Typeset by Asco Trade Typesetting Ltd., Hong Kong.

9 8 7 6 5 4 3 2 1

ISBN-13:978-1-4613-8750-3 e-ISBN-13:978-1-4613-87948-0
DOI: 10.1007/978-1-4613-87948-0

Preface

This book is based on a 1981 German language edition published by Springer-Verlag, Vienna, under the title *Bioprozesstechnik*. Philip Manor has done the translation, for which I am deeply grateful.

This book differs from the German edition in many ways besides language. It is substantially enlargened and updated, and examples of computer simulations have been added together with other appendices to make the work both more comprehensive and more practical.

This book is the result of over 15 years of experience in teaching and research. It stems from lectures that I began in 1970 at the Technical University of Graz, Austria, and continued at the University of Western Ontario in London, Canada, 1980; at the Free University of Brussels, 1981; at Chalmers Technical University in Göteborg, Sweden; at the Academy of Sciences in Jena, East Germany; at the "Haus der Technik" in Essen, West Germany, 1982; at the Academy of Science in Sofia, Bulgaria; and at the Technical University of Delft, Netherlands, 1986.

The main goals of this book are, first, to bridge the gap that always exists between basic principles and applied engineering practice, second, to enhance the integration between biological and physical phenomena, and, third, to contribute to the internal development of the field of biotechnology by describing the process-oriented field of bioprocess technology.

To achieve these goals, I have attempted to unify the many interdisciplinary fields that continue to become ever more specialized. For better comprehensiveness, quantitative facts are often followed by a qualitative discussion but without complete derivation of the final equations—these can be found in the original references. Since a major goal is to contribute to the development of a biotechnical methodology by integrating several fields of science into one, details are often omitted in that they are thought to be not significant for the overall concept of integrating strategy. This approach aims to foster new orientation, a unified, process-oriented strategy of "bioprocess technology" that follows a systematic method of thinking and working. This strategy, as an expression of research philosophy, includes four working principles: (a) working with simplifications, that is—making distinctions between the important and the unimportant, (b) quantifying, (c) process (regime) analysis, that is, separating biological from physical phenomena, and (d) thinking and working in terms of mathematical models.

In conjunction with the irreplaceable element of intuition, which always serves as the starting point, mathematical models should be considered as working hypotheses that can assist process development. Within the framework of adaptive model building, these models can be compared with, and fitted to, the experimental reality of biological processes.

The book is organized into chapters that seem sensible from a pedagogical viewpoint. After the definition of bioprocess technology and its delineation with respect to the whole area of biotechnology in Chapter 1, and description of principles of thinking and working to be used in the integration in Chapter 2, Chapters 3 through 5 discuss bioprocess analysis. Chapter 3 characterizes bioreactors quantitatively, as does Chapter 4, which also covers the general utilization of bioreactors for obtaining kinetic data in the analysis of specific processes. In Chapter 5 the use of mathematical models for simulating the kinetics of biological processes is described. Chapter 6 discusses the synthesis of biological data (kinetics) and physical data (transport phenomena) in order to estimate bioreactor performance, that is, the conversion and productivity of the most important types and operations of reactors.

This book does not primarily provide an exhaustive survey of the literature, but concentrates on the most significant facts. Rather, its principal feature is its description of a generally valid approach to calculations for bioprocess technology. The problems of achieving a quantitative understanding are primary, especially the problems of understanding kinetics as a cycle of engineering considerations and calculations. The biological synthesis of the products of intermediary metabolism and all biochemical/biological aspects are excluded; they are well described in other books, for example, volumes 1 through 8 of *Biotechnology—A Comprehensive Treatise* (Rehm, H.J., and Reed, G., eds., 1982ff) and *Comprehensive Biotechology* (Moo-Young, 1985). Discussion of the technical aspects of quantifying reactors and process development is limited in favor of description of working procedures for the analysis and synthesis of production-scale biological processes. As a consequence of the novel strategy being developed here, I hope to close a gap I perceive in the literature, especially for students and for people involved in industry.

The title "Bioprocess Technology" for this edition has been chosen in accordance with the semantic definition of "technology" as the "science of industrial arts" This title was preferred to the "working title" "Bioprocess Engineering." (Another possibility was "Bioprocess Engineering Sciences.") I believe that the conception of engineering basically means the area of applied sciences, that is, execution of the science of industrial arts, including mechanical aspects, auxiliary equipment, and measurement and control techniques, perhaps even being dominated by them (cf. title of the journal *Bioprocess Engineering*).

This book attempts to present the formal kinetic macro-approach following Einstein's dictum that "Everything should be made as simple as possible, but not simpler." And I firmly believe that "Nothing is more practical than theory, and—at the same time—nothing is more theoretical than practice." as a typical example if the "Yin & Yang" principle, which is at present stressed by Fritjof Capra in order to achieve a holistic, ecological paradigm.

Graz, Austria

ANTON MOSER

Contents

Preface v
Nomenclature (List of Symbols) xv
Subscripts xxv
Abbreviations xxvii
Greek Symbols xxviii

Chapter 1 Introduction 1

1.1 Biotechnology: A Definition and Overview 1
1.2 Bioprocess Technology 5
Bibliography 12

Chapter 2 The Principles of Bioprocess Technology 13

2.1 Empirical Pragmatic Process Development 13
2.1.1 Production Strains 13
2.1.2 Starting Points 13
2.1.3 Different Modes (or Strategies) in Process Development 15
2.1.4 Process Development Without Mathematical Models 15
2.2 Basics of Quantification Methods for Bioprocesses 18
2.2.1 Concepts of a Uniform Nomenclature for Bioprocess Kinetics 19
2.2.2 The Rates of a Bioprocess 20
2.2.3 Stoichiometry and Thermodynamics 25
2.2.4 Productivity, Conversion, and Economics (Profit) 38
2.3 Systematic, Empirical Process Development with Mathematical Models 41
2.3.1 An Integrating Strategy—A Basis for Biotechnological Methodology 41
2.3.2 Working Principles of Bioprocess Technology 46
2.4 Mathematical Modeling in Bioprocessing 49
2.4.1 General Remarks 49
2.4.2 Model Building 51
2.4.3 Different Levels and Types of Kinetic Models 54
Bibliography 62

Chapter 3 Bioreactors 66

3.1 Overview: Industrial Reactors 66
3.1.1 Microbiological Reactors (Fermenters, Cell Tissue Culture Vessels, and Waste Water Treatment Plants) 66
3.1.2 Enzyme Reactors 68
3.1.3 Sterilizers 68
3.2 Systematics of Bioreactors 69
3.2.1 Homogeneous Versus Heterogeneous Systems 70
3.2.2 Mixing Behavior 70
3.3 Quantification Methods 73
3.3.1 Residence Time Distribution (RTD)—Macromixing 74
3.3.2 Micromixing 81
3.3.3 Oxygen Transfer Rate (OTR) 90
3.3.4 Degree of O_2 Utilization, η_{O_2} 99
3.3.5 Degree of Hinterland, Hl 99
3.3.6 Power Consumption, P 100
3.3.7 O_2 Efficiency (Economy) E_{O_2} 101
3.3.8 Heat Transfer Rate, $H_v TR$ 101
3.3.9 Characteristic Diameter of Biocatalytic Mass $\bar{d}_p$ 105
3.3.10 Comparison of Process Technology Data for Bioreactors 105
3.3.11 Biological Test Systems 110
3.4 Operational Modes and Bioreactor Concepts 112
3.5 Bioreactor Models 118
3.5.1 Model 1: The Ideal Discontinuous Stirred Tank Reactor (DCSTR) 119
3.5.2 Model 2: The Ideal Continuous Stirred Tank Reactor (CSTR) with V = Constant 119
3.5.3 Model 3: The Ideal Semicontinuous Stirred Tank Reactor (SCSTR) with V = Variable 119
3.5.4 Model 4: The Ideal Continuous Plug Flow Reactor (CPFR) or Tubular Reactor 121
3.5.5 Model 5: The Real Plug Flow Reactor CPFR with Dispersion 122
3.5.6 Model 6: The Discontinuous Recycle Reactor (DCRR) 123
3.5.7 Model 7: The Continuous Recycle Reactor (CRR) 123
3.5.8 Multiple Phase Bioreactor Models 124
3.6 "Perfect Bioreactors" in Bench and Pilot Scale for Process Kinetic Analysis 126
Bibliography 130

Chapter 4 Process Kinetic Analysis 138

4.1 Kinetic Analysis in Different Types of Reactors 138
4.2 Regime Analysis—General Concept and Guidelines 141
4.3 Test of Pseudohomogeneity 146
4.4 Parameter Estimation of Kinetic Models with Bioreactors 151
4.4.1 Integral and Differential Reactors 151
4.4.2 Integral and Differential Reactor Data Evaluation Methods 154

4.4.3 Results of Differential and Integral Analysis: Linearization Diagrams 157
4.5 Modeling Heterogeneous Processes 168
4.5.1 External Transport Limitations 171
4.5.2 Internal Transport Limitations 175
4.5.3 Combined Internal and External Transport Limitations 183
4.5.4 Transport Enhancement 188
4.5.5 Concluding Remarks 192
Bibliography 193

Chapter 5 Bioprocess Kinetics 197

5.1 Temperature Dependence, $k(T)$, Water Activity, a_W, and Enthalpy/Entropy Compensation 198
5.2 Microkinetic Equations Derived from the Kinetics of Chemical and Enzymatic Reactions 204
5.2.1 The Dynamic Flow Equilibrium Approach to Life Processes 204
5.2.2 Contribution of Enzyme Mechanism to Bioprocess Kinetic Models 206
5.2.3 Contribution of Chemical Kinetic Laws to Bioprocess Kinetic Modeling 214
5.3 Basic Unstructured Kinetic Models of Growth and Substrate Utilization (Homogeneous Rate Equations) 216
5.3.1 $\mu = \mu(s)$: Simple Model Functions of Inhibition-Free Substrate Limitation (Saturation-Type Kinetics) 217
5.3.2 $\mu = \mu(x)$: Influence of Biomass Concentration on Specific Growth Rate 224
5.3.3 $\mu = \mu(t)$: Extensions of Monod-Type Kinetics to Stationary and Lag Phase 225
5.3.4 Negative Biokinetic Rates—The Case of Microbial Death and Endogenous Metabolism 227
5.3.5 Kinetic Model Equations for Inhibition by Substrates and Products 231
5.3.6 Kinetic Model Equations for Repression 237
5.3.7 $\mu = \mu(\mathrm{pH})$ 237
5.3.8 Kinetic Pseudohomogeneous Modeling of Mycelial Filamentous Growth Including Photosynthesis 237
5.3.9 Kinetic Modeling of Biosorption 238
5.4 Kinetic Models for Microbial Product Formation 240
5.4.1 Metabolites and End Products 240
5.4.2 Heat Production in Fermentation Processes 247
5.5 Multisubstrate Kinetics 250
5.5.1 Sequential Substrate-Utilization Kinetics 251
5.5.2 Simultaneous Substrate-Utilization Kinetics 253
5.5.3 Generalizations in Multisubstrate Kinetics 254
5.6 Mixed Population Kinetics 259
5.6.1 Classification of the Types of Microbial Interactions 260
5.6.2 Kinetic Analysis of Microbial Interactions 261
5.7 Dynamic Models for Transient Operation Techniques (Nonstationary Kinetics) 272

5.7.1 Definitions of Balanced Growth and Steady-State Growth 272
5.7.2 Mathematical Modeling of Dynamic Process Kinetics 274
5.8 Kinetic Models of Heterogeneous Bioprocesses 283
5.8.1 Biofilm Kinetics 283
5.8.2 Unstructured Models of Pellet Growth 288
5.8.3 Linear Growth 290
5.9 Pseudokinetics 290
5.10 Kinetics of Sterilization 292
5.10.1 Basic Kinetic Approaches in Sterilization Kinetics 292
5.10.2 Multicomponent Systems in Food Technology 294
Bibliography 295

Chapter 6 Bioreactor Performance: Process Design Methods 307

6.1 The Ideal Single-Stage, Constant-Volume Continuous Stirred Tank Reactor, CSTR (Pseudohomogeneous L-Phase Reactor Model) 308
6.1.1 Performance of the CSTR with Simple Kinetics 308
6.1.2 Performance of the CSTR with Complex Kinetics 311
6.1.3 Stability Analysis and Transient Behavior of the CSTR 318
6.2 Variable Volume CSTR Operation (Fed-Batch and Transient Reactor Operation) 325
6.3 Multistage Single and Multistream Continuous Reactor Operation .. 329
6.3.1 Classification 329
6.3.2 Potentialities of Multistage Systems 330
6.3.3 Single-Stream Multistage Operation 331
6.3.4 Multistream Multistage Operation 334
6.4 Continuous Plug Flow Reactors (CPFR) 337
6.4.1 Performance Equations 337
6.4.2 Potential Advantages of CPFR Operation 339
6.4.3 Principal Properties and Design of CPFRs Compared with CSTRs .. 340
6.4.4 Applications of CPFR 347
6.4.5 One-Phase (Liquid) Reactors with Arbitrary Residence Time Distribution and Micromixing 347
6.5 Recycle Reactor Operation 351
6.5.1 Performance Equations of Recycle Reactors 351
6.5.2 Application of CRR 356
6.6 Gas/Liquid (Two-Phase) Reactor Models in Bioprocessing 357
6.7 Biofilm Reactor Operation 358
6.7.1 Potentialities of Biofilm Reactors 359
6.7.2 Performance Equations of Biofilm Reactors 360
6.7.3 Application of Biofilm Reactors 370
6.8 Dialysis and Synchronous Culture Operation 371
6.8.1 Dialysis (Membrane) Reactor Operation 371
6.8.2 Synchronous Culture Operation 378
6.9 Integrating Strategy as General Scale-Up Concept in Bioprocessing . 381
6.9.1 Stoichiometry (Balancing Methods) Applied in Bioprocess Design ... 382
6.9.2 Interactions Between Biology and Physics via Viscosity of Fermentation Media 385

6.9.3 Influence of Mycelium—The Morphology Factors ("Apparent Morphology") 389
6.9.4 Structured Modeling of Bioreactors (OTR) 392
6.10 Final Note 395
Bibliography 396

Appendix I Fundamentals of Stoichiometry of Complex Reaction Systems 406
Appendix II Computer Simulations 412
Appendix III Microkinetics: Derivation of Kinetic Rate Equations from Mechanisms 436

Index 441

Nomenclature (List of Symbols)

ARABIC

Symbol	Unit	Meaning
A	[m^2]	surface area
	[A]	Ampère
A, B, C, D	—	compounds (general)
{A}	—	activated state of compound A
a	[m^{-1}]	specific area
a, b, c, d, e, f	—	coefficients (esp. in element-species matrix, cf. Equ. II.6)
a, d, c, d	[$kg \cdot m^{-3}$]	concentration of compounds
a_{L1}	[$m^2 \cdot m^{-3}$]	
a_W	—	water activity (Equ. 5.1)
BOD	[$kg \cdot m^{-3} \cdot t^{-1}$]	biological oxygen demand
B_v	[$kg \cdot m^{-3} \cdot h^{-1}$]	volumetric mass loading rate (Equ. 6.129)
B_x	[h^{-1}]	mass loading rate per unit sludge biomass (Equ. 5.212)
Bi	—	Biot number ($Bi = D_{L2} \cdot \delta_{L2}/D_s \cdot \delta_s$)
Bo	—	Bodenstein number ($Bo = v \cdot L/D_L$)
C	[kg]	amount of CO_2
C	[$]	costs
C	—	integration constant

CTD	—	circulation time distribution
c, c_i, c_j	$[kg \cdot m^{-3}]$	concentration (general) of component *i* or *j*
c	$[kg \cdot m^{-3}]$	concentration of CO_2
c_μ	—	constant in Equ. 5.218
c_P	$[kJ \cdot kg^{-1} \cdot {}^\circ C^{-1}]$	specific heat at $p = \text{constant}$
c_R	$[kg \cdot m^{-3}]$	concentration of ribosomes
c_{st}	$[n_i \cdot m^{-3}]$	sterilization level
D	$[m^2 \cdot s^{-1}]$	diffusion or dispersion coefficient
D	$[h^{-1}]$	dilution rate
D_c	$[h^{-1}]$	critical dilution rate where washout occurs
D_M	$[h^{-1}]$	dilution rate with maximum productivity
D_{10}	[h]	decimal reduction time (Equ. 5.8)
Da (Da_0, $Da_{L\|S}$, $Da_{G\|L}$)	—	Damkoehler number (Equs. 4.3, 4.5, 4.7)
Da_I	—	Damkoehler number of first degree (Equ. 6.89)
Da_{II}	—	Damkoehler number of second degree (Equ. 4.74)
d	[m]	diameter (general term)
d	—	constant
$\bar{d}$	[m]	mean diameter
d_B	[m]	bubble diameter
d_i	[m]	impeller diameter
d_P	[m]	particle diameter
$\tilde{d}_p$	[m]	active particle diameter (Equ. 5.258)
d_T	[m]	tank diameter
E	$[m^2 \cdot s^{-1}]$	convection coefficient (Equ. 3.30d)
E (E_{OTR})	—	enhancement factor (Equ. 4.115) of OTR
E_a, E_{Gr}, E_d	$[J \cdot mole^{-1}]$	activation energy, general (growth or death)

E_{O2}	$[kg\ O_2 \cdot kWh^{-1}]$	oxygen economy (Equ. 3.66) (oxygen efficiency)
E, ES, EP, EI	[kg]	amount of enzyme, enzyme-substrate, enzyme-product, enzyme-inhibitor complex
{ES}	—	activated state of ES-complex
e, es, ep, ei	$[kg \cdot m^{-3}]$	concentration of components
F	$[m^3 \cdot h^{-1}]$	volumetric flow rate
F_B	$[m^3 \cdot h^{-1}]$	bleed rate
F_F	$[m^3 \cdot h^{-1}]$	feed rate
F_p	$[m^3 \cdot h^{-1}]$	pumping capacity
F_s	$[kg \cdot m^{-3} \cdot h^{-1}]$	substrate feed rate (Equ. II.30)
Fr	—	Froude number ($Fr = v^2/g \cdot L$)
$F(t)$	—	mathematical RTD function acc. to step method
$f(t)$	—	mathematical RTD function acc. to pulse method
$f_E(t)$	—	singular $f(t)$ (Equ. 3.14)
f_1	—	kinetic term for overlapping multi-substrate degradation
f_2	—	kinetic term for repression of substrate degradation by other substrates
f_{O^*}	—	correction factor for oxygen saturation calculation (Equ. 3.40)
G		Material in "G compartment" proteins and DNA)
Ga	—	Galilei number

ΔG	$[J \cdot mole^{-1}]$	free reaction enthalpy
g	$[cm \cdot s^{-2}]$	acceleration of gravity (980)
g_c	$[cm^3 \cdot kg^{-1} \cdot s^{-2}]$	universal gravitational constant ($6.671 \cdot 10^{-8}$)
H	[m]	height
H	$[m^3\ atm \cdot mol^{-1}]$	Henry distribution coefficient
He	—	dimensionless Henry distribution coefficient
Ha	—	Hatta number (Equ. 4.112)
Hl	—	degree of hinterland (Equ. 3.59)
ΔH_R, $\Delta H_R^{(0)}$	$[J \cdot mole^{-1}]$	reaction enthalpy (molar)
$\Delta H_R^{(X)}$, $\Delta H_R^{(S)}$	$[J \cdot kg^{-1}]$	reaction enthalpy per unit mass of X or S
ΔH_i	$[J \cdot mole^{-1}]$	heat of combustion of components
H_v	$[J \cdot m^{-3}]$	amount of volumetric heat of fermentation
h	$[J \cdot s]$	Planck constant ($6.6 \cdot 10^{-34}$)
h (k_H)	$[J \cdot m^{-2} \cdot h^{-1} \cdot K^{-1}]$	heat transfer coefficient
h^+	$[kg \cdot m^{-3}]$	concentration of H^+ ions
h_s	—	crowding factor (Equ. 6.188)
h_v	$[J \cdot m^{-3}]$	"concentration" of volumetric heat
I	[kg]	amount of inhibitor
i	$[kg \cdot m^{-3}]$	concentration of inhibitor
i	—	increments in Equ. 3.34
J	—	degree of segregation (Equ. 3.1)
K		material in "K compartment" (RNA mainly)

K	$[kg \cdot m^{-3}]$	constant of equilibrium or saturation (general term)
K_{adapt}	—	constant of metabolic adaptation in Equ. 5.227
K_c	$[(Nsm^{-2})^{1/2}]$	constant in Casson law (Equ. 6.166), "Casson-viscosity"
K_D	—	constant diffusional resistance in Equ. 5.43
K_e	—	constant for endogenous metabolism in Equ. 5.87
K_{eq}	—	constant of thermo-dynamic equilibrium (Equ. 5.22) "Haldane relationship"
K_d	—	constant for microbial death (Equ. 5.86)
K_{IS}, K_{IP} (K_{ISX}, K_{IPX})	$[kg \cdot m^{-3}]$	inhibition constant of growth by substrate or product
K_m, K_S, K_L, K_F, K_H	$[kg \cdot m^{-3}]$	saturation constant for substrate in kinetic equation acc. to Michaelis–Menten, Monod, Langmuir, Freundlich, and Hill
K_O, K_P	$[kg \cdot m^{-3}]$	saturation constant in Monod-type kinetics for oxygen or product
K_0	—	constant in Equ. 3.34
K_R (K_{ISP})	$[kg \cdot m^{-3}]$	constant for repression
K_1, K_2	—	constants for pH function (Equ. 5.107)
k	$[(m^3 \cdot mol^{-1})^{n-1} \cdot s^{-1}]$	rate constant (general term)
k	—	number of elements for elemental balancing method (Sect. 2.4.2.2)
k_B	$[J \cdot K^{-1}]$	Boltzmann constant ($1.38 \cdot 10^{-23}$)
k_{cat}	$[\text{mole P} \cdot \text{mole E}^{-1} \cdot s^{-1}]$	catalytic constant or turnover number
k_d	$[h^{-1}]$	specific dead rate

k_E	[s^{-1}]	rate constant of electrode response
k_e	[s^{-1}]	rate constant of environmental change
k_H (h)	[$J \cdot m^{-2} \cdot h^{-1} \cdot K^{-1}$]	true rate coefficient of volumetric heat transfer
k_{HTR}	[$J \cdot h^{-1} \cdot K^{-1}$]	rate coefficient of volumetric heat transfer (phenomenological)
k_m	[s^{-1}]	first-order rate constant for mixing (Equ. 3.23)
$k_{L1}a$ (k_La)	[h^{-1}]	volumetric OTR coefficient (G|L interface)
k_{L2}	[$m \cdot s^{-1}$]	transfer coefficients at L|S interface
k_P	[h^{-1}]	rate constant of product formation
$k_{P,d}$	[h^{-1}]	specific rate constant of product degradation
k_{P1}, k_{P2}	[h^{-1}]	rate constant acc. to Kono equation (Equ. 5.126)
k_r (k_i, k_n)	[t^{-1} (mole/vol)$^{-n+1}$]	reaction rate constant (ith reaction or for reaction order n)
k_S	[$m \cdot s^{-1}$]	transport rate coefficient in solid phase
k_{TR}	[s^{-1}]	transport rate constant (phenomenological)
$k_{V,G}$	[h^{-1}]	transport rate constant of gas phase (Equ. 3.38)
k_x	[h^{-1}]	constant for biomass in Equ. 5.63 or 6.170
k_0, k_1, $k_{1/2}$	[!]	rate constants for reactor order $n = 0, 1, 1/2$
k_1, k_2, k_3	[!]	rate constants in BRE (Equ. 5.239), esp. Equ. 4.25 with 5.240
k_{+1}, k_{-1}	[!]	rate constants for forward or backward reaction
k_τ	[h^{-1}]	decay coefficient
k_∞	[$m^3 \cdot mole^{-1} \cdot s^{-1}$]	preexponential factor in Arrhenius equation (Equ. 5.3)

L	[m]	length (general term)
L_e, L_t, L_u, L_H,	[m]	morphology factors (e.g., Equ. 6.182, 5.109)
M	[kg]	mass
M_F		morphology factor (Equ. 6.183)
m_S, m_O, m_{H_v}	[h^{-1}]	specific maintenance rate coefficient for S, O_2, and H_v
m	[%]	degree of mixing (Equ. 3.17)
m	—	power factor in Equ. 6.164, "flow index" (see also Equ. 6.171)
N	—	number of significant components in balancing method (Sect. 2.4.2.2); number of vessels in series
N_A	—	aeration number (Equ. 3.75)
N_i		number of molecules of component i
N_M	—	mixing number (Equ. 3.73)
N_{OTR}	—	OTR number (Equ. 3.76)
N_P	—	power number (Equ. 3.74)
N_{Re} (Re)	—	Reynolds number ($Re = v \cdot L/\nu$)
N_{Sc} (Sc)	—	Schmidt number ($Sc = \nu/D$)
N_{Sh} (Sh)	—	Sherwood number ($Sh_L = k_L \cdot L/D$)
n	[rps]	rotational speed, stirrer speed
n (n_s, n_O)	—	reaction order (with respect to S or O_2)
n	—	power index
n		number of microbial cells
n_i	—	number (of moles of component i)

n_B	—	bed expansion index (Equ. 6.139)
n_H	—	Hill coefficient (Equ. 5.29)
n_i	$[kg \cdot m^{-3} \cdot s^{-1}]$	volumetric flux of mass (transport rate)
n_i'	$[kg \cdot m^{-2} \cdot s^{-1}]$	(mass) flux through area
O	kg	amount of O_2
OTR	$[kg \cdot m^{-3} \cdot h^{-1}]$	O_2 transfer rate
o (resp. o*)	$[kg \cdot m^{-3}]$	O_2 concentration (resp. saturation value)
obs	—	observed value
OUR	$[kg \cdot m^{-3} \cdot h^{-1}]$	O_2-uptake rate
P	[kg]	amount of product
P	$[W \text{ or } J \cdot s^{-1}]$	power
P_{mb}	$[m^3 \cdot h^{-1} \cdot kN^{-1}]$	membrane permeability (Equ. 6.146a)
p	$[kg \cdot m^{-3}]$	concentration of product
p	[bar]	pressure
Q	—	product formation activity function (Equ. 5.220)
q	$[J \cdot m^{-3} \cdot s^{-1}]$	rate of heat transfer
q_i ($q_s, q_o, q_c, q_{H_v}, q_P$)	$[h^{-1}]$ resp. $J \cdot kg^{-1} \cdot h^{-1}$ (for q_{H_v})	specific rate of formation or consumption of component *i* (substrate, O_2, CO_2, heat, and product)
q_{int}	$[h^{-1}]$	rate of internal production in biomass compartment
R	$[J \cdot mol^{-1} \cdot K^{-1}]$	gas constant ($8.314 \cdot 10^7$)
R	[m]	radius
R, R_ν, R_β	—	rank of matrix (general, of stoichiometrical coefficients and element-species), see App. I
Re	—	Reynold number (N_{Re})
Re_t	—	terminal settling velocity, Re-number of bioparticles (Equ. 6.139)

Re_ε	—	Re-number based on energy dissipation (Equ. 4.1b)
RTD		residence time distribution
r	—	excretion rate
r	—	correlation coefficient
r	—	recycle ratio
r_i	$[kg \cdot m^{-3} \cdot h^{-1}]$	rate of production or consumption (volumetric)
r_C	—	rate of CO_2-production
r_P	—	rate of production
r_S	—	rate of S-consumption
r_X	—	rate of growth
r_O	—	rate of O_2-consumption
r_H	$[J \cdot m^{-3} \cdot h^{-1}]$	rate of heat production
r_i'	$[kg \cdot m^{-2} \cdot s^{-1}]$	rate per unit area
r_i^*	$[s^{-1}]$	related reaction rate (general term)
S	[kg]	amount of substrate
Sc	—	Schmidt number (N_{Sc})
Sh	—	Sherwood number (N_{Sh})
ΔS	$[J \cdot mole^{-1} \cdot K^{-1}]$	reaction entropy
s	—	crowding factor (Equ. 6.183b) for space filling
s	$[kg \cdot m^{-3}]$	concentration of substrate
s	$[s^{-1}]$	surface renewal rate (Equ. 3.30)
s	—	standard deviation
$\hat{s}$ (s_{ads})	$[kg \cdot m^{-3}]$	"sludge loading" = adsorbed S concentration (biosorption) (Equ. 5.114)
s_c	$[kg \cdot m^{-3}]$	critical threshold concentration (Equ. 5.95)
s_E, s_I	$[kg \cdot m^{-3}]$	concentration of essential and enhancing substrates (Equ. 5.168)
${}^1s$	$[kg \cdot m^{-3}]$	"washout state" S concentration (cf. Fig. 6.1a, 6.4, and 6.5)

^2s	$[kg \cdot m^{-3}]$	"non washout state" S concentration (steady state)
^3s	$[kg \cdot m^{-3}]$	unstable S concentration (Fig. 6.11)
T	$[°C, K]$	temperature
T_β	$[K]$	isokinetic temperature (Equ. 5.11)
TOC		total organic carbon
t	$[s]$	time (general term)
$\bar{t}$	$[h]$	mean residence time
t_c	$[s]$	circulation time
$\bar{t}_G$	$[s]$	mean residence time of gas phase
$t_{H,TR}$	$[s]$	characteristic time of heat transfer (Table 4.3a)
t_L	$[s]$	lag time
t_M	$[s]$	maturation time (Equ. 5.136)
t_m ($t_{m,95}$)	$[s]$	mixing time (at m = 95%)
t_{max}	$[s]$	time of maximum production rate (Equ. 5.124)
t_{OTR}	$[s]$	characteristic time of OTR (Table 4.1)
t_P	$[s]$	period of oscillation (Equ. 5.208)
t_{q_i}	$[s]$	characteristic time of consumption or formation rates (cf. Table 4.1)
t_r	$[s]$	characteristic reaction time (cf. Table 4.1)
t_{St}	$[s]$	sterilization time
V	$[m^3]$	volume
	$[V]$	Volt
v	$[m \cdot s^{-1}]$	velocity (general term)
v (see r)	$[kg \cdot m^{-3} \cdot s^{-1}]$	reaction rate, esp. of enzyme kinetics
v_{SL}, v_{SG}	$[m \cdot s^{-1}]$	superficial velocity of liquid or gas phase
v_t	$[m \cdot s^{-1}]$	terminal settling velocity

v_{tip}	$[m \cdot s^{-1}]$	peripheric velocity of stirrer (stirrer top velocity)
v_{mf}	$[m \cdot s^{-1}]$	velocity of minimum fluidization
v_{ms}	$[m \cdot s^{-1}]$	velocity of minimum spouting
W	[kg]	weight
	[W]	Watt
w	—	mass fraction of component
X	[kg]	amount of biomass
x	$[kg \cdot m^{-3}]$	biomass concentration
x_i	—	measurement points or significant process variables or arbitrary quantity
${}^1x$, ${}^2x$	$[kg \cdot m^{-3}]$	"washout" and "non-washout state" of biomass concentration (Fig. 6.1a)
${}^3x$	$[kg \cdot m^{-3}]$	unstable steady state (Fig. 6.11) biomass concentration
Y, $Y_{i\|j}$	—	yield coefficient (general term) (see Table 2.14)
Z		Z value (Equ. 5.9)
Z_i	—	electric charge (Equ. 3.33)
z	—	coordinate

SUBSCRIPTS

ads, a	adsorption
agit	agitation
app	apparent
ass	assimilation
asym	asymptotic
adap	adaptation
av	available
ave	average
B	bulk or bubble or bleed
BC	bubble column
bio	biological
C	CO_2
c	circulation
c	cold

calc	calculated
cat	catalytic
charact	characteristic
crit, c	critical
d	death or decay
dc (DC)	discontinuous
degr, degrad, d	degradation
dist	distributor
E	electrode or enzyme or single
e	endogenous or environmental or electrons
eff	effective
elim, el, e	elimination
end	end
est	estimated
eq	equivalent
evap	evaporation (heat loss)
ex	exit
exp	experimental
ext	external
F	feed
f	film, biofilm
ferm	fermenter
G	gas phase, gas
G	G compartment
gl	glucose
gr	growth
H	heat, general
H_v	volumetric heat
i, j	number, components
in	inlet
ingest	ingestion
int	internal
intersect	intersection
K	K compartment
L	liquid phase
L	longitudinal
L1	liquid film at G|L interface
L2	liquid film at L|S interface
M	maturation
m	mixing degree, medium, membrane
mm	maximum mixedness
max	maximum
mb	membrane
min	minimum
mf	minimum fluidization
ms	minimum spouting
N	number of stages
N resp. Ni	nucleotide resp. internal nucleotide
O	oxygen
o	initial value
obs	observed
opt	optimal
OTR	oxygen transfer rate
Π	product sum
p	product
P	power
pr	production
R	reactor or ribosomes
r	reaction or resistant or recycle
rad	radiation
rds	rate-determining step
rel	relative
repr	repression
res	reservoir
Σ	sum
S	solid phase or substrate
ST	stirred tank

s	sensitive
salt	salt
sim	simulated
st	stirrer or sterilisation
T	tank
TR	transfer
t	terminal
tot	total, global
ts	total segregation
V	volumetric
v	viable
W	wall
w	warm
X	biomass
z	coordinate
$\neq$	value of activated complex
$\wedge$	normalized value
$-$	mean value (steady state)

ABBREVIATIONS

ADH, adh	alcohol dehydrogenase
ADP	adenosine diphosphate
AMP	adenosine mono-phosphate
ATP	adenosine triphosphate
ATPase	adenosine tripho-sphatease (enzyme)
BFF	biological film fermenter
BOD	biological oxygen demand
BRE	biological rate equation
CMMFF	completely mixed microbial film fermenter
COD	chemical oxygen demand
CPFR	continuous plug flow reactor
CRR	continuous recycle reactor
CSTR	continuous stirred tank reactor
CTCR	cycle tube cyclone reactor
CTD	circulation time distribution
DCRR	discontinuous recycle reactor
DCSTR	discontinuous stirred tank reactor
DNA	desoxyribonucleic acid
ES	enzyme substrate complex
EFB	European Federation of Biotechnology
E, ΔE_{260}	extinction esp. at 260 μm
FBBR	fluidized bed biofilm reactor
FDP	fructose diphosphate
G	G-compartment in structured modeling of cells
gapdh	glyceraldehyde-3-phosphate dehydro-genase
H_vTR	volumetric heat transfer rate
HK	hexokinase
idCPFR	ideal continuous plug flow reactor
idCSTR	ideal continuous stirred tank reactor

K	K-compartment in structured modeling of cells
mm	maximum mixedness
NAD	nicotinamide adenine dinucleotide
NADH	nicotinamide adenine dinucleotide, reduced form
NCSTR	cascade of CSTR with N stages
OTR	O_2 transfer rate
OUR	O_2 uptake rate
$\overline{\Delta O}$	log mean of O_2 concentration
PATR	pneumatically agitated & aerated tubular reactor
pfk	pyruvate fructokinase
PGK	phosphoglycerate kinase
Ph	Phosphorus
PK	pyruvate kinase
PHB	polyhydroxy butyric acid
Pr	productivity
pyk	pyruvate kinase
qss	quasi-steady-state
q	heat (rate)
rds	rate-determining step
rCSTR	continuous stirred tank reactor with cell recycling
RF	rotor fermenter
rH	redox potential
RNA	ribonucleic acid
RTD	residence time distribution
S	Solid
s	substrate or solid
SE	substrate enzyme complex
Su	Sulfur
SCR	semicontinuous reactor
SCSTR	semicontinuous stirred tank reactor
SOV	sum of variances
S_LTR	liquid-phase substrate transfer rate
S_STR	solid-phase substrate transfer rate
TLR	tubular loop reactor
ThLTBFF	thin-layer tubular biofilm fermenter
ThLTR	thin-layer tubular reactor
ThLTFF	thin-layer tubular film fermenter
TOC	total organic carbon
ts	total segregation
$\overline{\Delta T}$	log-mean of temperature
TOC	total organic carbon
TR	transport, general

Greek Symbols

α (var α)	—	variance
$\alpha, \beta, \gamma, \delta,$	—	coefficients
β	—	concentration factor of settling device
β (ϕ_0)	—	modulus (Equ. 4.85)
γ	[!]	factor in Equ. 4.123
$\dot{\gamma}$	$[s^{-1}]$	shear rate

γ_S, γ_X	—	degree of reductance (reduction)
δ	[m]	thickness of film acc. to two-film theory
δ^*	—	morphology factor (Equ. 6.178)
ε	$[m^2 \cdot s^{-3}]$	energy dissipation
ε	—	ratio of K_S values (Equ. 5.137)
ε_G	—	gas hold-up
ε_P	—	volume fraction of particles
ε_x	—	biomass hold-up (Equ. 6.134)
ζ_i	—	(relative) conversion of component *i* (Equ. 2.45)
η	—	effectiveness factor (general term)
η	$[N \cdot s \cdot m^{-2}]$	dynamic viscosity
$\tilde{\eta}$	—	overall effectiveness factor (Equ. 4.2)
η_{app}	$[N \cdot s \cdot m^{-2}]$	apparent viscosity
η_D	—	factor in Equ. 3.62
η_E	—	energy efficiency coefficient of growth (Sect. 2.2.3.4)
$\eta_{G\|L}$	—	effectiveness factor of G|L—mass transfer
$\eta_{L\|S}$	—	effectiveness factor of L|S—mass transfer
η_M	—	factor in Equ. 3.60
η_0	$[N \cdot s \cdot m^{-2}]$	viscosity of suspending fluid (for $\phi_s \rightarrow \phi$)
η_{O2}	[%]	O_2 utilization (Equ. 3.58)
$\eta_{r,A}$	—	surface-area-related effectiveness factor of reaction (Equ. 4.98)
η_r ($\eta_{r,V}$)	—	effectiveness factor of reaction (volume related)
η_r^*	—	effectiveness factor of reaction (Equ. 4.108)
$\hat{\eta}_r$	—	effectiveness factor of reaction (Equ. 4.110b)
η_{th}	—	thermodynamic efficiency
η_{TR}(E)	—	effectiveness factor of transport (enhancement factor)
θ	[!]	factor in Equ. 3.63
θ_x	[h]	sludge age (Equ. 6.122)
κ	—	dimensionless factor for bulk reaction rate in Fig. 5.75
Λ	[h]	cell age (Equ. 5.134)
λ	$[\Omega^{-1} \cdot cm^{-1}]$	conductivity
μ	$[h^{-1}]$	specific growth rate
$\dot{\mu}$	$[h^{-2}]$	first derivation of specific growth rate on time
μ_d	$[h^{-1}]$	specific microbial death rate (Equ. 5.86)
μ_1	—	first moment (average value) of distribution function (Equ. 6.191)
ν	$[m^2 \cdot s^{-1}]$	kinematic viscosity
ν	$[s^{-1}]$	specific doubling rate of cell number
ν_i	—	stoichiometric coefficients (Equ. I.4)
ξ	[mole]	extent of reaction (Equ. I.3)
ξ_{ads}	—	sorption capacity (Equ. 5.115)

Π	—	product sum
ρ	$[kg.m^{-3}]$	density
Σ	—	sum
σ	$[mole \cdot m^{-3}]$	ionic strength (Equs. 3.33, 5.93)
σ	—	selectivity (Equ. 2.43)
σ_i^2	—	deviation (Equ. 4.47)
σ_1	—	spread of distribution function (Equ. 6.191)
σ^2	—	variance of distribution function
τ	$[N \cdot m^{-2}]$	shear stress
τ	—	normalized time $(t/\bar{t})$
τ_E	[s]	characteristic time of electrode response
τ_e	[s]	characteristic time of environmental change
τ_r	[s]	response to signal
$\tau_{1/e}$	[s]	characteristic time of response acc. to moment method with first-order lags in series (Equ. 351)
τ_L	[h]	lag time (transients), Equ. 5.213
τ_o	$[N \cdot m^{-2}]$	yield stress
ϕ (ϕ_{int})	—	Thiele modulus (Equs. 4.80 and 4.77) (internal limitation)
Φ	—	modulus Equ. 5.251
ϕ_0 (β)	—	modulus for zero-order reaction (Equ. 4.85)
ϕ_1	—	modulus for first-order reaction (Equ. 4.83)
ϕ_P	—	generalized Thiele modulus of particles
ϕ_δ	—	Thiele modulus based on biofilm thickness δ (Equ. 4.111)
ϕ_{ext}	—	Thiele modulus for external S limitation (Equ. 4.63)
ϕ_{limit}	—	limiting value of ϕ_{ext} (Equ. 4.65)
ϕ	—	consumption coefficient (Equ. 5.48)
ϕ_c	—	Carman factor in Equ. 4.1a
ϕ_s	—	volume fraction of suspended solids (Equ. 6.184)

CHAPTER 1
Introduction

1.1 Biotechnology: A Definition and Overview

In a general sense, biotechnology is a multi- and interdisciplinary field of activities dealing with biological and biochemical processes carried out on a broad range of scale (10 l to 10^6 m^3) as a technique for production.

Biotechnology has been defined differently in different countries and continents. Based on a definition by the European Federation of Biotechnology (EFB), a broader sense is discussed in Europe recently:

"Biotechnology is the integrated use of natural sciences (biology incl. molecular biology, biochemistry, but also chemistry and physics) and engineering sciences (chemical reaction engineering, electronics) in order to achieve the application of (technological, industrial) organisms, cells (microbes, plant and animal cells), parts thereof and molecular analogues in order to provide biosociety desirable goods and services"

Thus, the main characteristics of biotechnology are integration, orientation towards applications but also a considerably impact on human society in the future ("Biosociety", FAST, 1980; Moser, 1988).

Significant fields of applications are

Industrial Biotechnology
Food- and Agricultural (Phyto-, Zoo-) Biotechnology
Healthcare (Pharma-) Biotechnology and
Environmental Biotechnology.

This definition excludes medical engineering, which is concerned with the construction of instruments for biological or health care use—for example, heart–lung machines. Biomedical engineering is also commonly referred to as biotechnology and/or bioengineering, and no clear-cut distinction is apparent from the name alone. Further, technological activities in the field of agriculture, for example, traditional breeding (phytotechnology and zootechnology), belong to biotechnology only in the broadest sense.

One systematic treatment of biotechnology results from a logical consideration of the catalysts responsible for the conversions in biochemical reactions and other biologically active agents. In this sense, biocatalysts are not only naturally or synthetically produced enzymes and biomolecules respectively analogues but also cells or subcellular fractions from microorganisms and from plant and animal tissues. The type of catalyst used to carry out a biological process delineates the individual technology: enzyme or (simple and complex) fermentation technology or tissue culture with suspended or immobilized catalysts. In addition to the production techniques, there is also a group of sterilization technique used to inactivate or kill biological materials.

For historical reasons, precise separation of biotechnology from other fields is often difficult. There is overlap among various areas, such as conventional food technology and biotechnology (Fig. 1.1). The production of yogurt or

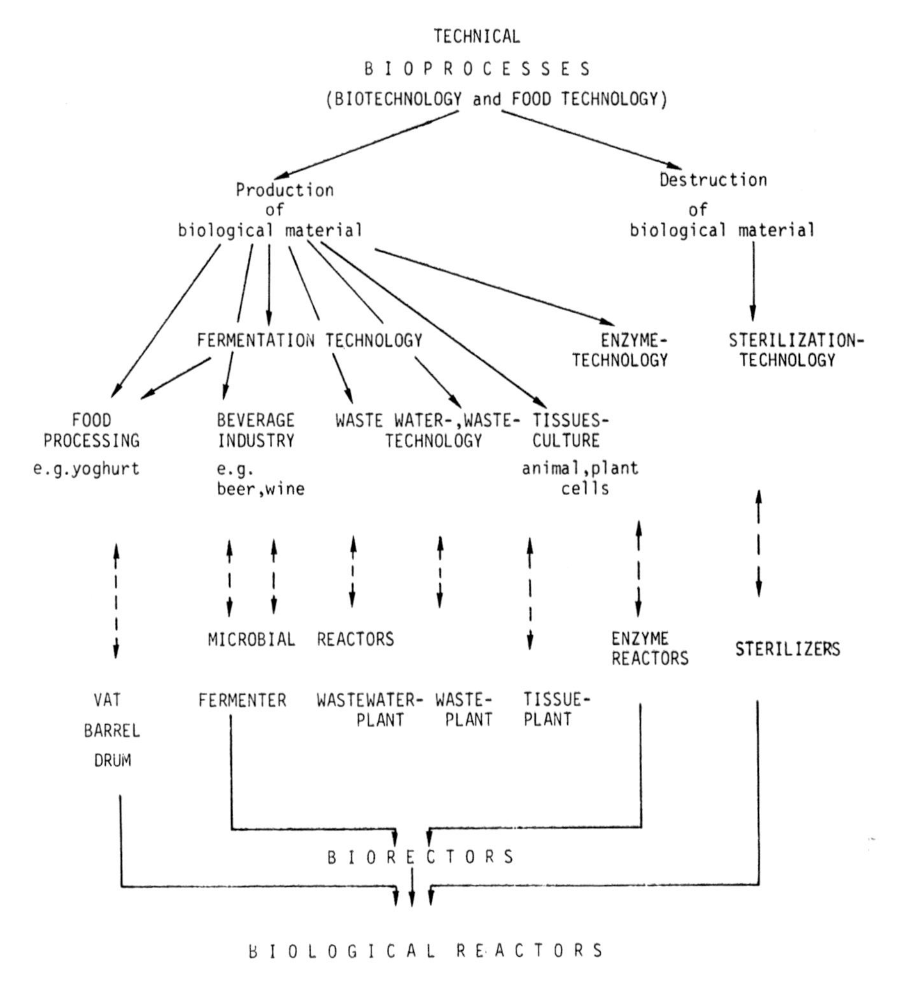

FIGURE 1.1. Biological reactors and their areas of application for "taming" many industrial-scale bioprocesses. (After Moser, 1978, 1981.)

cheese, for example, would be considered as bioprocesses of food technology. The high degree of complexity of bioprocessing in food technology and the difficulty of obtaining quantitative measurements (a difficulty shared by many other divisions of fermentation technology) usually retard or inhibit a systematic approach. In addition, special expectations exist for the quality of the foodstuff resulting from bioprocessing and the techniques of large-scale production are only slowly being introduced. Despite the fundamentally different orientation of food technology, however, progress is being made in introducing scientific methods. Because of commercial considerations, sooner or later the working principles of biotechnology will be applied in complex processes.

Precisely the same fate can be seen in neighboring area, the biological treatment of waste water and solid wastes (Fig. 1.1). The pressure of environmental problems has grown so great that here too the methods of thinking that have been successfully used in conventional biotechnical processes are being applied. One can now speak of the biotechnology of waste water treatment.

Figure 1.1 shows the correspondence between the various technologies of bioprocessing and the containers in which the reactions are carried out on an engineering scale. These "bioreactors" range from sterilizers to proper bioreactors (microbiological reactors such as fermenters, waste water treatment units, vessels for cultivation of plant & animal cells, enzyme reactors) to miscellaneous containers such as bottles, barrels, and so on used primarily for carrying out "arts and crafts"—type complex biological processes.

What are the future prospects of biotechnology? The range of products is being extended from antibiotics, steroids, vitamins, diagnostics, vaccines, enzymes, and polysaccharides to such fine chemicals as organic acids, alcohols, and single-cell protein (Rehm and Reed, 1982ff). Figure 1.2 shows the course of development of prices in relation to the scale of production facilities for the period 1970 to 2000 (Hines, 1980). From this figure one can conclude that, at present, biotechnical methods are more economical than chemical ones only in the case of pharmaceuticals that are complex molecules. The prediction is that soon single-cell proteins can be economically produced by biotechnical methods, and that by 1995 even inexpensive chemicals may be produced in the same way. Economic feasibility in processes for producing simple materials is first reached only in large-scale production units. At present, small-scale, high-price processes are being carried out, mainly on a discontinuous basis. For the future, one looks toward large-scale, low-price processes carried out on a continuous basis. Additional facts concerning economics are discussed by Hamer (1985).

Some general conclusions can now be drawn from experience in fermentation, food, and waste water treatment engineering:

1. The development of processes from the laboratory stage to the stage of technical maturity where they can be of service to mankind must be accelerated and made more reliable. Three stages are involved in "taming" a bioprocess.

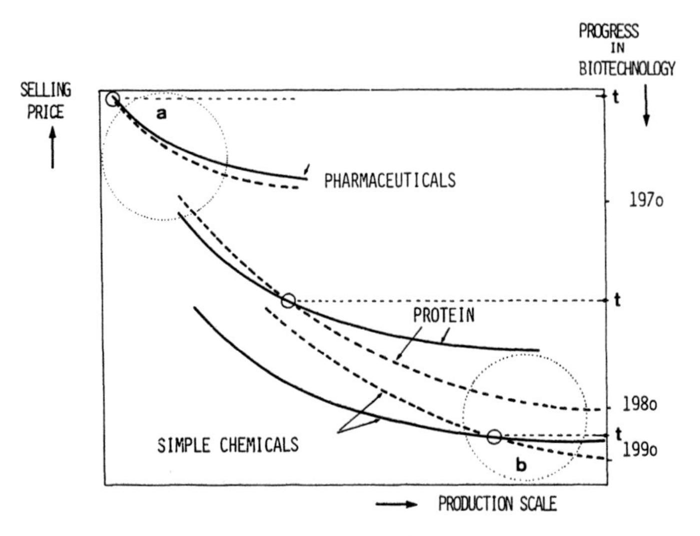

FIGURE 1.2. Predicted development of biotechnical processes (---) compared with development of chemical processes (——) for three classes of products (pharmaceuticals, single-cell protein, simple chemicals) as examples of products with different molecular complexity. The date (*t*) shows when it is predicted that the bioprocess will be more economical than the chemical process. Region a: small-scale, high-price processes. Region b: large-scale, low-price processes. (After Hines, 1980. With permission of Butterworth scientific Ltd.)

The first is that of nature itself. In the earth, in organisms, and in water, biological processes operate on their own. They are often first discovered when human intervention disturbs them. This stage was, and is, the school in which human beings learn through trial and error to reproduce the processes and use them for their own ends. The protocols for doing this are strictly observed and are often kept secret. This was, and is, the stage of the artist and craftsperson in, for example, the making of beer, wine, and sauerkraut. In part, it is still the foundation upon which modern technology builds to create complex products. Modern technology goes beyond this foundation to ensure reproducibility of products through quantitative methods. In the field of chemical engineering, such reproducibility was accomplished several decades ago.
2. Developing completely new methods of production demands a new organization of scientific and technical work. In recent years the problems of raw material supplies, energy, and the environment have posed new questions. With processes based on biotechnology, previously unused or underused natural sources of energy and raw materials can be used or used more intensively, and disturbed ecological cycles can be stabilized (for example, CO_2, sunlight, cellulose, waste materials, and the cycles of the self-cleaning rivers).

Industrial use of biotechnology requires new technical knowledge, new working methods, and valid principles of operation. The new processes extend well beyond the framework of traditional technology in both complexity and scale. Neither simple fermentation technology nor waste water treatment technology, nor the technologies of food processing, pharmaceuticals, or water purification, provide a model adequate to serve as the basis for organization of the new types of production. Traditional technologies developed largely in isolation; they were product oriented. The current existence within each field of independent terms describing technical measures and quantities common to them all is a sign of this product orientation.

To foster the rapid and reliable development of processes based on biotechnology, and especially to foster new production methods applicable on a usable scale, a more unified view is needed. The many similar biotechnical methods used by all of the scientific disciplines represented and by all of the industries involved must be placed on a more universal basis, both from necessity and from consideration of the advantages of universality. The product-oriented way of thinking should be woven into a "process-oriented" viewpoint that unites processes having similar working principles (Dechema, 1974; A. Moser, 1977; Ringpfeil, 1977). In the chemical industry, such process orientation led to the development of chemical engineering. Figure 1.1 shows the process-oriented way of thinking with regard to the food, beverage, and waste water treatment industries. The different bioreactors must be reduced to a few basic types. Finally, general working principles are needed. It is necessary and advantageous to universalize concepts, definitions, and nomenclature.

1.2 Bioprocess Technology

The circumstances just described are the starting point of this book and the initiative for the book's objectives. The term "bioprocess technology," created by the author in 1972 includes the reaction engineering of biochemical processes (Reuss, 1977). The term is now increasingly accepted for the integrated disciplines of engineering sciences in biotechnology (i.e., kinetics, reactors, measurement and control, upstream and downstream processing). Thus, it includes the general working principles for introducing, executing, and optimizing biological processes on a technical scale. The term "biotechnology" fulfills its comprehensive role only when it is understood to mean the scientific discipline of general applicability associated with engineering-scale bioprocesses made *reproducible* through the utilization of *systematic*, *quantitative*, and *system-analytical* methods. The inter and trans-disciplinary character of biotechnology that is reflected in the hybrid name is also found in other areas of engineering. No doubt, the purely technical component of reactors and their layout is fundamentally the same as in chemical engineering. The special feature of biotechnology lies in the nature of the biological processes, in the interaction of biochemistry and biology with technical capabilities. This book reflects this special feature in that it discusses less the problems of reactors and more the problems involved in integrated quantification of bioprocesses. The emphasis here, then, will be on bioprocess kinetics, because kinetics represents, so to speak, the heart of bioprocess technology including interactions with physics.

A further limitation on the contents of this book is the interdisciplinary structure of biotechnology. Complete coverage of all of the component areas of biotechnology exceeds the capabilities of one author. Such a book would also duplicate information already presented in readily available textbooks. Among these are

- Aiba, S., Humphrey, A., and Millis, N.F. (1973, 1976). *Biochemical Engineering*. New York, Academic Press.
- Atkinson, B. (1974). *Biochemical Reactors.* London: Pion, Ltd.
- Atkinson, B., and Mavituna, F. (1983). *Biochemical Engineering and Biotechnology Handbook.* New York: Nature Press; London: MacMillan.
- Bailey, J.E., and Ollis, D.F. (1977). *Biochemical Engineering Fundamentals.* New York: McGraw-Hill
- Bergter, F. (1983). *Wachstum, von Mikroorganismen—Experimente und Modelle.* Jena: G. Fischer.
- Biryukov, V.V. (1985). *Optimization of Periodic Processes for Microbial Synthesis.* Moscow. (in russ.)
- Kafarow, W.W., Winarow, A.J., and Gordiejew, L.S. (1979). *Lesnaja Promyshlenost* (*Modeling of Biochemical Reactors*). Moscow. (in russ.)
- Malek, I., and Fencl, Z. (1966). *Theoretical and Methodological Basis of Continuous Culture of Microorganisms.* New York: Academic Press.
- Moo-Young, M. (ed.-in-chief) (1985). *Comprehensive Biotechnology*, Vol. 1–4. Oxford: Pergamon Press.
- Moser, A. (1981) Bioprozeßtechnik, Springer-Verlag, Vienna, New York.
- Pirt, S.J. (1975). *Principles of Microbe and Cell Cultivation.* Oxford: Blackwell Scientific.
- Präve, P., Faust, U., Sittig, W., and Sukatsch, D.A. (1982). *Handbuch der Biotechnologie.* Wiesbaden: Akademische Verlagsgesellschaft.
- Rehm, H.J., and Reed, G. (eds.) (1982ff). *Biotechnology—A Comprehensive Treatise*, Vol. 1–8, Deerfield Beach, Fl. and Basel: Verlag Chemie Weinheim.
- Schügerl, K. (1985). *Bioreaktionstechnik.* Vol. 1. Aarau Salle u Sauerlander.

Figure 1.3 schematically illustrates the procedures and different fields that contribute to process development from concept to technological maturity. The important component parts (in circles in the figure) are

1. The biochemical-biological basis for the preparation of the catalyst (isolation, culture of pure cell lines, maintenance of strains, enzyme preparation), all the problems of raw materials (the composition of the culture media), and all the problems of analysis (chemical analysis; biochemical, microbiological, and physical methods).
2. The stage of "upstream" processing dealing with the sterilization of raw materials, with cell culture, and with equipment and instrumentation for

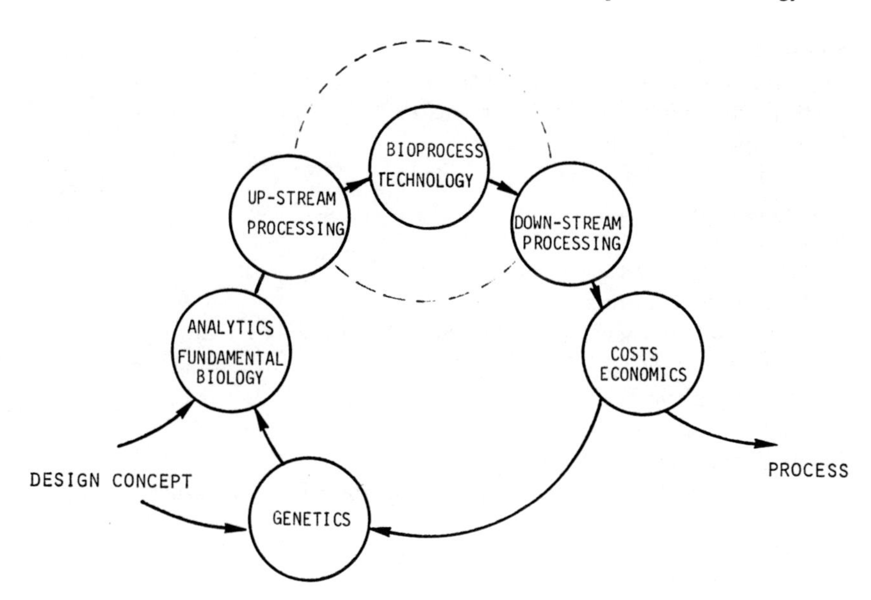

FIGURE 1.3. Areas of biotechnology potentially contributing to the development of a process. (From Moser, 1977.)

process control including techniques for carrying out the process on an experimental scale (measurement and control).

3. Bioprocess technology and its working principles used for planning, evaluating, and calculating (modeling) bioprocesses (kinetics and reactors).
4. "Downstream" processing, which includes methods for the separation of cells, cell rupture, product isolation, and purification.
5. Costs and other economic calculations.
6. Genetics, used for the development of biocatalysts with new characteristics through mutation, selection, and adaptation.

Understanding of the extent of the problems involved in biotechnology can best be reached through examining the professional literature, found in various journals and review series. The most important of these are

Journals
Acta Biotechnologica
Applied Microbiology and Biotechnology
Biochemical Engineering Journal
Bioengineering News
Biofuture
Bioprocess Engineering
Biosensors
Biotech News
Biotechnologie in Nederland

Bio/Technology
Biotechnology Abstracts (Derwent)
Biotechnology and Applied Biochemistry
Biotechnology and Bioengineering
Biotechnology Laboratory
Biotechnology Letters
Biotechnology News
Biotechnology Newswatch
Biotechnology Progress
Chemical and Biochemical Engineering Quarterly
Chemical Engineering Journal
Chemie-Ingenieur-Technik
Current Biotechnology Abstracts
Enzyme and Microbial Technology
European Journal of Applied Microbiology and Biotechnology
Forum Biotechnologie
Journal of Applied Chemistry and Biotechnology
Journal of Biotechnology
Journal of Chemical Technology and Biotechnology
Journal of Fermentation Technology
Journal of Industrial Microbiology
Pascal Biotechnology—Bulletin Signaletique
Practical Biotechnology
Process Biochemistry
Swiss Biotech
Trends in Biotechnology

Review Series
Advances in Biochemical Engineering
Advances in Biotechnological Processes
Annual Reports on Fermentation Processes
Biotechnology Advances
Biotechnology and Bioengineering Symposia
Developments in Industrial Microbiology
Methods in Microbiology
Progress in Industrial Microbiology

The process orientation of this book is illustrated by a comparison between chemical processes and bioprocesses such as simple fermentation or waste water treatment as a case of complex fermentation process from a process engineering viewpoint (Table 1.1). The advantages and disadvantages of the two processes are based on the characteristics of the catalysts. In comparison with the usual chemical catalysts, enzyme catalysts are both highly active and highly selective. In the last decades, even more active chemical catalysts have been developed; however, this has been costly. Since isolation and purification of enzymes also have costs, enzymes in pure form are also not inexpensive.

TABLE 1.1. Comparison of bioprocessing and chemical processing.

Criterion of comparison	*Chemical process*	*Bioprocess*
Catalyst	More or less active and selective Expensive if selective Expensive regeneration	Enzymes: highly active and selective Regeneration easy due to microbial growth "Construction" of new catalysts with genetic engineering
Reaction conditions	Mostly high temperature and high pressure	Mainly ~25°C and ~1 bar pressure Biological demands of nutrients, biological optimum (cf. Fig. 5.1), maintenance, mutation
Raw materials	Pure if poorly selective catalyst	Impure, inactive, and diluted ("unconventional" raw materials: cellulose, starch, oil)
Process	Often multistage processing with recovery of intermediates, resulting in low yields Environmental crisis Fast reaction rates at high concentration level and high yields	One-stage possible without intermediary product recovery (e.g., steroid transformation) Potentially better for environment Mainly slow reaction rates Low concentrations High demand in sophisticated apparatus/equipment, supplementary education "Demon of nature" (biological material, infections, mutations)

Today, use of enzymes in the whole cell as the catalysts is preferred; in general, fermentation technology is still the least costly option. In addition, catalyst regeneration is easily attained through cell growth and division.

All of the facts noted in Table 1.1 result from the nature of the catalyst. A quantitative treatment of the kinetics of biochemical-biological systems, given hereinafter, will mirror these process engineering characteristics. Thus, biological process kinetics will have fundamental formulas for processes of growth and product accumulation which operate with a typically biological optimum in a relatively narrow range of concentrations, pH and temperature. The formulas must, of course, also take into account the metabolism, the effects of the interaction between organism and environment, and the nutritional requirements of the cell.

The process-engineering comparison between simple fermentation and a complex bioprocess such as that used for waste water treatment, shown in detail in Table 1.2, brings out the problems involved in quantifying practices used in complex cases: multiple substrate kinetics operating in either sequential or parallel form; mixed populations; dependence on pH and temperature; the influence of homogeneous or heterogeneous reactor operation in discontin-

TABLE 1.2. Process engineering comparison between simple fermentation and complex fermentation bioprocess (e.g., biological waste water treatment).

Criterion of comparison	*Simple fermentation (lab scale)*	*Complex fermentation bioprocessing (technical scale)*
Catalyst (population)	Mainly pure cultures	Mixed population "biocoenosis"
Raw materials	Synthetic media with S limitations known (optimized media)	Complex media with unknown and even toxic components
Analytical methods	Expensive, modern (e.g., enzymatic analyses, research oriented)	Cheap and simple, often global (e.g., chemical, biological oxygen demand); need for stability (long time) and reproducibility
Temperature, pH	Constant	Gradients
Reactors, mode of operation	Mainly homogeneous type (e.g., stirred tanks) in batch or continuous mode of operation Pseudohomogeneous approach preferred	Heterogeneous types dominate (gradients in homogeneous reactors, transport limitation in three-phase systems, biofilm reactors) Batch, semicontinuous operation
Inflow	Constant	Changes
Kinetics	Idealized conditions: 1-S limitation	Multi-S limitation Microbial interactions (multi-X) Transient operation (dynamics) Interactions between biology and physics Structured approach needed (for biological cell as well as for reactors)

From A. Moser, 1981.

uous, semidiscontinuous, or fully continuous operations; unstable behavior with respect to changes in the quantity or quality of input materials, and special effects such as multiphase operation (heterogeneities) and the problems of applying global methods of measurement that reflect only overall kinetics.

Identification of the separate tasks in the area of bioprocess technology clearly shows that in every case the reaction is closely coupled to the reactor. The quantification of processes involves the kinetics, the transport phenomena, and the interactions between them. This fact influences the whole way of thinking and the whole mode of working. Figure 1.4 illustrates a general strategy for bioprocess kinetic analysis on the basis of this mode of thinking—an "integration strategy" (A. Moser, 1982)—which will be outlined in more detail in Sect. 2.3.1. At the same time, the scheme shown in Fig. 1.4 can serve as a guide to this book: The sequential steps represent the concepts that will be developed in the following chapters.

The strategy followed in the book is in agreement with the theory of deduction (Popper, 1972, 1976) which recently was stressed again for economic-

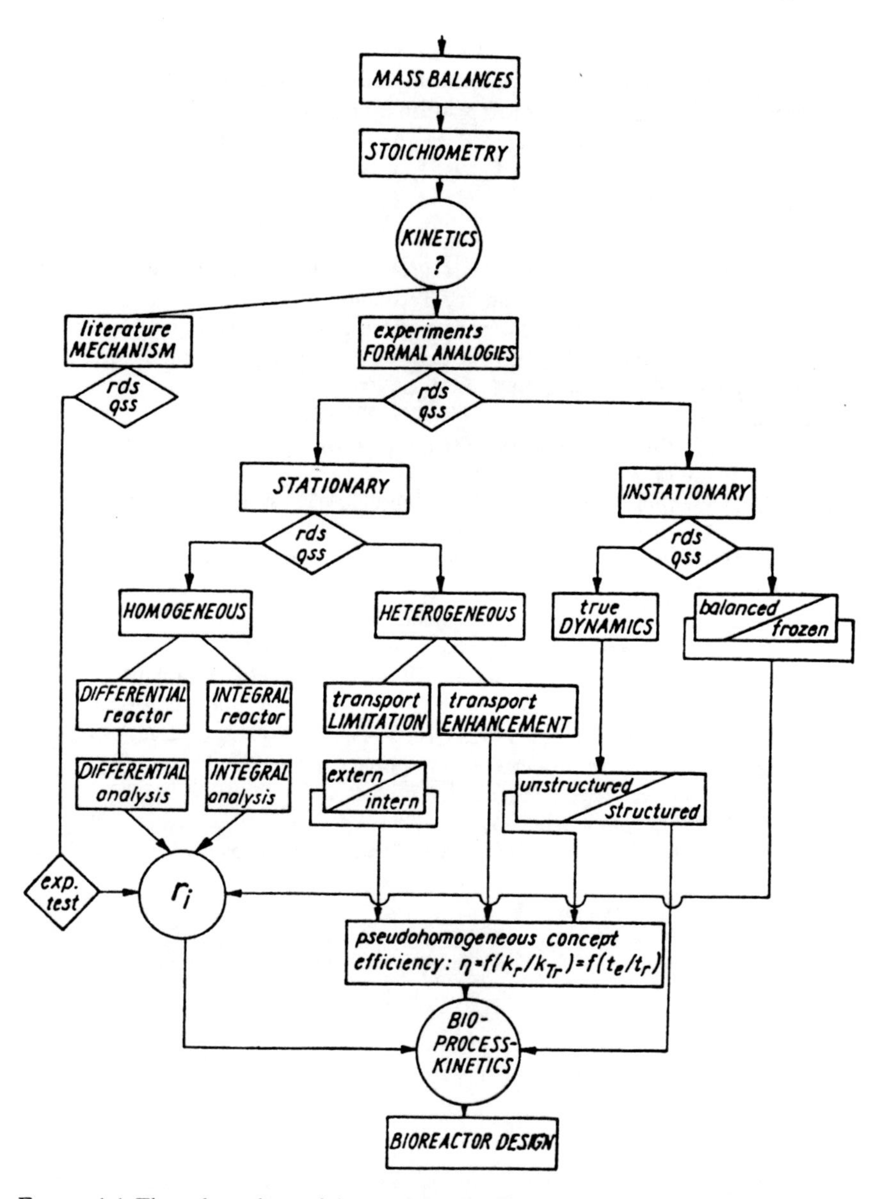

FIGURE 1.4. Flow chart sheet of the strategy of bioprocess kinetic analysis for different process situations: stationary/instationary, homogeneous/heterogeneous, differential/integral, and true dynamic/balanced (frozen) reactor operation. rds, rate-determining step; qss, quasi-steady-state, (From Moser, A. 1983.)

ecological evaluation (F. Moser, 1981, 1983). This scientific strategy is most helpful to analyze and synthetize the complex network of integrated phenomena of biology and physics. As integration means more than summing-up, the success of engineering design work depends on the elucidation of interactions.

Bibliography

Dechema (1974). *Biotechnologie, Studie über Forschung und Entwicklung.* Bonn: Bundesministerium für Forschung und Technologie.

FAST (1980) Sub-programme Bio-society, Research activities Commission of the European Communities EUR 7105. Brussels (M. Cantley)

Hamer, G. (1985). *Trends Biotechnol.*, 3, 73.

Hines, D.A. (1980). *Enzyme Microb. Technol.*, 2, 327–329.

Moser, A. (1977). *Habilitationsschrift*, Technical University, Graz, Austria.

Moser, A. (1978). *Ernährung/Nutrit.*, 2, 505.

Moser, A. (1981). *Bioprozesstechnik.* Vienna and New York: Springer-Verlag.

Moser, A. (1982). *Biotechnol. Lett.*, 4, 73.

Moser, F. (1981). Paper presented at 2nd World Conference of Chemical Engineering, 4–9 Oct., Montreal.

Moser, A. (1988). *Trends in Biotechnology*, Vol. 6, No. 8, August.

Moser, F. (1983). *Chem. Eng. (Lond)*, August/September, 13.

Popper, K.R. (1972). *The Logic of Scientific Discovery.* London: Hutchinson.

Popper, K.R. (1976). *Die Logik der Forschung.* 6th ed. Mohr Studienausgabe.

Rehm, H.J., and Reed, G. (eds.) (1982ff). *Biotechnology—A Comprehensive Treatise.* Deerfield Beach, Fla. and Basel Vol. 1–8 Verlag Chemie Weinheim.

Reuss, M. (1977). *Fort. Verfahrenstechnik*, 15F, 549–566.

Ringpfeil, M. (1977). *Chem. Tech.*, 29, 424–428.

Moo-Young, M. (ed.-in-chief) (1985). *Comprehensive Biotechnology*, Vol. 1–4, Oxford: Pergamon Press.

CHAPTER 2
The Principles of Bioprocess Technology

2.1 Empirical Pragmatic Process Development

Chapter 1 mentioned the central importance of the catalyst in biotechnological processes. A culture suitable for production purposes—that is, a cell culture with sufficiently high population density and capacity for economic production of the product—must be available.

2.1.1 Production Strains

Only about 1/10th of all the types of microorganisms present in nature are known. In principle, a production strain can be isolated from natural sources or can be constructed by genetic manipulations. Experience has shown that the potential of a strain to produce a product can be increased by a stepwise progression from the laboratory scale to the technical scale through screening stages (Aiba, Humphrey, and Millis, 1976). The complete stepwise screening program is shown schematically in Fig. 2.1. The steps for preparing the inoculum for a production-scale reactor begin most often with flasks, and the volume is increased in steps of about 1:10 or more until the size of the production plant is reached (Metz, 1975; Rehm, 1980). The point is to ensure that the large production reactor will have, from the very beginning, a sufficiently high cell concentration to minimize the danger of infection from foreign strains and to permit reaching the final cell density desired for production purposes (normally ca. 2–50 g/l) in an acceptably short period of time.

2.1.2 The Starting Points

In general, four situations are conceivable in starting a program of process development. In practice, a combination of all four usually exists.

1. Establishment of the Catalyst. Chemical and enzyme catalysts offer alternatives to biological fermentation (cf. Table 1.1). Enzyme technology offers the advantages of specificity and selectivity: The multiple-substrate-media

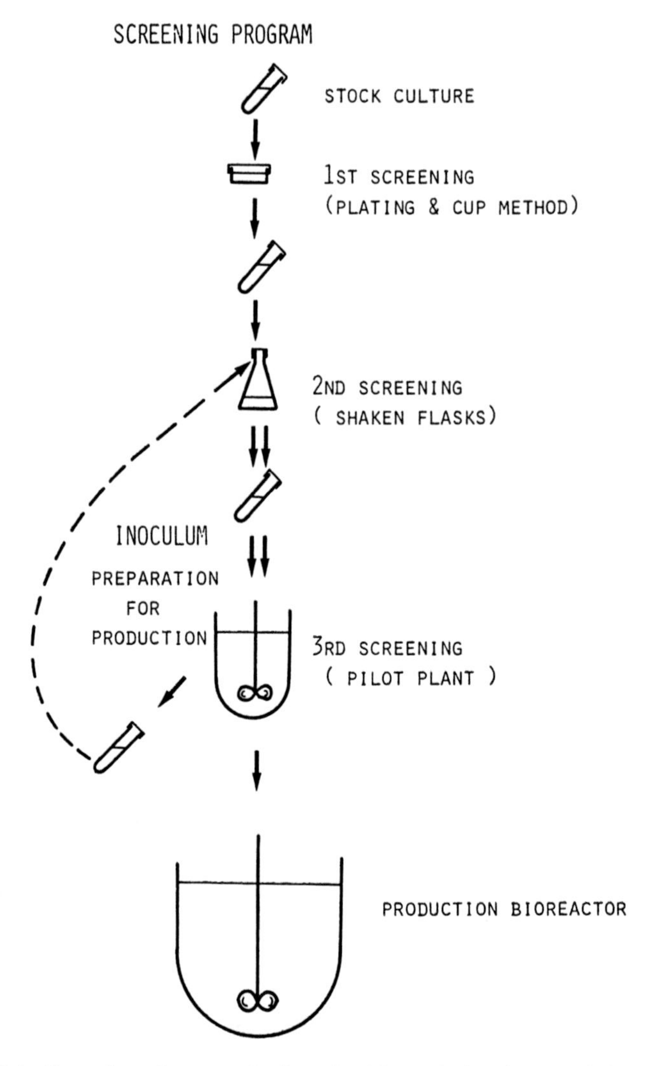

FIGURE 2.1. Procedure for transferring the biocatalytic characteristics of microorganisms from the laboratory to an industrial-scale operation. (Adapted from Aiba, Humphrey and Millis, 1976.)

processes of fermentation technology is replaced with a one-component solution. At the same time, downstream processing, which is normally labor intensive and costly, can be simplified. For a more detailed study of enzyme technology the reader is referred to the literature (e.g., Rehm and Reed, 1982, Vol. 7; Wingard, 1972; Wingard, Katchalski–Katzir, and Goldstein, 1976; Zaborsky, 1973).

2. Work with New Substrates. In the future, one expects that the basic feature

of enzymes (i.e., metabolizing of almost any substrate) may contribute to solving problems involving the supply of raw materials, the environment, and the energy supply. When these biotechnological processes might gain a foothold in industry depends on economic, ecological, and political considerations.
3. Manufacture of New Products. Information on the impressive possibilities for products and the large array of them already being produced may be found in almost every introductory level publication on biotechnology (e.g., Bogen, 1976; Moo-Young, 1985; Rehm, 1980; Rehm and Reed, 1982ff).
4. Application of New Types of Process Techniques. As will be further explored in Chap. 3, a great number of mixing and aeration systems have been developed that could be of industrial interest. All of these reactors can be operated in discontinuous, semicontinuous, or fully continuous modes. The semicontinuous mode of operation offers some advantages (Pickett, Topiwala, and Bazin, 1979). However, a fully continuous operation seems to be limited (infections, strain stability, continuous flow).

2.1.3 Different Modes (or Strategies) in Process Development

In the development of a process, precise information from the basic sciences is joined with information from the applied engineering sciences. From the standpoint of process technology, the problem area of every technology is the need to increase the scale of measurements by which the kinetic phenomena are coupled with transport phenomena. Findings about the behavior of systems of cells interacting with their environment, studied in the microbiological laboratory using 100-ml to 1-l Erlenmeyer flasks, must be extrapolated to large plants that might work with volumes in excess of 100 m^3.

The principal routes available in developing a process (Moser, 1977a) can be summarized as follows:

1. Trial and error. This has been successful in the past but requires a long time and entails much loss.
2. Pure theoretical route via the solution of all differential equations. This seems to be a utopian path for the future.
3. Engineering approach based on quantification, including pilot plant work and work with or without mathematical models. Pilot plants are generally 10 times larger than the laboratory reactors but are smaller than production-scale plants. In the first step, the volume of the pilot plant reactor is typically about 50 to 500 l; a second step extends this to 500 to 5000 l.

2.1.4 Process Development Without Mathematical Models

The individual steps of a process development procedure not involving a mathematical model are shown in Fig. 2.2. The important characteristic

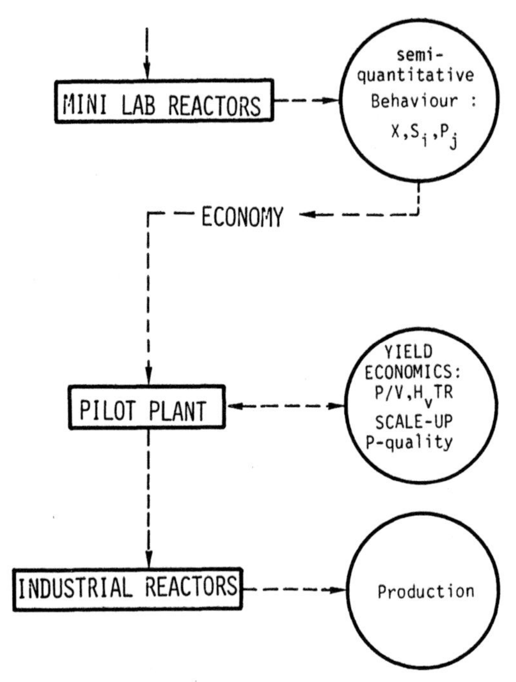

FIGURE 2.2. Process development involving no mathematical models: the empirical pragmatic approach. X, biomass; S_i, substrate, P_j, products; P/V, power/volume; H_vTR, volumetric heat rate.

feature of this route is that the type of reactor to be employed is determined pragmatically or intuitively on the basis of information from the literature or from experiments carried out according to more or less accidental considerations, such as availability.

The first phase, exploratory research, takes place mainly in the microbiological, biochemical, and chemical laboratories and yields mainly qualitative results concerning the reaction components—the cells and the nutrient medium. Typical results from experiments in Erlenmeyer flasks include

- Characterization of the cell culture in terms of, for example, the shape of the cells and the population (morphology).
- Characterization of the nutrient media with regard to temperature, pH, aerobic or anaerobic characteristics, and composition of a "minimal medium" (the limiting substrate concentration).
- Type of metabolism and range of products (physiology).

Optimization of the media can be done in two ways: (a) balancing the nutrient solution, the cells, and the remaining fluid (Dostalek et al., 1972; Häggström, 1977; Pirt, 1974), or (b) experiments in a chemostat to determine the influence on growth of different concentrations of the components of the growth medium (Kuhn, Friedrich, and Fiechter, 1979; Mateles and Batat, 1974; Tsuchiya, Nishio, and Nagai, 1980).

After a preliminary estimate of costs to see whether production is economically justifiable, that is, whether production costs are low enough that the product can be profitably marketed (see Sect. 2.2.4), a pilot plant is built. Because of the desire for more rapid development and minimized costs, the pilot plant is usually constructed on the smallest possible scale. The pilot plant gives results concerning "space–time–yield" relationships (see Sect. 2.2.3) and some economic factors (H_VTR, P/V), and provides indications about product quality and the efforts required in the final phases of separating and purifying the product. The pilot plant also aids in selecting the criteria for a scale-up (Hockenhull, 1975; Oosterhuis, 1984). This choice, like the choice of reactor type, results most often from pragmatic considerations and experience with the theory of similarity. In the case of presumed geometric similarity, the following quantities are used as pragmatic scale-up criteria, if a systematic scale-up is executed at all:

- P [W], resp. P_G/V [$kW \cdot m^{-3}$]
- F_P, pumping capacity of stirrer, resp. F_L [$m^3 \cdot s^{-1}$] or F_G/V_L
- v_{tip} [$m \cdot s^{-1}$] or $\dot{\gamma}$ [s^{-1}]
- Re_{st} [$-$]
- k_La [h^{-1}]
- r_O [$kgO_2 \cdot m^{-3} \cdot h^{-1}$]
- $v_{S,L}$ [$m^3 \cdot m^{-2} \cdot s^{-1}$]
- $t_{m,L}$ or $t_{c,L}$ [s]
- $D_{eff,L}$ [$m^2 \cdot s^{-1}$] in case of continuous plug flow reactor (CPFR)

All of these criteria are dependent variables. The variables available for a theoretical calculation are v_L, ρ_L, d_i, n, d_T, and F_G.

At the moment there is no way to attain a proper scale-up, although in the recent past there have been important developments in this field (Kossen and Oosterhuis, 1984; Moser 1977a; Ovaskainen, Lundell, and Laiho, 1976; Reuss et al., 1980).

With this type of process development, it is more interesting to establish the special conditions of an essentially given mode of operation in an also given reactor than to attempt optimization. The construction of a mathematical model, and any later systematic optimization of processes in the individual reaction steps, are in practice not done. The time and expense associated with model building are often thought to be economically unjustified: Quantification of the experimentally measured conversion rate, or productivity (yield), is sufficient, without separation of phenomena into biological and physical components. In practice, the space–time–yield mode of thought will continue to be justified in most cases.

A more systematic process development strategy extends beyond simple quantification of yield based on gross quantities (see Sect. 2.3.2). It involves mathematical models that are, among other things, useful for later optimization or computer-coupled systems operation (Stephanopoulos and San, 1984). In the future, models will be used together with systematic process design whenever (a) competition with other industries or the crisis in the environment, in raw materials supply, or in energy supply make optimization of the process and (b) improved methods of data acquisition, handling, and evaluation are available for key variables.

2.2 Basics of Quantification Methods for Bioprocesses

Figure 2.3, a typical schematical diagram for a bioprocess, represents the activities of microbial cells according to the macroscopic principle (see Sect. 2.4.1). In most cases the organisms live in the liquid phase (L) and the nutrients S_i and gases (such as O_2 in the case of an aerobic process) must be transported through phases and interfacial surfaces. The products P_j, CO_2 as a gas and heat (H_v = volumetric heat of reaction, $J \cdot l^{-1}$) must be transported away. The possible limiting steps involving transport are shown in Fig. 2.3 along with the steps numbered 1 to 4a, which will be important later (see Sect. 4.2).

Quantitative treatment is possible only for variables that are measurable. Such variables should be easily, reliably, and rapidly measurable in practice. Further, the variables measured should be key variables. The future development of measuring instrumentation will certainly bring further progress in this area. In practice, data collection is still dominated by the process variables listed in Table 2.1.

The quantitative treatment of such a simplified process scheme with six variables is derivable from the following gross equation with stoichiometric

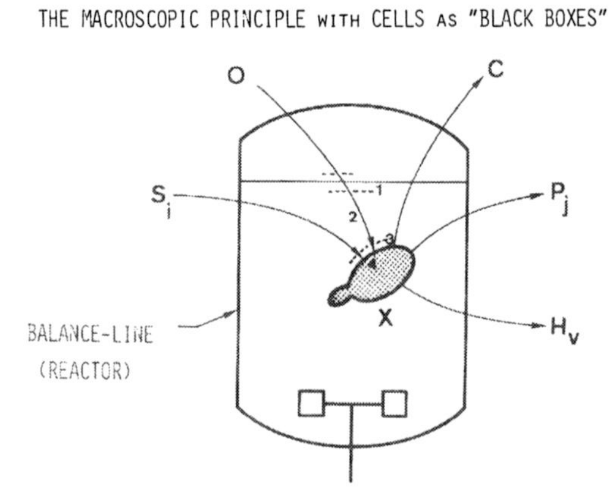

FIGURE 2.3. The macroscopic principle applied to bioprocessing expressed with pseudohomogeneous observable process variables in the liquid phase (L): biomass X, substrates S_i, oxygen O, products P_j, carbon dioxide C, and volumetric heat H_v. Pseudohomogeneity is checked by considering a series of mass transfer steps: L film at the gas phase–L interface (1), L bulk (2), L film at the L–Solid phase (S) interface (3), cell wall and membranes (4), resp. S-phase cell mass with cytoplasm.

TABLE 2.1. Conventional process variables and their measurements.

Symbol	*Process variable*	*Measuring method*
X	Biomass	Cell dry weight, turbidity, cell number
S	Substrates	Enzymatic analyses—global methods (e.g. biological, chemical oxygen demand)
P	Products	Enzymatic analyses or special method
O	Oxygen	P_{O_2}-electrode, gas analysis
C	Carbon dioxide	P_{CO_2}-electrode, gas analysis
H_v	Heat of fermentation	Temperature, heat balance

coefficients ν_i:

$$\nu_S S_i + \nu_O O_2 + X_0 \xrightarrow[T,\mathrm{pH}]{} \nu_X X + \nu_P P_j + \nu_C CO_2 + \nu_H H_v \tag{2.1}$$

This equation holds in general for discontinuous fermentations operating aerobically with the production of CO_2. In some cases, of course, such as reactions with algae, CO_2 is used and O_2 is produced. In principle, all fermentation processes are autocatalytic over a wide range—the presence of cells promotes the reaction. This is expressed in Equ. 2.7.

A quantitative treatment of bioprocesses includes the following basic concepts:

- Stoichiometry and thermodynamics
- Rate of the biological process
- Productivity, conversion, yield
- Costs, profits

2.2.1 Concepts of a Uniform Nomenclature for Bioprocess Kinetics

1. There is a distinction among symbols for the absolute rate of a reaction (r_i), the rate of transport of a substance (n_i), and the rate of heat (q). Lower case italic letters are used; the dot indicating the first derivative is omitted.
2. The specific, or relative, rates of growth (u) and of synthesis (q_p or q_c or q_{H_v}) and consumption (q_s or q_o) are distinguished from the absolute rates of consumption (r_s or r_o) or formation (r_x, r_p, r_c, r_{H_v}).
3. Rate constants are always designated with k, uniformly for both reactions and transport: k or k_r or k_{TR} with k_L, k_S, k_G, and k_{H_v}. The rate of reaction, with n the order of the reaction is generally

$$r_i = k_r \cdot c_j^n \tag{2.2a}$$

The dimensions of the reaction rate constant, k_r, are $[t^{-1}(\mathrm{mol/V})^{1-n}]$ thus, dependent on the order of the reaction. For the rate of mass transfer (with Δc = the concentration gradient):

$$n_i = k_{TR} \cdot \Delta c \tag{2.2b}$$

TABLE 2.2. Basic concept of unified nomenclature in bioprocess engineering.

Variable	*Biokinetics*	*Mass transport*	*Heat transport*
Rates			
Absolute (e.g., for discontinuous stirred tank reactor)	$r_i = \pm\frac{dc_i}{dt}$ [$g \cdot l^{-1} \cdot h^{-1}$]	$n_i = \pm\frac{dc_i}{dt}$ [$g \cdot l^{-1} \cdot h^{-1}$]	$q = \frac{dh_v}{dt}$ [$kJ \cdot l^{-1} \cdot h^{-1}$]
	$r_X, r_S, r_O, r_P, r_C, r_{H_v}$	n_O, n_C	
Relative (specific)	$\frac{dc_i}{dt} \cdot \frac{1}{x}$		
	$\mu, q_S, q_O, q_P, q_C, q_{H_v}$		
Rate constants (coefficients) $r_i = k \cdot (c_j)^n$	k_r $k_P, k_d, \ldots$	k_{Tr} $k_{L1}, k_{L2}, k_S, k_{tot}, k_g$	k_{H_v}
Stoichiometry			
Microscopic: stoichiometric coefficients $\nu_A \cdot a \rightarrow \nu_B \cdot b + \nu_C \cdot c$	$\nu_{ji}(\nu_A, \nu_B, \ldots)$		
Macroscopic: yield coefficients	$Y_{j\mid i}$		
$r_i = \pm\frac{1}{Y_{j\mid i}} \cdot r_j$	$Y_{X\mid S}, Y_{X\mid P}, Y_{H_v\mid X} \ldots$		
Equilibrium constants	K $K_m, K_S, K_I \ldots$		

and for the heat transfer rate (ΔT = temperature gradient):

$$q = k_{HTR} \cdot \Delta T \tag{2.2c}$$

Here, k_{TR} and k_{HTR} are phenomenological transfer coefficients, which are further interpreted using true transfer coefficients (k_{L1}, k_{L2}, k_H) and the corresponding interfacial areas ($a_{G\mid L}, a_{L\mid S}, a_H$).

4. The symbol Y is proposed for yield coefficient (see Equ. 2.13).
5. The stoichiometric coefficients are designated as ν_i, according to Equ. 2.1.
6. The concentrations of substances are indicated with lowercase letters, in accord with the nomenclature of the enzyme literature. Capital letters are used for quantities of materials (e.g., $X = x \cdot V$).
7. The symbol K is exclusively used as the equilibrium, saturation, or inhibition constant in a kinetic equation.

Tables 2.2 and 2.3 summarize the nomenclature used in this book and compare it with the nomenclature used elsewhere.

2.2.2 The Rates of a Bioprocess

The concentration–time profile for the six process variables shown schematically in Fig. 2.3 is shown in Fig. 2.4. This is for the case of a discontinuous process with constant reactor volume V_R. Conditions that can be observed and

TABLE 2.3. Symbols used for rates in bioprocessing.

			specific rates $[h^{-1}]$				
			this textbook:				
Varible	*concentration* $(g \cdot l^{-1})$	*Absolute rate* $(g \cdot l^{-1} \cdot h^{-1})$	*positive*	*negative*	*Pirt* (*1975*)	*Moser* (*1981*)	*IUPAC* *proposal* (*1982*)
Enzyme, *E*	e	r_E	k_{cat}				
Cell mass, *X*	x	r_X	μ	(μ_d, k_d, k_e)	μ	μ	μ
Cell number, *N*	n	r_N	ν			ν	
Substrate, *S*	s	r_S	q_S	(m_S)	$q_S(m_S)$	$\sigma(\sigma_e)$	q_S
Oxygen, *O*	o	r_O	q_O	(m_O)	q_{O_2} [mmol/gX · h]	$\sigma_O(\sigma_{O,e})$	q_{O_2}
Product, *P*	p	r_P	q_P		q_P	π	q_P
Carbon dioxide, *C*	c	r_C	q_C	(m_C)	q_{CO_2}	π_C	q_{CO_2}
Heat, H_v	h_v $[kJ \cdot l^{-1}]$	r_{H_v} $[kJ \cdot l^{-1} \cdot h^{-1}]$	q_{H_v}	(m_{H_v})		π_{H_v}	

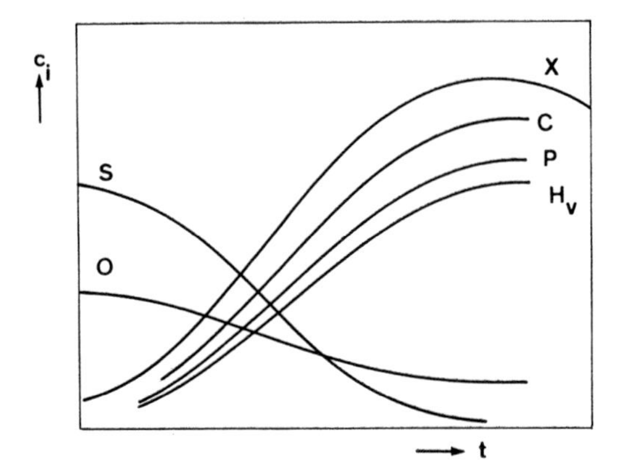

FIGURE 2.4. Typical concentration–time curves for fermentation process in discontinuous culture, with reactor volume constant. S, substrate; O, oxygen; X, biomass; C, carbon dioxide; P, product; H_v, heat.

measured are growth of biomass X, consumption of S_i and O_2, and accumulation of P_i, CO_2, and H_v. The rates involved are of prime interest for reactor balancing and kinetics. In formulating the rate equations, one must derive or define them from the conservation of mass equation (see Sect. 2.3.1). This equation contains two principal terms: (a) sources and sinks, where component i is made or used, and (b) the mass flux, n_i' [g/cm^{2}s], of component i through the surface volume element under consideration.

The flow of mass $n_i' = d(N/A)/dt$ can be due to various kinds of transport phenomena, and it is described by various terms (for example, the convective flow velocity v; or conduction through a gradient, that is, diffusion, D; or dispersion, D_{eff}; or interfacial transport, k_{TR} (see Fig. 2.3).

A change in the concentration c_i with time is expressed in Equ. 2.3. In general, the equation can be written using a vector operator ∇ (grad = partial derivative with respect to a direction coordinate) as follows:

$$\frac{\partial c_i}{\partial t} = -\nabla n_i' \pm r_i \tag{2.3a}$$

Separating the various transport terms with $n_i' = c_i \cdot v$ and $v = F/A$, this equation can be written in vector notation as

$$\frac{\partial c_i}{\partial t} = [-\mathrm{div}(c_j \cdot v) + \mathrm{div}(D \text{ grad } c_j) + k_{TR} \cdot \Delta c_i] \pm r_i \tag{2.3b}$$

or, in one-dimensional form (for example, for tubular reactors, cf. Equ. 3.4), as

$$\frac{\partial c_i}{\partial t} = -v_z \frac{\partial c_i}{\partial z} + D \frac{\partial^2 c_i}{\partial z^2} + k_{TR} \cdot \Delta c_i \pm r_i \tag{2.3c}$$

The rates of formation and/or consumption of component i are intensive variables [$g \cdot l^{-1} \cdot h^{-1}$], and they differ from each other in that the formation rate is positive and the consumption rate is negative. Equation 2.3c can be reduced to Equ. 2.3d for the rate of formation or consumption in a discontinuous stirred tank reactor:

$$\frac{dc_i}{dt} = \pm r_i \tag{2.3d}$$

Historically, this was the first rate equation, and later it was arbitrarily used as the definition of a "true" reaction rate. However, Equ. 2.3d is valid only with V_R = constant, T = constant, and ideal mixing. Furthermore, it represents only the rate of formation or consumption of a component. Other cases, with other reactor configurations, will be discussed in Chapter 3; in that Chapter, due to the introduction of a transport term, more complex formulas define rates.

We can now express the rates associated with the model process of Fig. 2.4 for the case of an ideally mixed discontinuous reactor with V_R = constant and T = constant. The rate of growth (biomass formation) is

$$r_X = \frac{dx}{dt} \tag{2.4a}$$

The rates of oxygen and substrate utilization are

$$r_S = -\frac{ds}{dt} \quad \text{and} \quad r_O = -\frac{do}{dt} \tag{2.4b}$$

The rates of synthesis for P, C, and H_v are

$$r_P = \frac{dp}{dt}, \quad r_C = \frac{dc}{dt}, \quad r_{H_v} = \frac{dh_v}{dt} \tag{2.4c}$$

These rates have dimensions of [$g \cdot l^{-1} \cdot h^{-1}$] or [$kJ \cdot l^{-1} \cdot h^{-1}$] and are intensive quantities that may be substituted directly into the mass balance equation. On the other hand, these absolute rates may take on any value and are therefore not characteristic of the system. Comparable variables, which are biologically representative, are the so-called specific rates of production or utilization, which refer to the catalytically active mass (to a first approximation, the concentration of biomass, x, is taken as the dry cell weight). With this, one has the definitions of the specific rates for bioprocesses of Equs. 2.5a–f where, in each case, the specific rate has the dimension [h^{-1}]. For growth, the specific rate is μ

$$\frac{1}{x} \cdot \frac{dx}{dt} = \mu \, [h^{-1}] \tag{2.5a}$$

or

$$\frac{1}{n} \cdot \frac{dn}{dt} = \nu \, [h^{-1}] \tag{2.5b}$$

For the consumption of substrate, the specific rate is q_S

$$q_S = \frac{1}{x} \cdot \frac{ds}{dt} \, [\mathrm{h}^{-1}] \tag{2.5c}$$

or for O_2

$$q_O = -\frac{1}{x} \cdot \frac{do}{dt} \, [\mathrm{h}^{-1}] \tag{2.5d}$$

The specific rate of product accumulation is q_P

$$q_P = \frac{1}{x} \cdot \frac{dp}{dt} \, [\mathrm{h}^{-1}] \tag{2.5e}$$

and for CO_2

$$q_C = \frac{1}{x} \cdot \frac{dc}{dt} \, [\mathrm{h}^{-1}] \tag{2.5f}$$

and for H_v

$$q_H = \frac{1}{x} \cdot \frac{dh_v}{dt} \, [\mathrm{kJ} \cdot \mathrm{g}^{-1} \cdot \mathrm{h}^{-1}] \tag{2.5g}$$

These definitions for specific rates are analogous to the definition of specific rates r_i^* in chemical kinetics, where the mass of the active catalyst, M_c in the case of heterogeneous catalysis, is used to define a specific rate

$$r_i^* = \frac{1}{M_c} \cdot \frac{dN_i}{dt} \tag{2.5h}$$

The reactor volume $V_R(V)$ is used predominantly in the case of homogeneous catalysis

$$r_i^* = \frac{1}{V} \cdot \frac{dN_i}{dt} \tag{2.5i}$$

In some cases other quantities are useful, for example, the external catalyst surface or the volume of the solid phase.

The quantity N_i in Equs. 2.5h and 2.5i represents the number of moles of component i ($c_i = N_i/V$). The connection between the definitions of Equ. 2.5i and Equ. 2.4 can be made by substituting this fact

$$r_i^* = \frac{1}{V} \cdot \frac{d}{dt}(c_i \cdot V) = \frac{dc_i}{dt} + \left(\frac{c_i}{V} \cdot \frac{dv}{dt}\right) \tag{2.5k}$$

Equation 2.5k is identical to Equ. 2.3d for the case of V = constant.

For the sake of completeness, Equ. 2.5i is often taken as the definition of the reaction rate in chemical kinetics. Nevertheless, the true reaction rate, r, is connected with the previously indicated and used rates of growth or consumption, r_i or r_i^*, through the stoichiometric coefficients ν_i

$$r = \frac{\pm r_i}{\pm \nu_i} \tag{2.6}$$

For consumption, $\nu_i < 0$ and for growth, $\nu_i > 0$. Naturally, a component can also have no reaction rate. A reaction rate can only be defined in cases where both consumption and formation occur as indicated by Equ. 2.6.

For bioprocesses, the stoichiometric coefficients are seldom known (Cooney, Wang, and Wang, 1977), so the reaction rate in the strict sense of Equ. 2.6 can seldom be used. However, rates of consumption or formation of the type given in Equs. 2.5a–g are well defined.

Finally, the autocatalytic nature of fermentation becomes evident from Equ. 2.7, which can be derived from Equ. 2.5a

$$r_X = \mu \cdot x \tag{2.7}$$

2.2.3 Stoichiometry and Thermodynamics

2.2.3.1 General Remarks

Considered as the study of the quantitative composition of chemical compounds and the quantitative conversion of chemical reactions, stoichiometry belongs to the fundamentals of reaction engineering, along with thermodynamics and kinetics. Industrial plants for bioprocessing on a large scale will be expensive to construct, and therefore maximum profit will have to be extracted from the process. Intensive studies of the quantification of metabolic reactions in different reactor configurations are required, studies that examine all factors making for maximum yield. Hence the principle of balancing plays a central role. One can distinguish between macrobalances and microbalances. For conservative properties, which are not altered during the reaction, only macrobalances are used; for nonconservative properties, microbalances are used as well (Kossen, 1979).

In contrast to kinetics, where laws are useful for predictions, in stoichiometry calculations have to be based on knowledge of the reaction mechanism, the balances of the initial values of the significant compounds ("key variables"). It must be remembered that, just as in the balancing of chemical processes, macroconversion in bioprocessing is always the sum of transport phenomena and biological transformation reactions. As a consequence, Equ. 2.3 must be applied in all special cases of reactor operation modes (see Fig. 2.5). Modifications are needed for complex reactions and complex reactor operations. In this basic equation, all facets of reaction engineering sciences are manifest: balancing, stoichiometry, kinetics of reaction rates, and physical transport. Thermodynamics are not directly included here because thermodynamics emphasizes only initial and final states of the system, whereas kinetics deals primarily with the time required for the transitions involved. Stoichiometry includes the rules linking changes in the composition of reaction components during the process. Thermodynamics gives information about the maximum

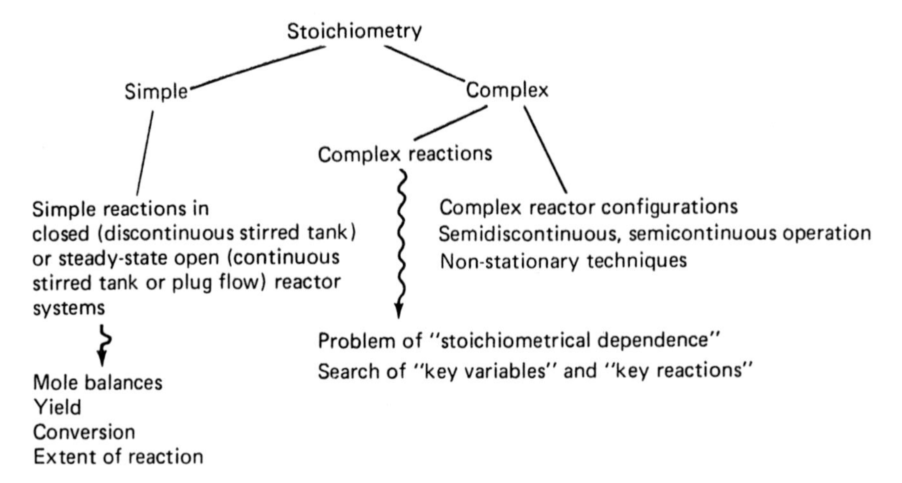

FIGURE 2.5. Overview of the problems involved in stoichiometry of technical processes.

possible extent of reaction and the heat liberated or absorbed and also permits calculation of the equilibrium constant from the standard free energy.

The activities of microbial cells in bioreactors can also be expressed by the balance equation for the chemical elements, instead of Equ. 2.1, considering the conservation of the atomic species (carbon C, hydrogen H, Oxygen O, nitrogen N):

$$\nu_S(C_aH_bO_cN_d) + \nu_O(O_2) + \nu_N(NH_3)$$
$$\rightarrow \nu_x(C_\alpha H_\beta O_\gamma N_\delta) + \nu_p(C_{\alpha'}H_{\beta'}O_{\gamma'}N_{\delta'}) + \nu_c(CO_2) + \nu_w(H_2O) + \nu_{H_v}\cdot H_v$$

Here, $C_aH_bO_cN_d$ is a generalized carbon source, $C_\alpha H_\beta O_\gamma N_\delta$ a cell unit, and $C_{\alpha'}, H_{\beta'}, O_{\gamma'}, N_{\delta'}$, a product. In special cases, the elements sulfur and phosphorus can also be included.

The overall biosynthetic equation, Equ. 2.8, is the net equation resulting from hundreds of metabolic reactions in the living cell. The various cycles and chains in the metabolic network, including the ATP system and other energy-handling systems that do not result in the output of new cells or products, do not contribute to the net reaction. Hence, detailed knowledge of these cycles is unnecessary in a macroscopic treatment using the net stoichometric equations. Bioprocess analysis, therefore, is made less complex with no significant loss of information.

The general form for the conservation of atomic species in the case of multiple reactions, expressed as a sum, is given by

$$\sum_j \nu_j \cdot r = 0 \tag{2.9}$$

Direct application of Equ. 2.9 to the stoichiometry of bioprocesses does, however, present some difficulty, because several hundred components take part in metabolism. This problem has thus far been avoided, and only a limited number of compounds have been considered in a so-called gross stoichiometric equation of growth. This approach has been theoretically justified (Roels, 1980a).

A practical implication of Equ. 2.3 using Equ. 2.6 in a case of multiple reactions in the steady state, with $\partial c_j/\partial t = 0$, written as a sum

$$\sum_j \nabla n_j' = -\sum_j v_j \cdot r \tag{2.10}$$

is that the stoichiometry of a reaction pattern can be readily studied experimentally in steady state by observations of the exchange flows ($\nabla n_j'$) with the environment. There is a group of extensive quantities, called conservative quantities, that have the property that in the reaction pattern of the system no net production take place (Roels and Kossen, 1978). The balance equation for conservative properties has no production term

$$\sum_j \nabla n_j' = 0 \tag{2.11}$$

In the steady-state case, the reasoning presented formally validates the treatment of microbial metabolism in terms of a gross stoichiometric equation of growth and product formation (Roels, 1980a). This formalism, however, results in complicated equations that offer little advantage over other equations. Therefore, only situations of limited complexity can be studied in detail.

The main problem in applying stoichiometric considerations to bioprocessing (beyond quantification in non-open-reactor systems) arises from the complex metabolic reaction network. In simple reactions stoichiometry is trivial, and complex reactions can only be handled with the aid of a formal mathematical approach analogous to the approach for complex chemical reactions (Schubert and Hofmann, 1975). In such a situation, an elementary balance equation must be set up. Due to complexity, it is not surprising that the approach first used in the quantification of bioprocesses was much simpler—the concept of yield factors Y. This macroscopic parameter Y cannot be considered a biological constant.

2.2.3.2 Definition of a "Mole" of Microorganisms

Several suggestions have been made as to how the molar concept may be applied to microorganisms. Elementary analysis, performed by standard Pregl microcombustion techniques on cells, has shown that

1. There is no unique empirical formula for a microorganism—the elementary composition varies with the growth rate and limiting nutrient. For yeast, for example, the following "formula" was given (Harrison, 1967):

$$C_{3.72}H_{6.11}O_{1.95}N_{0.61}S_{0.017}P_{0.035} \quad \text{or} \quad C_6H_{11}O_3N \tag{2.12a}$$

   For aerobic microorganisms, the four principal elements are present in a ratio of $C_5H_7O_2N$ (Hoover and Porges, 1952), which is very similar to that obtained for anaerobic microorganisms (Speece and McCarty, 1964). A widely used formula that includes phosphorus is (McCarty, 1970)

$$C_{60}H_{87}O_{23}N_{12}P \tag{2.12b}$$

2. There is, consequently, no unique elementary balance equation for microbial growth; rather, there is a range of such equations with different numerical coefficients.

All quantities considered (e.g., number of organisms containing 1 g · atom of total nitrogen or of protein-nitrogen, or of DNA phosphorus) vary with growth rate and limiting substrate. It would seem logical to derive the concept of a "mole" of microorganisms from the empirical formula that (neglecting sulfur, phosphorus, and ash) can be expressed in the general form $C_aH_bO_cN_d$. Finally, by dividing all the subscripts by *a*, giving the formula $CH_{b/a}O_{c/a}N_{d/a}$, the definition of 1 "*C-mole*" represents the quantity of microorganisms containing 1 g · atom (12.011 g) of carbon (McLennan et al., 1973). The general use of this formula for expressing quantities of microorganisms on a molar basis is strongly recommended (Herbert, 1975).

2.2.3.3 The Concept of Yield Factors (Yield Constants)

A parameter that is of great relevance to the economic evaluation of the potential of, for example, a carbon source for biomass or product formation is the macroscopic yield factor $Y_{i|j}$. This yield factor was originally defined by Monod (1942) in mass units by the quotient

$$Y_{X|S} = -\frac{\Delta X}{\Delta S} = \frac{X_t - X_0}{S_0 - S_t} \approx -\frac{dx/dt}{dS/dt} = -\frac{r_X}{r_S} = \frac{dX}{dS}\left[\frac{\text{g cells formed}}{\text{g substrate used}}\right] \tag{2.13}$$

In Fig. 2.6, a schematic evaluation of the yield constant is shown, the value for $Y_{i|j}$ can be taken from the slope of the line, representing, for example, the data r_s versus r_x. The definition of yield factor used by Monod is of a purely macroscopic nature and can be applied to all compounds, in analogy to Equ. 2.13, to quantify cell or product yields including heat, by a relation between

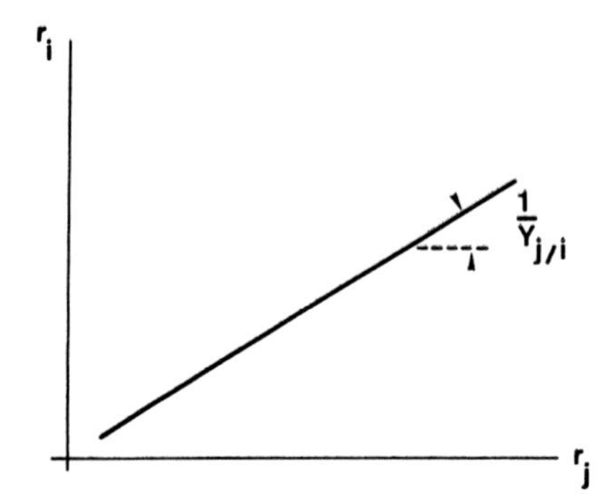

FIGURE 2.6. Evaluation plot for yield coefficients: Rate of consumption of production of component i versus rate of production of consumption of component j.

TABLE 2.4. Summary of macroscopic yield coefficients in bioprocessing.

Yield coefficient $Y_{i\|j}$*	*Reaction or relation between components*	*Definition*	*Equ. no. 2.14a–o*
$Y_{X\|S}$	$S \rightarrow X$	$r_S = -(1/Y_{X\|S}) \cdot r_X$	2.14a
$Y_{X\|O}$	$O_2 \rightarrow X$	$r_O = -(1/Y_{X\|O}) \cdot r_X$	2.14b
$Y_{P\|S}$	$S \rightarrow P$	$r_S = -(1/Y_{P\|S}) \cdot r_P$	2.14c
$Y_{C\|S}$	$S \rightarrow CO_2$	$r_S = -(1/Y_{C\|S}) \cdot r_C$	2.14d
$Y_{P\|O}$	$O_2 \rightarrow P$	$r_O = -(1/Y_{P\|O}) \cdot r_P$	2.14e
$Y_{H_v\|S}$	$S \rightarrow H_v$	$r_S = -(1/Y_{H_v\|S}) \cdot r_{H_v}$	2.14f
$Y_{C\|O}$	$O_2 \rightarrow CO_2$	$r_O = -(1/Y_{C\|O}) \cdot r_C$	2.14g
$Y_{H_v\|O}$	$O_2 \rightarrow H_v$	$r_O = -(1/Y_{H_v\|O}) \cdot r_O$	2.14h
$Y_{X\|P}$	$P \sim X$	$r_P = (1/Y_{X\|P}) \cdot r_X = (Y_{P\|X}) \cdot r_X$	2.14i
$Y_{X\|C}$	$CO_2 \sim X$	$r_C = (1/Y_{X\|C}) \cdot r_X$	2.14j
$Y_{X\|H_v}$	$H_v \sim X$	$r_{H_v} = (1/Y_{X\|H_v}) \cdot r_X$	2.14k
$Y_{S\|O}$	$O_2 \sim S$	$r_O = (1/Y_{\|O}) \cdot r_S$	2.14l
$Y_{C\|P}$	$P \sim CO_2$	$r_P = (1/Y_{C\|P}) \cdot r_C$	2.14m
$Y_{P\|H_v}$	$H_v \sim P$	$r_{H_v} = (1/Y_{P\|H_v}) \cdot r_P$	2.14n
$Y_{C\|H_v}$	$H_v \sim CO_2$	$r_{H_v} = (1/Y_{C\|H_v}) \cdot r_C$	2.14o

* Dimension in the case of mass components: [g component i/g component j]; dimension in the case of heat components: [kJ/g component j].

the amounts consumed and the amounts formed. The same concept of yield can be applied to quantify the relation between consumed or produced amounts. The definitions of yield factors are summarized in Table 2.4 and represented by Equs. 2.14a–o.

In some instances, however, it is more advantageous to express yields in the units of moles:

$$\left[\frac{\text{g cells formed}}{\text{mole substrate used}}\right] = Y_{X|S}^{\text{mol}} \tag{2.15a}$$

or

$$\left[\frac{\text{g cells formed}}{\text{g}\cdot\text{atom substrate used}}\right] = Y_{X|S}^{\text{g}\cdot\text{atom}} \tag{2.15b}$$

These units are always used for yields of cells in grams per mole of ATP formed during growth

$$Y_{ATP} = \frac{\Delta x}{\Delta_{atp}} \tag{2.16}$$

This yield factor plays an important role when the molecular energetics of biochemical pathways are considered. Bauchop and Elsden (1960) found that Y_{ATP} was about 10.5, independent of the nature of the organism and the environment. According to Stouthamer and Bettenhaussen (1973), an equation can be assumed for the consumption of ATP in a metabolizing cell,

$$Y_{ATP} = \frac{1}{Y_{ATP,max}} \cdot r_X + m_{ATP} \cdot x \tag{2.17}$$

that has the same form as the linear equation often used for biomass growth

$$r_S = \frac{1}{Y_{X|S}} \cdot r_X + m_S \cdot x \tag{2.18}$$

where m_S and m_{ATP} are the maintenance requirement for substrate and ATP.

Although the mathematical structures of Equs. 2.17 and 2.18 are the same, their status in terms of the system's description is totally different. Equ. 2.18 is a macroscopic balance equation concerning the flow of substrate entering the microbial system. Equ. 2.17 is a microscopic assumption about the internal functioning of the regulation of the energy transformation processes in the cell, describing the fate of the energy carrier ATP (Roels, 1980b). Stouthamer and Bettenhaussen (1973) pointed out that the effect of maintenance energy on Y_{ATP} has been underestimated, and it is one factor responsible for the wide range of Y_{ATP} values found (5–32 g dry biomass per mole ATP, depending on the nature of the carbon). The concept of the "C-mole of biomass" allows yields to be expressed in molar units. This can be done in two ways

$$\left[\frac{\text{g}\cdot\text{atom cell C formed}}{\text{mole substrate used}}\right] = Y_S^C \tag{2.19}$$

and

$$\left[\frac{\text{g}\cdot\text{atom cell C formed}}{\text{g}\cdot\text{atom substrate C used}}\right] = Y_C^C \tag{2.20}$$

While both may be useful, the latter is usually preferred, because it is independent of the number of carbon atoms in the substrate molecule (Herbert, 1975).

Applying similar concepts to the important constants Y_{ATP} and Y_O, these are now expressed as

$$\left[\frac{\text{g}\cdot\text{atom cell C formed}}{\text{mole ATP formed}}\right] = Y_{ATP}^C \tag{2.21}$$

and

$$\left[\frac{\text{g}\cdot\text{atom cell C formed}}{\text{g}\cdot\text{atom } O_2 \text{ used}}\right] = Y_O^C \tag{2.22}$$

The quantity Y_O^C is of practical importance in the derivation of elementary balance equations and of theoretical importance in connection with oxidative phosphorylation (Tempest and Neijssel, 1975).

In addition to cell yields from carbon substrates and oxygen, cell yields from other nutrients perhaps are best expressed in the units of Equ. 2.20, as for example

$$Y_N^C \quad Y_{Mg}^C \quad Y_K^C \quad Y_{Ph}^C \quad Y_{Su}^C$$

Other important yield factors and their significance are reviewed in articles by Payne (1970), Bell (1972), and Atkinson and Mavituna (1983). Various proposals for macroscopic efficiency measures have appeared in the literature, and they are all more or less related. The yield can be based on electrons initially available in the substrate for transfer to oxygen during combustion of the substrate:

$$\left[\frac{\text{g substrate}}{\text{electrons}}\right] = Y_{\text{av e}^-} \tag{2.23}$$

The concept of "available electrons" was first developed by Mayberry et al. (1968) and was later applied to biomass by Minkevich and Eroshin (1973). The number of available electrons is a quantitative measure of the energy content of growth substrates (Eroshin, 1977).

A second measure proposed by Payne (1970) is the yield of biomass dry weight per kilocalorie heat of combustion of the substrate, ΔH_S:

$$\left[\frac{\text{g cell dry matter}}{\text{kcal heat of combustion of S}}\right] = Y_{\text{kcal}} \tag{2.24a}$$

which can be deduced for aerobic fermentations, for example, as

$$Y_{\text{kcal}} = \frac{Y_{X|S}^{\text{mol}}}{\Delta H_S}\left[\frac{\text{g}\cdot\text{mol}^{-1}}{\text{kcal}\cdot\text{mol}^{-1}}\right] \tag{2.24b}$$

or (Prochazka et al., 1970)

$$Y_{\text{kcal}} = \frac{Y_{X|S}}{\Delta H_S - Y_{X|S}\cdot\Delta H_X} \tag{2.24c}$$

where ΔH_X is the heat of combustion of cell mass (kcal/g).

These two quantities of $Y_{\text{av e}^-}$ and Y_{kcal} are related, because the heat of combustion of an organic compound, ΔH_S, is known to be proportional to the electrons available for transfer to oxygen (Minkevich and Eroshin, 1973).

2.2.3.4 Some Thermodynamic Explanations of Yield Factors

The concept of available electrons allows the energetic yield of growth to be expressed as a fraction of the available electrons of the substrate being changed into biomass. The rest of the available electrons are transferred to oxygen, and their energy is dissipated. The energy efficiency coefficient of growth η_E is the part of the available electrons that are transferred to biomass. $1 - \eta_E$ is the energy consumed during the process of biosynthesis and released as heat. This efficiency measure, η_E, is proportional to $Y_{\text{av e}^-}$ as well as to Y_{kcal} and seems to be very promising in applications, although Roels (1980a) has suggested that the correct formulation of the second law of thermodynamics for open systems is missed.

The various efficiency measures have been reviewed by Nagai (1979) and are used in studying the energetics of growth and product formation (Erickson et al. 1978; Roels, 1980a). In the case of, for example, aerobic growth on a single source of carbon and energy without product formation, the efficiency factor for oxygen of Minkevich and Eroshin (1973) is shown to be related to the "degrees of reduction" of substrate γ_S and of biomass γ_X:

$$\eta_E = \frac{Y'_{X|S} \cdot \gamma_X}{\gamma_S} \tag{2.25}$$

with

$$Y'_{X|S} = \frac{Y_{X|S}}{a} = \frac{\nu_X}{\nu_S/a} \tag{2.26}$$

where a is the number of carbon atoms in one molecule of substrate. Equ. 2.26 is a somewhat modified definition of the yield factor in Equ. 2.13, which is more convenient for the treatment to be presented (Roels, 1980b). $Y'_{X|S}$ expresses the mole dry weight of biomass per carbon equivalent substrate consumed.

The correlation in Equ. 2.25 is numerically different for different nitrogen sources because γ_X depends on the nitrogen source used (Roels, 1980a). These generalized degrees of reduction are analogous modifications of the original definition of the "reductance degree" (Minkevich and Eroshin, 1973) defined by

$$\gamma_S = 4a + b - 2c - 3d \tag{2.27a}$$

and

$$\gamma_X = 4\alpha + \beta - 2\gamma - 3\delta \tag{2.27b}$$

The reductance degree of biomass, γ_X, is the number of equivalents of O_2 required per quantity of biomass containing 1 g · atom of carbon; and γ_S, the reduction degree of substrate, is the number of equivalents of available electrons in that quantity of organic substrate that contain 1 g · atom carbon, according to the general balance equation for aerobic microbial growth on a single carbon and energy source with NH_3 as the nitrogen source.

The number of equivalents of available electrons is taken as 4 for carbon (a, α), 1 for hydrogen (b, β), -2 for oxygen (c, γ), and -3 for nitrogen (d, δ).

It is interesting to note that a relationship exists between $Y'_{X|S}$ and $Y_{av\,e^-}$ (Roels, 1980a)

$$Y_{av\,e^-} = \frac{Y'_{X|S}}{\gamma_S} \cdot M_X \tag{2.28}$$

where M_X is the molecular weight of the biomass [g/C-mole]. In place of the efficiency factor η_E, which was based on an incorrect statement of the second law of thermodynamics, Roels (1980a) developed the concept of the "thermodynamic efficiency" η_{th}

$$\eta_{th} = \frac{Y'_{X|S}}{(Y'_{X|S})_{max}} \tag{2.29}$$

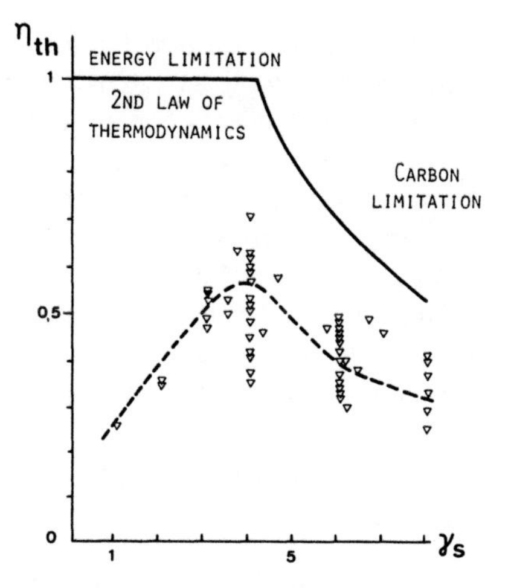

FIGURE 2.7. Thermodynamic efficiency of aerobic growth η_{th} with one carbon source as a function of the substrate's degree of reduction γ_s, and theoretical limits of the efficiency (Roels, 1980a. With permission of John Wiley & Sons, Inc.).

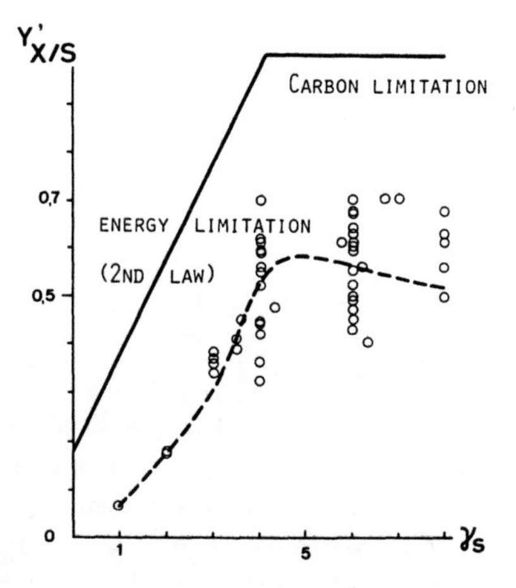

FIGURE 2.8. Relationship between substrate yield factor $Y'_{X|S}$ and substrate degree of reduction γ_S, and the theoretical limits to $Y'_{X|S}$ (Roels, 1980a. With permission of John Wiley & Sons, Inc.).

where the maximum value of the yield is calculated in full accordance with the second law. This equation has been used in the analysis of some recent yield data from the literature. The data are presented graphically in Fig. 2.7: The thermodynamic efficiency η_{th} is plotted against the degree of reduction γ_S of the substrate for growth, with NH_3 as a nitrogen source. Figure 2.7 shows appreciable variation; the trend seems to be adequately represented by the dotted line. The tendency for η_{th} to be low for highly reduced substrates (e.g., methane) becomes clear, and probably it is also low for highly oxidized substrates (e.g., oxalic and formic acids). Due to carbon limitation, $Y'_{X|S}$ can never exceed 1; hence, the value of η_{th} is subject to the restrictions given by the right part of the solid line in Fig. 2.7. The left part of the solid line is the restriction due to the second law; η_{th} can never exceed unity. Figure 2.8 shows the same results as Fig. 2.7 but plotted in a somewhat different way. The carbon conversion into biomass $Y'_{X|S}$ is plotted against substrate degree of reduction γ_S. Again, the solid line gives the restrictions mentioned.

The tendencies observed in Figs. 2.7 and 2.8 can be summarized as follows: For substrates up to a degree of reduction of about 4.2, the degree of reduction of biomass (the energy content of the substrate) is insufficient to allow all substrate carbon to be converted to biomass, even if the thermodynamic efficiency were unity. For substrates with a degree of reduction higher than 4.2, as far as energy requirements are concerned, all carbon can be converted into biomass, and even CO_2 could be fixed; the extent to which the energy can be stored in biomass becomes limiting. Hence, the thermodynamic efficiency must decrease with growth on substrates of a high degree of reduction. For unknown reasons, the carbon conservation efficiency $Y'_{X|S}$ never exceeds 0.7.

2.2.3.5 Relationship Between Yield Factors and True Stoichiometric Coefficients

On the basis of the general stoichiometric equation of microbial growth, given in Equ. 2.8, the following balance equations for each of the main elements [C, H, O, N, neglecting S, P, and ash] can be written

Starting material Products

$$\text{g·atom C: } \nu_S \cdot a = \nu_X \cdot \alpha + \nu_C + \nu_P \cdot \alpha' \quad (2.30a)$$

$$\text{g·atom H: } \nu_S \cdot b + 3\nu_N = \nu_X \cdot \beta + \nu_P \cdot \beta' + 2\nu_W \quad (2.30b)$$

$$\text{g·atom O: } \nu_S \cdot C + 2\nu_O = \nu_X \cdot \gamma + \nu_P \cdot \gamma' + 2\nu_O + \nu_W \quad (2.30c)$$

$$\text{g·atom N: } \nu_S \cdot d + \nu_N = \nu_X \cdot \delta + \nu_P \cdot \delta' \quad (2.30d)$$

In Equs. 2.30a–d, the quantities of substrate (ν_S) and of NH_3 (ν_N) required to make 1 C-mole of cells can be measured directly, and the composition of cells and products can be determined by elementary analysis, giving α, β, γ, δ and α', β', γ', δ'. In the case of no nitrogen in the substrate and product, $\nu_N = \nu_X \cdot \delta$ and the determination of NH_3 is not strictly necessary. However, it provides a useful check on overall accuracy. O_2 and CO_2 can be determined with gas analyzers, giving ν_O and ν_C. The only quantity that cannot be conveniently measured is ν_W, the moles of H_2O produced. This value, however, can be obtained by difference from Equs. 2.30b and 2.30c, and the values so obtained should balance. From Equs. 2.8 and 2.30a–d, a number of important yield constants can be derived, e.g.,

$$Y_{X|S} = \frac{\nu_X(12\alpha + \beta + 16\gamma + 14\delta)/(1-r)}{\nu_S(12a + b + 16c + 14d)} \left[\frac{\text{g cells formed}}{\text{g substrate used}}\right] \quad (2.31)$$

where r is the fraction of cell mass represented by ash ($r \approx 0.07$–0.10) (Wang et al., 1979). Other yield factors, valid for cases without product formation, can be expressed as follows (Herbert, 1975)

$$Y_S^C = \frac{1}{\nu_S}\left[\frac{\text{g·atom cell C formed}}{\text{mole substrate used}}\right] \quad (2.32)$$

$$Y_{C}^{C} = \frac{1}{\nu_S \cdot a} \left[\frac{\text{g} \cdot \text{atom cell C formed}}{\text{g} \cdot \text{atoms substrate C used}} \right] \tag{2.33}$$

$$Y_{CO_2}^{C} = \frac{\nu_C}{\nu_S \cdot a} \left[\frac{\text{mole } CO_2 \text{ formed}}{\text{g} \cdot \text{atom substrate C used}} \right] \tag{2.34}$$

$$Y_{O}^{C} = \frac{\nu_X \cdot \alpha}{2\nu_O} = \frac{1}{2\nu_O} \left[\frac{\text{g} \cdot \text{atom all C formed}}{\text{g} \cdot \text{atom } O_2 \text{ used}} \right] \tag{2.35}$$

In a similar manner, the yield of product becomes

$$Y_{P|S} = \frac{\nu_P(12\alpha' + \beta' + 16\gamma' + 14\delta')}{\nu_S(12a + b + 16c + 14d)}. \tag{2.36}$$

In practice, it is usually convenient to calculate the values of these yield coefficients from experimental data and to use them to obtain the values of ν_S, ν_O, and ν_C for use in the cell balance equation.

2.2.3.6 Yield Coefficients in Real Fermentation Processes

Yields in real fermentations are not constant, so that the concept of "yield constants" must be modified. Yield factors exhibit a significant dependence on various biological and physical parameters. It must be remembered that, although the yield factor is defined for a given strain with respect to a particular substance, yield is not only a function of the substrate. Generally, the following dependence is valid:

$$Y = f(\text{strain, substrate}; \mu, m, S; \bar{t}, t_m, \text{OTR}, C/N\text{-ratio}, P/O\text{-ratio}). \tag{2.37}$$

Earlier studies on $Y_{X|S}$ (Humphrey, 1970; Wang, 1968) and other reports have also illustrated the dependence of $Y_{X|O}$ and Y_{kcal} on $Y_{X|S}$ (Guenther, 1965; Mateles, 1971). Pirt (1966) proposed the following function to demonstrate the relationship of Y to the specific growth rate μ, as realized in a CSTR operating at different dilution rates D ($D = \mu$), and to the maintenance coefficient m

$$\frac{1}{Y_{X|S}} = \frac{1}{Y_{X|S}^{max}} + \frac{m_S}{\mu} \tag{2.38}$$

or its equivalent form, shown as Equ. 2.18. $Y_{X|S}^{max}$ is the value of $Y_{X|S}$ as μ approaches infinity. Parameter estimation can be carried out, as shown in Fig. 5.27 (cf. Fig. 2.7). Equation 2.18 can be used in modified form to calculate the influence of μ and m on $Y_{X|S}$ (Abbott and Clamen, 1973).

$$Y_{X|S} = \frac{\mu \cdot Y_{X|S}^{max}}{\mu + Y_{X|S}^{max} \cdot m_S} \tag{2.39}$$

Figure 2.9 shows the general behavior of the dependence of Y on μ. Beyond $Y_{X|S}$ the values of Y_{kcal} and $Y_{X|O}$ are also shown according to Abbott and Clamen (1973). Simultaneously, another dependence is shown in Fig. 2.9,

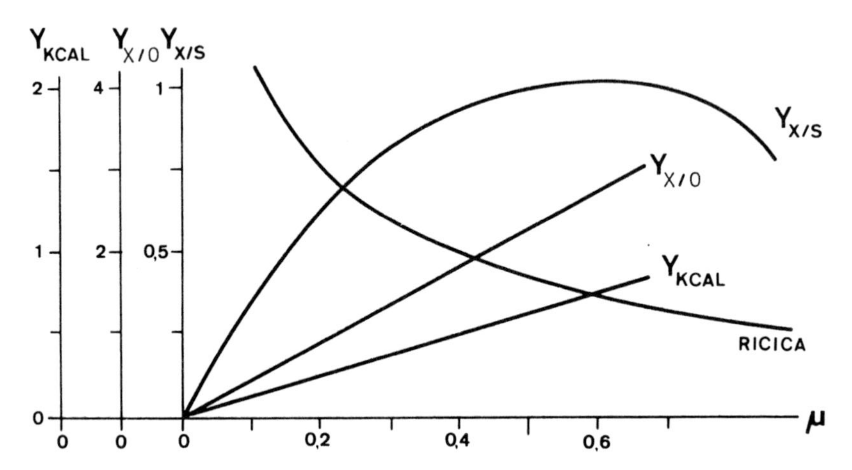

FIGURE 2.9. Dependence of yield factors Y_{kcal}, $Y_{X|O}$, and $Y_{X|S}$ on specific growth rate μ. "Ricica" refers to data from *Escherichia coli* with a nitrogen source as the limiting substrate. (Adapted from Abbott and Clamen, 1973; data for *E. coli* from Ricica, 1969. With permission of John Wiley & Sons, Inc.).

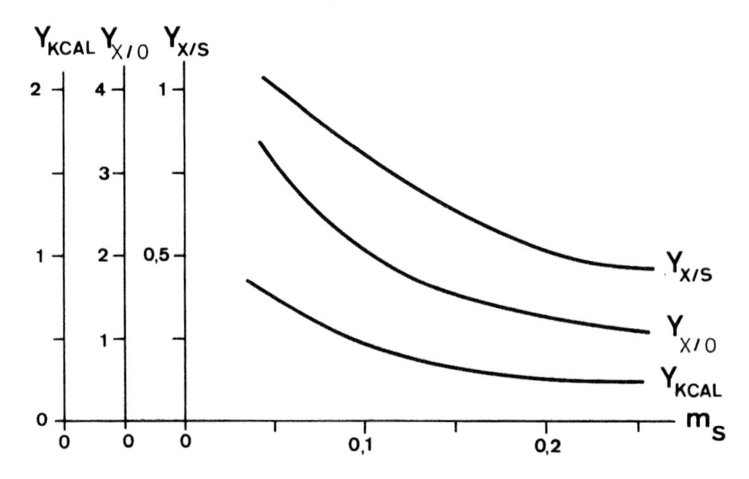

FIGURE 2.10. Dependence of yield coefficients Y_{kcal}, $Y_{X|O}$, and $Y_{X|S}$ on substrate maintenance coefficient m_s. (Results according to Abbott and Clamen, 1973. With permission of John Wiley & Sons, Inc.).

derived from data from *Escherichia coli* with a nitrogen source as limiting substrate (Ricica, 1969).

The rates at which the yield coefficients change as μ changes depend on the value of the maintenance coefficient m_S as shown in Fig. 2.10.

In a number of cases this approach is not satisfactory, because it looks at substrate rather than ATP. Therefore, an equation with formal similarity to

Equ. 2.38 has been proposed for the specific production of ATP (cf. Equ. 2.17, where q_{ATP} is a linear function of μ (Stouthamer, 1980; Stouthamer and Bettenhaussen, 1973). More detailed investigations show that, for example, the maximum specific growth rate is determined by the rate of energy production in the cells. As ATP production is correlated with substrate consumption, the growth behavior can be formally represented at the level of mass, that is, $\mu = f(s)$.

The same background to formal modeling is evident in the several attempts that have been made to explain the biomass yield variation with specific growth rate based on Equ. 2.18 (Reuss et al., 1974; Rock et al., 1978). Such an explanation does not appear to be fully justified. Equation 2.38 cannot explain why the yield changes during the exponential phase of batch cultures when μ is constant. Furthermore, it is not true that, in the case of C_1-compound utilizers, $Y_{X|S}$ increases monotonically with μ (or D for steady-state CSTR). Papoutsakis and Lim (1981) stressed the carbon-flow-branching concept to illustrate the general possibility of a highly variable biomass yield:

$$Y_{X|S} = \left(\frac{M_X}{\nu_X \cdot M_S}\right)\frac{1}{1 + r_1/r_2} \tag{2.40}$$

where r_1 and r_2 are the rates of branching into metabolic pathways 1 and 2 for the carbon source, M_X and M_S are the molecular weights of the biomass and substrate, and ν_X is the stoichiometric coefficient of the reaction $S \rightarrow X$. In the case of methylotrophs, two distinct carbon-flow processes exist: assimilation (r_2) and oxidation (r_1). Cells produce only as much NADH and/or ATP through oxidation as may be required for assimilation. This fine tuning of the two processes produces maximal possible biomass yields. According to Equ. 2.40, the yield will change only when r_1 or r_2 changes, and this can happen if the concentration of any substrate or of any other nutrient or the culture condition such as T and pH changes. This concept shows that the unusual behavior of biomass yields of methylotrophs is a kinetic rather than a biosynthetic question.

An alternative concept has been proposed by Agrawal et al. (1982), who achieved the desired variability of stoichiometry by defining a yield factor that increases linearly with substrate concentration:

$$Y_{X|S} = a + b \cdot s \tag{2.41}$$

where a, and b are constants.

The influence of physical parameters on yield coefficients can be quantified on the basis of the scheme that is valid for yeast metabolism:

$$S_1 + O \begin{cases} \xrightarrow{\mu_1} X \\ \xrightarrow{K_R} P \ldots\ldots S_2 + O \xrightarrow{\mu_2} X. \end{cases} \tag{2.42}$$

The "selectivity," σ, of this process with simultaneous glucose and oxygen effects can be written

$$\sigma = \frac{X_{max}}{P_{max}} = \frac{Y_{X|S}}{Y_{P|S}} \tag{2.43}$$

Yield, that is, the selectivity, depends on OTR, t_m, t_c, or CTD. Thus *Saccharomyces cerevisiae* works as a test organism for micromixing if OTR is held constant (Moser, 1977b).

2.2.3.7 Temperature Dependence of Yield Factors

As a stoichiometric coefficient, the yield factor is not expected to vary significantly with temperature (Moletta et al., 1978; Peters, 1976; Shiloach and Bauer, 1975; Topiwala and Sinclair, 1971). Temperature seems to have only a minor influence on $Y_{X|S}$. In the case of the yield factor $Y_{X|O}$ based on BOD, for example, a decreasing value with increasing temperature is recorded (Peters, 1976).

2.2.4 Productivity, Conversion, and Economics (Profit)

The productivity, r_j, of a process is generally defined as the mass of the product, j, produced per unit time and per unit volume; it has the dimensions $[\mathrm{kg \cdot m^{-3} \cdot h^{-1}}]$. To clarify this concept, the curve (Fig. 2.11) demonstrates r_j for the example of the discontinuous process whose course was diagrammed in Fig. 2.4. With discontinuous processes, a certain lag time is also necessary between production cycles due to harvesting, emptying, cleaning, and refilling operations. This lag time, t_0, is shown on the negative time axis in Fig. 2.11.

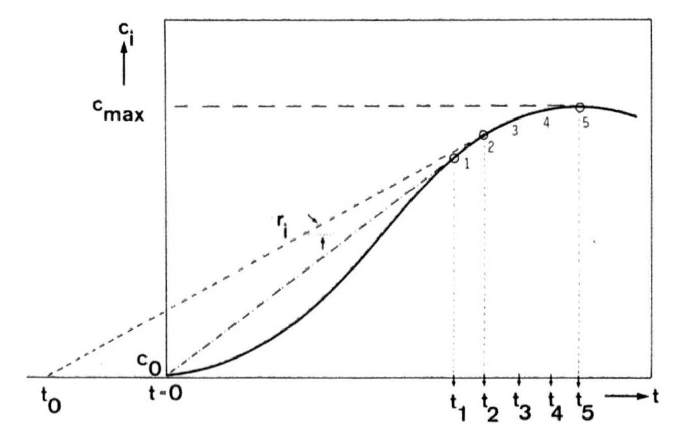

FIGURE 2.11. Schematic representation of optimal operating point for a process based on different economic criteria. Productivity may be obtained from the tangent to the curve. Reaction times: t_1, maximum productivity in a continuous process; t_2, maximum productivity in a discontinuous process with t_0 = deadtime; t_3, maximum profit; t_4, minimum cost; t_5, maximum conversion—maximum product concentration as $S \rightarrow 0$.

Drawing a tangent from this point to the concentration/time curve, one obtains at point 2 the value of the product concentration that can be reached in the whole production time ($t_{tot} = t_0 + t_r$). The maximum productivity attainable with a discontinuous process can be calculated from

$$r_{j,dc,max} = \frac{c_{max,j} - c_{0,j}}{t_0 - t_r} \tag{2.44}$$

Point 1 in Fig. 2.11 gives, for comparison, the maximum productivity of an equivalent continuous process. Since a continuous process has no dead time, the slope of the tangent and therefore $r_{j,max}$ is greater than in a discontinuous process (see also Sect. 6.1).

Figure 2.11 illustrates another principle that is an issue in the question of economic optimization. At point 5, productivity is zero. This point may be of interest when very expensive substrates are being used. In these cases the process may be run to complete substrate utilization.

The conversion ζ_i is defined (V = constant)

$$\zeta_i = \frac{c_{i,0} - c_{i,t}}{c_{i,0}} \tag{2.45a}$$

and can also be given as a relative quantity

$$\zeta = \frac{\zeta_i}{\zeta_{i,max}} \tag{2.45b}$$

The yield, $Y_{i|j}$, can be determined by Equ. 2.46 (V = constant); it compares the total amount of a product j yielded from an amount of material i consumed

$$Y_{i|j} = \frac{c_{j,t} - c_{j,0}}{c_{i,0}} \tag{2.46a}$$

(cf. yield coefficient, Equ. 2.13) with

$$Y_{i|j,rel} = \frac{Y_{i|j}}{Y_{i|j,max}} \tag{2.46b}$$

The output of a reactor, with the dimensions [tons/day], can be calculated from the relation

$$\text{Output} = \zeta_i \cdot n_i \qquad \text{or} \qquad Y_j \cdot n_i \tag{2.47}$$

where n_i is the mass flux of component i $[t \cdot d^{-1}]$. In the case $\zeta_i \to 1$ or $Y_i \to 1$, the output will be equal to n_i.

Point 4 in Fig. 2.11 represents the point where minimum costs (C_{min}) are reached. The accumulated cost at any time can, in general, be formulated according to Equ. 2.48 (Richards, 1968a, b)

$$C_{tot/kg} = \frac{C_B}{W} + \frac{C_R}{W} \tag{2.48}$$

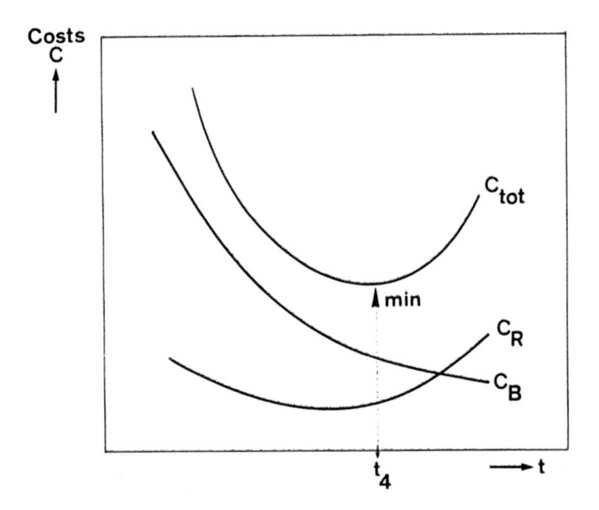

FIGURE 2.12. Typical change in costs for a discontinuous process over time (C_{tot} = total costs, C_R = running costs independent of batch size, C_B = variable batch costs, which depend on batch size). t_4, minimum total costs (see Fig. 2.11). (Adapted from Richards, 1968a,b).

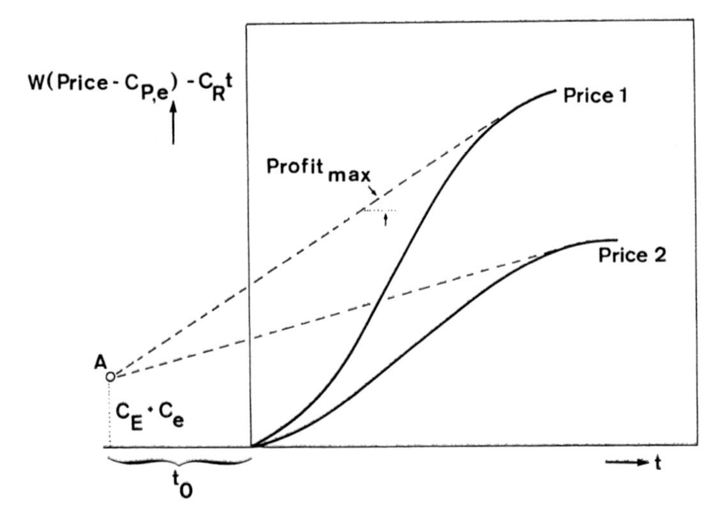

FIGURE 2.13. Graphic method for evaluating the maximum profit at a given selling price and starting point, A. (Fixed cost = $C_B + C_e$ for deadtime t_0) (Adapted from Geyson and Gray, 1972. With permission of John Wiley & Sons, Inc.).

The total costs, $C_{tot/kg}$, consist of the sum of the batch costs C_B (sterilization, materials, and costs independent of the amount produced) and the operating or running costs, C_R (for the power used for stirring and aeration). A typical time course for the accumulation of costs in a discontinuous fermentation process is shown in Fig. 2.12. From this, the time when costs are minimal (t_4) may be calculated.

The last remaining operating point, point 3 of Fig. 2.11, is that of maximum profitability. By experience, this point lies between the point of maximum productivity for a discontinuous process and the point representing minimum costs.

Profit is calculated according to Equ. 2.49 (Geyson and Gray, 1972)

$$\text{Profit [\$/t]} = W(\text{price} - C_{P,e}) - C_R \cdot t - (C_B + C_e) \qquad (2.49)$$

$C_{P,e}$ is the extraction cost for product isolation and C_e is the extraction running cost that is independent of the amount produced. The maximum profit can be obtained graphically from Equ. 2.49. In Fig. 2.13, the maximum profit can be obtained from the slope of the tangent to the curve drawn from starting point A. The use of economic analysis of processing costs is a powerful tool in establishing priorities for process improvements (e.g., Swartz, 1979; Trilli, 1977).

2.3 Systematic, Empirical Process Development With Mathematical Models

2.3.1 An Integrating Strategy—A Basis for Biotechnological Methodology

Research work is often governed in such a way that biology and physics are elaborated independently of each other. As a consequence, results from the basic sciences are difficult to apply on an engineering scale for process design. An integrating strategy has been proposed to bridge this gap (Moser, 1982). It represents a systematic approach to empirical bioprocess analysis and design, in which the kinetics of biological reactions play the central role.

The working concept of the macroscopic principle, recently adapted from chemical engineering (see Roels, 1980; Roels and Kossen, 1978), operates with macroscopically observable variables that are thought to be closely related to significant phenomena of the process (cf. Table 2.1).

While simplifications are needed, one essential feature of bioprocessing cannot be neglected: the interaction between microbial metabolism and physical transport in bioreactors. On the basis of macroscopic process variables, a systematic analysis of experimental data must include these interactions. As shown in Fig. 2.14 the fundamental equation of this approach is the law of the conservation of mass (cf. Equ. 2.3). Macroconversion, r_{eff}, must be elucidated to elaborate the kind and degree of interactions between metabolism (kinetic r_i) and reactor (physical transport, n_i). Thus, it is necessary to quantify bioreactors and their physical transport with the aid of biological systems having known kinetics ("biological test systems," Fiechter, 1978). It is also necessary to quantify biokinetics using bioreactors with known physical transport properties ("perfect bioreactor," Moser, 1978a). Characterization of

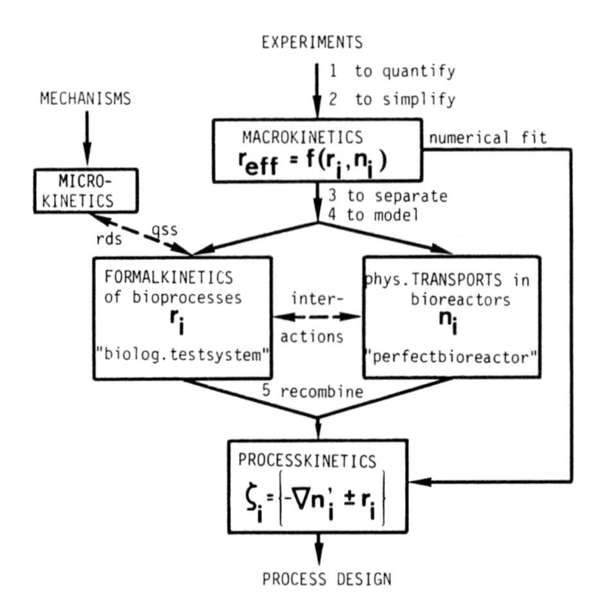

FIGURE 2.14. Strategies in bioprocess design based on the macroscopic principle and using the formal kinetic concept as part of an integrating strategy. The strategy incorporates the spatial change of mass flux through area $\nabla n_i'$ (kg/m^2 · h) and the rate of consumption or formation r_i (kg/m^3 · h). rds, rate-determining step; qss, quasi-steady-state. (From Moser, 1983b.)

biological test systems is mainly gained by quantification: concentration/time and specific rate/time curves (Dechema, 1982). In a strict sense, these curves should be elaborated at the level of kinetic modeling.

The characterization of bioreactors is not limited to the standardization of stirred tank type reactors (Dechema, 1982): It also includes ideal reactors as model types in a wider framework. The problem of defining a "perfect bioreactor," including tests of pseudohomogeneity (Moser, 1983a) will be discussed later.

The interactions of physical transport phenomena and biokinetics, formulated in a sum of process kinetic equations, is qualitatively represented in Fig. 2.15. The bioreactor behavior on macroscopic level, represented by the broken line in the figure as a symbol for the area of the balance considered, can be characterized by calculating derived quantities such as the specific rates of consumption $q_{S,i}$ growth μ, and production $q_{P,j}$ and formulating the interactions of the kinetic data with the transport of mass, energy, and momentum. Some other physical properties of the extracellular environment may be incorporated very usefully as "missing links": viscosity ν, shear rate γ, density ρ, gas hold-up ε_G, apparent morphology η^* (or M_F), morphology index δ^*, and so on. This approach has been successfully used for the scale-up of, for

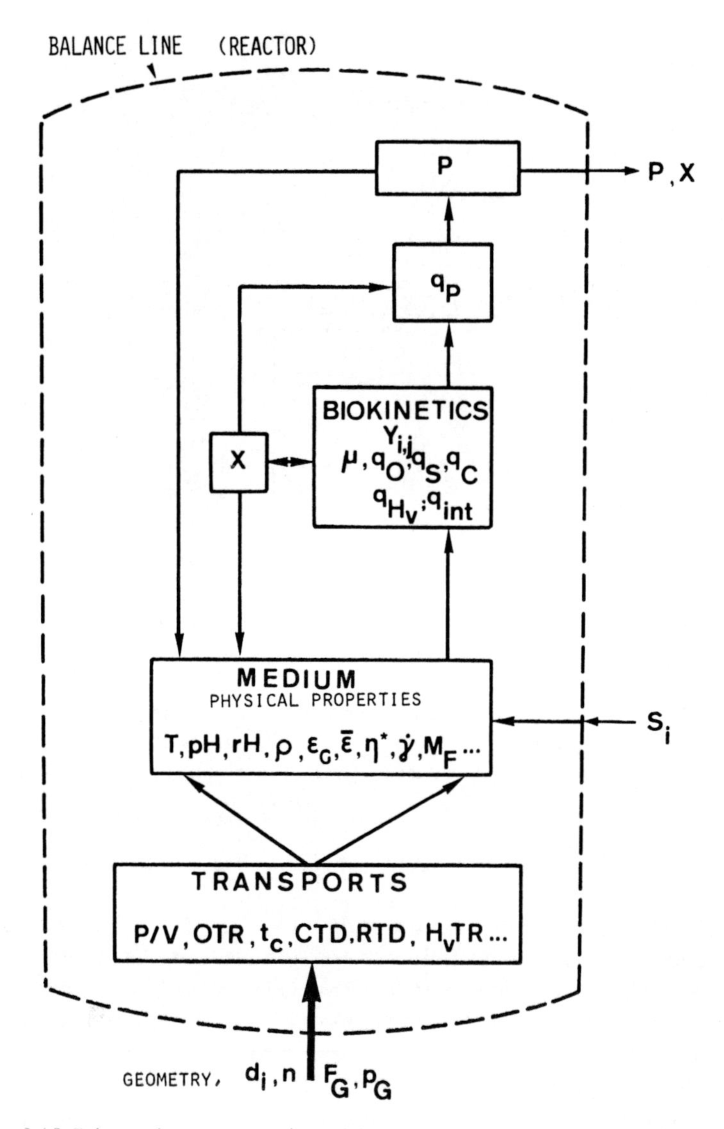

FIGURE 2.15. Schematic representation of the concept of interactions between physical transport phenomena (oxygen transfer rate, OTR; heat transfer rate, H_vTR; power consumption, P/V; circulation time, t_c, and distribution, CTD; residence time distribution, RTD) and biokinetics (specific rates of growth μ; of consumption of substrates q_S, resp. consumption of oxygen, q_O; of production of heat, q_{H_v}, of CO_2 q_C, of product q_P; and of all internal compartments q_{int}). The physical properties of the medium (temperature T, pH value, redox potential rH, density ρ, gas hold-up ε_G, mean energy dissipation $\bar{\varepsilon}$, shear rate $\dot{\gamma}$, resp. "specific viscosity" $\eta^* = \eta \cdot x^{-1}$ as a quantitative estimation of "engineering morphology," and M_F = morphology factor) represent the missing link for modeling the interaction concept. (Adapted from Reuss et al., 1980.)

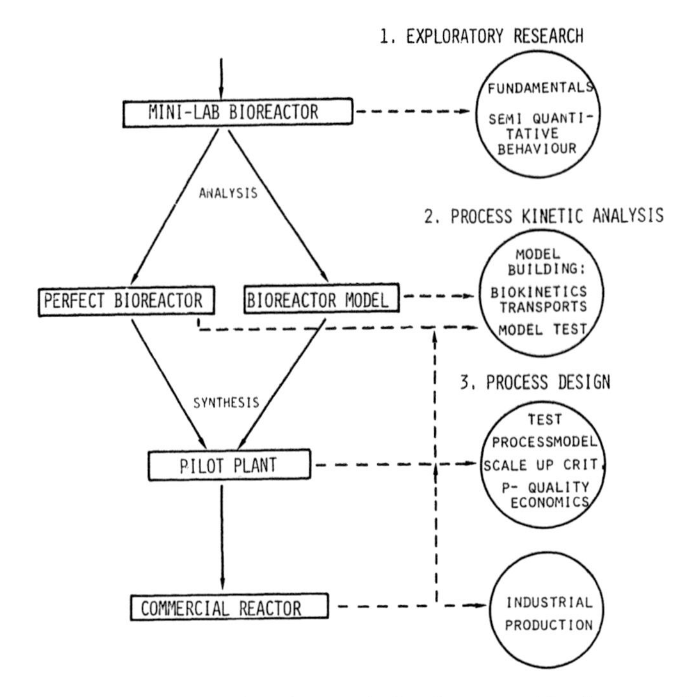

FIGURE 2.16. Strategy for engineering analysis of process kinetics according to a systematic empirical approach. Bioreactors of different scales are used to obtain process engineering data at different stages (1–3) of development (From Moser, 1978a.)

example, penicillin production (Reuss et al., 1980) as will be described in Sect. 6.7.

The interactions between physics and biology govern large-scale microbial production and also heavily influence reaction rates on the laboratory scale. This fact is to be considered in the case of process kinetics analysis. All methods used for quantification in bioprocessing (cf. Chapters 3 and 4) should be critically considered, and standards would be desirable (cf. EFB, Dechema 1984) for development of a general methodology for biotechnology.

In contrast to Fig. 2.2, which demonstrates a process development protocol without modeling, Fig. 2.16 demonstrates the steps involved in an *empirical, systematic process development* that does operate with mathematical models.

Three main areas are associated with the flow of further work: bioreactors, kinetics, and conversion. This pathway will also be followed in this book (Chapters 3–6). (Chapter 4 is a detailed look at the problems associated with the coupling of kinetic and transport phenomena in the formulation of an analysis of process kinetics.) The goal in each of the three main areas is the establishment of a mathematical model. In correspondence with the main objective of this book (see Sect. 1.2), the hallmark of this procedure lies in the careful formulation of the biokinetics. Basic research is primarily interested in clarifying mechanisms and in discovering *why* a process runs in a particular way. Engineering research is primarily interested in questions of *know-how*—how a process runs and how it can be influenced so that it runs optimally. Whether the measured values reflect the microkinetics depends not on the scale (laboratory or pilot) but rather on the technique and on whether transport phenomena are at work.

The term "microkinetics" is understood to mean the kinetics of a reaction that are not masked by transport phenomena and to refer to a series of reaction steps. For the investigation of intermediary metabolism, idealized conditions are chosen that often do not correspond to the real conditions of engineering processes. This fact makes it difficult to transfer microkinetic data to technical processes. For the purposes of technologically oriented research and the development of a process to technical ripeness, it is often sufficient to know quantitatively how a process runs without necessarily knowing why. (Macrokinetics, however, must be avoided, as they are scale dependent). Mathematical formulations are needed that reproduce the kinetics adequately for the purpose but are as simple and have as few parameters as possible. Today, even when electronic computers greatly reduce the labor of computation, the criterion of simplicity remains important due to the problem of experimental verification. The iterative nature of the process of building an adequate model is an important point that will be considered in greater detail in Sect. 2.4.

The successive working stages of analysis and synthesis characterizing the flow of work in a systematic process development are shown schematically in Fig. 2.16. A distinction is made among three levels:

- Fundamental research in small laboratory reactors (Erlenmeyer flasks)
- Process kinetic analysis in laboratory-scale reactors ($V = 10-50$ l)
- Process design in pilot plant reactors

The process kinetic analysis is carried out in two different types of reactors. The so-called "perfect bioreactor" is used for obtaining "true" kinetic data. This laboratory reactor must therefore meet certain requirements with regard to all possible transport phenomena (see Sect. 4.2). A so-called "bioreactor model" is a scaled-down bioreactor with some geometrical similarity to the production unit. Here, significant transport phenomena (cf. Chap. 3) can be studied and quantified with a physical system in a first step.

Pilot plant bioreactors serve one purpose supplementary to those mentioned before (see Sect. 2.1.4): test of the process model (process kinetics), including separate tests of kinetics and physical transport phenomena.

Finally, there is a precautionary rule: Kinetics should always be measured in two different reactors to identify any hidden variables or parameters. In addition to experimenting in a perfect bioreactor, one should also evaluate a trial in the pilot plant. This testing has another purpose as well, which might be called "kinetic similarity" (Moser, 1978b) (as exists, e.g., between DCSTRs and CPFRs). Real biological processes are so complex that altered flow conditions cause changes in the concentrations of materials and corresponding changes in the metabolism and constitution of the mixed population of cells (biocoenosis). The advantage of kinetic similarity is that it guards against the eventuality of unexpected factors not considered in the model. Perhaps with further progress in building reliable models this step may no longer be necessary.

2.3.2 Working Principles of Bioprocess Technology

Based on the strategy of systematic process development, an engineering approach to bioprocessing includes several stages, as illustrated in the scheme of Fig. 2.14. As already stated in Fig. 1.4, the concepts of a rate-determining step (rds) and of the quasi-steady-state (qss) are very useful (see Sect. 2.4.1). Some basic principles play an important role: to simplify, to quantify, to separate, to model, and to recombine the separated phenomena to form a total process model.

2.3.2.1 Principle of Simplification: The "Formal Macroapproach"

The engineer responsible for planning must always strike a balance between involved complications and practicality; in addition every model will be characterized by the number of variables and the measuring techniques used for them. The predictive power of the model will thus be limited.

The following statement will serve as a general guideline for the first working principle: Simplification includes compressing the complex structures of a process into those few schemes that may be regarded as significant, that is, that are reflected in key variables. In this way both the experimental and theoretical research may remain limited to just that degree of precision necessary for the purpose at hand.

Due to the complexity of both aspects—the biological and the physical—simplifications must be made without loss of essential information. This approach, well known in chemical engineering, is a consequence of the macroscopic principle (Glansdorff and Prigogine, 1974), which has recently been applied to bioprocessing (Kafarow et al., 1979; Moser, 1978a, 1981; Roels, 1980a, b; Roels and Kossen, 1978; Votruba, 1982). In other words: "Everything should be made as simple as possible, but not simpler," as Einstein said.

According to Fig. 2.3, microbial cells suspended in the liquid phase of the bioreactor are treated as black boxes; this does not mean that they are neglected. Their macroscopic behavior, manifested in the changes of the concentrations in the liquid phase, is taken into consideration. As a consequence of this approach, the bioreactor itself is not treated as a black box. The core of the formal macroapproach is that formal analogies are used (see Sect. 2.4.3). The utility of the principle of simplification can also be demonstrated in the case of modeling the dynamics of bioprocesses (see Sects. 3.5.3 and 5.7).

2.3.2.2 Principle of Quantification

Quantitative data collection is widely accepted as the basis of every modern, effective technology. Section 2.2 laid out the concepts for data collection for bioprocesses. Analytical methods for measuring process variables are often faulty, that is, they may be falsified by interactions between physics and biology. The question of the extent to which the measured variable is a useful, representative variable is generally of far greater importance, thus the analytical methods as used, and this is especially so in evaluating the value of a mathematical model. Examples where this is an issue include (a) the dry cell mass used as a substitute for the catalytically active cell mass and (b) the biological oxygen demand as a measure of the degree of pollution in waste water treatment.

2.3.2.3 Principle of Separation

Separation deals with the accurate design of experiments for obtaining data on biological and physical phenomena in circumstances where the rates of the biological and physical processes are independent of one another. Macroconversion (macrokinetics) is an ill-defined basis for scale-up, and efforts must be made to avoid the appearance of such data. This is done with the aid of the test of pseudohomogeneity (see Sect. 4.3) and of regime analysis in general (see Sect. 4.2).

In connection with the problem of pseudohomogeneity, it is clear that a reaction occurring inside a solid phase is not directly measurable. The internal flux of a metabolic reaction can thus only be checked by measuring the external fluxes in the liquid medium by means of computer simulation. Barford and Hall (1979) found that an external overall flux does not reflect, even in an approximate manner, the internal fluxes. Moreover, in vitro examination of an isolated section of metabolism is inadequate for quantification of the coordinated and integrated biochemical control of an intact living system due to in vivo interactions between different parts of the metabolism.

Another application of the principle of separation is the case of truly heterogeneous processes (see Sects. 4.5, 6.6, and 6.7). Here, cases of external and internal transport limitation of a bioprocess must be distinguished from the case where the physical transport may be enhanced by the biological reaction. A simplified treatment in all these cases is possible through introduction of the concept of efficiency η_r, defined by

$$\eta_r = \frac{r_{\text{eff}}}{r_{\text{ideal}}} \tag{2.50}$$

The efficiency factor of the reaction η_r can be interpreted as

$$\eta_r \propto f \frac{k_r}{k_{Tr}} \tag{2.51}$$

being a function of the interaction between rate constant of the reaction k_r and transport k_{Tr} (see Sect. 4.5). With the concept of efficiency, the real heterogeneous situation can sometimes be handled as a pseudohomogeneous case.

The principle of separation constitutes an important prerequisite for the sensible application of mathematical models, and vice versa.

2.3.2.4 Principle of Mathematical Modeling

Mathematical models are mathematical formulations able to represent the behavior of a process on a simplified level and adequately for a given purpose. Such models serve to generalize singular results and provide a basis for the deduction of further properties of the system. Process and system engineering use mathematical models and a high degree of theoretical background; this is often admired, but also often rejected by biologists. However, Louis Pasteur's admonition should be kept in mind: "Without theory, practice is but routine born of habit." This statement is related to deductive research methodology, which is accepted as the basis for scientific work (Popper, 1934, 1972). The steps of deduction in bioprocess analysis and design are illustrated in Fig. 2.17. As theories are always fallible, the only way to circumvent failure is to compare hypotheses with experiments. To be sure that no undetected interaction is disturbing the measured kinetics, a supplementary run should be made in a pilot plant reactor. Here transport phenomena are normally quite different from those of the bench-scale reactor, so that interactions are easier to detect and modifications to the model can be made, such as the addition of new model functions f, other process variables x_i, or other model parameters k. All we can do is to apply the principle of analogy, describing the behavior of the system under certain conditions ("if, then"). These "black-box" models, however, are not only descriptive but also predictive when the deductive method is applied.

Even if some mechanistic background allows a deeper interpretation of the parameters so that they have more biological significance, with such an approach "gray boxes" still remain; the approach is still, more or less, an optical illusion (Roels and Kossen, 1978), since the final subsystems are always black boxes. In conclusion, the level of modeling to be considered as adequate depends on the principle of analogy, in which the model parameters are adapted to experiments ("adaptational parameters," according to Hofmann, 1975).

To summarize the principal aims of mathematical models:

1. Models are built to allow prediction of the conversion or productivity of any system.
2. Models are built to examine the nature and behavior of a plant's operation under a variety of operating conditions and to examine the regions where the model is valid, including how far from this region extrapolation is permitted.

3. Models allow generalization to other situations within the boundaries of their validities.
4. Models are mathematical formulas that can be manipulated to optimize processes.
5. Computer simulations are types of mathematical models.
6. Models are used to identify unknown or previously disregarded process variables and parameters that may be significant variables.
7. Models serve as controls in examining whether an effective separation of biological and physical phenomena has been achieved.
8. As an indirect result, models may also help in clarifying reaction mechanisms.

2.4 Mathematical Modeling in Bioprocessing

Because of the central meaning of working with mathematical models, their general nature, objectives, and construction will be discussed here.

2.4.1 General Remarks

In the literature there are several ways of classifying different types of models (cf. Roels and Kossen, 1978, and Sect. 2.4.3). From the viewpoint of process engineering, in this book distinctions will be made among macrokinetic, microkinetic, formal kinetic, and process kinetic models (see Fig. 2.14). The formal models that will be discussed individually in Chap. 5 are sometimes referred to in the literature as "unsegregated" or "unstructured models" that have a "descriptive" or "predictive" nature.

Classification by the concepts of "segregation" and "structuring" (Tsuchiya, Fredrickson, and Aris, 1966) is predominant in the area of microkinetic models, but it is also useful for making broad distinctions among formal kinetic models. *Unsegregated models* are those that consider the biomass of each cell as identical in terms of physiology, morphology, and genetics. The number of cells involved in most fermentation processes of technological importance is very large. Within the limits of the precision desired for an engineering process, the macroscopic behavior can be treated as constant, representing an average value throughout the whole population. On the other hand, *unstructured models* neglect the fact that the cell mass is not uniform but rather consists of many different compounds (DNA, RNA, proteins, etc.) that, kinetically, behave quite differently. The problem of structuring mathematical models will doubtless become more important, especially in the case of nonstationary processes such as batch and fed batch cultures (cf. Sect. 5.7.2.2).

It cannot be expected that any kinetic model will be directly applicable to a real process situation. Mathematical modeling must start with the simplest type, but it must be reiterated, modified, and extended until eventually it leads to an adequate process kinetic model (see Fig. 2.17). In modeling, great value

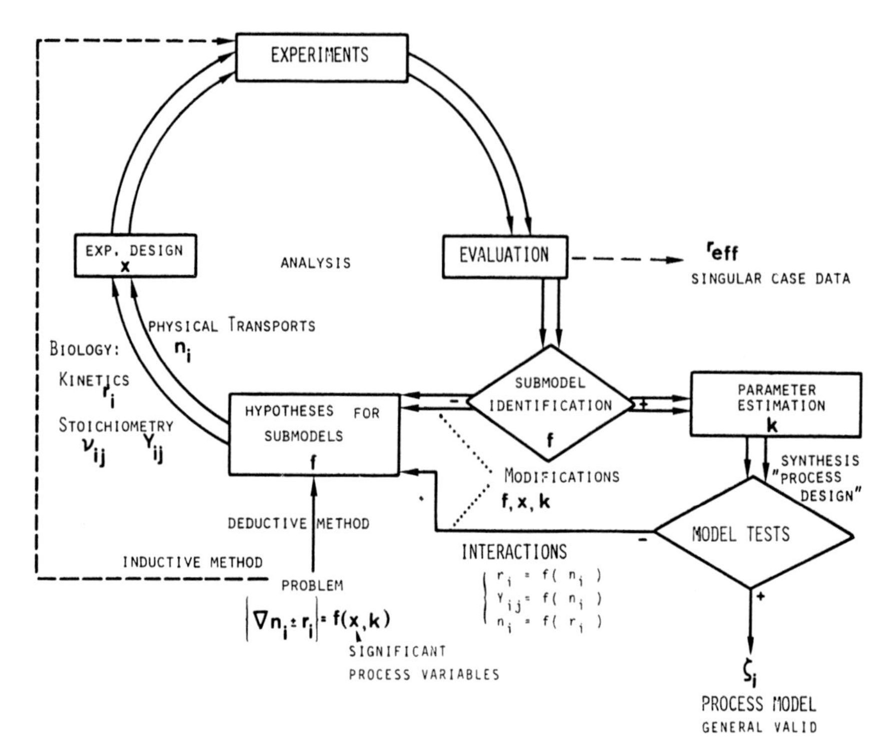

FIGURE 2.17. Flowchart of adaptive modeling in the integrating strategy, applying deductive research methodology in which mathematical models (for physical transports n_i, biokinetics r_i, and stoichiometry on the macroscopic level $Y_{i|j}$ or the microscopic level $\nu_{i|j}$) serve as working hypotheses to be compared consequently with carefully designed and evaluated experiments. Special attention concerns the interactions between n_i, r_i, and $Y_{i|j}$, as indicated. (From Moser, 1981).

should be placed on establishing the boundaries within which the mathematical functions are valid and the actual values to be used for the parameters in concrete instances of application.

In connection with this discussion it might be said that a rather simple method of structuring, such as, for example, distinguishing between the quantity of protein and the quantity of nucleic acids present, can lead to a great number of otherwise inaccessible parameters (Reuss, 1977). This can make an important contribution to the understanding of intracellular events. For engineering calculations, however, the primary claim of simplicity must also be satisfied. Furthermore, within the scope of the analytical methods presently available, distinctions among models are often simply not possible (Boyle and Berthouex, 1974).

The important characteristics of models to be considered for engineering purposes are

1. The model presents a simplified picture of the complex structure of a process in terms of a few key variables.
2. The model is developed to serve a particular need (generalizations can be achieved by "putting the model in jeopardy").
3. The model forms a bridge across the gap between microscopic and macroscopic phenomena.
4. The model has been "verified"—that is, it has been compared with experimental results and adapted to them.
5. The model is valid only within certain boundaries, which correspond to the models purposes. No model can completely replace experimentation. Models can, however, contribute to saving some of the time and money necessary for experiments. "Parameter sensitivity analysis" is obligatory.

2.4.2 Model Building

The mathematical formalism of Equ. 2.2a presents one case of a method for reproducing kinetics. The general form of a mathematical model is

$$r = f(x_i, k) \tag{2.52}$$

In accord with both theory and experience, the total function can be separated into a function of temperature and a function of the concentration variables:

$$r = k(T) \cdot f(c_x) \tag{2.53}$$

This very general equation is the starting point for developing an analysis of process kinetics (see Chap. 4).

2.4.2.1 General Procedure

Model building is concerned with the establishment of equations of this type. Important considerations in doing this are

- The choice of the significant process variables, x_i.
- The choice of an appropriate mathematical function for the model, f, and of the appropriate fitting parameters, k.
- Comparison of the first model with experimental results and the expansion or modification of the model functions.

The stages of this procedure are shown in Fig. 2.18. The initial motivation is always the existing concept of the purpose of the model. On the basis of information from the literature and/or exploratory experiments (and with, of course, the element of intuition), the problems can be recognized and a hypothesis formulated in mathematical terms. A model is selected by first selecting the significant process variables and then preparing a qualitative structure diagram in the form of a simplified process flowchart. The next phase can be described as one of testing or verification. It consists of using estimated test data to obtain a mathematical solution to the system of equations, to

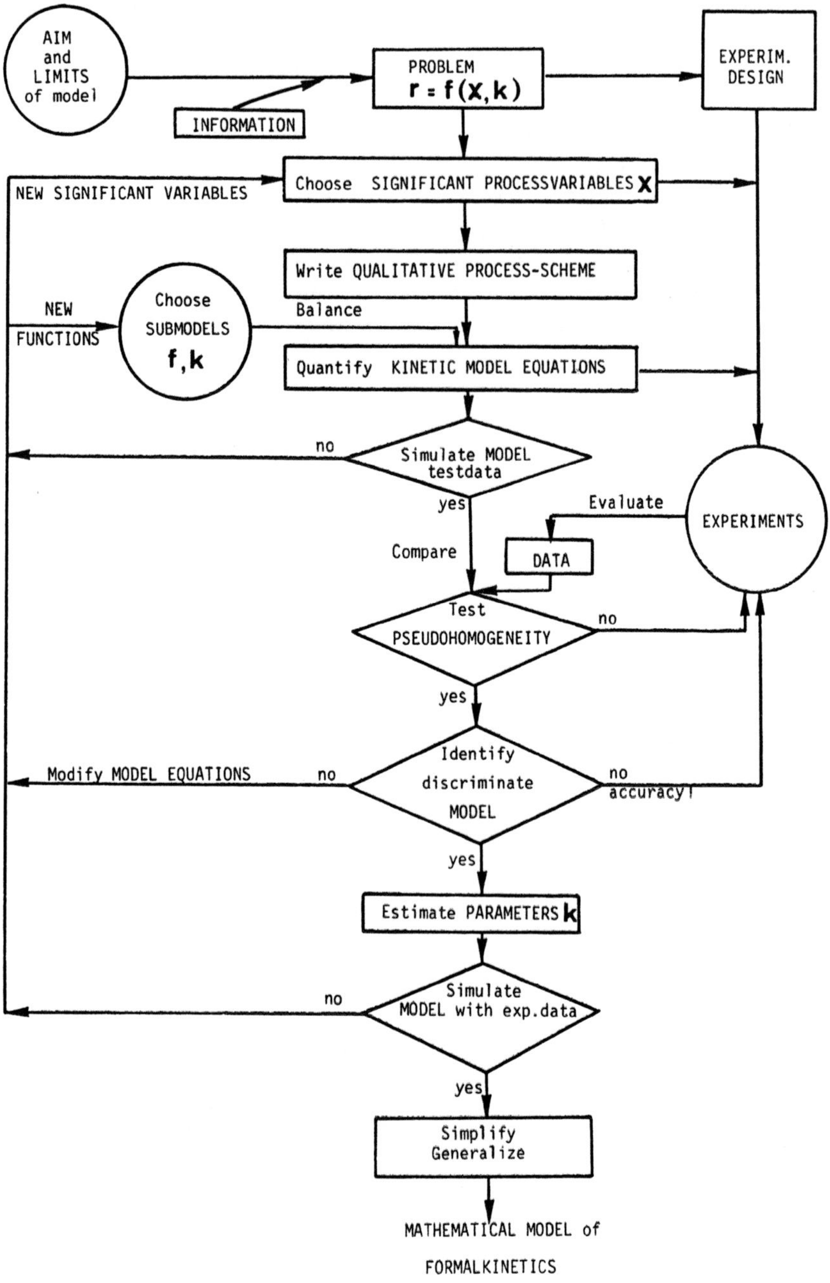

AIM and LIMITS of model
INFORMATION
PROBLEM $r = f(X, k)$
EXPERIM. DESIGN
NEW SIGNIFICANT VARIABLES
Choose SIGNIFICANT PROCESSVARIABLES X
Write QUALITATIVE PROCESS-SCHEME
NEW FUNCTIONS
Choose SUBMODELS f, k
Balance
Quantify KINETIC MODEL EQUATIONS
no
Simulate MODEL testdata
yes
Evaluate
EXPERIMENTS
Compare
DATA
Test PSEUDOHOMOGENEITY
no
yes
Modify MODEL EQUATIONS
no
Identify discriminate MODEL
no accuracy!
yes
Estimate PARAMETERS k
no
Simulate MODEL with exp.data
yes
Simplify Generalize
MATHEMATICAL MODEL of FORMALKINETICS

FIGURE 2.18. Procedure for constructing a formal kinetic mathematical model of a bioprocess (Moser, 1977c.) Explanation see text.

determine whether the model appears, in principle, to reproduce the course of the process.

Identification with the model parameters is accomplished in most cases with use of graphically estimated linearizations or of nonlinear regression analysis. After model identification, parameterization may be undertaken. In the beginning the control of parameterization should not be delegated to the computer operating blindly with prescribed criteria. Deviations can suggest a previously unknown influence that should be incorporated into the model.

A "sensitivity-analysis" has to show finally how the model reacts to changing values of parameters. An efficient biokinetic experimental design has been demonstrated by Johnson and Berthouex (1975a), and multiresponse data are used for parameter estimation by the same authors (1975b).

Three factors are possible sources of error in constructing a model (Moser, 1978b):

1. Process variables not adequately considered, for example, O_2, CO_2, viscosity, T temperature pH, inhibitory metabolites.
2. Choice of inadequate submodels of the kinetic processes, for example, unknown influences from endogenous metabolism, inhibitors, the use of homogeneous rather than heterogeneous models, multisubstrate limitations, mixed population interactions.
3. Experimental requirements or assumptions concerning transport not fulfilled, for example, the mixing behavior of fluids and gases, deviation from the ideal residence time distribution, critical dependence on particle size, the thickness of layers of cells, the response lag of the p_{O_2} electrode.

Searching for the source of error is mainly an intuitive process, but one that is aided by close observation during the course of experiments and close adherence to protocols. Model building has to be repeated several times. The iterative nature of the "adaptive model building" process is of special importance not just for effective parameterization (Johnson and Berthouex, 1975a, b) but also for the whole process of conceptualization and working with mathematical models of kinetics.

2.4.2.2 Integration of Balance Methods and Kinetics in the Construction of Unstructured Models

To construct a simple unstructured model for bioprocesses, at least one of the reactions taking place in the culture must be specified in kinetic terms. Generally a complete set of constitutive equations for each of the N chemical reactions taking place in the culture can be written in the form of a sum or a matrix (Roels, 1980a; Schubert and Hofmann, 1975). The net conversion rate of each of the components present follows with the aid of $r_j = v_j \cdot r$. For a system in steady state, the net production rate is equal to minus the flow into the system, as is clear from Equ. 2.10. Furthermore, the elemental balance principle (according Equ. 2.11) specifies k relationships between the flows F_j

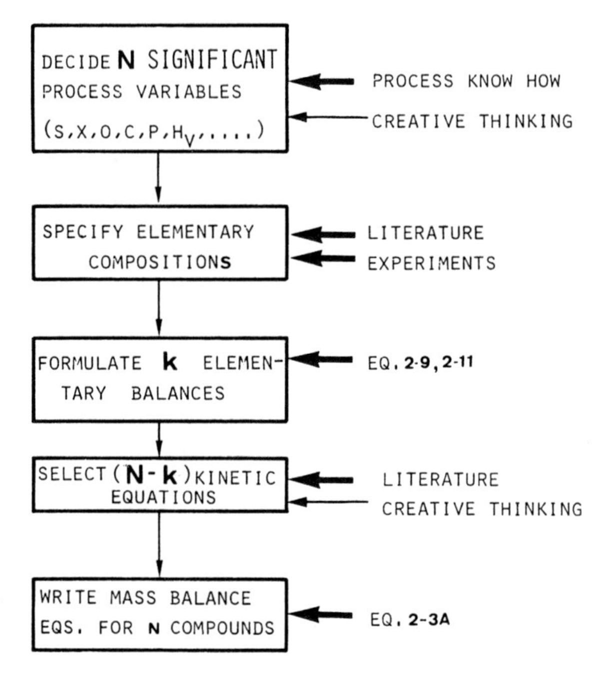

FIGURE 2.19. Flowchart of the construction of unstructured biokinetic models as a consequence of joint activity between kinetic work and stoichiometry. (Adapted from Roels, 1980a. With permission of John Wiley & Sons, Inc.).

and hence only $(N - k)$ net conversion rates can be chosen independently. The number of independent kinetic equations to be postulated cannot be chosen at will—it is completely specified by the number of elemental balances (k) and the number of components in the system (N). The procedure for the construction of a simple unstructured process model is outlined in individual steps in Fig. 2.19.

From the foregoing it is clear that the theory of elemental balance is an invaluable tool in the description of the systems commonly encountered in bioengineering. It is as fundamental as stoichiometry in chemical reaction engineering systems. The theory seems to be well developed, and the field is open to the development of specific applications. A significant problem is, however, associated with the application of the theory. It applies to instances in which more flows are measured than are minimally needed to calculate the remaining ones. In this case, a statistical procedure can be applied to obtain a more optimal estimate of all measured and unknown flows.

2.4.3 Different Levels and Types of Kinetic Models

Some remarks should be made here to clarify discrepancies observed in the kinetics literature. Classification of mathematical models of microbial processes normally involves contrasting pairs, for example: continuous time–discrete time, deterministic–stochastic, unsegregated–segregated, unstructured–structured, lumped parameter–distributed parameter (Harder and Roels, 1981; Kossen, 1979; Reuss, 1977; Roels and Kossen, 1978), to use terminology-from Tsuchiya et al. (1966). The kinetics of biological systems may be expressed at four different system levels: (a) molecular or enzyme, (b) macromolecular or cellular component, (c) cellular, and (d) population (Humphrey, 1978). Each level of expression has a unique characteristic that leads to a rather specific kinetic treatment, similar to the situation with chemical reactions. The similarity to chemical processing should be emphasized even more when the high complexity of bioprocessing at the engineering scale is considered. In complex systems, the mutual relations among phenomena are clarified by means of systems analysis (Kafarow, 1971). Knowledge of Macrokinetics—as named by van Krevelen in his introduction to the first symposium on chemical reaction engineering (1957)—is a sine qua non for studying conversions and developing reactor operations. It is only by understanding the relation and the interaction between microkinetics and macrokinetics that the real phenomena occurring in a chemical or biochemical/biological reactor can be understood.

On the basis of empirical data from reactor operations, and with use of systems analysis, a general research methodology can be derived (Fig. 2.20). The process need not be examined in its infinite complexity but can be hierarchically subdivided into five different levels. The fourth level of bioreactors is the most essential, according to the macroscopic principle. But all other levels contribute to the development of bioplants according to the special aims of the investigation. Generally, model building can proceed in various ways:

1. Derivation of an equation from the microkinetics of the reaction mechanism
2. Simulation of experimental curves by means of purely numerical methods
3. The establishment of quantitative formulas with the help of formal analogies

Each of these three ways of constructing models will be discussed separately.

2.4.3.1 Mechanistic Models of Microkinetics

In principle it is possible to investigate the biochemical mechanisms of the metabolic pathways (enzyme induction and metabolic repression). In many cases a good beginning is described in the literature (Moreira et al., 1979). From this start, a mathematical function usually can be formulated that possesses such a high degree of flexibility that it can be brought into agreement with the experimental facts. In most cases, however, good ideas about the biochemical regulation of metabolism are all too few, and exact values for the parameters of the model are not available. One must then deal with estimates, often estimates taken from the literature.

The complex efforts of modeling also serve another end, namely that of process control and special process optimization (Calam, Ellis, and McCann, 1971). For this, naturally, special methods of analysis must be available. In

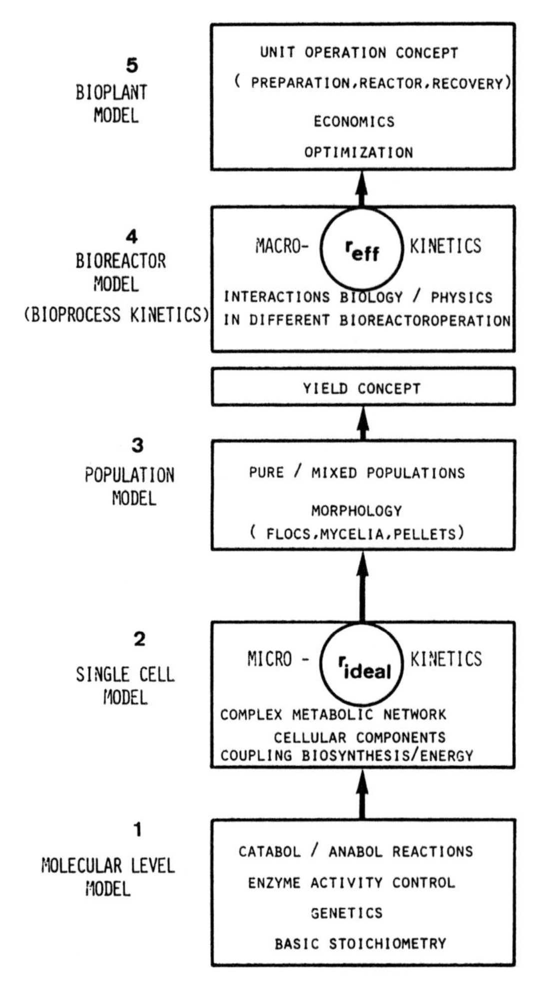

FIGURE 2.20. Overview of the five levels of process research and mathematical modeling of bioprocessing and bioreactor operation.

the future, the microkinetic approach will no doubt contribute a great deal to the formulation of process kinetics (see App. II, especially the use of rds and qss).

The derivation of kinetic equations from postulated mechanisms can, for instance, show the analogy between enzyme kinetics and chemical kinetics for heterogeneous catalysis. Poison-free enzyme kinetics follow the form of a rate equation (Michaelis–Menten type)

$$r_S = r_{S,\max} \frac{s}{K_m + s} \tag{2.54}$$

while for heterogeneous catalysis, the Langmuir type is useful (with $c = s$)

$$r = r_{max}\frac{K_L c}{1 + K_L c} \tag{2.55}$$

Evidently, both approaches are identical when

$$K_m = \frac{1}{K_L} \tag{2.56}$$

(i.e., the Michaelis–Menten constant is inversely proportional to the adsorption equilibrium constant).

Engineers are mainly interested in manipulating a population of cells, and fundamental biology can contribute significantly to deriving population models from single-cell models, this has been pointed out by Shuler and Domach (1982) at a symposium on kinetics and thermodynamics in biological systems (Blanch et al., 1982). Single-cell models can be formulated that mimic very closely the responses of living cells. Such models are convenient tools for testing the plausibility of basic biochemical mechanisms. They may be particularly attractive for testing the potential in vivo compatibility of mechanisms postulated on the basis of in vitro enzymology. Fig. 2.21 shows an idealized sketch of a single-cell model (the Cornell single-cell model, Shuler and Domach, 1982), representing a modification of a previous simple mathematical model for the growth of an individual bacterium (Ho and Shuler, 1977). The simple mathematical model could predict the growth pattern for a cell of a given shape (filamentous, bacillus, or spherical), and it initially included four components (NH_4^+, glucose, precursors, and macromolecules). This number was later enlarged to 14 components (Shuler et al., 1979). The Cornell single-cell model makes reasonable predictions about the dependence of cell size, cell shape, cell composition, growth rate, and timing of cellular events on external concentrations of glucose.

More complex models with a larger number of parameters are able to describe real processes better than simple models. For practical engineering, however, models should be as simple as possible but no simpler (i.e., with a minimum number of parameters).

The field of kinetic modeling of nonstationary bioprocess situations with the aid of structured and mechanistic models will be discussed in more detail in Chap. 5.

2.4.3.2 Numerical Fitting

Equ. 2.57 shows a purely mathematical function that is well suited for reproducing the time-dependent growth curve of a discontinuous process. It contains both an exponential term and a polynomial, which makes it very flexible (Edwards and Wilke, 1968).

Type 1: $r_i^* = f(t)$, Example

$$r_i^* = \frac{K}{1 + \exp(\alpha_0 + \alpha_1 t + \alpha_2 t^2 + \cdots)} \tag{2.57a}$$

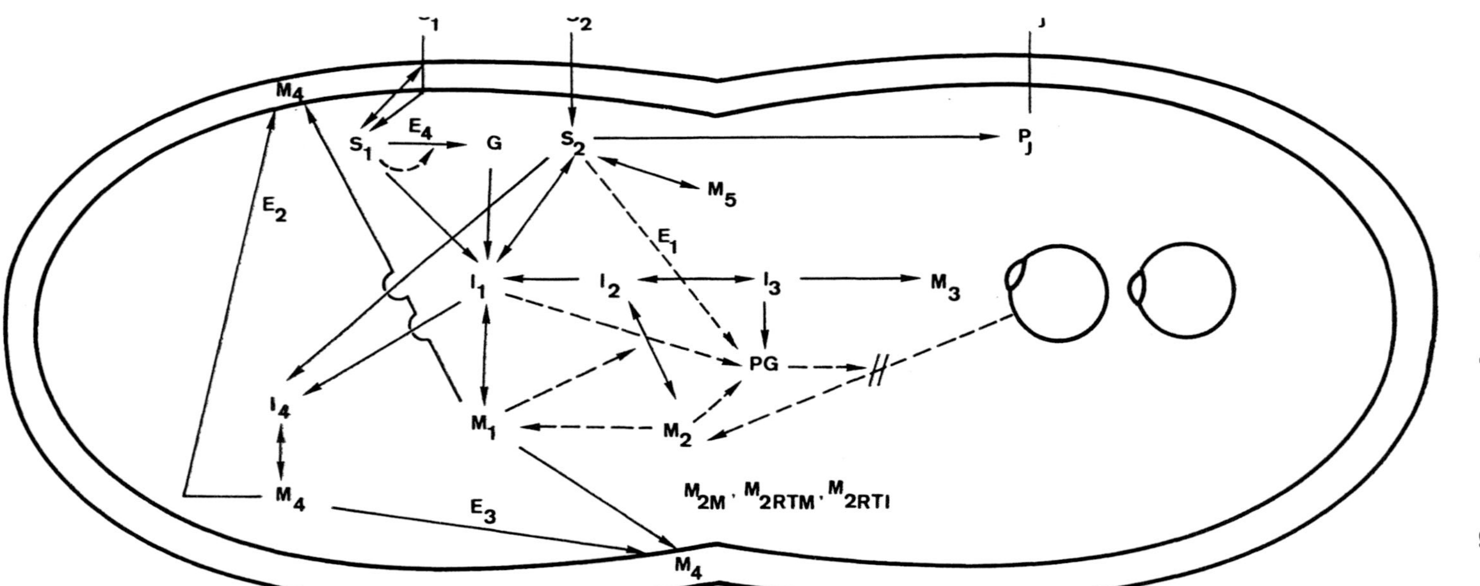

FIGURE 2.21. An idealized sketch of the model for *Escherichia coli* growing in a glucose–ammonium salts medium with glucose or ammonia as the limiting nutrient. At the time shown, the cell has just completed a round of DNA replication and initiated cross-wall fermentation and a new round of DNA replication. Solid lines indicate the flow of material, while dashed lines indicate flow of information. S_1, ammonium ion; S_2, glucose (and associated compounds in the cell); P_i, waste products (CO_2, H_2O, and acetate) formed from energy metabolism during aerobic growth; I_1, amino acids; I_2, ribonucleotides; I_3, desoxyribonucleotides; I_4, cell envelope precursors; M_1, protein (both cytoplasmic and envelope); M_{2RTI}, immature "stable" RNA; M_{2RTM}, mature "stable" RNA (r – RNA and t – RNA; assuming 85% r – RNA throughout); M_{2M}, messenger RNA; M_3, DNA; M_4, nonprotein part of cell envelope (assuming 16.7% peptidoglycan, 47.6% lipid, and 35.7% polysaccharide); M_5, glycogen; PG, guanosine-tetra phosphate; E_1, enzymes in the conversion of P_2 to P_3; E_2, E_3, molecules involved in directing cross-wall formation and cell envelope synthesis (the approach used in the prototype model was used here but more recent experimental support is available); G, glutamine; E_4, glutamine synthetase. An * indicates that the material is present in the external environment. (From Shuler and Domach, 1982. Reprinted with permission from Kinetics and Thermodynamics in Biological Systems, ACS-Winter Symposium. 1982 American Chemical Society).

with $r_i^* = \mu$, $K = x_{max}$, $\alpha_0 = x_{max}/x_0$, and $\alpha_1 = \mu_{max}$. The disadvantage of the usually anonymous functions is that the parameters are mainly devoid of any biological meaning. Other simple functions are also used to describe the time-dependent curve, for example

$$r_i^* = k \cdot \exp[-\alpha_1(t - t_0)] \tag{2.57b}$$

The same type can be used to describe the curves for the concentration dependence $r = f(c)$ and/or $r = f(t)$. In principle, the same function as that in Equ. 2.58, or even a simple polynomial, may be used.
Type 2: $r_i^* = f(c)$: Example

$$r_i^* = \alpha_0 + \alpha_1 c + \alpha_2 c^2 \cdots \tag{2.58}$$

The limitations that apply to this type of analysis do not necessarily diminish its value. Applications will remain limited to cases of very difficult to analyze processes and to the initial phases of process development (e.g., Grm, Mele, and Kremser, 1980). Sooner or later, when economic incentive is adequate, mathematical models with parameters interpretable in clear biological terms will be preferred. The well-known kinetic approach of Kono (see Chap. 5.3) is also a numerical fitting procedure of great engineering value.

2.4.3.3 Formal Kinetic Approach

Here, thinking in terms of analogies arises as a fundamental principle very useful for process development but also having dangers. A formal kinetic model contains analogies to known precedents that are represented by mathematical formulas, formula that allow a simple (and therefore usable) reproduction of the course of a process, at least those aspects regarded as significant. A formal kinetic approach binds the user to a somewhat distorted and mechanistically causal interpretation. This is not expressly stated in the approach and may be of no importance with regard to the primary purpose at hand. However, the implication is that the microkinetic reality is somehow mirrored in the formal kinetic analysis (cf., e.g., Fig. 5.47). Despite the progress of fundamental sciences in the field of biology, especially in the development of models with a mechanistic background, the useful mechanistic interpretation of a bioprocess with even normal complexity is still a demanding task. In this situation, real experiments should be the starting point of process kinetic analysis (Moser, 1978a) as in Fig. 2.14.

The special definition of formal kinetics can be summarized as follows (Moser, 1983b):

1. The experimental data and not postulated mechanisms are the starting point.
2. The mathematical function to be chosen as the model is not of the purely numerical form but is taken from analogies. Analogy means that known mathematical functions, generally from conventional mechanistic work with a similar phenomenology, are selected (e.g., the Monod equation as an analogy to enzyme kinetics).

3. A representative quantification of the process is reached by adapting the values of the model parameters to experimental data (so-called adaptational or fitting parameters e.g. K_S).
4. The meaning of the parameters of such formal kinetic models is different from that of real mechanistic models. A degree of mechanistic interpretation can be attained only with supplementary work. This fact is the main difference between formal kinetics and microkinetics; the mathematical structure used in both approaches often is the same (e.g. $K_S \neq K_m$).

This formal kinetic approach is also successfully used in chemical kinetics (Hofmann, 1975), and it is advantageously used for complex chemical reactions (Schmid and Sapunov, 1982).

A systematic formal kinetic analysis starts with measured concentration–time curves (e.g., in batch processes, as illustrated in Fig. 2.4 for substrate concentrations). From these data a reaction scheme can be extracted. At this point a clear differentiation must be made between reaction scheme and reaction mechanism. Due to the fictitious character of a mechanism, it may be disproven but never proven. A reaction scheme, on the other hand, can be more or less definitely established and may be extended later only if there is evidence of additional steps. From the shape of the concentration–time curves several conclusions can be made (Moser, 1983b) concerning the interpretation of apparent reaction orders *n*. Linearity can be a sign for transport limitation or can indicate the presence of a biosorption effect resulting in a reaction order of zero. Half- and first-order reaction can be interpreted as internal transport

TABLE 2.5. Significant phenomena in bioprocessing accord to the formal macroapproach, their quantification, and pseudo-homogeneous modeling.

Macroscopic phenomena	*Observed quantity*	*Derived quantity*	*Model parameters*
Kinetics			
Growth	x	r_x, μ	μ_{max}, K_s
S consumption	s	r_s, q_s	$q_{s,max}, K_s, Y_{x\|s}$
O_2 consumption	o	r_o, q_o	$q_{O,max}, K_o$
Product formation	p	r_p, q_p	$q_{p,max}, Y_{P\|X}, k_p$
CO_2 formation	c	r_c, q_c	$q_{c,max}$
Heat formation	h_v	r_{H_v}, q_{H_v}	$q_{H_v,max}, k_x \cdot E_a$
Maintenance	c_i		k_d, m_s, k_e, μ_d
Stoichiometry	c_i		$Y_{X\|S}, Y_{X\|P}, \ldots$
Transports			
L phase			
O_2	o	n_o (OTR)	$^{o}k_{L1}a, {}^{o}k_{L2}a$
CO_2	c	n_c (CTR)	$^{c}k_{L1}a \ldots$
Heat	h_v, T	q (H_vTR)	$k_{H_v}a_{H_v}$
Micromixing	c_i	t_m, t_c, CTD	t_m, t_c
Macromixing	c_i	$\bar{t}$, RTD	s^2 (Bo, N_{eq})
S phase	d_p	d_{crit}	$\bar{d}_p(D_{eff})$

Adapted from Moser, 1981.

limitations (see biofilm kinetics, Chap. 5.5), while a changing order from zero to one is to be expected for enzymatically controlled bioprocesses. However, the magnitude of the model parameters, especially of the K_s value, can be taken as an indicator for different overlapping phenomena (Fiechter, 1982; Moser, 1981; see Chap. 5.9). Table 2.5 summarizes the measurable process variables considered significant. The working principles discussed in this chapter are also apparent in this table. The last column identifies the parameters that are incorporated in the case of a mathematical model of the kinetic and transport processes.

Finally, the field of scientific activities for the set-up of mathematical models is shown by illustrating the real situation in biopressing (see scheme of Fig. 2.22). The explanation of details given in the legend to this figure is intended to serve as an introduction to the next chapters.

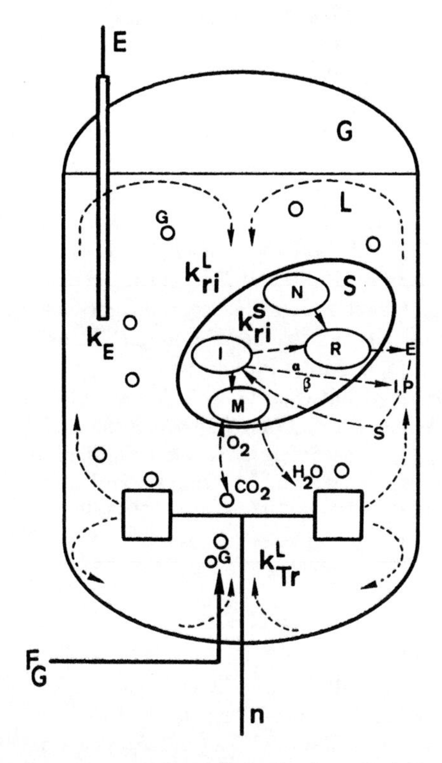

FIGURE 2.22. Schematic representation of the basic experimental situation in bioprocess/bioreactor analyses, where the interactions between physical transports (k^L_{Tr}) and biokinetic rates (k^L_{ri}) in the liquid phase are thought to be representative for the process rates in the solid phase of cell mass (k^S_{ri}). At the same time, response lags of measuring electrodes (k_E) have to be taken into account. G, gas phase; L, liquid phase; S, solid phase or substrate; E, enzyme or electrode; I, intermediary metabolites or products; P, end product; N, nucleus; R, ribosomes; M, mitochondria; α, anabolism; β, catabolism; F_G = gas flow rate; n = agitators rotational speed.

Bibliography

Abbott, B.J., and Clamen, A. (1973). *Biotechnol. Bioeng.*, 15, 117.

Agrawal, P., et al. (1982). *Chem. Eng. Sci.*, 37, 453.

Aiba, S., Humphrey, A.E., and Millis, N.F. (1976). *Biochemical Engineering.* New York: Academic Press.

Atkinson, B., and Mavituna, F. (1983). *Biochemical Engineering and Biotechnology.* The Nature Press and McMillan Publ. Ltd.

Bailey, J.E. (1973). Chem. Eng. Commun., 1, 111.

Bajpaj, R.K., and Reuss, M. (1981). *Biotechnol. Bioeng.*, 23, 717.

Barford, J.R., and Hall, R.J. (1979). In Proc. *7th Australian Conf. Chem. Eng.* p. 21.

Bauchop, T., and Elsden, S.R. (1960). *J. Gen. Microbiol.*, 23, 457.

Bazin, M.J. (ed.) (1982). *Microbial Population Dynamics.* Boca Raton, Fla.: CRC Press.

Bell, G.H. (1972). *Proc. Biochem.*, 7 (April), 21.

Blanch, H.W. et al. (eds.) (1982) ACS winter symp. "Kinetics and Thermodynamics in Biological Systems" Boulder, Colorado.

Bogen, H.J. (1976). *Buch der Biotechnik.* Knaur Verlag, no. 3478. Munich-Zurich:.

Boyle, W.C., and Berthouex, P.M. (1974). *Biotechnol. Bioeng.*, 16, 1139.

Budde, K., Budde, H., and Rückauf, H. (1981). *Stoichiometrie chemischtechnologischer Prozesse*, Berlin: Akademie-Verlag.

Calam, C.T., Ellis, S.H., and McCann, M.J. (1971). *J. Appl. Chem. Biotechnol.*, 21, 181.

Cooney, Ch.L. (1979). *Proc. Biochem.*, 14 (May), 31.

Cooney, Ch.L., and Alcevedo, F. (1977). *Biotechnol. Bioeng.*, 19, 1949.

Cooney, Ch.L., Wang, H.Y., and Wang, D.I.C. (1977). *Biotechnol. Bioeng.*, 19, 55.

Dechema Monograph (1982). *Arbeitsmethoden für die Biotechnologie.* Biotechnology working group (chairman H.J. Rehm). Frankfurt: Dechema.

Dechema Monograph (1984). *Process Variables in Biotechnology.* Bioreactor Performance EFB-working party (chairman W. Crueger) Frankfurt: Dechema.

Dostalek, M., Häggström, L., Molin, N., and Terui, G. (1972). In *Ferm. Technol. Today,* Proc. 4th Intern. Ferm. Symp. (Terui G., ed.) Society of Ferm. Technol. Japan, 497.

Edwards, H.V., and Wilke, Ch.R. (1968). *Biotechnol. Bioeng.*, 10, 205.

Endo, I., and Inoue, I. (1976). 5th International Fermentation Symposium, Berlin, Abstract 5.01. Dellweg H., ed., Inst. f. Gärungsgewerbe und Biotechnologie.

Erickson, L.E., et al. (1978a). *Biotechnol. Bioeng.*, 20, 1595.

Erickson, L.E., et al. (1978b). *Biotechnol. Bioeng.*, 20, 1623.

Erickson, L.E., et al. (1979). *Biotechnol. Bioeng.*, 21, 575.

Eroshin, V.K. (1977). *Proc. Biochem.*, 12 (July/August), 29.

FAST (1980). Sub-programme Bio-society research activities, Commission of the European Communities, EUR 7105, Brussels (M. Cantley).

Fiechter, A. (1974). In Proc. 4th Intern. Symp. Yeasts, Klaushofer H. and Sleytr U. (eds.), Univ. of Bodenkultur Vienna, Part II, 17.

Fiechter, A. (1978). In *Proceedings of the 1st European Congress on Biotechnology,* Interlaken, Switzerland *Dechema Monograph* 82, 17. Dechema 1D.

Fiechter, A. (1982). In Rehm, H.J., and Reed, G. (eds.). *Biotechnology—A Comprehensive Treatise,* Deerfield Beach, Fla and Basel: Verlag Chemie Weinheim. Vol. 1, Chap. 7.

Geurts, Th.G., et al. (1980). *Biotechnol. Bioeng.*, 22, 2031.

Geyson, H.M., and Gray, P. (1972). *Biotechnol. Bioeng.*, 14, 857.

Glansdorff, P., and Prigogine, I. (1974). *Thermodynamics of Structure, Stability and Fluctuations.* New York: John Wiley.

Grm, B., Mele, M., and Kremser, M. (1980). *Biotechnol. Bioeng.*, 22, 255.
Guenther, K.R. (1965). *Biotechnol. Bioeng.*, 7, 445.
Häggström, L. (1977). *Appl. Environ. Microbiol.*, 33, 555.
Harder, A., and Roels, J.A. (1981). *Adv. Biochem. Eng.*, 21, 56.
Harrison, J.S. (1967). *Proc. Biochem.*, 2, 41.
Heijnen, J.J., and Roels, J.A. (1981). *Biotechnol. Bioeng.*, 23, 739.
Heijnen, J.J., et al. (1979). *Biotechnol. Bioeng.*, 21, 2175.
Herbert, D. (1975). In Dean, A.C.R., et al. (eds.). *Continous Culture*, Vol. 6. London: SCI, E. Horwood Ltd, Chichester England p. 1.
Ho, S.V., and Shuler, M.L. (1977). *J. Theoret. Biol.*, 68, 415.
Hockenhull, D.J.M. (1971). *Progr. Ind. Microb.*, 9, 133.
Hockenhull, D.J.M. (1975). *Appl. Microbiol.*, 19, 187.
Hofmann, H. (1975). *Chimia*, 29, 159.
Hoover, S.R., and Porges, N. (1952). *Sewage Indus. Wastes*, 24, 306.
Humphrey, A.E. (1970). *Proc. Biochem.*, 5 (June), 19.
Humphrey, A.E. (1977). Chem. Eng. Prog., May, 85.
Humphrey, A.E. (1978). *ACS Symposium Series*, 72. (American Chemical Society)
IUPAC-proposal (1982) list of symbols for use in Biotechnology Pure & Appl. Chem. 54, 1743.
Johnson D.B., and Berthouex, P.M. (1975a). *Biotechnol. Bioeng.*, 17, 557.
Johnson, D.B., and Berthouex, P.M. (1975b). *Biotechnol. Bioeng.*, 17, 571.
Kafarow, W.W. (1971). *Kybernetische Methoden in der Chemie und Chemischen Technologie.* (K. Hartmann, ed.) Verlag Chemie und Akademie Verlag, Berlin (German) edition).
Kafarow, W.W., et al. (1979). *Modelling of Biochemical Reactors.* Moscow: Lesnaja Promyshlenost. (Russ.)
Kobayashi, T. (1972). Group Training Course, Osaka University, Osaka.
Kossen, N.W.F. (1979). *Proc. Soc. Gen. Microbiol. Symp.* 29, (Bull A.T. et al., eds.) 327.
Kossen, N.W.F., and Oosterhuis, N.M.G. (1985). In Rehm, H.J., and G. Reed (eds.). *Biotechnology—A Comprehensive Treatise*, Vol. 2. Deerfield Beach, Fla., and Basel: Verlag Chemie Weinheim, Chap. 24.
Kuhn, H., Friedrich, U., and Fiechter, A. (1979). *Europ. J. Appl. Microbiol. Biotechnol.*, 6, 341.
Mateles, R.I. (1971). *Biotechnol. Bioeng.*, 12, 581.
Mateles, R., and Battat, E. (1974). Appl. Microbiol. 105, 51.
Mayberry, W.H., et al. (1968). *J. Bacteriol.*, 96, 1424.
McCarty P.L. (1970). In Proc. Wastewater Reclamation Reuse Workshop, Lake Tahoe, Calif., 226.
McLennan, D.G., et al. (1973). *Proc. Biochem.*, 8, 22.
Metz, H. (1975). *Chem. Ing. Techn.*, 43, 60.
Minkevich, I.G., and Eroshin, V.K. (1973). *Folia Microbiol.*, 18, 376.
Moletta, R., et al. (1978). *Arch. Microbiol.*, 118, 293.
Monod, J. (1942). *Recherches sur la croissance des cultures bacteriennes.* Paris: Hermann & Cie.
Moo-Young, M. (ed.-in-chief) (1985). *Comprehensive Biotechnology*, Vol. 1–3. Oxford: Pergamon Press.
Moreira, A.R., van Dedem, G., and Moo-Young, M. (1979). *Biotechnol. Bioeng. Symp.*, 9, 179.
Moser, A. (1977a). *Habilitationsschrift*, Technical University, Graz, Austria.
Moser, A. (1977b). *Chem. Ing. Tech.*, 49, 612.

Moser, A. (1977c). *Chimia*, 31, 116.
Moser, A. (1978a). In preprints 1st Eur. Congress on Biotechnology, Interlaken, Switzerland Dechema/D Part 1, p. 88.
Moser, A. (1978b). *Gas-Wasserfach, Wasser/Abwasser*, 119, 242.
Moser, A. (1980). Paper presented at 6th *International Fermentation Symposium*, London, Ontario, July 20–25 (Zajic J.E et al., eds.) abstract no F.11.1.7
Moser, A. (1981). *Bioprozesstechnik*. Vienna and New York: Springer-Verlag.
Moser, A. (1982). *Biotechnol. Lett.*, 4, 73.
Moser, A. (1983a). In *Adv. in Fermentation 83* (Suppl. Process Biochemistry), p. 202. Wheatland J. Ltd, England (ed.).
Moser, A. (1983b). In *Proceedings of "Biotech 83", International Conference on Commercial Applications and Implications of Biotechnology*. London: Online Publications, p. 961.
Moser, A. (1984). *Acta Biotechnol.*, 4, 3.
Moser, A. (1985). In *Proc. Advances in Bioreactor Engineering*. Lodz, Poland, September 25–27 1985. (Michalski H., et al., eds.)
Moser, A., and Lafferty, R.M. (1976). Paper presented at 5th Intern. Fermentation Symposium, Berlin. (H. Dellweg, ed.) Institut für Gärungs gewerbe und Biotechnologie, Berlin.
Nagai, S. (1979). *Adv. Biochem. Eng.*, 11, 49.
Oosterhuis, N.M.G. (1984). Ph.D. thesis, Technical University, Delft, Netherlands.
Ovaskainen, P., Lundell, R., and Laiho, P. (1976). *Proc. Biochem.*, 10 (May), 37.
Papoutsakis, E., and Lim, H.C. (1981). *Ind. Eng. Chem. Fundam.*, 20, 307.
Payne, W.J. (1970). *Annu. Rev. Microbiol.*, 17.
Peters, H. (1976). In Hartmann, L. (ed.). *Karlsruher Berichte zur Ingenieurbiologie*, Vol. 9. Univ. Karlsruhe, Inst. t. Ingenieurbiologie und Biotechnologie des Abwassers, Karlsruhe.
Pickett, A.M., Topiwala, H.H., and Bazin, M.J. (1979). *Proc. Biochem.*, 13 (November), 10.
Pirt, S.J. (1966). *Proc. Roy. Soc.*, (B) 163, 224.
Pirt, S.J. (1975). *Principles of Microbe and Cell Cultivation*, Blackwell Scientific Publications, Oxford.
Popper, K. (1934,1972). *The Logic of Scientific Discovery*. London: Hutchinson.
Prigogine, I. (1962). *Introduction to Nonequilibrium Thermodynamics*. New York: Wiley-Interscience.
Prochazka, G.J., et al. (1970). J. *Bacteriol*, 15, 117.
Reardon, K.F., Scheper Th.H., Bailey J.E. (1987). Biotechnol. Prog. 3, Sept, 153.
Rehm, H.J. (1980). *Technische Mikrobiologie*. Berlin: Springer-Verlag.
Rehm H.J., and Reed, G. (eds.) (1982ff). *Biotechnology—A Comprehensive Treatise*, Vol. 1–9. Deerfield Beach, Fla., and Basel: Verlag Chemie Weinheim.
Reuss, M. (1977). *Fort. Verfahrenstechnik*, 15F, 549.
Reuss, M., et al. (1974). *Chem. Ing. Techn.*, 46, 669.
Reuss, M., et al. (1980). *6th International Fermentation Symposium*, London, Ontario.
Richards, J.W. (1968a). *Proc. Biochem.*, 3 (May), 28.
Richards, J.W. (1968b). *Proc. Biochem.*, 3 (June), 56.
Ricica, J. (1969). In Perlman, D. (ed.). *Fermentation Advances*. New York: Academic Press, p. 427.
Rock, J.S., et al. (1978). *Biotechnol. Bioeng.*, 20, 1557.
Roels, J.A. (1980a). *Biotechnol. Bioeng.*, 22, 2457.

Roels, J.A. (1980b). *Biotechnol. Bioeng.*, 22, 23.
Roels, J.A., and Kossen, N.W.F. (1978). In Bull, M.J. (ed.). *Progress of Industrial Microbiology*, Vol. 14. Amsterdam: Elsevier, p. 95.
Romanovski, J.M., et al. (1974). *Kinetische Modelle in der Biophysik.* Jena: Fischer. (German translation by W.A. Knorre and A. Knorre)
Schmid, R., and Sapunov, V.N. (1982). *Non-Formal Kinetics*, Deerfield Beach, Fla., and Basel: Verlag Chemie Weinheim.
Schubert, E., and Hofmann, H. (1975). *Chem. Ing. Techn.*, 47, 191.
Shiloach, J., and Bauer, S. (1975). *Biotechnol Bioeng.*, 17, 227.
Shuler, M.L., and Domach, M.M. (1982). In (Blanch, H.W., et al. (eds.). *Kinetics and Thermodynamics in Biological Systems.* ACS winter symposium, Boulder, Colorado. (American Chemical Society)
Shuler, M.L., et al. (1979). *Ann. N. Y. Acad. Sci.*, 326, 35.
Speece, R.E., and McCarty, P.L. (1964). *Adv. Water Poll. Res.*, 2, 305.
Stephanopoulos, G., and San, K.Y. (1984). *Biotechnol. Bioeng.*, 26, 1176.
Stouthamer, A.H. (1980). *Vierteljahrsschrift der Naturforschenden Gesellschaft in Zürich*, 125/1, 43.
Stouthamer, A.H., and Bettenhaussen C.W. (1973). *Biochim. Biophys. Acta*, 301, 53.
Stouthamer, A.H., van Versefeld H.W. (1987). Trends in Biotechnol., 5, 149.
Sukatsch, D.A., and Faust, U. (1977). *Proceedings of the Tutzing Symposium, Dechema Monograph* 81, 197.
Swartz, R.W. (1979). *Annu. Rep. Ferm. Proc.*, 3, 75.
Tanner, R.D. (1970). *Biotechnol. Bioeng.*, 12, 831.
Tempest, D.W., and Neijssel, O.M. (1975). In Dean, A.C.R., et al. (eds.). *Continuous Culture*, Vol. 6. London: SCI, p. 283.
Toda, K. (1981). *J. Chem. Tech. Biotechnol.*, 31, 775.
Topiwala, H.H. (1973). In Norris, J.R., and Ribbons, D.W. (eds.). *Methods of Microbiology*, Vol. 8. London-New York: Academic Press, p. 35.
Topiwala, H., and Sinclair, C.G. (1971). *Biotechnol. Bioeng.*, 13, 1975.
Trilli, A. (1977). *J. Appl. Chem. Biotechnol.*, 27, 251.
Tsuchiya, H.M., Fredrickson, A.G., and Aris, R. (1966). *Adv. Chem. Engng.*, 6, 125.
Tsuchiya, Y., Nishio, N., and Nagai, S. (1980). *Europ. J. Appl. Microbiol. Biotechnol.*, 9, 211.
van Krevelen, D.W. (1957). In Proceedings of the First *Symposium on Chemical Reaction Engineering.* Oxford: Pergamon Press. (van Krevelen D.W., ed.)
Votruba, J. (1982). *Acta Biotechnol.*, 2, 119.
Wang, D.I.C. (1968). *Chem. Eng.*, 75, 99.
Wang, D.I.C., Cooney Ch.L., Demain A.L., Humphrey A.E. and Lilly M. (1979). "*Fermentation and Enzyme Technology*" John Wiley & sons, New York.
Wingard, L.B. Jr. (ed.) (1972). *Enzyme Engineering. Biotechnol. Bioengng. Symp.* 3. Interscience-Publ., J. Wiley & sons, New York.
Wingard, L.B., Katchalski-Katzir, E., and Goldstein, E. (eds.) (1976). *Immobilized Enzyme Principles.* New York: Academic Press.
Zaborsky, O. (1973). *Immobilized Enzymes.* Cleveland, Ohio: CRS Press.

CHAPTER 3
Bioreactors

3.1 Overview: Industrial Reactors

Bioreactors exist to "tame" biological systems on an industrial scale (see fig. 1.1), and they should present the optimum conditions for serving the needs of biological processes. A large number of reactor types are found corresponding to the large number of different industrial processes (e.g., Moo-Young, 1985; e.g. Rehm and Reed 1982ff). Reviews with detailed discussions are to be found in the literature and also in almost all symposium volumes on biotechnology (Atkinson, 1974; Atkinson and Kossen, 1978; Fiechter, 1978; Ghose and Mukhopadhyay, 1979; Moo-Young, 1985; Mosen 1985a; Schügerl, 1979, 1980; Sittig, 1977; Sittig and Heine, 1977). Here some of the most important types will be briefly mentioned.

3.1.1 Microbiological Reactors (Fermenters, Cell Tissue Culture Vessels and Waste Water Treatment Plants)

A more or less clear overview of the enormous number of very different types of designs in fermentation, cell tissue culture, food, and waste water and waste treatment technology is possible. The designs cover technological application, the stirring and aeration system, and the phase of the main substrates.

The basic types of aeration systems are the submerged sparger, with or without mechanical stirrer, the surface aerator, and the film reactor. The simplest type of reactor, a vessel or vat with no special movable stirring or aeration system, is well suited for carrying out liquid phase, anaerobic cooking-type operations, such as in brewing. Containers with certain shapes (egg shaped or cylindrical with tapered top and bottom parts) are preferable due to improved fluid dynamic characteristics. Vats, together with suitable horizontal-type reactors such as channels with floating covers, are used in producing biogas (CH_4), in agricultural installations or large-scale community waste treatment plants. When used on a small scale, such reactors make few technical demands and may be constructed from fiberglass, reinforced polyester, concrete, or steel. In this class would also be a rotating horizontal drum that works like a cement mixer and is suitable for use in fermentations involving solid substrates (animal wastes, or the fermentation of grain for use in antibiotic production) (Hesseltine, 1977a,b). Similar designs are used in, for example, making yogurt (Driessen, Ubbels, and Stadhouders, 1977).

Aerated, stirred vessels can be called standard or reference bioreactors (Dechema, 1982): Equipped with various stirrers, they are generally suitable for most uses (Zlokarnik, 1972). Additional mixing of the fluid can be obtained by building in various baffles and fins. A new development is the so-called "totally filled bioreactor" (Karrer, 1978; Puhar et al., 1978).

Multiple stirring devices are advantageous for mixing highly viscous media such as myceliae fermentations used in producing antibiotics. Column-type reactors with multiple stirrers and sieve base plates are highly developed reactors that require a great deal of energy and whose use is justified only in the case of special problems such as oil–water emulsions. A cascade of stirred reactors is used, naturally, only for continuous processes. From a technical viewpoint, the cascade may be thought of as a substitute for a genuine tubular reactor. The so-called "paddle wheel reactor" is a horizontal container that has good aeration but can only be built to a limited size (Zlokarnik, 1975). The self-priming aeration systems using low-pressure air are energy efficient but are limited to low-viscosity media. A new development is the submerged tower system with injector aeration ("towerbiology," Leistner et al., 1979), also called the "bio-high reactor." Systems that have been used for a comparatively long time in water purification are oxidation ponds, ditches, and lagoons, which can be aerated in various ways. New types of systems worth mentioning are the agitated and aerated tubular reactors (Moser, 1973b, 1977a) and the scraped-tube reactors (Moo-Young et al., 1979).

The various types of plug flow bioreactors were recently surveyed by Moser (1985a). They utilize surface aeration by means of a variety of rotating brushes, rotors, cone aerators, or gas or fluid jets such as are found in biological waste water treatment plants. Beyond all mechanically driven systems, reactors can also be both aerated and mixed pneumatically, or one pump can serve for both mixing and hydrodynamic stirring.

There is a great deal of interest in bubble columns with their countless designs: They have no moving parts and are very energy efficient (Deckwer, 1977; Schügerl, Oels, and Lücke, 1977; Schügerl et al., 1978). The air lift fermenter uses additional fins to achieve the mixing of the liquid (Wang and Humphrey, 1969). Another new design is the "Andritz" reactor (Paar et al. 1988). The "pressure cycle fermentor" from Imperial Chemical Industry ICI (Cow et al., 1975) and the "deep shaft reactor" (Hines et al., 1975) also represent the types of reactors in which the rising column of gas bubbles is used to circulate the liquid. They have the same characteristics as the loop reactors, of which there are a large number of variants with internal or external circulation, most often with jets as the aeration system (Blenke, 1979; the tubular loop reactor, Ziegler et al., 1977; the torus or cycle ring reactor,

Läderach, Widmer, and Einsele, 1978; the cyclone reactor, Dawson, 1974; the cycle tube cyclone reactor, Liepe et al., 1978). The plunging jet reactor developed by Vogelbusch, Vienna, in collaboration with the Engineering Center, Böhlen G., East Germany, is another type of special design that works with a two-phase pump and a foam-like gas/liquid mixture (Schreier, 1975; Steiner et al., 1977). The last type of bioreactor to be mentioned is the thin-film type. Here the liquid and/or solid phase is in the form of a thin layer, and this promotes the reaction.

Further classifications of biofloc and biofilm bioreactors may include thin-layer and adhesive fermenters, horizontal trays, the biodisk, and packed bed and fluidized bed bioreactors (Moser, 1977a). Most reactors of this type are still used for research at the laboratory scale. Simple trays have long been used for tissue culture. The fixed bed reactor or percolating filter has long been used for producing acetic acid, while the biodisk is used in waste water treatment. Bubble column and air lift reactors are especially suitable for tissue culture applications, which are generally very sensitive to sheer forces (Katinger, Scheirer, and Krömen, 1979). Finally, photobioreactors should be mentioned; they are used to achieve high photosynthetic activity with algae (Jüttner, 1982; Märkl and Vortmeyer, 1973; Pirt, 1980; Pirt et al., 1983). Integrated reactor systems using membranes for separation purposes are also being developed.

3.1.2 Enzyme Reactors

The biologically active enzymes can be used as catalysts either in a soluble, dispersed form or in a carrier-bound form. Because of the need for the isolation step and losses of enzyme activity, enzyme processes are sparingly used at the present time. Immobilized enzymes show promise for minimizing these activity losses and for facilitating enzyme recovery (Pitcher, 1978).

Usable types of laboratory- or engineering-scale enzyme reactors include packed bed columns, hollow fiber membrane reactors, and reactors containing rolled sheets of membrane catalyst. Almost all have some of the characteristics of plug flow reactors. The so-called "spinning basket" reactor is an analogy to chemical reactors in which the catalyst is directly attached to the stirring apparatus (Carberry, 1964). Enzyme reactors currently are being introduced on a large scale in, for example, the production of amino acids (separation of d- and l-forms, Chibata et al., 1972). Surely they will play a larger role in the future (Coughlin et al., 1975; Moo-Young, 1985; Mulcahy and La Motta, 1978; Nelböck and Wandrey, 1978; Pye and Wingard, 1974ff; Wandrey, 1983; Wang et al., 1979).

The problems of working with carrier-bound enzymes and cells in biofilm reactors and the biofloc reactors are compared in Table 3.1.

3.1.3 Sterilizers

Sterilizers are used to kill biological materials. Use of sterilizers is often necessary to produce infection-free, sterile nutrient media that can be used in pure cultures.

TABLE 3.1. Biofilm operation compared with floc bioprocessing.

Criterion	*Floc*	*Biofilm*
Mode of operation	Discontinuous Continuous (wash out)	Continuous (no wash out)
Process control	Multiple, difficult	Simple
Product recovery	Expensive	Easy and inexpensive
Kinetics	Homogeneous Heterogeneous	Heterogeneous Resp. pseudohomogeneous
Transport problems	Yes	Yes
At G|L interface	Yes	Yes
In L phase	Yes	Yes
At L|S interface	Nearly undetectable	Yes (easier to handle)
In S phase	Difficult to model	Yes
Particle size	Size distribution uncontrolled in STRs	Film thickness uncontrolled but possibilities for controlled and uniform film growth

From Moser, 1981.

Sterilizers can be operated discontinuously or continuously. For engineering-scale processes, sterilization is usually done with heat (i.e., steam) on economic grounds. Chemical and physical processes also exist (Aiba, Nagai, and Nishizawa, 1976; Richards, 1968). The stirred processes differ from each other in method of heat exchange. They are far surpassed for continuous operations by tube-type sterilizers, a consequence of the formal first-order kinetics of sterilization (see Chap. 5). Continuous steam sterilizers have been used for a long time in food processing technology (e.g., milk production).

3.2 Systematics of Bioreactors

A classification scheme for bioreactors can be devised with the help of the following criteria:

1. The geometry of the reactor, or the way the mass of catalyst is distributed throughout the reactor volume: biofloc or biofilm reactors. Associated with this is the question of whether the reactor should be considered a "homogeneous" or a "heterogeneous" system, that is, whether it has or does not move significant physical transport limitations.
2. The mode of operation of the reactor—discontinuous, completely continuous, semicontinuous, or semidiscontinuous—including transient or steady-state operation techniques.
3. The state of mixing in the reactor—uniform distribution ("lumped parameters") or uneven distribution ("distributed parameters"). Associated with this is the question of whether the reactor is maximally mixed or totally segregated, that is, whether the reactor is to be considered an ideal stirred vessel or an ideal tube reactor.

3.2.1 Homogeneous Versus Heterogeneous Systems

The concept of homogeneous versus heterogeneous used as basis for systematization refers to the relationship of the extent or size of the solid mass (S phase) to the extent of the reaction phase (L phase).

Genuine homogeneous reactions are found only when enzymes in soluble form are used. The "normal" situation of fermentation technology, in which biological cells are agglomerated to flocs representing a solid phase is a heterogeneous system. Nevertheless, such a system can be treated as pseudohomogeneous (see Sect. 4.3).

With increasing floc diameter (or in the case of biofilm processes increasing film thickness), the system becomes truly heterogeneous. Transport within and between different phases is significant, and differential equations must be formulated for the different situations. These complex cases often can be reduced to pseudohomogeneous ones by introducing an algebraic factor, and this is the concept of reaction rate efficiency (cf. Sect. 4.5).

3.2.2 Mixing Behavior

Process engineering characteristics of reactors deal with the mixing conditions of the main reaction phase, the L phase. The degree of segregation (Danckwerts, 1958) can be used as an example. The two extreme conditions are referred to as maximal mixing, mm, and total segregation, ts (see Fig. 3.1).

Consider a particular reactor space, for example a tube in which there is a flow of velocity v_z in the direction z. The entering fluid stream is thought of as separate layers, and the fate of these layers is observed as they pass through the reactor space. In the case of mm, there is complete mixing of the layers over the cross section of the tube. In the case of ts, the layers leave the reactor unchanged. These extreme conditions are realizable in the reactor conformations shown in Fig. 3.1c (Zwietering, 1959).

In practice, in the case of stirred reactors, the mixing time t_m is used for characterizing mixing; its experimental measurement is described in Sect. 3.3.2. That this one-dimensional quantity can represent the actual three-dimensional time course of mixing is due to the condition of mm, which occurs at the same rate in all three directions ("lumped parameters" in a stirred vessel). The state of mixing in reactors with concentration profiles (tubular reactors or reactors with internal and external circulation) is experimentally more difficult to observe (Hartung and Hiby, 1973; Hiby, 1972).

For understanding of the general concept of the degree of mixing in reactors it is important to recognize that the degree of mixing is made up of two equally important components: (a) micromixing, characterized by the time needed for mixing in the L phase inside the reactor (stirred reactors), and (b) macromixing, characterized by the residence time distribution (RTD) (continuous reactors), the reactor being considered a black box.

The two characteristic parameters of micromixing and macromixing, as

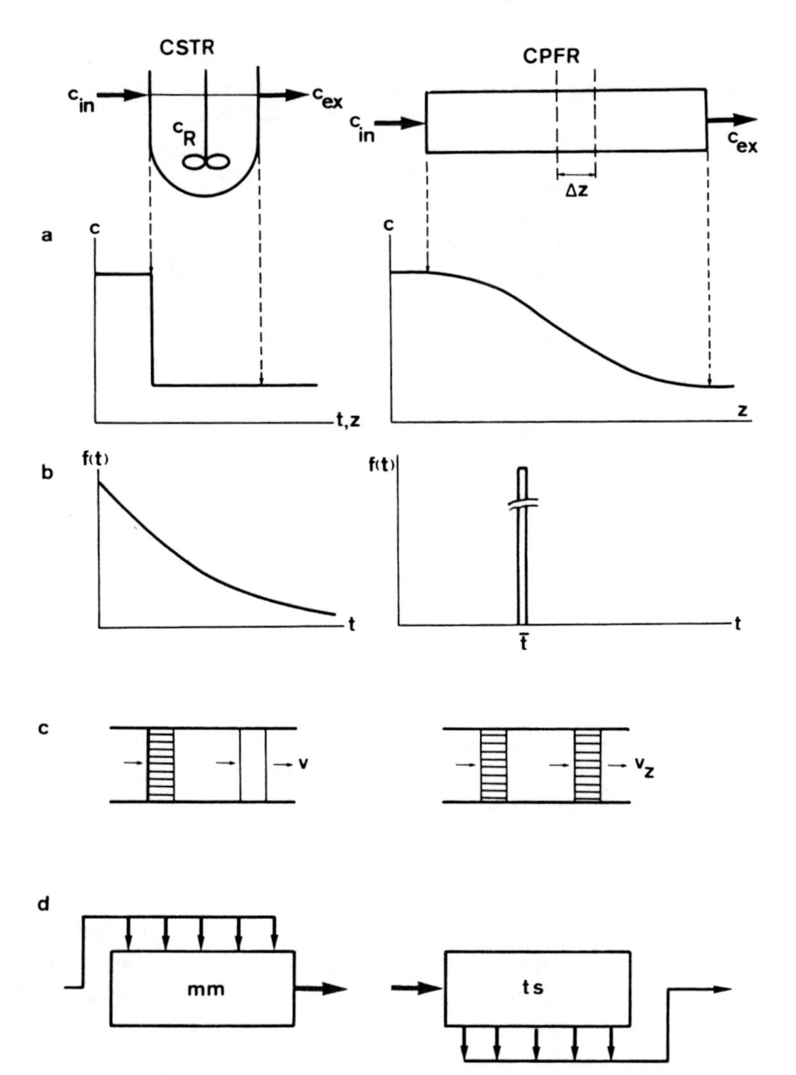

FIGURE 3.1. Continuous stirred tank reactor (CSTR) versus continuous plug flow reactor (CPFR). Comparison of (a) concentration profiles, (b) residence time distribution according to impulse function $f(t)$, representing macromixing, (c) visualization of micromixing flow behavior with extreme cases of maximum mixedness (mm) and total segregation (ts) (according to Weinstein and Adler, 1967, Pergamon Journals Ltd.) and (d) realization of the flow behavior (according to Zwietering, 1959, Pergamon Journals Ltd.) v, velocity; z, direction.

shown in Fig. 3.1, are actually coupled. The simplification implies, however, that this model using CSTR and CPFR as extreme cases is not appropriate to adequately represent transition states. The degree of mixing can be especially well studied in loop reactors, because here the intermediate states appear in clearer form (Fu et al., 1971; Lehnert, 1972; Moser, 1985a; Moser and Steiner, 1974, 1975a,b; Rippin, 1967).

Because of the restricted applicability of the concept of mixing time, alternatives are sought to characterize mixing more generally. The degree of segregation previously mentioned, and also referred to as the "inhomogeneity," J (Dankwerts, 1958; Zwietering, 1959), is defined by Equ. 3.1

$$J = \frac{\operatorname{var} \alpha_P}{\operatorname{var} \alpha} = 1 - \frac{\operatorname{var} \alpha_i}{\operatorname{var} \alpha} \tag{3.1}$$

and has a range

$$0 \leq J \leq 1$$

with boundary values

$J = 0$ for mm (maximum mixedness)

$J = 1$ for ts (total segregation)

The quantities $\operatorname{var} \alpha_i$, $\operatorname{var} \alpha_P$, and $\operatorname{var} \alpha$ are variances (sum of squares) of the age distributions at a single "point" (α_i), between two points (α_P), or of the whole system (α). The point (Danckwerts, 1958) is defined as an element of volume that is small in relation to the volume of the reactor but nevertheless large enough to hold many molecules. At this point the process of mixing, along with its two extremes, can be conceived of as shown in Fig. 3.1a and c. The problem with this concept is that the quantity J is—in real bioprocessing—experimentally almost unmeasurable. It can nevertheless be used as a model in thinking and working. The cited references demonstrate the fact that the states of micromixing and macromixing are directly dependent on the recycle flow ratio r of a loop reactor (Dohan and Weinstein, 1973; Moser and Steiner, 1975a; Rudkin, 1967). The behavior of loop reactors often approximates that of tube-type reactors, with total segregation when the value of the recycling ratio is low. When the recycling ratio is high, the behavior changes to that of a stirred reactor with maximal mixing (Fig. 3.2). The consequence is important when, for example, circulating reactors are used to obtain kinetic data (see Sect. 3.5 and 4.3), and it is also important in obtaining the optimal configuration and conversion in a reactor (see Chap. 6).

The same model of inhomogeneity has also been used to model conversion in various combinations of reactors (Tsai et al., 1969,1971; Wen and Fan, 1975; see Chap. 6). Finally, a systematic picture of the various types of bioreactors on the basis of the criteria discussed is shown in Fig. 3.3. Of course, in practice a strict classification is not possible since the intermediate states (especially with regard to micro and macromixing) are often dominant.

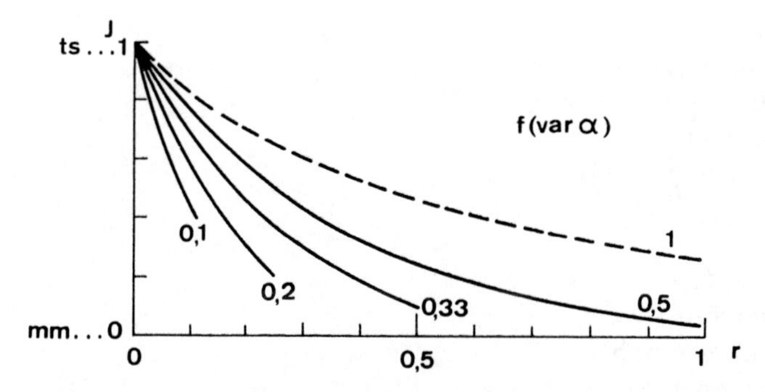

FIGURE 3.2. Relation between degree of segregation J (Danckwerts, 1958) and recycle ratio r, in a recycle reactor with the variance α (proportional to Bo_{tot} as a measure of residence time distribution) as parameter (From Dohan and Weinstein, 1973. With permission from Ind. Eng. Chem. Fundam., 12, 64. Copyright American Chemical Society.)

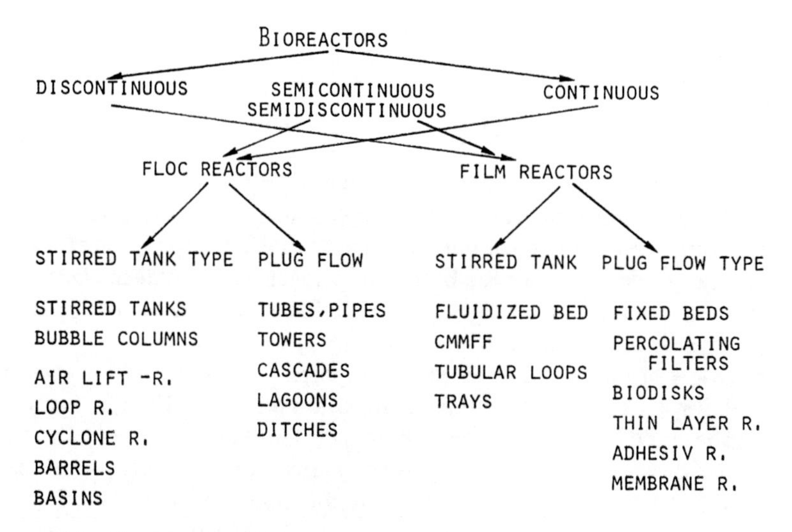

FIGURE 3.3. Classification of bioreactor types on the basis of technical criteria, with examples of known reactor designs. (From A. Moser, 1977a).

3.3 Quantification Methods

A complete list of all process variables necessary for a detailed characterization of bioreactors includes the following (Dechema 1984,1987):

- Temperature (T)
- Stirrer speed (n)
- Foam

- Gas flow (F_G)
- Weight, level (W)
- Addition of feed stock
- O_2 partial pressure (p_{O_2})
- Redox potential (rH)
- pH
- Sterility
- Biomass (x)
- Gas hold-up (ε_G)
- O_2/CO_2 gas exchange
- Power input (P)
- Rheology (ν, η_{app}, or K_C)
- Shear ($\dot{\gamma}$, τ_0)
- Mixing (t_m, t_C, CTD)
- O_2 transfer ($k_{L1}a$)
- Morphology (δ^* resp. η^*)

The appropriate methods of measurement and evaluation of the most essential variables will be briefly discussed here.

3.3.1 Residence Time Distributions (RTD)—Macromixing

The experimental measurement and typical results for different residence time distributions in a continuous reactor are summarized in Fig. 3.4. The same arrangements used for determining the mixing time are appropriate for determining the residence time distribution in a reactor. A signal in the form of a pulse or step function or a periodic function is formed at the input, and the response is measured at the output.

The shape of the response curve for an impulse function $f(t)$ is shown in Fig. 3.3b for various residence time distribution functions. Figure 3.4c, shows the response curves $F(t)$ for a step function. The limiting cases of an ideal stirred vessel and an ideal tubular reactor are shown in both b and c. To quantify RTD curves, two fundamentally different models are used: (a) the so-called "one-dimensional dispersion model," primarily used for tubular reactors with low backmixing, and (2) the so-called "cell model" ("tanks in series model") which was primarily intended for stirred vessel reactors but is of general validity.

3.3.1.1 One-Dimensional Dispersion Model (1-parameter model)

The picture involved in the one-dimensional dispersion model is the one-dimensional process of flow in a tube. There is a flow velocity v_z in direction z, which, in the ideal case, is constant over the reactor cross section a_T. Because of molecular diffusion, turbulent convection, and the parabolic velocity profile that results from boundary friction (roughness, ε), there are large deviations from a uniform flow front. The effective longitudinal dispersion coefficient

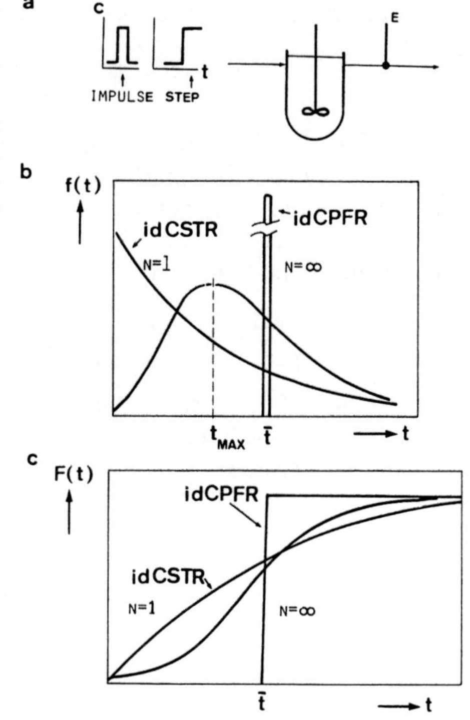

FIGURE 3.4. Measurement set-up (a) for the characterization of the residence time distribution of continuous reactors as exemplified by the fluid phase: (b) pulse method data with the pulse function $f(t)$. (c) Step function data with step function $F(t)$. The extreme cases of ideal reactor behavior is indicated by idCSTR and idCPFR.

$D_L(D_{eff})$ is used as a measure of this effect. It is an analogy to a genuine diffusion coefficient

$$D_L = f(v_z, D, d_T, \rho, \nu, \varepsilon) \tag{3.2}$$

Applying Equ. 2.3, the equation for the one-dimensional dispersion model can be written

$$\frac{\partial c}{\partial t} = -v_z \frac{\partial c}{\partial z} + D_L \frac{\partial^2 c}{\partial z^2} \tag{3.3a}$$

To facilitate a solution, this equation can be written in dimensionless form

$$\frac{\partial c/c_0}{\partial t/\bar{t}} = -\frac{\partial c/c_0}{\partial z/L} + \left(\frac{D_L}{v_z \cdot L}\right) \frac{\partial^2 c/c_0}{\partial (z/L)^2} = 0 \tag{3.3b}$$

In this equation there is a dimensionless number, the so-called Bodenstein number, Bo (sometimes also called Peclet number, especially when a diameter is taken as characteristic length):

$$\mathrm{Bo} = \frac{v_z \cdot L}{D_L} \tag{3.4}$$

The exact solution of Equ. 3.3b, after carefully choice of the appropriate boundary conditions, (see Levenspiel and Smith, 1957) is

$$f(\tau) = \bar{t} \cdot f(t) = \sqrt{\frac{\text{Bo}}{4\pi\tau}} \cdot \exp\left[-(1-\tau)^2 \frac{\text{Bo}}{4\tau}\right] \tag{3.5}$$

In this equation, $\tau = t/\bar{t}$, with $\bar{t}$ being the mean residence time. The Bodenstein number is therefore the parameter of the dispersion model used in quantifying the residence time distribution, and it may be obtained from the experimentally measured curves using Equ. 3.6a with particular boundary conditions (see Levenspiel and Smith, 1957):

$$\frac{1}{\text{Bo}} = \frac{1}{8}\left(\sqrt{\frac{8s^2}{\bar{t}^2} + 1} - 1\right) \tag{3.6a}$$

Concerning boundary conditions, it should be stated here that two of the many possible conditions for a flow vessel are called "closed" or "open" boundaries, when there is or is not a change of flow, respectively, at the vessel boundary.

Thus, for "closed" boundaries Equ. 3.6b is used rather than Equ. 3.6a

$$\frac{s^2}{\bar{t}^2} = 2\,\text{Bo} - 2\,\text{Bo}^2(1 - e^{-1/\text{Bo}}) \tag{3.6b}$$

and for "open" boundaries

$$\frac{s^2}{\bar{t}^2} = 2\,\text{Bo} + 8\,\text{Bo}^2 \tag{3.6c}$$

Because the treatment of end conditions is full of mathematical subtleties (so that the additivity of variances can become questionable), one should always state clearly the assumptions about what is happening at the vessel boundaries before using one of the solutions given in the literature.

In these equations $s^2/\bar{t}^2$ is the total variance of the distribution function shown in Fig. 3.4a. It is connected with the spread of the distribution s^2 (the second moment), which can be obtained directly from the measured values $f(t)$ versus t when Equ. 3.7 is used with all summations

$$s^2 = \frac{\sum t^2 \cdot f(t)}{\sum f(t)} - \left[\frac{\sum t \cdot f(t)}{\sum f(t)}\right]^2 \tag{3.7}$$

The second term on the right side of Equ. 3.7 is the mean residence time, $\bar{t}$ (first moment of the distribution, the average value of t).

The numerical value of the Bodenstein number can, in principle, range between ∞ for an ideal tube reactor and 0 for an ideal stirred vessel. One should remember here that, in the first instance, the dispersion model is thought of as appropriate for the region of tubular flow. Although the model may be extrapolated to Bo = 0, one should be very careful in using the dispersion model when backmixing is great, particularly if systems are not closed. The use of the cell model is recommended for this region.

The analysis of residence time distribution curves thus far refers to situations in which the signal is a single pulse. The method of measurement involving a step function (see Fig. 3.4c) gives curves that are referred to by the symbol $F(t)$ and, logically, have the following relationship to $f(t)$

$$F(t) = \int_0^t f(t) \cdot dt \tag{3.8}$$

3.3.1.2 Tanks in Series Model (Cell Model)

Besides the dispersion model, the tanks in series model is the other one-dimensional model widely used to represent non-ideal flow. Here the fluid is thought to flow through a series of equal-size ideal stirred tanks, and the parameter in this model is the number of tanks in the cascade (N_{eq}). The RTD curves and moments of this model are easy to obtain, since problems of proper boundary conditions and method of tracer injection and measurement do not intrude. N_{eq} need not be an integer for curve-fitting purposes. It is strictly empirical, and no theoretical justification, such as Taylor diffusion, or theoretical estimates of the model parameter, are generally possible. This model starts from the mass balance equation for a series of i stirred vessels with $1 \leq i \leq N$, and N, the number of vessels in the series (or the number of equivalent stages, N_{eq})

$$\frac{1}{N} \frac{dc_i}{dt} = \frac{1}{\bar{t}}(c_{i-1} - c_i) \tag{3.9}$$

The general solution for the case of a stirred vessel series with N tanks is, according to Levenspiel (1972),

$$f(t) = \frac{N^N \cdot t^{N-1}}{(N-1)!\bar{t}^N} \cdot e^{-N \cdot t/\bar{t}} \tag{3.10}$$

This equation is simplified in the case of one ideal stirred vessel ($N = 1$) to

$$f(t) = \frac{1}{\bar{t}} \cdot e^{-t/\bar{t}} = D \cdot e^{-Dt} \tag{3.11}$$

where D is the rate of dilution that is inversely proportional to $\bar{t}$.

Evaluation of the model parameter N is done in a way analogous to Equ. 3.6a using the variance of the measured distribution s^2 according to Equ. 3.6d:

$$N = \frac{\bar{t}^2}{s^2} \tag{3.6d}$$

Recently, an alternative method of estimating N in the range of transition from plug flows to stirred tank behavior ($1 \leq N \leq 10$) was found with the aid of a computer simulation of Equ. 3.10. For the given case, the time value of maximum $f(t)$ value, t_{max}, of the individual RTD function in Fig. 3.5 can be taken as a measure of N_{eq} (Bauer and Moser, 1985). The result of this computer evaluation is plotted in Fig. 3.5, which can easily be used directly, to estimate N_{eq} in the range $1 \leq N \leq 3$

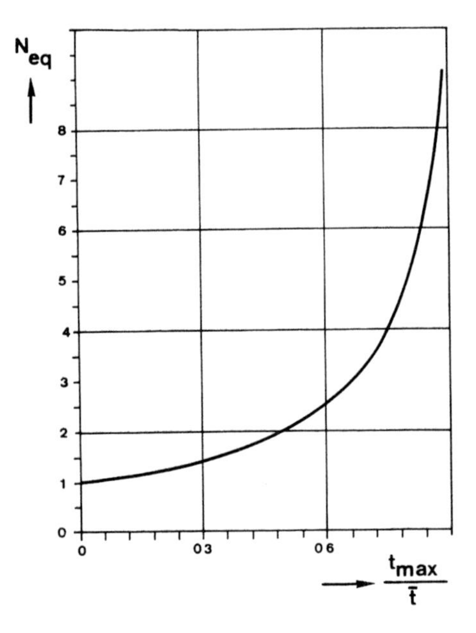

FIGURE 3.5. Computer simulation of the relationship between equivalent number of tanks in series, N_{eq}, and the time of maximum $f(t)$ function, t_{max} (see Fig. 3.4b), normalized by mean residence time $\bar{t}$ (Bauer and Moser, 1985).

$$N = \frac{\bar{t}}{\bar{t} - t_{max}} \tag{3.12}$$

There is no exact way to compare the two models (Bo and N), since the RTD curves are never identical. However, some useful relations are obtained from equating the variances of the two models (e.g., Kramers and Alberda, 1953):

$$N_{eq} = \frac{\text{Bo}}{2} + \frac{1}{2} \qquad \text{for Bo} > 2 \tag{3.13a}$$

or for other special cases (according to Pawlowski, 1962)

$$N_{eq} = 1 + 1/2\sqrt{\text{Bo}^2 + 1} \tag{3.13b}$$

This type of comparison of the moments of RTD curves of two models has wide applicability. For large N_{eq} the RTD curve becomes increasingly symmetrical and approaches the normal curve of the dispersion model and a comparison of these two curves allows one to relate the two models. The range where these conversion equations are valid is determined by the range of the validity of each of the models.

In evaluating the residence time distributions in a continuous system operating without recycling, for the case of the ideal discontinuous stirred vessel the curve with an exponential decay of the dilution process normally appears. But it appears in such a way that the pulse functions do not overlap (see Fig. 3.6 and also Blenke, 1979). This means that mixing and dilution processes are superimposed, and that in reactors that deviate from the ideal continuous

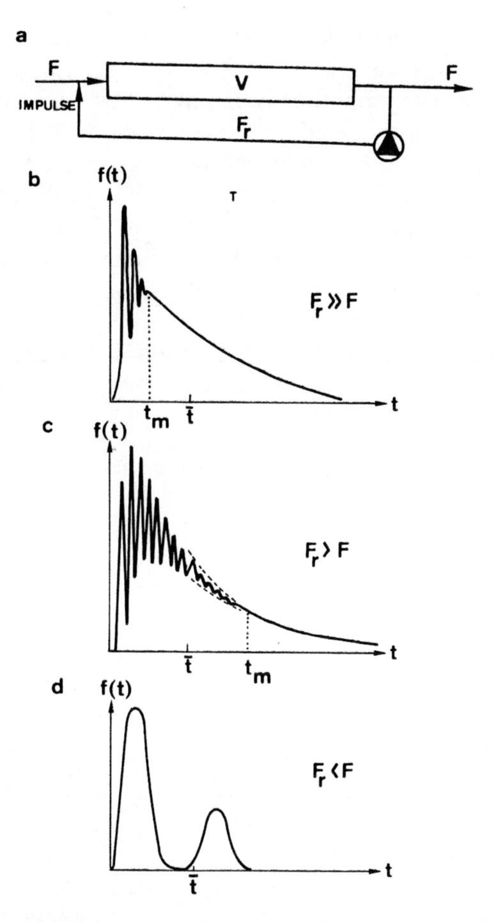

FIGURE 3.6. Measurement set-up (a) for determining the mixing time, t_m, and inhomogeneity, J, for a continuous loop reactor with differing degrees of mixing as a function of the strength of the recycled stream, F_r. (b) $CSTR_{mm}$ behavior with $t_m < \bar{t}$. (c) Intermediate case with $t_m \sim \bar{t}$. (d) $CPFR_{ts}$ behavior with $t_m \gg \bar{t}$. (Adapted from Moser and Steiner, 1975b).

stirred reactor, such as loop reactors, the mixing and residence time distribution behaviors are manifest in the curves.

The curves obtained experimentally in the case of loop reactors (cf. Fig.3.6b–d,) are interpretable as a special case quantitatively using Equ. 3.14 (Moser and Steiner, 1975a)

$$f(\tau) = \sum_{i=1}^{i_{end}} f_{E,i}(\tau)$$

$$= \sum \left[\left(\frac{1}{r+1} \right) \left(\frac{r}{r+1} \right)^{i=1} \sqrt{\frac{Bo}{4\pi t/\bar{t}}} \right] \cdot \exp\left[-\left(i - \frac{t}{\bar{t}} \right)^2 \frac{Bo}{4t/\bar{t}} \right] \tag{3.14}$$

The mathematical function describing this type of pulse signal is the sum of a number, i_{end}, of single functions, $f_{E,i}(t)$, which represent the individual passes from $i = 1$ to i_{end}. The amplitude is diminished on each pass by a "washing out" effect in addition to the effect of mixing behavior. The factor preceding the term containing the Bo number takes into account this dilution effect ($r = F_r/F$). The mathematical description of the individual functions uses the concept of the Bo number, as presented in Equ. 3.5 and Bo_E. The first and second moments of the whole distribution function (residence time distribution plus mixing) can be obtained in the same way as Equs. 3.6 and 3.7; accordingly, this combined function fulfills all the requirements of a normal distribution function. The mean residence time in the whole system $\bar{t}_{tot}$ can be calculated from

$$\bar{t}_{tot} = \frac{V}{F} \tag{3.15a}$$

and the mean residence time of a single pulse function $\bar{t}_{int}$ can be found from Equ. 3.15b

$$\bar{t}_{int} = \frac{V}{F + F_r} = \frac{\bar{t}_{tot}}{1 + r} \tag{3.15b}$$

The second moment of the whole distribution function is the Bo number, Bo_{tot}, representing the time distribution in the entire reactor system. In the

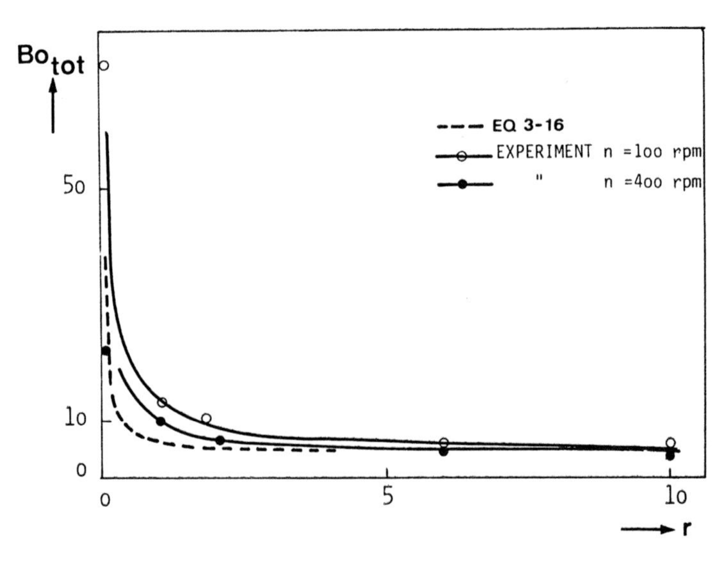

FIGURE 3.7. The longitudinal Bodenstein number of the total pulse function, Bo_{tot} for the liquid phase of a tube-type reactor with recycling as a function of the recycle ratio, r. The evaluation of the experimental results are compared with the theoretical calculation, Equ. 3.15b. (From Moser and Steiner, 1974 and 1975).

case of a loop reactor, it is dependent on the recycle ratio, r. As shown in Fig. 3.7, the plug flow characteristics dominant at the beginning are rapidly dissipated with an increasingly strong recycle flow F_r, and the system goes over to the residence time distribution characteristic of a stirred reactor vessel. Figure 3.7 shows experimental values (Moser and Steiner, 1974 and 1975a), and a theoretically derived equation for the variance of the system according to Fu et al., (1971):

$$\frac{s_{tot}^2}{\bar{t}_{tot}^2} = \frac{1 + N \cdot r}{N(1 + r)} \tag{3.16}$$

3.3.2 Micromixing

3.3.2.1 Mixing Time, t_m, and degree of mixing, m

The experimental set-up for the measurement and a typical result are shown in Fig. 3.8. A measuring device is installed in a reactor. The device is sensitive to a change of some property: conductivity, pH, color, optical density, "Schlieren methods," O_2, T, radio pill, fluorimetry, radioactivity (Beyeler et al., 1981,1983; Bryant and Sadeghzadeh, 1979; Einsele, 1976b; Käppel, 1976; Kipke, 1984; Middleton, 1979; Moser, 1987; Schneider et al. 1986; Zlokarnik, 1967). The "response function" of such a measurement often has the typical appearance shown in Fig. 3.8.

The so-called "degree of mixing," m, is sufficient to determine the mixing time, t_m, necessary to reach a particular value of m. Representing the asymptotic value of the concentration by c_∞, the degree of mixing m can be defined with Equ. 3.17

$$m = \frac{c - c_\infty}{c_\infty - c_0} \cdot 100 \tag{3.17}$$

and it represents the residual deviation from the final concentration in percent. Usually m is given as $\pm 5\%$ or 1%; that is, at a value of 95% or 99% of the totally mixed concentration. In quoting the mixing time, the degree of mixing should also always be quoted. Figure 3.9 illustrates a typical plot of dependence of t_m on the degree of mixing.

The mixing time, which was primarily intended as a measurement for discontinuous stirred reactors, can also be applied in the case of loop reactors. As shown in Fig. 3.8b the deviation from c_∞ can be taken as a direct measure in the definition of an inhomogeneity, I (Lehnert, 1972).

In the case of a continuously operated loop reactor, one can also apply the concept of mixing time over a wide range. Figure 3.6 brings together the different results obtainable by superimposing a mixing process such as that to be found in a discontinuous, closed system on a process of removing fluid in an open, continuous reactor (Moser and Steiner, 1975a,b). The different curves in the figure, labeled b through d, result from different ratios, r, of the

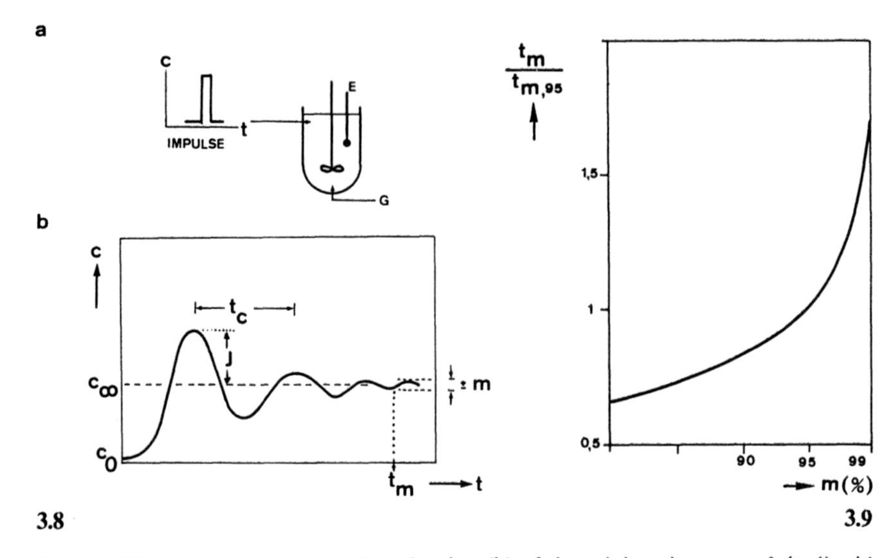

FIGURE 3.8. Measurement (a) and evaluation (b) of the mixing time, t_m, of the liquid phase in a discontinuous stirred tank reactor using the degree-of-mixing parameter m, Equ. 3.17. In reactors with constant recirculation, the circulation time, t_c, may also be determined as indicated. The degree of inhomogeneity J is also shown. (From Dechema, 1982).

FIGURE 3.9. General form of the dependence of mixing time t_m (normalized with t_m at $m = 95\%$) on the degree of mixing m. (From Kipke, 1984).

recycled flow to the output flow ($r = F_r/F$). This occurs simultaneously with changes in the average value of the time spent in the reactor, $\bar{t}$, relative to the mixing time, t_m. The envelopes to the curve, indicated by the dotted line, can be used to evaluate the mixing time, as shown in Fig. 3.6c. Recent work on structured mixing models using circulation time and CTD is shown in Sects. 3.3.2.3 and 6.9.4.

3.3.2.2 Mixing Models of Bioreactors (Structured Models)

It should be noted that the RTD character of a whole recycle reactor system, quantified with Bo_{tot}, depends not only on the recycle ratio r, but also on Bo_{int}, the internal RTD characteristic of the reactor (Moser, 1985a).

In practice, situations are even more complicated due to the simultaneous effect of r on Bo_{tot} and Bo_{int}, so that computer simulations fail to describe experiments as they operate with constant Bo_{int}. Experimental evaluation of Bo_{int} from RTD functions shows that the value goes through a maximum at about $r = 10$, when using a tubular reactor with recycling (Moser and Steiner, 1975a,b).

Another factor to be considered in analyzing RTD functions on this basis can be observed by comparing curve b and c in Fig. 3.7. Obviously, at $F_r \gg F$, the RTD curve approaches that of an ideal CSTR and simultaneously the micromixing behavior tends in the direction of maximum mixing. Thus, the degree of segregation J can be expressed as a function of recycle ratio r and σ^2_{tot} and σ^2_{int}, according to Dohan and Weinstein (1973)

$$J = 1 - \frac{\frac{r}{r+1}\left(\sigma^2_{tot} + 1 - \frac{r}{r+1}\right)}{\sigma^2_{int}} \tag{3.18}$$

Especially for technical-scale reactors empirical correlations were developed (Equs. 3.19 and 3.20 and Tables 3.2 and 3.3) that include the diameter of the tank d_T, or the ratio d_i/d_T. This fact indicates that high-volume vessels are far from the well mixed (cf. also Sect. 6.9.4). Thus, reactor models must take into account imperfect mixing; several approaches are used, and they are summarized in Fig. 3.10.

TABLE 3.2. Correlations between t_m and n, resp. d_i and d_T, for liquid systems.

Correlation	*Reference*	*Equ. No.*
$t_m \sim n^{-1}$	Kramers and Alberda, 1953	3.19a
$t_m \sim n^{-1} \cdot d_i^{\frac{1}{4}}$	Khang and Levenspiel, 1979 van de Vusse, 1955	3.19b
$t_m \sim (n \cdot d_i^2)^{-\frac{2}{3}}$	Norwood and Metzner, 1960	3.19c
$t_m \sim n^{-1}\left(\frac{d_i}{d_T}\right)^{-2}$	Holmes et al., 1964	3.19d
$t_m \sim n^{-\frac{2}{3}}$	Biggs, 1963	3.19e
$t_m \sim n^{-1}\left(\frac{d_i}{d_T}\right)^{-2\text{ to }-2.5}$	Prochazka and Landau, 1961	3.19f
$t_m \sim n^{-1}\left(\frac{d_i}{d_T}\right)^{-5/3}$	Mersmann et al., 1976	3.19g
$t_m \sim n^{-1\text{ to }-0.8}$	Stenberg, 1984	3.19h
$t_m \sim n^{-0.86\pm0.07} \cdot d_T^{-0.23\pm0.14}$	Stenberg, 1984	3.19i

From Moser, 1987.

TABLE 3.3. Correlations between t_m in case of aerated reactors.

Correction	*Reference*	*Equ. No.*
$\frac{t_{m,G}}{t_m} = 1 + 7.5 \cdot v^{0.27} \cdot \varepsilon_G$	Einsele and Finn, 1980	3.20a
$t_m = a \cdot n^b \cdot v_S^c \; (0.52 < b < 0.63; c \to 0)$	Stenberg, 1984	3.20b
$\frac{t_{m,G}}{t_m} = d + e \cdot v_S \; (e \to 0)$	Stenberg, 1984	3.20c

From Moser, 1987.

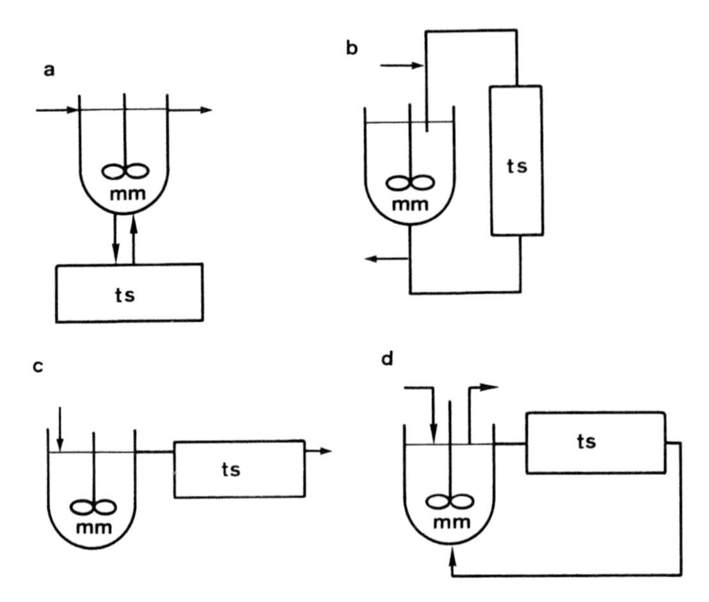

FIGURE 3.10. Graphic presentation of mixing models: (a) two-region mixing model, (b) two-environment mixing model, (c) reversed two-environment mixing model, and (d) combined backmix–plug flow configuration. In all cases maximum mixed zones (mm) are combined with zones of total segregation (ts). (Adapted from A. Moser, 1985a, Pergamon Books Ltd.).

A two-region mixing model was successfully employed by Sinclair and Brown (1970) to explain the experimental deviations observed in a CSTR (Herbert et al., 1956). Basically, this model is identical to the so-called "two-environment model of micromixing" first introduced by Ng and Rippin (1964) and then successfully applied to bioprocessing in the more empirical reversal of the original arrangement of mm region and ts environment. The "reversed two-environment" model (Tsai et al., 1971) involves an entering region with mm and a leaving environment of ts. (See, e.g., Dŭdŭkovic, 1977) Similar attempts have been made by Toda and Dunn (1982) by simulating several combinations of backmix–plug flow units, representing the flow behavior of mm–ts sequences (see Fig. 3.10d). Brown et al. (1979) adapted a multiloop recirculation model previously proposed by van de Vusse (1962), shown in Fig. 3.11, to quantify imperfect mixing in CSTRs. A series of mixing modules with feed rates F_i between them are assumed, and unsteady-state mass balances of each module in sequence are written and solved. This results in a collection of simultaneous equations relating the performance of each module with that of its neighbor. Results of varying performance as a consequence of imperfect mixedness have been described by Sinclair and Brown (1970) and by Toda and Dunn (1982) (see Fig. 3.12), even though different model approaches were used. Bajpaj and Reuss (1982) applied the two-environment model to

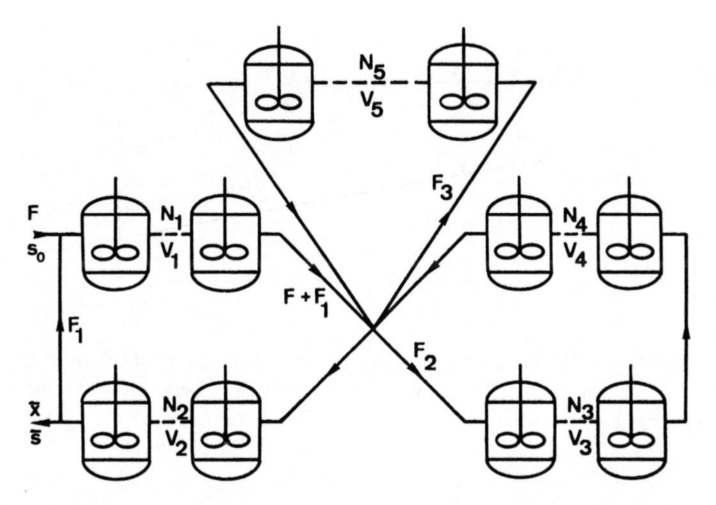

FIGURE 3.11. Multiloop recirculation mixing model (according to van de Vusse, 1962) for continuous culture systems. N_i, number of mixing modules in series; V_i, volume of modules; F_i, feed rate to one mixing module [$m^3 \cdot h^{-1}$]; F, feed rate to complete system. (From Brown et al., 1979).

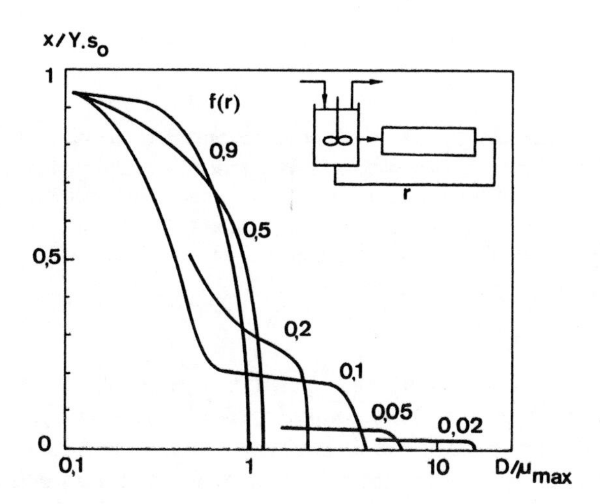

FIGURE 3.12. Dimensionless cell mass concentration x versus dilution rate D in a backmix–plug flow configuration (according to Fig. 3.10d) at varying recycle ratio r. (From Toda and Dunn, 1982. With permission of John Wiley & Sons, Inc.).

bioprocessing by coupling micromixing and microbial kinetics in the case of growth of *Saccharomyces cerevisiae*. They achieved good agreement with experimental observations when the circulation time distribution model is used (Bryant, 1977; Mukataka et al., 1981). This model permits application to other reactor operation modes, such as batch and fed-batch cultures, in which,

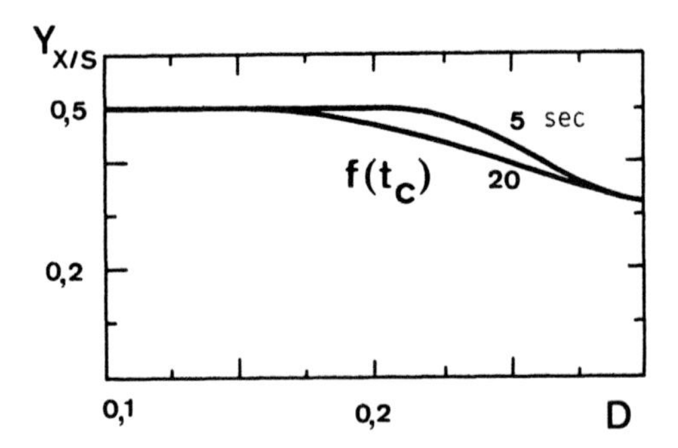

FIGURE 3.13. Yield coefficient $Y_{X|S}$ in dependence on dilution rate D in a continuous stirred tank reactor as a result of modeling the coupling between biokinetics in case of yeast fermentation and circulation time t_c. (From Bajpaj and Reuss, 1982).

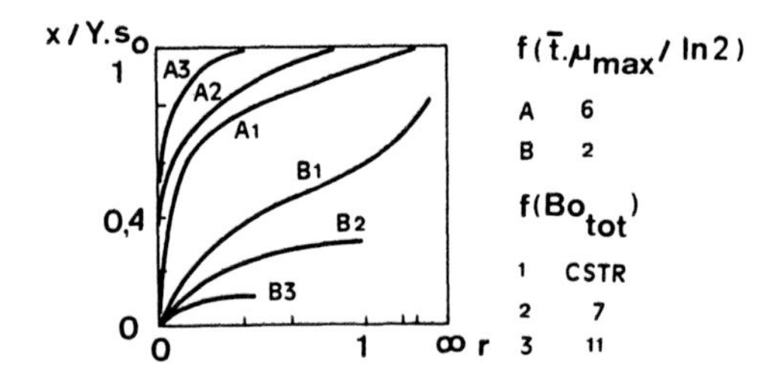

FIGURE 3.14. Microbial growth in dependence on micromixing and macromixing: dimensionless cell mass concentration x as a function of internal recycle ratio r in a recycle reactor at different values of total Bo number (Bo_{tot} as a measure of macromixing) and varying mean residence time $\bar{t}$. (Adapted from Dohan and Weinstein, 1973. With permission from Ind. Eng. Chem. Fundam., 12, 64. Copyright American Chemical Society).

in the case of continuous reactor operation, previously mentioned models cannot be used. Figure 3.13 is a computer simulation of cell yield $Y_{X|S}$, which varies with changing mean circulation time $\bar{t}_c$ (Bajpaj and Reuss, 1982).

In conclusion, it is clear that a bioreactor needs a high degree of micromixing to operate properly. However, the large amount of macromixing necessary to allow sufficient micromixing is not favorable to conversion, because the variable apparent order of reaction is always positive. Clearly an optimum depends on the coupling between the two components of the mixing. Figure 3.14 supports this by showing computed cell mass concentration versus recycle ratio in reactors with different values of var α (i.e., different Bo_{tot}). Mean residence time is made dimensionless (τ) by multiplication with the factor $\mu_{max}/\ln 2$. As can be seen, complete conversion cannot be approached in a CSTR (var $\alpha = 1$) for any state of micromixing, even for infinite τ. Conversion in a reactor with var $\alpha = 0.5$ (Bo $\sim$ 7) increases monotonically with r until r_{max} is reached. In this case, even at moderate t and at an intermediate level of segregation, complete conversion can be approached. A further increase in Bo_{tot} (curve 3 with $Bo_{tot} \sim 11$) causes the same tendency as with curve 2 to be exhibited, but a much longer mean residence time is needed to approach complete conversion because the maximum amount of micromixing permitted is quite small. The optimum value for macromixing and micromixing found in this simulation study (Dohan and Weinstein, 1973) was experimentally verified in a CPFR with recycling in the case of beer brewing and was called "mixed plug flow" (Moser, 1973b,1977b). A greater maximum production rate of cell mass than that achievable in a CSTR was realized in a piston-flow loop reactor by theoretical analysis (Grieves et al., 1964). Similar experimental findings of an optimum recycle rate, which depends on Bo and Da (Damkoehler) number are reported for bubble columns (Schügerl, 1977, 1982) and enzyme reactors (Wandrey and Flaschel, 1979) and represent alltogether a fact quantified in Chap. 6.

Thus, recycle reactors represent an advantageous device for experimental verification of mixing models, as emphasized here. Recycle reactors can be operated in either batch or continuous mode for this purpose, and they also exhibit the property of short cycle time distribution (CTD), which is very favorable compared with the CTD of normal stirred tanks.

In practice, in the designing of reactors it will no doubt be advantageous to use methods utilizing both biological and physical test systems (see Sect. 3.3.11). It will also be worth testing the extent to which the strategies can be combined. For example, the relevance of mixing time for bioprocessing seems somehow clear at first but becomes less clear when the meaning of the degree of mixing is considered. Einsele et al. (1978), by examining mixing times and comparing conventional measurements (method 1) according to Fig. 3.8 with measurements obtained with a fluorometer (method 2), showed that the response time of biological cells seems to be always the same (4.3 s). It should be remembered that method 1 contains transport resistances (cf. no. 2, according to Fig. 2.3), while the culture fluorescence measurements include mass transfer steps 2 through 5, in the legend to Fig. 2.3. Thus, bulk mixing is particularly important when t_m is greater than cell response time of 4.3 s. This is the case in pilot plants and technical scale reactors. This is further discussed in Sect. 3.3.11 and in the next section.

3.3.2.3 Experimental Verification of Structured Mixing Models

Additional data are required for the experimental verification of structured mixing models. The concept of mixing time in liquid phase reactors is widely used (e.g., Kipke, 1984). The simple method often fails in the case of real complex and gassed fermentation media, and it also fails in situations in which the mixing behavior cannot be represented by a single parameter (t_m), something which occurs in large-scale reactors. Normally problems arise in the application of basic tracer injection methods such as pH- or conductivity change: Neither property yields a significant response in real fermentation media due to the media's buffer capacity and high ion concentration. Einsele and Finn (1980) described a series of experimental results in aerated fermenters using a pH-electrode. The flow-follower method ("radio pill") was developed by Bryant and Sadeghzadeh (1979). A flow follower consists of a small encapsulated radio transmitter; it has been used by Middleton (1979), Mukataka et al. (1980), Cooker et al. (1983), and by Oosterhuis (1984). Detection of the pill outside the impeller area is achieved by mounting two aerials, each consisting of an isolated steel cable loop, concentrically around the impeller plane. Limitations in the use of the radio pill are imposed by the relatively large size of the pill (1–3 cm) in comparison with liquid flow pattern. Conductivity methods have been developed by Stenberg (1984). Disturbances resulting from gas bubbles passing a conductivity electrode were significantly reduced with the aid of a series of modified conductivity electrodes (Stenberg 1984). Each electrode was made of 10 platinated wire ends of 0.5 mm diameter. The 10 wire ends were separated from each other by approximately 1 cm. The general advantage of conductivity measurements is the small amount of chemicals needed and the small-scale electrode. Another simple alternative using a T-method was recently developed (Schneider et al., 1986). Furthermore, in production-scale bioreactors the cells will always be exposed to a circulation time distribution (CTD) and not to a mean value of $\bar{t}_c$. A typical plot of CTD would show that the first part of the distribution is better described than the tail. However, the tail of the distribution is of especially great importance for microbial processing. A first attempt to scale down a CTD was made out by Katinger (1976) simulating S concentration in a loop reactor. Later, Bajpaj and Reuss (1982) coupled mixing and microbial kinetics using a model of CTD to evaluate the performance of a bioreactor for yeast production. A number of research workers (Bryant, 1977; Bryant and Sadeghzadeh, 1979; Middleton, 1979) measured CTD in stirred tanks and reported it as being log normal, that is, a two-parameter distribution described by the first and second moment ($\bar{t}$ and σ^2) can be used adequately:

$$f(t)dt = \left(\frac{1}{\sqrt{2\pi}} \cdot \sigma_1 \cdot t\right) \exp\left[-\frac{1}{2}\left(\frac{\ln t - \mu_1}{\sigma_1}\right)^2\right] dt, \tag{3.21}$$

where $f(t)dt$ is the fraction of circulations whose circulation time lies between t and $t + dt$. Here μ_1 and σ_1 are the mean and the standard deviation of normally distributed variables that are related to $\bar{t}$ and σ^2.

In Figure 3.12 and 3.13 (and later Fig. 3.29) present typical results of the influence of varied circulation times on bioprocessing. An advanced evaluation strategy for mixing time, in addition to the methods presented in Sect. 3.3.1, which contributes to easier verification in experiments, was developed by Khang and Levenspiel (1979), showing that the liquid mixing process may be described as a first-order process with a single decay rate constant, k_m. If a salt pulse is added to the system, the concentration at a given point may be written as

$$C(x, y, z, t) = C^* + f(x, y, z, t) \cdot e^{-k_m t} \tag{3.22a}$$

where $f(x, y, z, t)$ is a bounded function with a mean value equal to zero. At $t = \infty$ the salt concentration is $C = C^*$ everywhere in the reactor. The degree of homogeneity as a measure for mixing is then described after intergration over the entire volume by considering the absolute deviation α

$$\alpha = \frac{1}{N} \sum_{n=1}^{N} |C_n - C^*| \approx e^{-k_m t} \frac{1}{V} \int_V |f(x, y, z, t)| dV \tag{3.22b}$$

or a quadratic deviation (variance var α or σ^2)

$$\sigma^2 = \frac{1}{N} \sum_{n=1}^{N} (C_n - C^*)^2 \approx e^{-2k_m t} \frac{1}{V} \int_V [f(x, y, z, t)^2] dV \tag{3.22c}$$

The exponential factors of Equs. 3.22a and b and are assumed to describe the main time dependence; consequently, the integrals are assumed to be constant in time. This indicates that plots of $\ln(\alpha)$ and $\ln(\sigma^2)$ versus t would yield straight lines, with a slope $-k_m$ and $-2k_m$, respectively. This yields, according to Equ. 3.22a

$$e^{-k_m t_m} \frac{1}{V} \int_V |f(x, y, z, t = t_m)| dV = 0.05 \frac{1}{V} \int_V |f(x, y, z, t = 0)| dV \tag{3.22d}$$

Since the integral is assumed to be independent of t, the mixing time for $m = 95\%$ is calculated according to

$$t_{m,i} = \frac{\ln 0.05}{k_m} \tag{3.23}$$

Mixing time thus can be evaluated in three different ways:

1. By fitting a straight line to $\ln(\alpha)$ versus t, giving $t_{m,1}$
2. By fitting a straight line to $\ln(\sigma^2)$ versus t, giving $t_{m,2}$
3. By determining the time for each electrode to lie within 5% from final value, giving $t_{m,3}$ according to conventional evaluation (cf. Sect. 3.3.1)

Mixing times evaluated according to α and var α are in good agreement with each other and also with the conventional method. Further, it can be shown that the approximation to a first-order system seems to fit quite well even when gas is sparged through the vessel.

An alternative method is the regular circulation of the liquid, which Holmes et al., 1964, introduced as the circulation time concept t_c. Generally, a relationship between t_c and t_m is valid

$$t_m = c \cdot t_c \tag{3.24}$$

where $7 < c < 3$ in most cases (Bruxelmane, 1983; Holmes et al., 1964; Prochazka and Landau, 1961; Stenberg, 1984). Experimentally, circulation time can be obtained using the same methods as for t_m (Mukataka et al., 1981), that is, ionic tracer pulse (conductivity), radio nucleotides (Merz and Vogg, 1978), and the radio pill (see also Reuss and Bajpaj, 1987, and Moser, 1987).

Joshi (1980) calculated the mean $t_{c,L}$ on the basis of t_m and the maximum length of circulation loop and found good agreement with experimental data. Thus, in practice, an exactly reverse procedure can be followed to calculate t_m from t_c data. Obviously, t_m is directly proportional to the longest loop length, which depends on the type of impeller and position. Furthermore, t_m is inversely proportional to the mean circulation velocity near the reactor wall (flat-blade turbine) or near the surface (upflow propeller) (Joshi et al., 1982)

$$t_c = \frac{\text{length of longest loop}}{\text{mean circulation velocity}} \tag{3.25}$$

3.3.3 Oxygen Transfer Rate (OTR)

There are several principal methods for measuring the OTR:

1. Sulphite oxidation
2. Gas in/gas out method (physical adsorption)
3. Gas analysis
4. Dynamic methods during the process using p_{O_2} electrodes
5. Glucose oxidase method
6. Combined methods using biological test systems
7. New methods (e.g., von Stockar and Stravs, 1983)

The Cu^{++}- or especially the Co^{++}-catalyzed oxidation of sulphite (Cooper, Fernstrom, and Miller, 1944; Reith, 1968) is thought to be suitable for estimating the OTR in comparing and designing gas–liquid (G|L) reactors only under similar physical conditions as bioprocessing. Mass transport occurring simultaneously with a chemical reaction is a theoretical problem that results; it is dealt with in Sect. 4.4.

Fortunately, the purely physical methods used in analyzing input or output gases can be applied directly in the medium, so that these methods reproduce the hydrodynamic behavior more accurately than is possible using sulphite oxidation. Analysis can be done using the laws of physical adsorption. Some methods, however, cannot be applied directly in the fermentation, during the run (sulphite and glucose oxidase), or due to problems with the p_{O_2} electrode (gas in–gas out and dynamic methods using p_{O_2} electrodes). Industry now prefers to operate with gas analyzers because of their superior long-term stability, and researchers at universities often use p_{O_2} electrodes in bench-scale reactors. Recently, combined methods have been introduced. The basic methodology will be outlined briefly here.

3.3.3.1 Theory of Methodology

According to the theory of mass transport, the OTR depends on the following quantities.

1. Specific exchange surface area, a

$$a = \frac{A}{V} \approx \frac{6\varepsilon_G}{\bar{d}_B} \tag{3.26}$$

where ε_G is the partial volume of the gas phase in the reactor (the so-called "gas hold-up"). The quantity ε_G can be calculated from

$$\varepsilon_G = \frac{V_G}{V_R} = \frac{V_G}{V_L + V_G} \equiv 1 - \varepsilon_L \tag{3.27}$$

and $\bar{d}_B$ is the average diameter of the gas bubbles (the Sauter diameter). Equation 3.28 can be used to calculate $\bar{d}_B$ from the number of bubbles, n, each of which has a diameter x

$$\bar{d}_B = \frac{\sum_i n_i \cdot x_i^3}{\sum_i n_i \cdot x_i^2} \tag{3.28}$$

Beyond this physical method using photography or light scattering for measuring a, the chemical method based on sulfite oxidation is often successfully used despite disadvantages due to different fluid properties. According to the theory of mass transfer with simultaneous chemical reaction, A can be calculated from

$$A = \frac{n_o}{\sqrt{\bar{o} \cdot (\frac{2}{3} D_L \cdot k_r) \cdot (1 + C)}} \tag{3.29}$$

where n_o = OTR and $\bar{o}$ = steady-state concentration (cf. Sect. 4.5.4).

2. Mass transport coefficient, k_L, a parameter defined and calculated according to various theories of mass transport, as given in Equ. 3.30. (As O_2 is a gas with very low solubility, the gas side mass transport coefficient can be ignored).

   Two-film theory, with δ = hypothetical film thickness (as $t_k \to \infty$):

$$k_L = \frac{D}{\delta} \tag{3.30a}$$

Penetration theory, with t_K = G|L contact time

$$k_L = \sqrt{\frac{4D}{\pi t_k}} \tag{3.30b}$$

Surface renewal theory, with s = rate of renewal of surface

$$k_L = \sqrt{D \cdot s} \tag{3.30c}$$

Convection theory (King, 1966; Kishinevskii, 1951; Moser, 1973c), with E = formal convection coefficient [cm²/s]

$$k_L = \sqrt{(D + E) \cdot s} \tag{3.30d}$$

(see also Philipps, 1969):

$$\mathrm{OTR} = k_L \cdot a(o^* - o) + 2 \cdot 10^{-7} a \cdot s \tag{3.30e}$$

The convection coefficient takes into account the adsorptive effects of the gas at the liquid surface; this is particularly important in the case of high

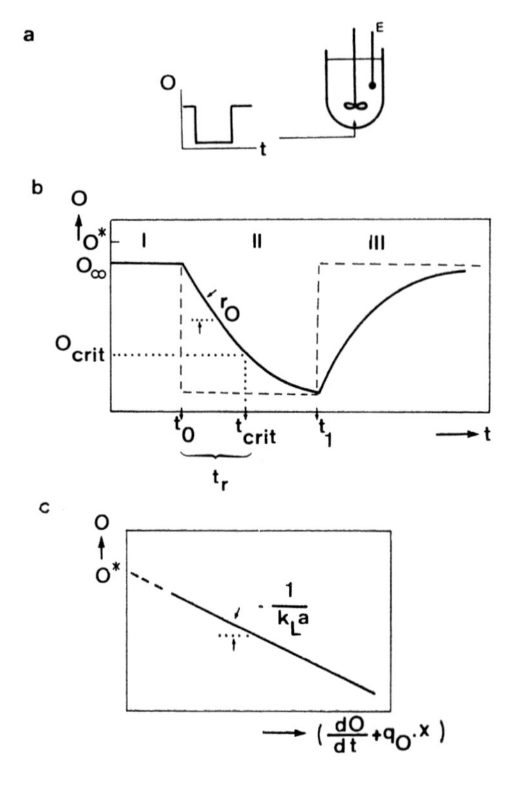

FIGURE 3.15. Experimental set-up and evaluation of the O_2 transfer rate characteristics of bioreactors using the dynamic method for measuring $k_{L1} \cdot a$. (a) Measuring set-up using a step change in inflowing O_2 concentration. (b) Response of the dynamic method. (c) Evaluation plot. Further explanation is found in the text.

rates of surface renewal (that is, as $t_K \to 0$). Normally k_L is determined as $k_L a$ (cf. Fig. 3.15) or using sulfite oxidation (see Sect. 4.5.4).

3. *Concentration difference*, $(o^* - o)$, between the saturation concentration, o^*, and the actual O_2 concentration, o (the "driving force").

3.3.3.2 O_2 Solubility

The saturation concentration is that concentration that is in equilibrium with the partial pressure of O_2 in the gas phase (p_G)

$$o^* = \frac{p_G}{\text{He} \cdot \text{RT}} \tag{3.31}$$

where He is the dimensionless Henry distribution coefficient (H = He · RT). The problem of O_2 solubility plays a central role in bioprocess engineering due to its importance for the analysis of O_2 consumption kinetics and OTR, process control, and scale-up calculations in aerobic processes. Many investigators still use the value for o^* in water as the first approximation because of a lack of a reliable method. Recently, Schumpe et al. (1982, 1985) reviewed the applicable methods: calibration techniques (Baburin et al., 1981; Käppeli and Fiechter, 1981; Lehmann et al. 1980; Liu et al., 1973), the classical saturation technique (Popovic, Niebelschütz, and Reuss, 1979; Quicker et al., 1981), and calculation methods based on the van Krevelen-Hooftijzer approach (1948):

$$\frac{\log o^+_{eff}}{o^*_{H_2O}} = -K_o \cdot \sigma \tag{3.32}$$

where σ is the ionic strength of a solution given by

$$\sigma = \tfrac{1}{2} \sum c_i z_i^2 \tag{3.33}$$

z being the electric charge of the ions. The proportionality factor K_O can be calculated from the three terms

$$K_O = i_+ + i_- + i_G \tag{3.34}$$

where $i_+ = i_{cations}$

$$i_- = i_{anions}$$

$$i_G = i_{gas}$$

Thus, oxygen solubility is not solely dependent on p and T. Even this relation is more complex, as shown by Mihaltz and Hollo (1980) in the following equation:

$$o^* = \frac{P_{tot} \cdot \sigma}{0.1353 \cdot 10^6 T^2 - 31.73 \cdot 10^6 T} \tag{3.35}$$

Significant differences between actual measurements and tables based on the CMEA recommendation (1968) initiated the derivation of Equ. 3.35. The strong influence of salts and of nutrients such as glucose in real fermentation broths was quantified by the following approach, taking into account the additive effect of salts and nutrients, Equ. 3.36 (Popovic et al., 1979):

$$\log \frac{o^*}{o^*_{eff}} = \log \frac{o^*}{o^*_{salt}} + \log \frac{o^*}{o^*_{gl}} \tag{3.36}$$

In this equation, the effect of a glucose solution of concentration c_{gl} is considered in terms of a linear equation of the form

$$o^*_{gl} = o^*(1 - 0.0012 - c_{gl}) \tag{3.37}$$

Equ. 3.38 gives the effect of the salt solution, which is represented in terms of the ionic strength as measured by the conductivity λ $[\Omega^{-1} \cdot cm^{-1}]$

$$\log \frac{o^*}{o^*_{salt}} = a_0 + a_1 \lambda + a_2 \lambda^2 \tag{3.38}$$

The coefficients a_i in Equ. 3.38 are not influenced by the glucose concentration.

Recently an experimental method was presented by Schneider and Moser (1984, 1985) that can be applied directly to the measurement of o^* in cultivation media during a process. It is based on a joint analysis of the gas and liquid phase using a paramagnetic analyzer, respectively a mass spectrometer and a p_{O_2} electrode, setting both values from gas and liquid analysis equal:

$$(q_O)_G \equiv (q_O)_L = \frac{\sigma_2 - \sigma_1}{t_2 - t_1} \cdot f_{o^*} \tag{3.39}$$

This identity equation gives the value for f_{o^*}, from which σ^*_{eff} can be calculated

$$\sigma^*_{eff} = \frac{\sigma^*_{H_2O}}{f_{o^*}} \tag{3.40}$$

3.3.3.3 Methods for $k_{L1}a$ Measurement

Method 1. Gas in/gas out

The mass transport equation can be written

$$\mathrm{OTR} = \frac{do}{dt} = n_o = k_L \cdot a(o^*_{eff} - o) \tag{3.41}$$

where o^*_{eff} represents the effective O_2 saturation and $k_L \cdot a$ is the volumetric mass transport coefficient $[h^{-1}]$, also known as the aeration constant. The solution to this differential equation is

$$\ln \frac{o^* - o}{o^*} = -k_L \cdot a \cdot t \tag{3.42}$$

and the value of $k_L \cdot a$ can be found by using the slope of a semilogarithmic plot. Recently an alternative equation for Equ. 3.41 was proposed (Sinclair, 1984).

Method 2. The dynamic OTR measurement

The dynamic method of OTR measurement is the one most often used in laboratories (Bandyopadhyay, Humphrey, and Taguchi, 1967; Taguchi and Humphrey, 1966). It is applicable during processing and uses a sterilizable p_{O_2} electrode (Lee and Tsao, 1979). The experimental arrangement is shown in Fig. 3.15a. A step-type concentration change is created in the fully assembled bioreactor by turning off the air supply at time t_0 and restarting it at time t_1. The response to this signal contains information reflective of the O_2 content. A typical dynamic method response curve is given in Fig. 3.15b. Incidentally, when $t \leq t_0$ there is a equilibrium between OTR and O_2 utilization, which results in a constant value of oxygen concentration o_∞. In phase I

$$\frac{do}{dt} = k_L \cdot a(o^* - o_\infty) - q_O \cdot x \tag{3.43}$$

in the stationary state, $do/dt = 0$, so that

$$k_L \cdot a = \frac{q_O \cdot x}{o^* - o_\infty} \tag{3.44}$$

If O_2 consumption is known (see Equ. 2.5d), one can, in principle, determine $k_L \cdot a$ from a measurement of o_∞. The accuracy of this type of calculation is low; therefore one also uses phase II and phase III for this type of calculation.

In phase II, with OTR = 0, Equ. 3.43 may be reduced to Equ. 2.5d, which describes the rate of respiration of the cell mass. From the slope of the curve, the parameters of the kinetic model can be obtained when x is known (for example, $q_{O,max}$ from the maximum slope and K_O from the half maximum; see Chap. 5).

Differential Equ. 3.43 is completely valid in phase III. With a slight rearrangement,

$$o = \frac{1}{k_L \cdot a}\left(\frac{do}{dt} + q_O \cdot x\right) + o^* \tag{3.45}$$

a graphic presentation is possible, so that with a given value for $q_O \cdot x$ one can obtain the volumetric mass transport coefficient $k_L \cdot a$. This type of graphic solution is shown in Fig. 3.15c.

An analysis according to this saturation procedure is, however, subject to many errors (Sobotka et al., 1982); Equ. 3.43 is a simplified representation of a dynamic process in which, in reality, not only respiration and aeration processes are at work. Neglected is the additional dependence of the gas phase dynamics on the separation from, or mixing of, the old or the fresh air or nitrogen in the reactor volume and the electrode behavior. The dynamic behavior of both influences can be accounted for by means of a first-order equation (Dunn and Einsele, 1975). For the gas phase

$$\frac{do_{ex}}{dt} = k_{V,G}(o_{in} - o_{ex}) \tag{3.46}$$

where $k_{V,G}$ is the dilution rate, or dilution constant, of the gas phase in the reactor [h^{-1}]. This constant is inversely proportional to $\bar{t}_G$, the mean residence time of the gas phase (V_G/F_G) in cases of maximal mixing

$$k_{V,G} = \frac{1}{\bar{t}_G} \tag{3.47}$$

For the electrode response, where the O_2 concentration read is o_E, one has

$$\frac{do_E}{dt} = k_E(o_L - o_E) \tag{3.48}$$

where k_E is the time constant of the electrode response [t^{-1}], which very often is formally of first order (Aiba and Huang, 1969). It is inversely proportional to the often-used time quantity t_E, which is the time necessary for the electrode response to reach 63.2% of its final value when the electrode is subjected to a sudden change (e.g., transferring it from a solution with $o_L = 0$ to $o_L = o^*$)

$$k_E = \frac{0.49}{t_E} \tag{3.49}$$

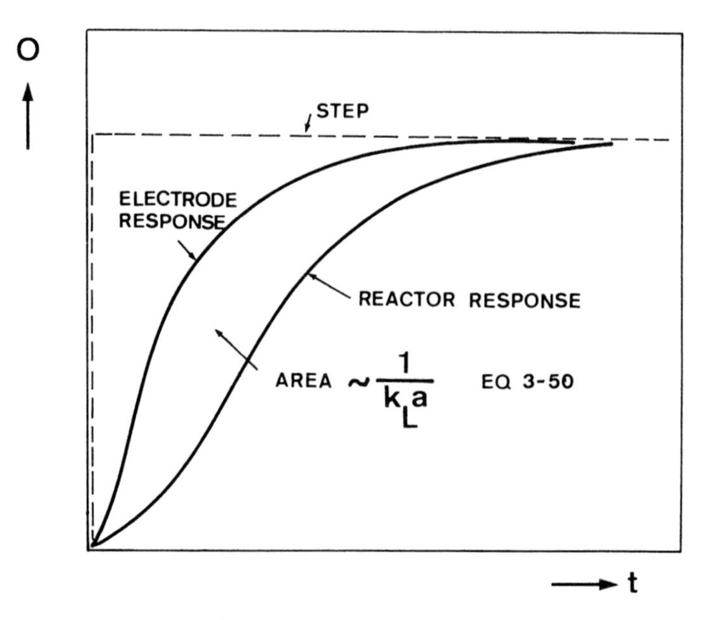

FIGURE 3.16. Measurement of $k_{L1} \cdot a$ using the momentum method (Dang et al., 1977; Nikolaev et al., 1976) assuming gas and electrode effects known from the response curve of the electrode, respectively the aeration system, and assuming a step function signal.

Available p_{O_2} electrodes are very slow in responding, and $k_L \cdot a$ measurement using the dynamic method results in significant errors. Many of the methods described in the literature to correct for gas and electrode dynamics are complicated and require a computer for calculations (Heineken, 1970; Lee and Tsao, 1979; Sobotka, Linek, and Prokop, 1973; Votruba and Sobotka, 1976). A simple method of evaluation that directly utilizes the response curve to a step function of the electrode and the aeration system is shown in Fig. 3.16 (Dang, Karrer and Dunn, 1977; Nikolaev et al., 1976). Use of this relative procedure allows the influence of the electrode to be eliminated even in viscous media. The influence of gas phase dynamics can be obtained from the area between the two curves, Equ. 3.50 (for the case of a maximally mixed gas phase in a well-stirred reactor vessel)

$$a = \frac{1}{k_L a} + \frac{1}{\mathrm{He}} \cdot \frac{V_L}{F_G} + \frac{1}{k_{V,G}} + \frac{1}{k_E} \tag{3.50}$$

In the case of a gas phase with plug flow, the area between the curves is inversely proportional to $k_L \cdot a$ (cf. Fig. 3.16). In all cases, fast response electrodes, if available, should be used ($k_E \gg k_L \cdot a$), which minimizes distortions in measuring OTR.

A simple method of correcting $k_L a$ values distorted by electrode lag was used by Heinzle (1978), following a formal kinetic approach, by a simultaneous

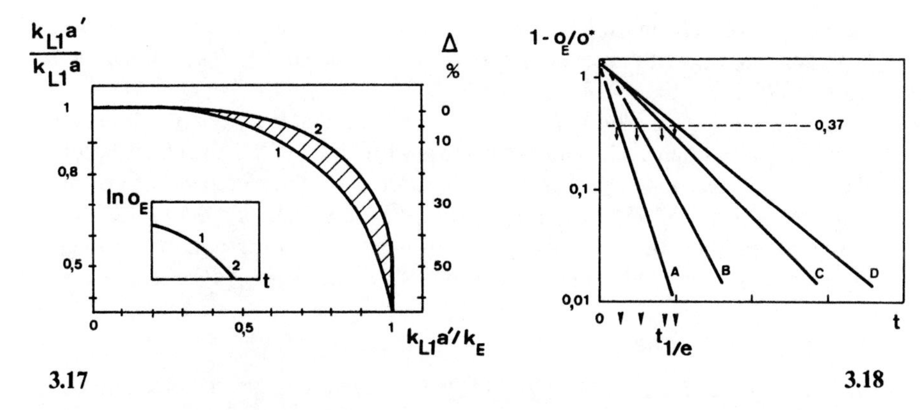

FIGURE 3.17. Graphic plot of the relation between actual (apparent) value of $k_{L1}a$, the electrode response constant k_E, and the true $k_{L1}a$ value. The axis on the right side indicates the deviation Δ from the true value. The two curves resulting from two different regions of the evaluation plot (see inserted semilogarithmic plot). (Adapted from Heinzle, 1978.)

FIGURE 3.18. Semilogarithmic plots of $(1 - o_E/o^*)$ versus time t of aeration curves at different conditions (curves B–D) and a step response experiment to determine k_E (curve A). Shifting the curves to intersect the ordinate at unit value yields the time constants $t_{1/e}$ at the 0.37 line, according to Equ. 3.51. (From Ruchti et al., 1981. With permission of John Wiley & Sons, Inc.).

solution of Equs. 3.41 and 3.48. The results are plotted in Fig. 3.17, showing the deviation in percent of actual k_La from the true value k_La with increasing value of k_E. As can be seen, no correction is needed if $k_La':k_E \leq 0.5$, and the measurements are seriously in error if this ratio is ≥ 0.5.

Another simplification of k_La calculations was proposed by Ruchti et al. (1981). This approach again uses the moment method, taking the form of four first-order lags in series, leading to a simple graphical estimation. The last dependent variable in the series is delayed by approximately the sum of the time constants. Thus, for this model, using normalized concentration of oxygen $\hat{o}$, the quantity $(1 - \hat{o}_E)$ attains the value of $1 - e^{-1}$ (or 0.623) at approximately

$$t_{1/e} = \frac{1}{k_La} + \frac{1}{k_{V,G}} + \frac{1}{k_E} \tag{3.51}$$

A further important property is that the $(1 - \hat{o}_E)$ versus t curve will be approximately exponential for large values of t. This characteristic is evident in Fig. 3.18, where $(1 - \hat{o}_E)$ from the response data from three experiments is plotted versus time on a semilogarithmic scale. Also shown in this figure is the response of the electrode to a step change in σ_L. It has been found empirically that if the lines are shifted but their slope is left unchanged, so that their intercept with the $(1 - \hat{o}_E)$ axis becomes equal to 1.0, then the k_La values can be calculated from Equ. 3.51. At highest $k_L a$ values, in the range of $10^3\ h^{-1}$, however, inaccuracies due to the error in $\bar{t}_G$ appear ($\pm 30\%$) (Bauer and Moser, 1985).

Finally, there is a method of OTR determination that is especially applicable in cases of very intensively aerated reactors with high OTR. For this situation, the dynamic method just mentioned is limited by the electrode's dynamic response. The so-called "stationary method" (Lücke, Oels, and Schügerl, 1977) works by passing the liquid phase, which is aerated in the reactor, through a circuit containing a de-aerator so that a stationary state is reached.

3.3.3.4 *Gas Analysis*

The basic gas exchange rate is generally correlated with gas balance and can be calculated by the equation

$$\mathrm{OTR} = (\sigma_{G,\mathrm{in}} - \sigma_{G,\mathrm{ex}}) \frac{F_G}{V_L} \tag{3.52}$$

where $o_{G,\mathrm{in}}$ and $o_{G,\mathrm{ex}}$ refer to the inlet and exit concentration of O_2, related to partial pressures around the reactor. In a steady state of OTR and O_2 uptake rate (OUR) during a bioprocess, the value of $k_L a$ can be estimated from Equ. 3.43 as

$$k_L a = \frac{q_o \cdot x}{\overline{\Delta o}} \tag{3.53}$$

where $\overline{\Delta o}$ is the logarithmic mean value of O_2 in the inlet and outlet gas stream, given by

$$\overline{\Delta o} = \frac{(\mathrm{p}_{O,\mathrm{in}} - \mathrm{p}_L) - (\mathrm{p}_{O,\mathrm{ex}} - \mathrm{p}_L)}{\ln(\mathrm{p}_{O,\mathrm{in}} - \mathrm{p}_L)/(\mathrm{p}_{O,\mathrm{ex}} - \mathrm{p}_L)} \tag{3.54}$$

where p_L is the partial pressure of O_2 in the liquid phase and $\mathrm{p}_{O,\mathrm{in}}$ and $\mathrm{p}_{O,\mathrm{ex}}$ are the partial pressures in the air stream. In small vessels where perfect mixing is approached, this averaging treatment is not required. Furthermore, the log mean value can only be used if the gas exhibits good plug flow and the liquid phase exhibits high backmixing.

The OUR term, $q_o \cdot x$, can be evaluated from O_2 partial pressures in the inlet and outlet gas from the relation (e.g., Lehmann et al., 1982)

$$q_o \cdot x = \frac{32}{22.4} \cdot 10^6 \left(\tilde{o}_{\mathrm{in}} - \tilde{o}_{\mathrm{ex}} \frac{\tilde{n}_{\mathrm{in}}}{1 - \tilde{o}_{\mathrm{ex}} - \tilde{c}_{\mathrm{ex}}} \right) \frac{F_{G,\mathrm{in}}}{V_L} \tag{3.55}$$

with $\tilde{o}$, $\tilde{c}$, and $\tilde{n}$ being the moles O_2, CO_2, and N_2 per moles of air mixture ($\tilde{o} = 0.2095$, $\tilde{c} = 0.0003$, and $\tilde{n} = 0.7902$). Van Meyenburg and Fiechter (1968) showed that Equ. 3.55 appears because normally the O_2 uptake and the CO_2 evolution differ in volume, and a correction of the apparent values is necessary to account for the differences in flow into and out of the reactor, on the basis of the balance of N_2, the usual inert component. Further correction for T and p is also needed.

Similarly, the CO_2 formation rate, $q_c \cdot x$, can be calculated from

$$q_c \cdot x = \frac{44.01}{22.4} \cdot 10^6 \left(\tilde{c}_{ex} \frac{\tilde{n}_{in}}{1 - \tilde{o}_{ex} - \tilde{c}_{ex}} - \tilde{c}_{in} \right) \frac{F_{G,in}}{V_L}. \tag{3.56}$$

Gas phase analysis is superior to p_{O_2} electrode methods due to better stability, and more and more it is being also accepted in fundamental research when gas flow rate or liquid volume is high enough for accuracy. In analogy to this approach, a "biological" method for OTR measurement has recently been recommended (Käppeli and Fiechter, 1980,1981b). In O_2-limited conditions, the overall maximum OTR of the bioreactor equals the OUR, which is calculated by Equ. 3.57

$$\mathrm{OTR}_{max} = k_L a (o_L^* - o_L)_{o_L \to 0} \equiv q_o \cdot x. \tag{3.57}$$

For this method it is essential to know the value of o_L^* (cf. previous section). The use of *Trichosporon cutaneum* is recommended by Käpelli and Fiechter as particularly attractive, since this organism is relatively easy to handle, although weekly subculture from the original strain is needed. It exhibits a simple and stable metabolism (strictly respiratory strain, even at O_2 limitation; glucose is converted only to biomass and CO_2, with some formation of polysaccharides).

Another new method for the dynamic measurement of $k_L a$ uses gas phase dynamics and consists of continuously measuring the composition of the outlet gas in response to a step input of a nonreactive tracer such as CO_2 in the inlet gas stream (André et al., 1981). This method is especially useful under particular conditions for application to high viscosity media and solid-substrate fermentations.

3.3.4 Degree of O_2 Utilization, η_{O_2}

In calculating the effectiveness of an aeration system or the economics of aeration, important use is made of the degree of oxygen utilization given in %

$$\eta_{O_2} = \frac{o_{G,in} - o_{G,ex}}{o_{G,in}} \tag{3.58}$$

This can be calculated from a measured balance of the gas phases.

3.3.5 Degree of Hinterland, Hl

The characteristic number Hl, the "hinterland" (Beek, 1969), also designated by the volumetric number B (Hirner and Blenke, 1974) or by the reciprocal "degree of volume utilization," f_V (Nagel, Kürten, and Sinn, 1972), can be calculated for a G|L reactor from

$$\mathrm{Hl} = \frac{V_L}{V_{L,film}} = \frac{V_L}{A \cdot \delta} = \frac{V_L \cdot k_L}{A \cdot D} = \frac{(1 - \varepsilon_G) k_L}{a \cdot D} \tag{3.59}$$

Hl represents the ratio of the whole liquid volume of the reactor, V_L, to that volume of fluid found at the gas–liquid exchange surface A in a film of thickness δ (see film theory, Equ. 3.30a). The dimensionless number Hl characterizes the size of the gas–liquid exchange surface of an aerator, and it will be especially significant in the case of so-called "fast reactions" (cf. Sect. 4.5.4).

3.3.6 Power Consumption, P

The electrical energy used for mixing and aeration in a reactor is determined as the quantity of power supplied over a period of time, expressed in kilowatts, [kW], or $[\mathrm{J \cdot s^{-1}}]$. Power input measurements of impellers depend on the scale of operation (Brown, 1977): On a plant scale, accurate data are obtained for alternating current from Watt-meter readings

$$P = V \cdot A \cdot \cos\phi \cdot \eta_M \tag{3.60}$$

where η_M is motor efficiency and $\cos\phi$ is power factor, which can be measured directly with a $\cos\phi$ meter or calculated by

$$\cos\phi = \frac{\text{true power}}{\text{apparent power}} = \frac{W}{V \cdot A} \tag{3.61}$$

Drive efficiencies η_D and losses of power transmission in the stirrer shaft gland and bearings occurring as heat to the belts due to excessive slip at the pulleys are to be taken into account for better accuracy (Brown, 1984). For a given stirrer speed, it is possible to formulate an equation for the power to the fluid

$$P = [(\text{load kW} \cdot \eta_N) - (\text{no-load kW} \cdot \eta_N)] \cdot \eta_D \tag{3.62}$$

As an alternative and more accurate method, measurements can be made by means of torsion dynamometers, which consist of a torsion bar coupling from which torque can be measured either by surface strain (Aiba et al., 1973) or by angular twist. However, relatively expensive equipment must be used, due to difficulties in interpretation of the torque signal. For laboratory-scale measurements ($P < 1$ kW or $V < 50$ l), the use of electrical techniques is not recommended, because circuit losses are greater than the power transmitted to the fluid. Arrangements are described where the motor is supported by a piano wire and is free to rotate in air bearings (Calderbank, 1958) or where the motor is mounted in two radial load bearings located on the motor shaft (Brown, 1967). Such arrangements are almost friction-free, and they allow the calculation of power

$$P = 2\pi \cdot n \cdot \Theta \tag{3.63}$$

where n is stirrer speed and Θ is radius of torque times the restraining force. Strain gauges represent another method for measuring agitator shaft power and are preferred because of greater accuracy (Aiba, Humphrey, and Millis, 1973; Einsele and Fiechter, 1974).

The power input to bioreactors from gas sparging can be calculated from

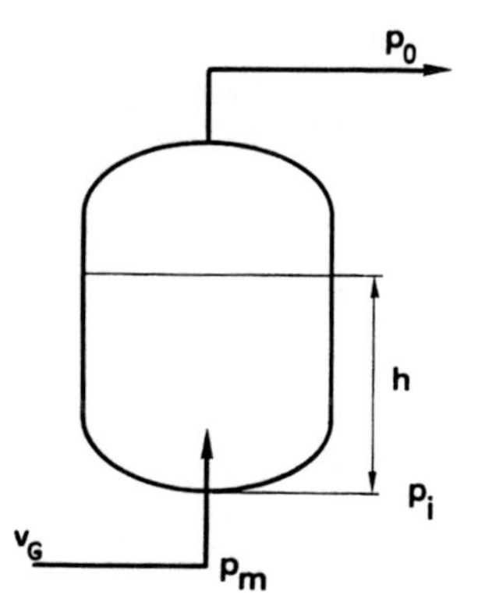

FIGURE 3.19. Graphic illustration of experimental situation when quantifying power consumption due to gas sparging, according to Equ. 3.64, respectively Equ. 3.65 (Adapted from Brown, 1977, 1984). For explanation see text.

$$P = \frac{F_G}{\rho_G}\left(M_G \cdot \eta_N \frac{v_G^2}{2} + RT \ln \frac{p_i}{p_0}\right) \tag{3.64}$$

The two terms in this equation represent the kinetic energy change of the gas due to the velocity difference of v_G in the pipe and in the vessel ($\Delta v_G = v_{G,\text{pipe}}$) and the isothermal expansion of the gas from a point in the liquid at the nozzle (p_i) to a point where the gas leaves the liquid (p_0). Useful values for P can be calculated neglecting the first term. In practice, the kinetic energy term can be included approximately as a pressure contribution by replacing p_0 by the pressure p_m, which must be measured as shown in Fig. 3.19. Thus, Equ. 3.64 becomes

$$P = \frac{F_G}{\rho_G} \cdot RT \cdot \ln \frac{p_m}{p_0} \tag{3.65}$$

3.3.7 O_2 Efficiency (Economy), E_{O_2})

This term refers to the quotient of the OTR and the power used per volume

$$E_{O_2} = \frac{\text{OTR}}{P/V}\left[\frac{\text{kg } O_2}{\text{kWh}}\right] \tag{3.66}$$

E_{O_2} can also be defined in reciprocal form.

3.3.8 Heat Transfer Rate, H_vTR

Because there is a biological optimum over a relatively narrow temperature range and because biological processes are exothermic, heat transfer problems play an important role, even though the amount of heat generated is comparatively less than that in chemical processes. An analogy to mass transfer may be used to formulate an equation and obtain model parameters for the H_vTR. The experimental set-up for measuring the heat transfer coefficient is given in Fig. 3.20. The reactor in which the average temperature should be T_R

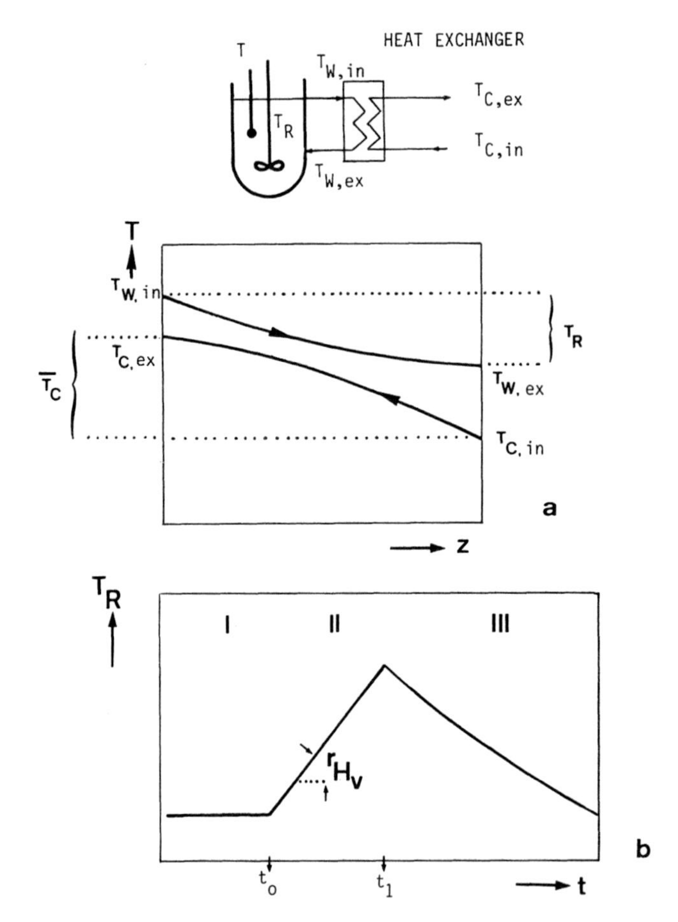

FIGURE 3.20. Quantification of heat exchange characteristics. The set-up of measurements contains a counter-current heat exchanger with indicated values of temperatures (T_w at warm sides, T_c at cool sides). The temperature profile shown (a) serves to calculate the logarithmic mean according to Equ. 3.67 in such real cases of greater linear extension. From the plot of reactor temperature T_R versus time t (b), in analogy to Fig. 3.15b, heat evolution rate r_{H_v} and heat transfer coefficient $k_{H_v}a_{H_v}$ can be calculated as shown in the text. At time t_0, the heat exchanger is stopped; and at T_1 it is reactivated.

is equipped with a temperature sensor and a heat exchanger in which the flow can be arranged in either a forward or a reverse direction. The appropriate temperature profiles for a counter-current heat exchanger are shown in part b of Fig. 3.20. They may be used for the definition and estimation of the "driving force" for the H_vTR, that is, the temperature difference ΔT. In the case of a real heat exchanger, where there are long surfaces with different temperatures, the logarithmic average, $\overline{\Delta T}$, is used to obtain ΔT. In analogy to Equ. 3.54, the $\overline{\Delta T}$ may be calculated from

$$\overline{\Delta T} = \frac{(T_{W,ex} - T_{C,in}) - (T_{W,in} - T_{C,ex})}{\ln[(T_{W,ex} - T_{C,in})/(T_{W,in} - T_{C,ex})]} \tag{3.67}$$

As an approximation suitable for small units that can be considered ideal (that is, without internal temperature gradients and with a constant transport coefficient) and in equilibrium situation, ΔT can be calculated from (Ebert, 1971)

$$\Delta T = T_R - T_C \tag{3.68}$$

The total heat balance, available on the basis of the directly accessible quantity of the volumetric heat of reaction h_V [$kJ \cdot l^{-1}$], is made up of several terms, q [$kJ \cdot l^{-1} h^{-1}$]

$$q_{tot} = \left(\frac{dh_v}{dt}\right)_{tot} = +\left(\frac{dh_v}{dt}\right)_r - q_{Tr} + q_{agit} - q_{rad} - q_{evap} - q_{Gas} \tag{3.69}$$

The various terms take into account the various "sinks" ($q < 0$) or "sources" ($q > 0$) of the heat balance. These include, specifically, the reaction itself, transport, agitation, radiation, and heat loss through evaporation and gas stream. In most cases the last four terms can be neglected (Bronn, 1971; Cooney, Wang, and Mateles, 1968; Mou and Cooney, 1976). The term for the reaction, the heat produced, is represented in Equ. 2.4c. The kinetics are, as always, introduced in the balance equation. All heats of reaction can be calculated from a series of T/t measurements using Equ. 3.70 when the specific heat capacity of the reactor system is known

$$\frac{dh_v}{dt} = \frac{M}{V} c_p \frac{dT}{dt} \tag{3.70}$$

This Uhlich approximation uses the value of the specific heat capacity c_p [$kJ \cdot kg^{-1} \cdot {}^\circ C^{-1}$] multiplied by the mass of the fermentation medium per unit volume, M/V/ [kg/l]. Typical values for fermentation processes are given in Table 3.4 (Conney et al., 1968).

The quantity refered to as the volumetric heat of reacton h_v is unconventionally defined from a thermodynamic viewpoint. The dimensions [$kJ \cdot l^{-1}$] are given for practical reasons. Strictly speaking, a (molar) reaction enthalpy has

TABLE 3.4. Typical data of specific heat and heat capacity with fermentations.

Medium	*Specific heat, c_p* [$kcal \cdot kg^{-1} \cdot {}^\circ C^{-1}$]	*Heat capacity, $M/V \cdot c_p$* [$kcal \cdot l^{-1} \cdot {}^\circ C^{-1}$]
Fermentation broth	1	1.01
Glass material	0.2	0.067
Steel material	0.12	0.0094
Total, global value	—	1.076

From Cooney, Wang, and Maleles, 1968).

dimensions [kJ · mol^{-1}]. In a modified form this is also used in kinetics as the specific, metabolic reaction enthalpy with the dimension [kJ · (gΔX)$^{-1}$] or

$$[\mathrm{kJ} \cdot (\mathrm{g}\Delta S)^{-1}] \qquad \text{(see Sect. 5.4.2).}$$

To evaluate the parameter for heat transfer (the heat transfer coefficient, k_{H_v} [kJ/m^2 · h · °C]), one can begin with Equ. 3.69, which refers to an exothermic bioprocess; neglecting the last four terms one obtains Equ. 3.71

$$\frac{dh_v}{dt} = -k_{H_v} \cdot a_H \cdot \overline{\Delta T} + r_{H_v} \tag{3.71}$$

By analogy to the OTR measurement (see Equ. 3.43), the dynamic method can also be used for heat transfer measurements. The bioreactor is then treated as an adiabatic calorimeter with constant internal heat production (Falch, 1968). A typical temperature curve of the dynamic calorimetry is shown in Fig. 3.20b in analogy to Fig. 3.15b. At time t_0, cooling is interrupted, and the rate of heat production r_{H_v} may be directly calculated from the linear increase in T (phase II). The quantity $k_H \cdot a_H$ is calculated from the slope dT/dt in phase III, which, in contrast to OTR measurements, is linear. Equation 3.71, together with Equ. 3.70, is used for this arithmetic calculation. A graphical method completely analogous to Fig. 3.15c is impossible, due to the large changes in the cooling temperature and the consequent necessity of calculating a logarithmic average.

An alternative approach for the quantification of H_vTR (which, however, is difficult to apply to bench-scale reactors, $V_R < 50$ l) follows the concept of Equ. 3.52. The energy balance on the cooling water is given by

$$\frac{dh_v}{dt} = \frac{F_{H_2O}}{V_R} \cdot c_{p,H_2O}(T_{ex} - T_{in}) \tag{3.72}$$

Although much of the earlier work involved the use of crude calorimeters, new techniques require relatively complicated apparatus and procedures for determining the heat of fermentation. Different methods are cited and are known as the principle of gradients, the isothermic and adiabatic principles (in the case of calorimetry), and the quasi-adiabatic method (Bayer and Fuehrer, 1982; Bronn, 1971; Cardosó-Duarte, et al., 1975; Eriksson, 1971; Eriksson and Home, 1973; Forrest, 1972; Imanaka and Aiba, 1976; Jones, 1973; Korjagin et al., 1978; Luong and Volesky, 1980, 1982; Marison and von Stockar 1984; Minkevich and Eroshin, 1973; Monk, 1978; Mou and Cooney, 1976; Schauerte, 1981; van Uden, 1971; Volesky et al., 1982; Wang et al., 1978).

Finally, Table 3.5 presents a summary of the heat transfer coefficient for a few types of heat exchangers (Bronn, 1971).

In practice, one finds that heat exchange poses engineering difficulties: Because of the low temperature difference, it is a given that the method of transferring heat will be somewhat ineffective and uneconomical. Therefore, attempts are now being made to culture thermophilic microorganisms that

TABLE 3.5. Typical data on heat transfer capabilities.

Type of heat exchanger	*Heat transfer coefficient*, k_{H_v} [$kcal \cdot m^{-2} \cdot h^{-1} \cdot °C^{-1}$]
External irrigation	500
Double jacket	900
Internal coils	1200
External plates	2000

will grow and produce optimally at higher temperatures, around 50°C and above. Further information on heat formation kinetics is given in Sect. 5.4.2.

3.3.9 Characteristic Diameter of the Biocatalytic Mass, $\bar{d}_p$

The concept and the importance of homogeneity and heterogeneity in characterizing bioreactors was mentioned in Sect. 3.2. In choosing the correct kinetic model, it is important also to include information on possible transport limitations in the solid phase, for example thick films or large particles.

Measuring this quantity is difficult in an operating plant. Methods involving a micrometer and phase contrast microscopy are described in the literature both for particles (Parker, Kaufmann, and Jenkins, 1971) and for films (Kornegay and Andrews, 1968). Biofilm operation is characterized in comparison with floc operation in Table 3.1.

No doubt in all real cases there will be a distribution of diameters d_p. This difficulty can be avoided by using Equ. 3.28 to calculate an average diameter $\bar{d}_p$ for a particular distribution function (Atkinson and Ur-Rahman, 1979). The further application of $\bar{d}_p$ for evaluating the extent of transport-limiting factors is described in Sect. 4.5.2. Biofilm reactors with mechanical or hydrodynamic control of film thickness exhibit a uniform $\bar{d}_p$ and thus are more suitable than floc bioreactors for process kinetic analyses (see Sects. 3.6 and 4.3).

3.3.10 Comparison of Process Technology Data for Bioreactors

Table 3.6 summarizes the types of process technology data and bioreactors thus far described. Establishing empirical correlations between particular systems generally is done with the aid of dimensionless quantities. The mixing time, t_m, can be made dimensionless by means of Equ. 3.73 (Zlokarnik, 1967)

$$N_M = \frac{t_m \cdot \nu}{D^2} \tag{3.73}$$

TABLE 3.6. Criteria of process engineering characterization of bioreactors/bioprocessing and range of typical data.

Quantity	*Symbol*	*Units*	*Numerical range*
O_2 transfer	OTR	$kg \cdot m^{-3} \cdot h^{-1}$	0.3–12 (50)
	$k_{L1}a$	h^{-1}	100–1500 (3000)
	k_{L1}	$m \cdot h^{-1}$	0.3–2
	a_{L1}	$m^2 \cdot m^{-3}$	10^2–10^6
OTR enhancement	E_{OTR}	—	≥ 1
O_2 utilization	η_{O_2}	%	5–40 (90)
O_2 efficiency	E_{O_2}	$kg\ O_2 \cdot kWh^{-1}$	0.2–3.5
		$kWh \cdot kgO_2^{-1}$	5–0.3
Hinterland	Hl	—	10^3–20
Power consumption	P/V	$kW \cdot m^{-3}$	0.3–2 (20)
Micromixing (L)	t_m	s	1–200 (500) as $f(V_R)$
Macromixing (L)	Bo, N_{eq}	—	CSTR: $N = 1$ (Bo $\rightarrow 0$)
			CPFR: Bo ≥ 7 ($N \geq 5$)
Biological data			
Floc size	V/A	mm	10^{-2}–5
Film thickness	d	mm	10^{-2}–10
Growth: max. rate:	μ_{max}	h^{-1}	0.02–2
Monod constant:	K_S	$mg \cdot l^{-1}$	2–50 (100)
Respiration	$q_{O,max}$	h^{-1}	0.05–3
Yield	$Y_{X\|S}$	—	0.05–0.5 (1)
Productivity	r_i	$kg \cdot m^{-3} \cdot h^{-1}$	0.5–20
Conversion	ζ	%	0–1

From Moser, 1981.

The power requirement, P or P_G, can be transformed into the power number N_p (Zlokarnik, 1967)

$$N_p = \frac{P}{\rho \cdot n^3 \cdot d^5} \tag{3.74a}$$

or can be referred to the gas flow velocity, F_G, (Schügerl, 1980) as

$$N_p' = \frac{P/F_G}{\rho(gv)^{2/3}} \tag{3.74b}$$

The aeration rate, F_G, can be given as a dimensionless number N_A

$$N_A = \frac{F_G}{n \cdot d^3} \tag{3.75}$$

and the volumetric mass transfer coefficient $k_L \cdot a$ can be made dimensionless using the following equation, which defines the sorption number (Zlokarnik, 1978, 1979)

$$N_{OTR} = \frac{k_L \cdot a}{F_G|V} \tag{3.76a}$$

or

$$N'_{\mathrm{OTR}} = k_{\mathrm{L}} \cdot a \left(\frac{\nu}{g^2}\right)^{1/3} \tag{3.76b}$$

In addition to the dimensionless numbers, there are well-known others, such as the Sherwood (Sh), Reynolds (Re), Schmidt (Sc), Froude (Fr), Bodenstein (Bo), and Weber (We) numbers. On the basis of these types of dimensionless numbers, empirical correlations for a large number of bioreactors have been made (for example, Blanch, 1979; Schügerl, 1980; Zlokarnik, 1979). The results of the experimental measurements of process engineering data are often presented in the form of a graph; they have the form of the relationships given in Equs. 3.77a and 3.77b. For the volumetric mass transport coefficient (Ryu and Humphrey, 1972) (see Fig. 3.21)

$$k_{\mathrm{L}} \cdot a = \alpha \cdot \left(\frac{P}{V}\right)^{\beta} \cdot v_{\mathrm{s}}^{\gamma} \cdot \left(\frac{\nu}{\rho}\right)^{\omega} \tag{3.77a}$$

For the specific G|L exchange surface, as shown in Fig. 3.22 (Nagel et al., 1972)

$$a = \alpha \cdot \left(\frac{P}{V}\right)^{\beta} \cdot v_{\mathrm{s}}^{\gamma} \tag{3.77b}$$

where α, β, γ, and ω in Equ. 3.77a are experimentally determined coefficients. For the power consumption of various stirring systems (Blanch, 1979; Rushton, Costich, and Everett, 1950) (see Fig. 3.23)

$$N_{\mathrm{p}} = f_1(N_{\mathrm{Re}}) \tag{3.78}$$

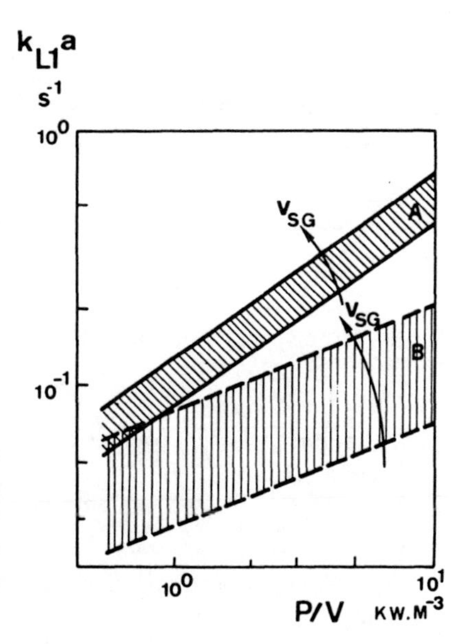

FIGURE 3.21. Typical plot of empirical engineering correlation between G|L transfer coefficient $k_{\mathrm{L1}}a$ and specific power consumption P/V as a function of superficial gas velocity $v_{\mathrm{S,G}}$. The shaded areas (A for pure water and B for ionic solutions) represent the range of validity of correlations.

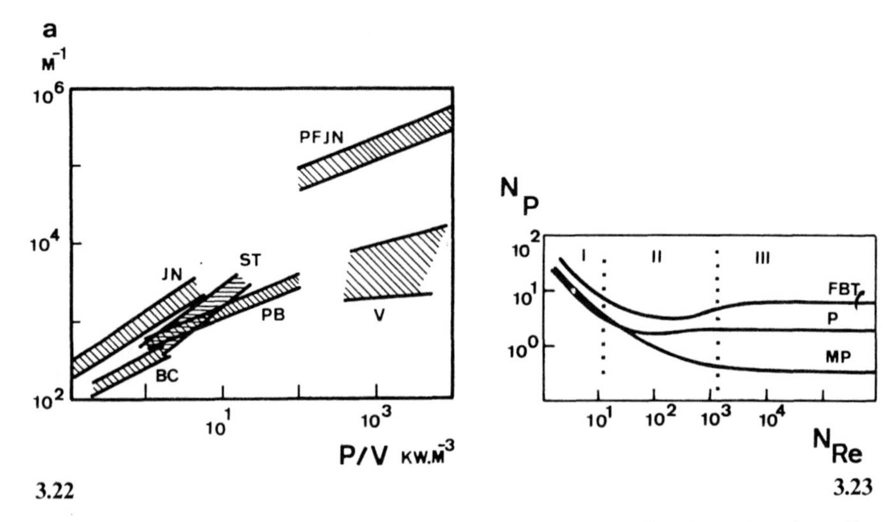

FIGURE 3.22. Comparison of G|L reactors of different constructions in a plot of specific interfacial area a versus specific power consumption P/V: stirred tanks (ST), bubble columns (BC), jet nozzles (JN), packed beds (PB), venturi washers (V) and plug flow-jet nozzles (PFJN). (Adapted from Nagel et al., 1972).

FIGURE 3.23. Typical graphical plot of dimensionless numbers of power consumption N_p versus N_{Re} in case of flat-blade turbine (FBT), paddles (P), and marine propellers (MP). The laminar transient, and turbulent, regions are indicated (I, II, III). (Adapted from Rushton et al., 1950).

and for the power consumption of various stirrers as aeration systems (Ohyama and Endho, 1955) as illustrated in Fig. 3.24

$$N_p = f_2(N_A) \tag{3.79}$$

For the mixing time of various mixing systems (Zlokarnik, 1967) represented in Fig. 3.25

$$N_M = f_3(N_p) \tag{3.80}$$

For the effectiveness of aeration in various bioreactors, see Fig. 3.26 (Zlokarnik, 1978)

$$N_{OTR} = f_4(N_p) \tag{3.81}$$

and for the special case of surface aeration (Zlokarnik, 1979), see Fig. 3.27

$$N_{OTR} = f_5(N_{Fr}) \tag{3.82}$$

For the mass transfer coefficients, as demonstrated in Fig. 3.28 according to Sano et al. (1974)

$$\mathrm{Sh}_{L2} = \alpha + \beta\, \mathrm{Re}_{\varepsilon}^{\gamma} \cdot \mathrm{Sc}^{\omega} \tag{3.83}$$

These correlations, among others, quantify and allow comparisons to be made. They make possible a system of generalizations in which the values from many systems can be organized. The mathematical functions f_i in Equ. 3.78 through Equ. 3.82 also represent correlations for scale-up purposes.

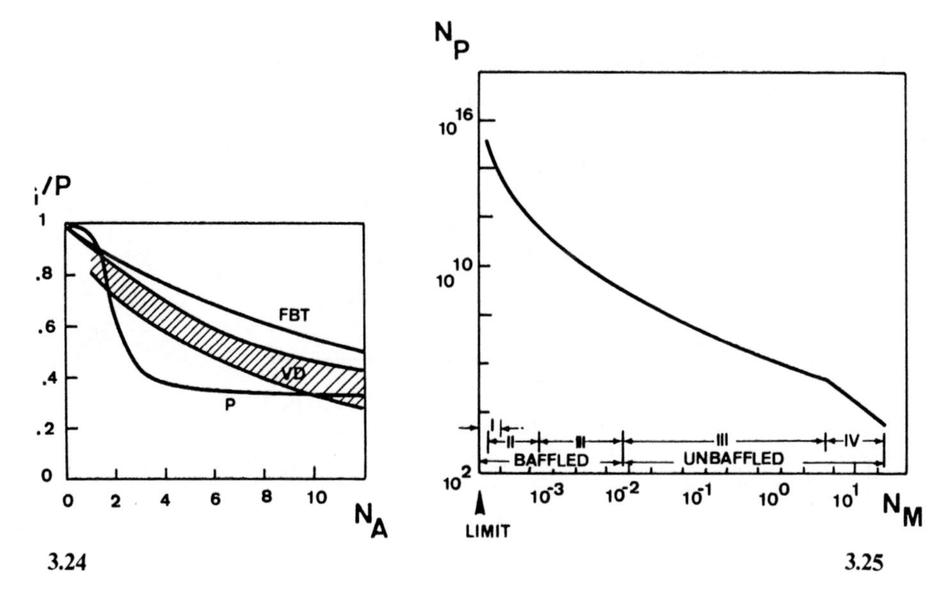

FIGURE 3.24. Typical plot of power requirement for agitation in gassed reactors, expressed as the degree of power decrease P_G/P versus aeration number N_A as a function of different types of impellers [flat-blade turbines (FBT; vaned disks (VD) with varying number; paddle (P)]. (Adapted from Ohyama and Endho, 1955.)

FIGURE 3.25. Plot of dimensionless numbers of power consumption N_P versus mixing time N_M with the range of different mixing devices with and without baffles [propeller (I), anchor (II), flat-blade turbine (III), and helical ribbon impeller (IV)]. (Adapted from Zlokarnik, 1967.)

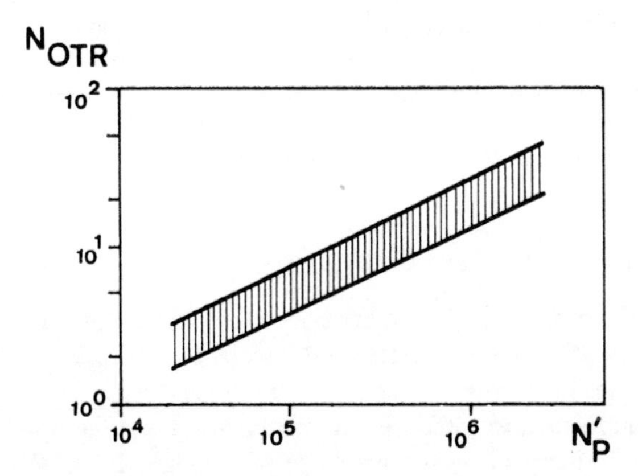

FIGURE 3.26. Typical plot of data showing the relation between oxygen transfer number N_{OTR} versus modified dimensionless power number $N_{P'}$ at different conditions. (Adapted from Zlokarnik, 1979).

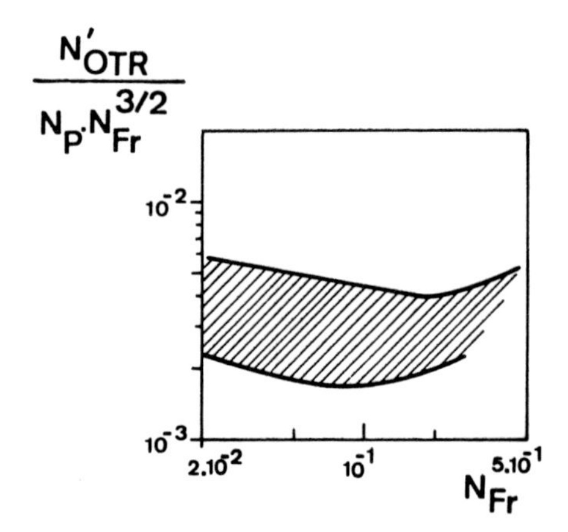

FIGURE 3.27. Typical plot of data of surface aerator efficiencies expressed as modified oxygen transfer number versus Froude number N_{Fr}. The shaded area includes the range of different tested types. (Adapted from Zlokarnik, 1979.)

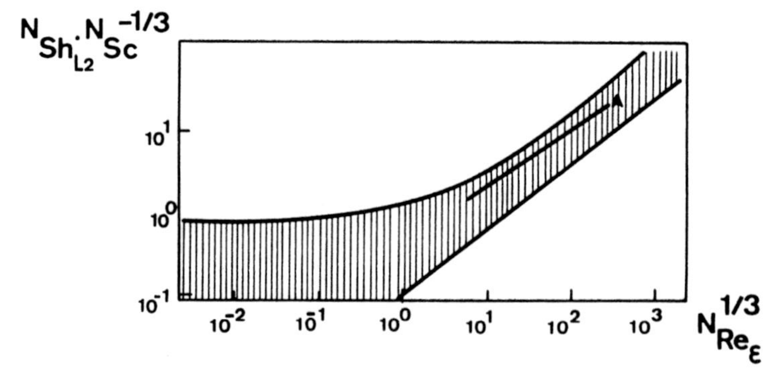

FIGURE 3.28. Typical plot of energy correlation for the L|S-mass transfer coefficient k_{L2} expressed as Reynolds number based on energy dissipation N_{Re_ε}, respectively Sherwood and Schmidt number $N_{Sh_{L_2}}$ and N_{Sc} (according to Sano et al., 1974). The shaded area includes all engineering correlations known in literature with varied N_{Sc}, while the correlation of Sano et al. is given with line A.

3.3.11 Biological Test Systems

Data about the physical characteristics of a bioreactor can provide only limited information. The complete characterization of a bioreactor requires additional studies involving biological test systems ("reference fermentations"). The fluid dynamics and rheological behavior of media are both directly and indirectly influenced by the presence of biological cells (Fiechter, 1978). Microbial processes whose growth or production kinetics are specifically dependent on changes in their medium or the reactor come into consideration as biological test systems (Karrer, 1978). Because of the central role of mass

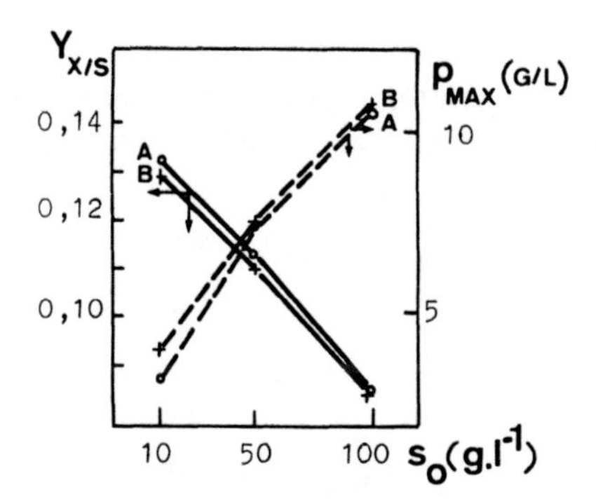

FIGURE 3.29. Graphic representation of the formal macroapproach to kinetic modeling of yeast metabolism, showing the dependence of yield coefficient $Y_{X|S}$, respectively maximum product concentration p_{max}, on initial glucose concentration in batch cultures. Compared are a conventional stirred tank (curve B) and a horizontal tubular loop reactor with mechanical agitation and aeration (curve A), both working at constant $k_{L1}a$-value (700 h^{-1}). (Adapted from A. Moser, 1977a.)

transport, biological test systems are particularly sensitive to micromixing, OTR, and shearing forces, as indicated in Fig. 2.15.

Naturally, good mixing is a necessary precondition for the mutual contact of all phases participating in the reaction (see Fig. 2.3). The adequacy of mixing is far more important for bioprocesses than for chemical or physical processes. Bioprocesses not only have a marked optimum but are also often very sensitive to concentration changes. A good example is bakers' yeast (*Saccharomyces cerevisiae*), which suffers from O_2 repression (O_2–Pasteur effect) and glucose repression (glucose or Crabtree effect). At glucose concentrations exceeding a critical value, $s_{crit} \approx 50$ mg/l (Fiechter, 1974), and despite aerobic conditions, alcohol is produced and cell yield is reduced (see Fig. 3.29). In the special case of bakers' yeast, there is, in practice, a special strategy for the introduction of fresh substrate—the fed-batch culture technique (Aiba et al., 1976)—that reduces glucose repression by minimizing glucose concentration. Despite this, when concentrated solutions are added in well-mixed reactors there are momentary nonuniformities so that the biological system experiences a change in concentration and a corresponding change in metabolism (Einsele, 1976a). Therefore, especially suitable biological test systems for reactors are those that are sensitive to, for example, glucose or O_2 (*Candida tropicalis*, *Trichosporum cutaneum*; *Bacillus subtilis*, *Escherichia Ecoli*, *Beauveria tenella*). These systems provide more information on the mixing conditions of bioreactors than do purely physical measurements (Cleland et al., 1984; Einsele, 1978; Fiechter, 1978; Griot et al., 1986; Karrer, 1978; Küng and Moser, 1986; Moes et al., 1984). Also, research on the shear force effects is carried out with the help of a physical (e.g., Grosse et al., 1985) or a biological test system, *Tetrahymena pyriformis* (Midler and Finn, 1966) to study the damage to cells. *Pullularia pullulans* is a shear-sensitive biological system. Reuss (1977) presented a quantification of this effect on the basis of investigations by Tanaka et al. (1975a,b), showing that the ratio between internal nucleotide concentration (c_{Ni}) and total concentration (c_N) is exponentially dependent on the excretion rate of nucleotides (r_N) which is measurable by measuring extinction (ΔE_{260})

$$c_{Ni}/c_N = k_i e^{-k_i r_N} \tag{3.84}$$

Excretion rates are correlated to peripheral stirrer speed v_{tip}, increasing linearly for *Mucor javanicus* and *Rhizopus javanicus* and exponentially for *P. pullulans*:

$$r_N = f(v_{tip}) = f(n \cdot d_i \cdot \pi) \tag{3.85}$$

where n is the rotational speed and d_i is the diameter of the impeller. A list of "reference fermentations" was recommended for the practical purpose of standardization, including the production of bakers' yeast, pentalenolacton (*Streptomyces arenae*), xanthan (*Xanthomonas campestris*), and gluconic acid with *Aspergillus niger* (Lehmann et al., 1982; 1985). Recently it has been emphasized that process kinetic modeling should be incorporated in these considerations, based on the integrating strategy (Küng and Moser, 1986a; Moser, 1985a). Meyer (1987) summarized to state of art of biotest systems.

3.4 Operational Modes and Bioreactor Concepts

Reactor operation includes discontinuous and continuous operations, steady-state and transient-state modes, and a group of intermediate modes that can be referred to collectively as semicontinuous operation. The modes of operation are compared in Fig. 3.30.

A discontinuous stirred tank reactor (DCSTR) with ideal mixing conditions has no concentration profile in space. The process is, however, time dependent and thus has a concentration/time profile (c/t).

A semicontinuous reactor (SCSTR) operation has most of the characteristics of a DCSTR in space, and the c/t profile is typically as shown in Fig. 3.30 (see also Figs. 3.32 and 3.33).

A continuous stirred tank reactor (CSTR) with ideal mixing has a uniform concentration both in time and space. As an important consequence, the concentration at the output (c_{ex}) is the same as that prevailing in the reactor ($c_{ex} = c_R$).

The ideal continuous plug flow reactor (CPFR) has no profile at any point of the tube in the steady state. The process, however, advances along the tube, and so shows a longitudinal concentration variation. The profile of a CPFR in space is identical to the profile of a DCSTR in time in case of a constant volume process; this fact is of great importance for process design ("kinetic similarity").

Finally, Fig. 3.30 shows the behavior of a *cascade of CSTRs* with a number,

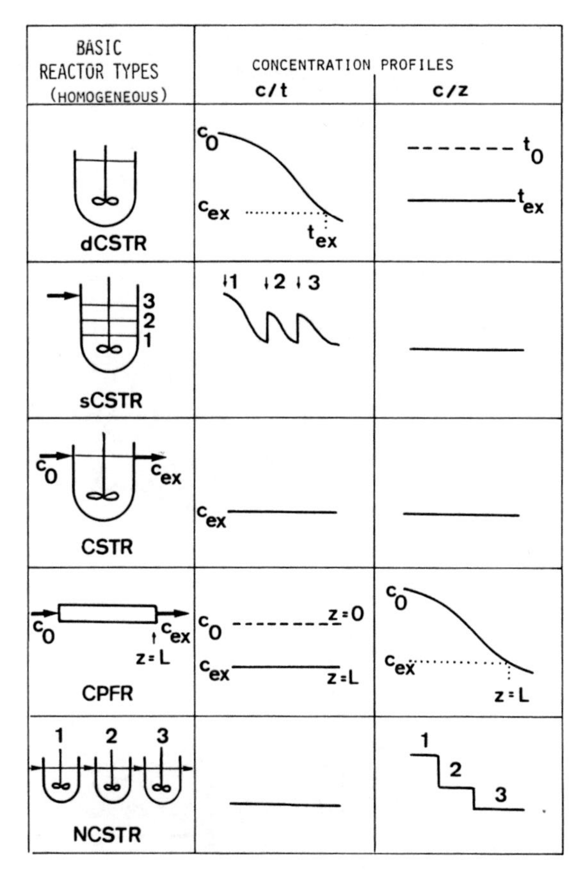

FIGURE 3.30. Basic reactor concept and concentration-versus-time and concentration-versus-space profiles. DCSTR, discontinuous stirred tank reactor; SCSTR, semicontinuous stirred tank reactor; CSTR, continuous stirred tank reactor; CPFR, continuous plug flow reactor; NCSTR, a cascade of N stirred vessels.

N, of units in a series (NCSTR). As with the CSTR mode of operation, the concentration profile in each tank is uniform in both space and time. However, over the entire length of the cascade, the space/concentration profile will show a typical step function curve. It is not difficult to see that this spatial behavior approximates that of a CPFR. As a rule, a cascade with $N \geq 5$ can actually be used as a process engineering substitute for a CPFR (see residence time distribution, Sect. 3.3.1).

The operational modes of bioprocess technology, especially the semicontinuous modes, can be categorized using a function describing the time dependence of the input. This is done in Fig. 3.31. The discontinuous and fully continuous processes are special cases with $F_{in} = 0$ and $F_{in} = F_{ex}$, respectively through a constant reactor volume. In the case of a CSTR there are practical difficulties; a constant, continuous input is not particularly easy to arrange.

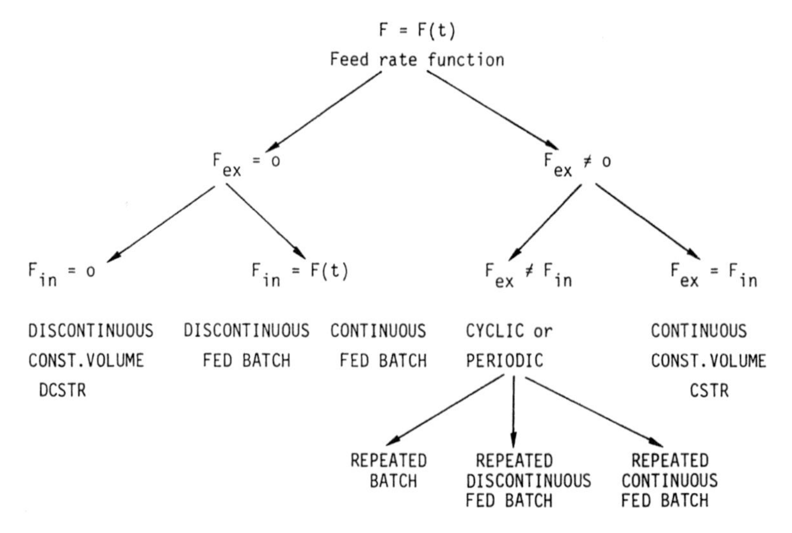

FIGURE 3.31. Systematic, engineering classification of fermentation processes based on the behavior in time of the input stream, F_{in}, and the exist stream, F_{ex}.

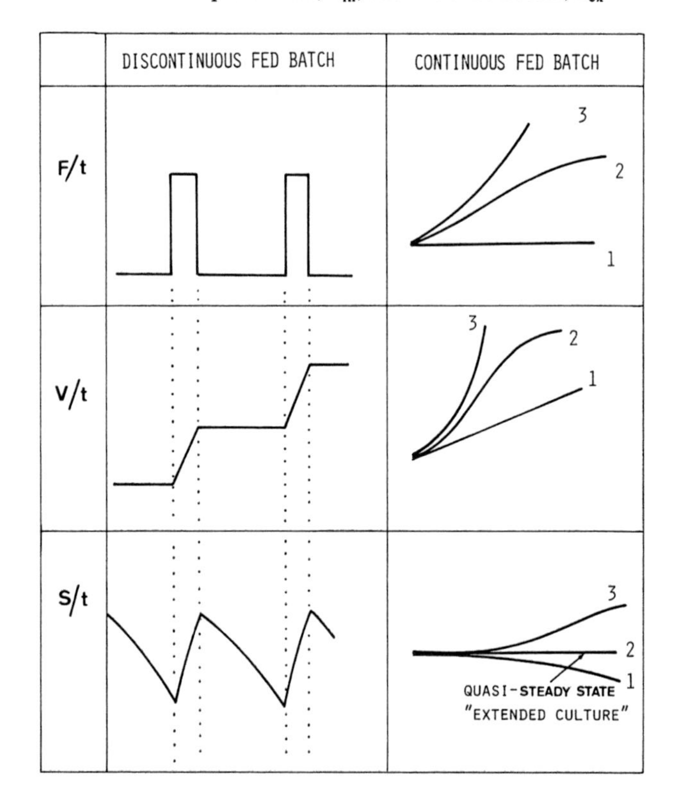

FIGURE 3.32. Discontinuous and continuous feed systems characterized by the time dependence of the input, F; the volume of the liquid phase, V; and the substrate concentration, s. The continuous feeding types are designated 1–3.

The processes with $F_{ex} = 0$ can be further subdivided into discontinuous input and continuous input processes.

Among the cyclic or periodic processes with $F_{in} \neq F_{ex}$, but $F_{ex} \neq 0$, one can distinguish cyclic processes with discontinuous or continuous input from simple, cyclic discontinuous processes. The time dependence of the addition of liquid, F/t, the volume of liquid in the reactor V_L, and a substrate concentration s are shown in Fig. 3.32 for substrate feeding operations in a fermentation process. They are shown in Fig. 3.33 for cyclic fermentation processes. The following simple equations are suitable for a mathematical representation of continuous feed operations.

$$F_L(t) = \frac{dV_L}{dt} = \alpha_1 \cdot t + \beta_1 \tag{3.86a}$$

or

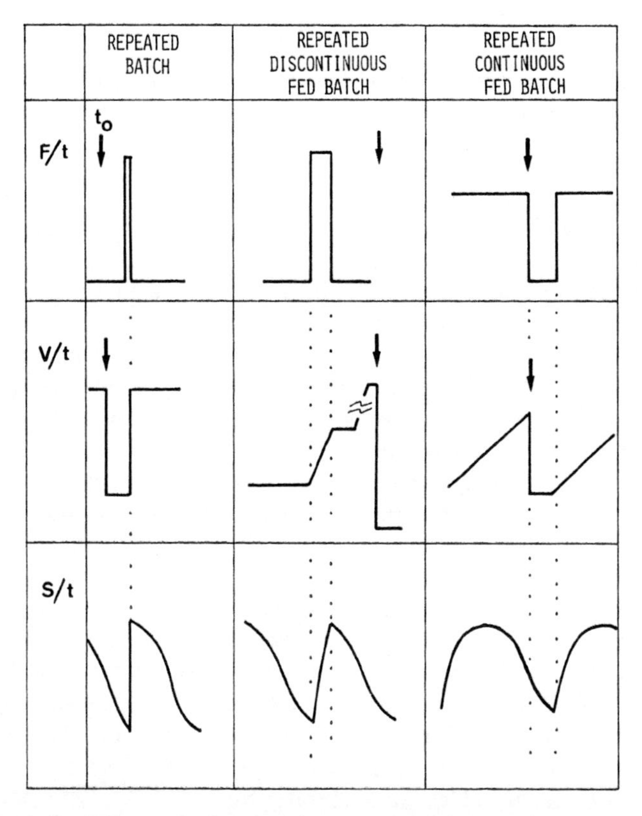

FIGURE 3.33. Periodic, cyclic fermentation processes characterized by the time dependence of the input, F; the volume of the liquid phase, V; and the substrate concentration, s. The arrow designates the moment when each cycle of the process begins.

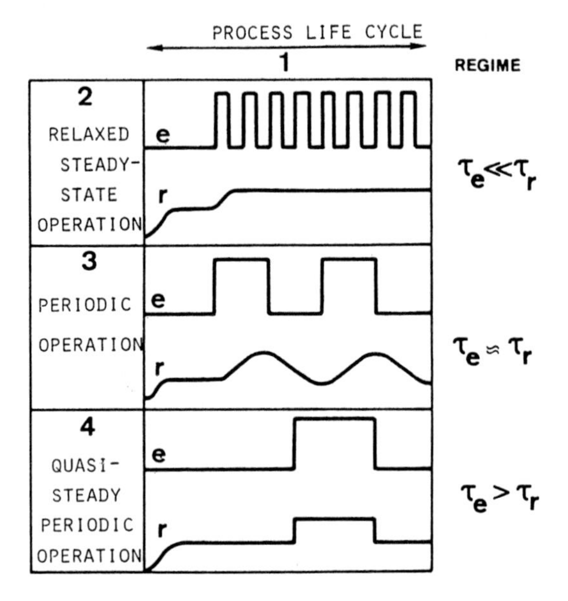

FIGURE 3.34. The four classes of periodic reactor operation as defined by Bailey (1973) represented by typical reactor responses (*r*) at different environmental changes (*e*) and characterized by the ratio of characteristic times τ_r, respectively τ_e, as indicated.

$$F_L(t) = \alpha_2 \cdot e^{\beta_2 \cdot t} \tag{3.86b}$$

The so-called "extended culture," which operates with constant substrate concentration, is a special case of a mode of operation with constant feed.

The criteria of the classification system recommended by Bailey (1973) involve a comparison of the time delay between a signal, for example, τ_e, and its response, τ_r. As shown in Fig. 3.34, transient operation techniques consist of true intermediate periodic operation ($\tau_e \sim \tau_r$), and relaxed steady-state ("frozen" systems with $\tau_e \ll \tau_r$), and quasi-steady-state ("balanced" systems with $\tau_e \gg \tau_r$) operations. With this system a great many bioprocesses fall into the category of cyclic continuous feed operations (Pickett, Topiwala, and Bazin, 1979).

The principal advantages of this type of cyclic system with transient operating techniques are apparent in bioprocesses whose maximum productivity is in a transient region. The products of secondary metabolism (Pirt, 1974) are a typical example of this group of processes. Another group consists of processes whose optimal operation requires an optimal substrate concentration—biomass production with bakers' yeast, for example (Aiba et al., 1976)—or where the process is subject to substrate inhibition. An important area of application for this is in biological waste water purification. These periodic modes of operation generally show increased productivity. More systematic and detailed study is needed in this area.

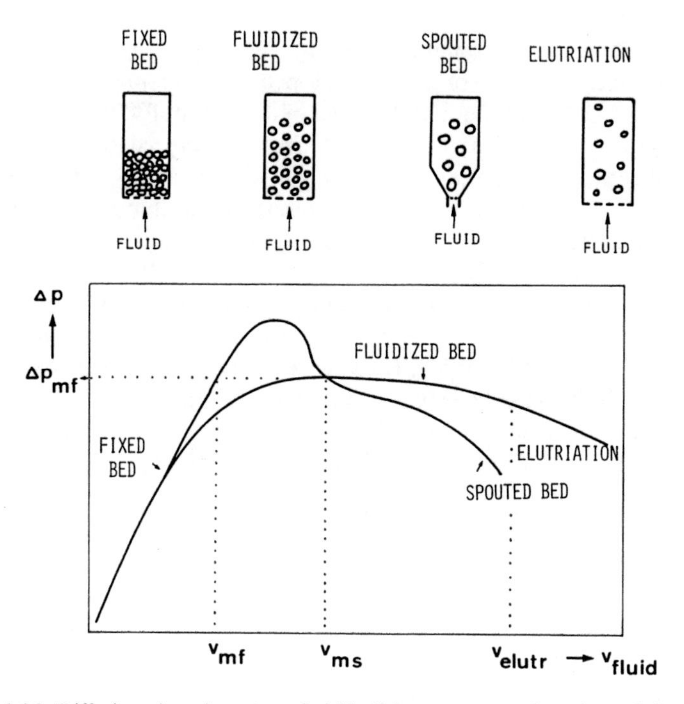

FIGURE 3.35. Differing situations in a fluid/solid reactor as a function of the velocity of the fluid phase, v_{fluid}, along with fluidization diagrams for the various reactor types. The pressure drop, Δp, increases in solid bed operations until the minimum velocity for fluidization is reached, v_{mf}. Above this velocity, fluidized bed conditions exist. In case of particles with diameters about 0.5 cm, the spouted bed can be operated in fluidized conditions above the minimum velocity of spouting (v_{ms}). A further increase in velocity finally leads to the elutriation of the solid phase.

Finally, the concept of the *fluidized bed* reactor should be briefly mentioned. Like so many others, this process technology stems from chemical reactor technology, and it has its roots in solid bed reactors. In the later, one can introduce a gas or liquid stream from below that can "fluidize" the particles as the velocity of the stream, v_{fluid}, increases.

The situation is diagrammed in Fig. 3.35. Above a particular flow velocity the small solid-phase particles become suspended. This point is called the "minimum fluidization velocity," v_{mf}, and it occurs at the point where the pressure drop in the column, Δp, remains constant. This working range is that of the "fluidized bed." With further increases in v_{fluid}, the solid phase would be prematurely "washed out" (elutriation). Another fluidization technology available for particles with diameters of about 5 mm is called a "spouted" or "whirling bed" reactor (Mathur and Epstein, 1974). A true fluidized bed has a porous plate at the bottom of the column. In the whirling bed mode of operation, the fluid is introduced directly through a small opening in the bottom of a cone-shaped column. A reactor type representing a cross between a fluidized and a spouted bed was developed for three-phase (GLS) fluidization, and it is called a "spouted fluidized bed" (Kono, 1980).

Because of the hydrodynamic effects in the fluidized reactor, mixing of all the phases involved (gas–liquid or gas–liquid–solid) is improved. The advantages of the fluidized bed are, among other things, improved mass and heat transfer coefficients, increased interfacial area, and high biomass concentration ($x > 80$ g/l).

The fluidized bed concept is now beginning to be applied in biological waste water treatment (e.g., Atkinson, 1980) and in food technology (e.g., Moser, Kosaric and Margaritis, 1980). Enzyme technology has already successfully operated with fluidized beds (Coughlin et al., 1975) even though, on a larger scale (even with chemical processes), all the problems are not yet solved (Andrews, 1982; Baker et al., 1980; Werther, 1977, 1980).

3.5 Bioreactor Models

There are three basic homogeneous reactor models (DCSTR, CSTR, and CPFR, Fig. 3.30) that can be considered ideal cases for calculating conversion. The equations for balancing all reactor models derive from the conservation of mass equation, Equ. 2.3. The equation for a balancing all types of ideal continuous stirred tank reactors (idCSTR) ($c_{ex} = c_R$) can uniformly be based on a consideration of the following (see Fig. 3.36):

Total change in mass in V_L = input mass with F_{in} − output mass with F_{ex} ± change in mass due to reaction (3.87a)

or, in mathematical form,

$$\frac{d(V \cdot c_{ex})}{dt} = F_{in} \cdot c_{in} - F_{ex} \cdot c_{ex} \pm r_i \cdot V \tag{3.87b}$$

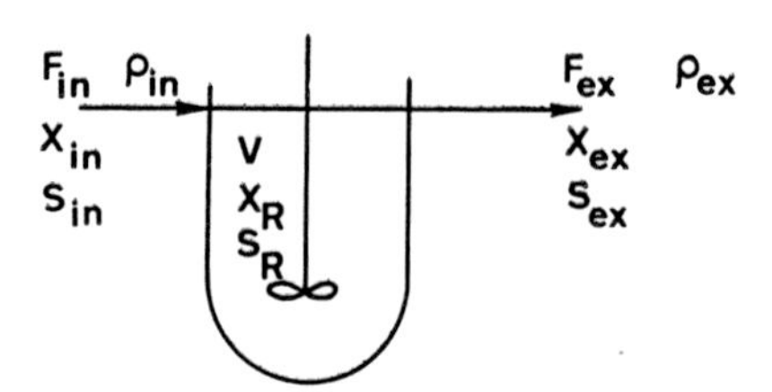

dCSTR $F_{in} = 0 = F_{ex}$

$F_{in} = F_{ex}$

sCSTR $F_{in} = F\ (t)$

FIGURE 3.36. Derivation of mass balance equations that can be based on an ideal stirred tank (condition $c_{ex} = c_R$, where c_R is the actual concentration in the vessel). All three configurations are included (DCSTR, CSTR, and SCSTR).

3.5.1 Model 1: The Ideal Discontinuous Stirred Tank Reactor (DCSTR)

The special conservation equation for the DCSTR, assumed to operate with ideal mixing and therefore with div grad $n_i = 0$ (see Equ. 2.3), follows from Equ. 3.87b, where $F_{in} = 0 = F_{ex}$ and V = constant, that is, $dV/dt = 0$:

$$\frac{dc_i}{dt} = \pm r_i \tag{2.3d}$$

This equation is not only the conservation of mass equation; it is also the definition of the rate of formation ($r_i > 0$) or consumption ($r_i < 0$) of a component. To calculate conversion, one integrates this equation.

3.5.2 Model 2: The Ideal Continuous Stirred Tank Reactor (CSTR) with V = Constant

Under the conditions $F_{ex} = F_{in} = F$ with V = constant and $c_{ex} = c_R$, the mass equation is

$$V\frac{dc_i}{dt} = F(c_{in} - c_{ex}) \pm r_i \cdot V \tag{3.87c}$$

If one uses the quantity $\bar{t}$ $[h]$ = mean residence time or the quantity D $[h^{-1}]$ = dilution rate of the fluid phase (cf. Equ. 3.47)

$$D = \frac{1}{\bar{t}} = \frac{F_L}{V_L} \tag{3.88}$$

one obtains from Equ. 3.87c

$$\frac{dc_i}{dt} = D(c_{in} - c_{ex}) \pm r_i \tag{3.89}$$

In the steady state ($dc_i/dt = 0$), one obtains the following mass equation for the CSTR

$$D(c_{in} - c_{ex}) = \pm r_i \tag{3.90}$$

If, for example, one is dealing with a microbial growth process in a CSTR, the $x_{in} = 0$, and one may apply Equ. 2.5a for the rate of growth and thereby obtain the well-known relations for the equilibrium state. This is the basis for the CSTR, the so-called "*chemostat*" or "*realstat*" mode of operation of a continuous fermentation process (see Sect. 6.1).

$$D = \mu \tag{3.91}$$

3.5.3 Model 3: The Ideal Semicontinuous Stirred Tank Reactor (SCSTR) with V = Variable

In this case, $dV/dt \neq 0$ (see Fig. 3.36) and therefore Equ. 3.87b remains entirely valid. For the calculation of a continuous feed operation as shown in Fig. 3.33,

F_{in} is a constant and F_{ex} is zero. In the case of an additional feed according to Equ. 3.86a, with $\alpha_1 = 0$ but $\beta_1 > 0$, then Equ. 3.87b can be rewritten to arrive at Equ. 3.92 as the mass conservation equation for the SCSTR with variable volume

$$\frac{d(V \cdot c_{ex})}{dt} = F_{in} \cdot c_{in} \pm r_i \cdot V \tag{3.92a}$$

or

$$F_{in} \cdot c_{ex} + \frac{dc_{ex}}{dt} \cdot V = F_{in} \cdot c_{in} \pm r_i \cdot V \tag{3.92b}$$

For the case of a growth process, using Equ. 2.5a for the rate of growth and with $x_{in} = 0$, one obtains (Dunn and Mor, 1975)

$$\frac{dx_R}{dt} = (\mu - D) \cdot x_R \tag{3.93}$$

This equation describes the "quasi-steady state" in which this type of addition process can be maintained for some time ($D = \mu$), as shown in Fig. 3.37 with simple Monod kinetics.

Equation 3.93 can be compared with Equ. 3.91 for the true steady state in a CSTR. The two are formally identical. The difference is in the real meaning of the factor $D \cdot c_{ex}$; in Equ. 3.90 it represents a dilution effect in the case of a chemostat. In the case of a semicontinuous stirred vessel with variable volume, the factor $D \cdot x_R$ in Equ. 3.93 describes the decrease in cell concentration due to the change of the volume in the reactor (Dunn and Mor, 1975). Similar equations can be formulated for all other types of fed-batch processes using the appropriate modification of Equ. 3.86.

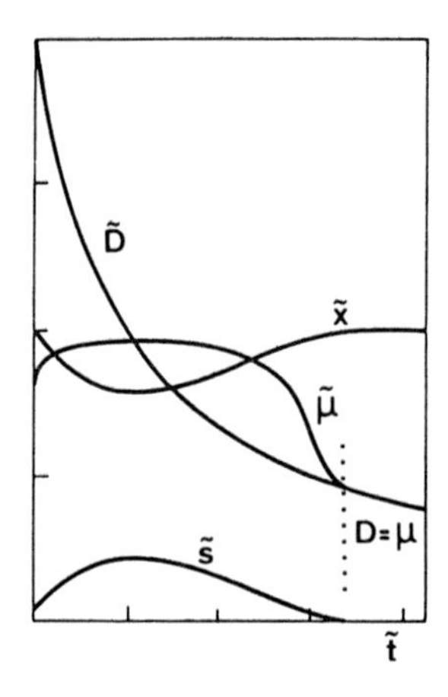

FIGURE 3.37. Computer simulation of fed-batch processes: approach to and attainment of the quasi-steady state with $D = \mu$ from the dimensionless kinetic model (see Equs. 3.92 and 3.93). The dimensionless parameters are $D = F_{in}/V \cdot \mu_{max}$, $\tilde{x} = x/Y \cdot s_{in}$, $\tilde{\mu} = \mu/\mu_{max}$, and $\tilde{s} = s/s_{in}$. (From Dunn and Mor, 1975. With permission of John Wiley & Sons, Inc.).

3.5.4 Model 4: The Ideal Continuous Plug Flow Reactor (CPFR) or Tubular Reactor

To quantify the balances in a CPFR, one may begin from the analogy between the space profile in a CPFR and the *c*/*t* profile in a DCSTR (Fig. 3.30). This implies a reformulation of Equ. 2.3d so that the time and space coordinates are interchanged. Dimensional analysis shows that the term representing length dependence must be multiplied with a velocity. Thus, for an ideal CPFR, with z is the length coordinate, the conservation equation is

$$v_z \cdot \frac{dc_i}{dz} = \pm r_i \tag{3.94}$$

In this equation, v_Z is the apparent flow velocity [mh^{-1}], obtained from the mass flow rate F [m^3h^{-1}] divided by the cross-sectional area of the tube A [m^2].

The same result is found by proceeding in the usual way, developing a differential conservation of mass equation for each volume element in the CPFR, as diagrammed in Fig. 3.38. This figure shows recycling in an amount F_r [m^3h^{-1}] for maintaining cell culture, so that the recycling ratio is $r = F_r/F$. For the stationary state and with $dV = A \cdot dz$, the equations for the conservation of cell mass, x, and substrate, s, are

$$F(1 + r) \cdot ds + r_s \cdot A \cdot dz = 0 \tag{3.95a}$$

$$F(1 + r) \cdot dx - r_x \cdot A \cdot dz = 0 \tag{3.95b}$$

These two equations, along with the equations for the boundary conditions, form a system of nonlinear differential equations that are very difficult to solve in closed form; recourse is therefore made to computer techniques (numerical methods). With certain simplifications, direct integration is possible. The appropriate boundary conditions are found in this case by considering the balance of mass at the point M where entering fresh medium meets with the recycled medium (see Fig. 3.38). For $z = 0$, one then has

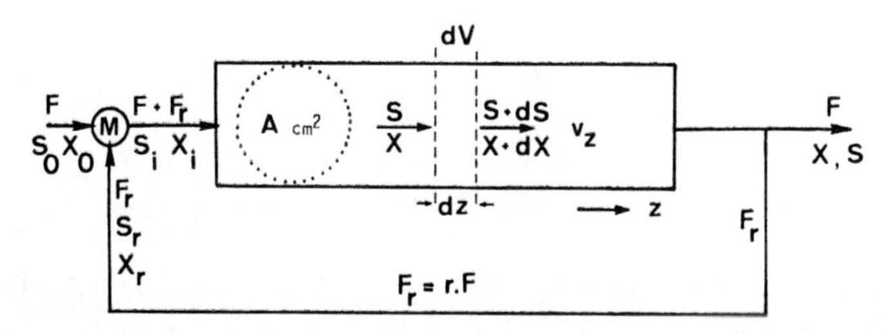

Figure 3.38. Derivation of mass balance equations in the case of an ideal tube reactor. Here, differential mass conservation $dV = A \cdot dz$ for s and x for a tubular reactor with recycle, leading to Equ. 3.85 at steady state.

$$s_i = \frac{s_0 + r \cdot s_r}{1 + r} \tag{3.96a}$$

and

$$x_i = \frac{x_0 + r \cdot x_r}{1 + r} \tag{3.96b}$$

From Equ. 3.95 and using the relationship $dV = A \cdot dz$, one can verify the validity of Equ. 3.94 for the case of no recycling.

The mass balance equation for the case of a cascade of N continuous stirred reactors, or for processing variants involving the return of concentrated solid-phase material, can be formulated with analogous considerations (Powell and Lowe, 1964).

3.5.5 Model 5: The Real Plug Flow Reactor CPFR with Dispersion

The quantification of real reactors, those with deviations from ideal plug flow, can now also be undertaken. In agreement with Equ. 3.3a, in addition to the convection term operating via the flow velocity v_z, the term with the dispersion coefficient D_L is also operative.

The conservation of mass equation for this situation, which is directly applied in modeling tube reactors (F. Moser, 1977) and bubble columns (Reuss, 1976), is thus identical to Equ. 3.3a. These types of one-dimensional one-phase models are not only necessary for calculating conversion: They are also very useful in, for example, calculating the $k_L \cdot a$ value of a reactor with a concentration profile:

$$\frac{do_L}{dt} = 0 = k_L a(o_L^* - o_L) - v_L \frac{do_L}{dz} + (1 - \varepsilon_G) D_L \frac{d^2 o_L}{dz^2} \tag{3.97}$$

In the steady state, for the actual determination of $k_L \cdot a$ one needs not only the experimental measurement of the O_2 concentration along the length of the reactor; one also needs to know the dispersion coefficient, which can be obtained from Equ. 3.4 in Sect. 3.3.1, and, for stationary method, Sect. 3.3.3, Lücke, Oels and Schügerl, 1977).

The types of models described thus far have already been characterized as one-dimensional one-phase models. This means that only a single coordinate (the z axis) is considered; in reality, in the case of, for example, flow in a tube, processes occurring in a radial direction (coordinate R) can also be of importance.

The formulation of a two-dimensional one-phase model for dispersion takes the following form, according to Equ. 3.3a:

$$\frac{\partial c_i}{\partial t} = -v_z \frac{\partial c_i}{\partial z} + D_z \frac{\partial^2 c_i}{\partial z^2} + \frac{D_R}{R} \frac{\partial}{\partial R}\left(R \frac{\partial c_i}{\partial R}\right) \tag{3.98}$$

In addition to having the dispersion coefficient D_z, this equation also has a term for the radial dispersion coefficient, D_R, which must also be measured to refine the model.

3.5.6 The Discontinuous Recycle Reactor (DCRR)

An adequate experimental device to avoid serious limitations in the sample size needed to analyze relatively slow reactions such as fermentation processes is illustrated in Fig. 3.39. This is the DCRR. The operating conditions are as follows:

- Differential operation in V_R
- Good mixing at high recirculation
- Accurate measurement of c_i in adequate time intervals

As a consequence, the concentration will be uniform. Hence $\nabla n_i' = 0$, and thus, from Equ. 2.3

$$\int_v \frac{dc_i}{dt} dv = \int_v r_i \, dv \tag{3.99}$$

Due to the differential performance of V_R, the result is

$$r_i = \frac{dc_i}{dt} \cdot \frac{(V_r + V_R')}{V_R} \tag{3.100}$$

Therefore, a high ratio of V_r/V_R reduces the problem of correcting the equation for the definition of reaction rates,* but at the same time it may prolong the time needed to achieve measurable conversion.

3.5.7 The Continuous Recycle Reactor (CRR)

Figure 3.40 shows a schematic diagram of a continuous recycle reactor, CRR. With high values of F_r, the whole reactor (balance line 2) approximates a CSTR, working at differential conversions with $r_{i,2}$ according to Equ. 3.90. Writing the balance for line 1, the equation for $r_{i,1}$ is

$$r_{i,1} = \frac{F + F_r}{V_R}(c_{ex} - c_{in}) \tag{3.101}$$

Since $r_{i,1} = r_{i,2}$, it follows that

$$c_1 = \frac{F_r \cdot c_{ex} - F \cdot c_{in}}{F + F_r} \tag{3.102}$$

Thus, if F_r is sufficiently high ($F_r \gg F$), as was first assumed, $c_1 \simeq c_{ex}$ and

* In the case of high sample volumes, it should be kept in mind that at each recycle the reactor behaves as a steady-state CPFR. Connecting tubes between reactor V_R and recycle reservoir V_r should be as short as possible.

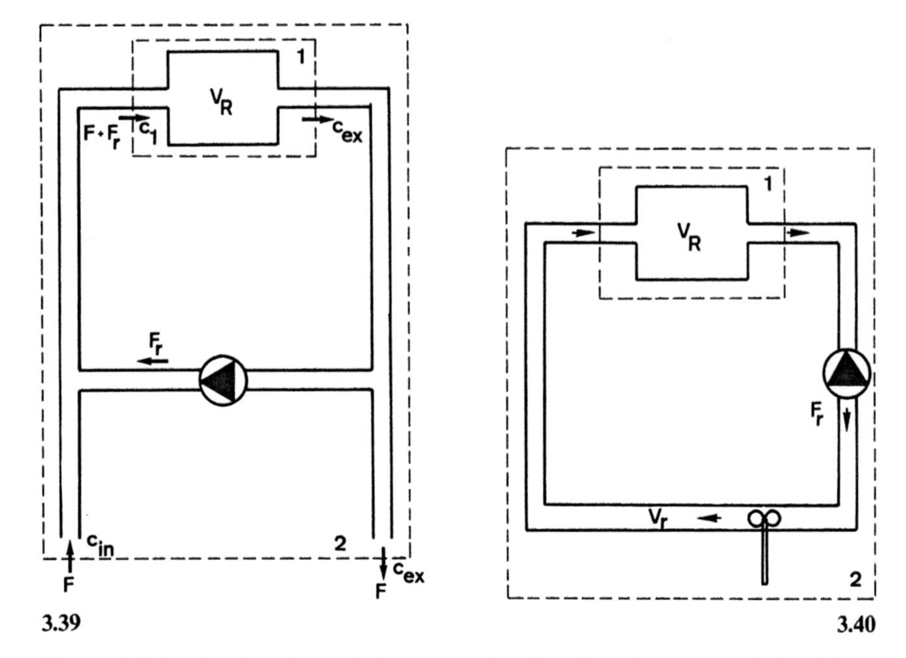

FIGURE 3.39. Derivation of mass balance equations based on the concept of closed-loop reactors. The total liquid volume in the recycle, V_r (reservoir) is recirculated with the aid of a pump with a flow rate F_r through the reactor volume V_R. The dotted lines indicate the balance lines of the reactor (1) and the whole system (2).

FIGURE 3.40. Derivation of mass balance equations based on the concept of open loop reactors with a stream F through the whole system (balance line 2) and a recycle flow rate F_r. The concentration in the reactor inlet (c_1) is different from c_{in} due to recycle.

under such conditions the global reactor system of a CRR can be treated as an ideal CSTR, according to Equ. 3.90. Logically, the condition $F_r \gg F$ or $r \approx 100$ is only an overall formulation, which can be written more accurately by considering the dependence on the reaction rate (Luft and Herbertz, 1969). Using a CRR is important mainly in the case of G|S or L|S catalytic reactions (e.g., Paspek et al., 1980) and in the case of reactions with immobilized cells or enzymes (e.g. Ford et al., 1972).

3.5.8 Multiple Phase Bioreactor Models

Figure 3.41 classifies the various types of bioreactor models according to the criteria floc–film, stationary–nonstationary, homogeneous–heterogeneous, and gradient-free–gradient bioprocessing. The types of models presented thus far are understood as single-phase reactors that consider the liquid phase and operate under pseudohomogeneous conditions. Section 4.3 deals with tests for whether and under what conditions, the pseudohomogeneous model is valid. Because gas, liquid, and solid phases are all present in bioreactors,

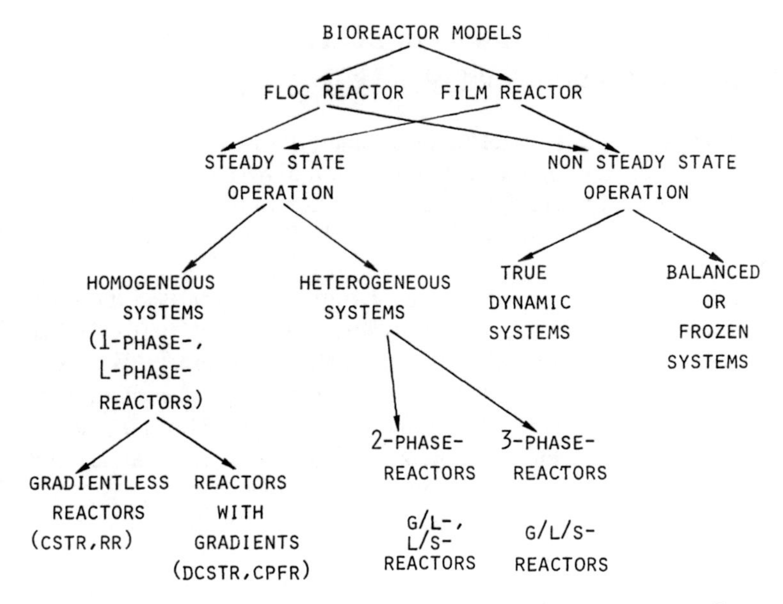

FIGURE 3.41. Classification of mathematical models of bioreactors according to engineering criteria. (Adapted from A. Moser, 1977a)

heterogeneous models in two or three phases should be more realistic. Due to their complexity, however, such models are only beginning to be applied. One such two-phase dispersion model has contributed to the successful clarification of oxygen limitation and inhibition in the case of a tall bubble column reactor. The two-phase model considers the gas and liquid phases, and it involves both mass conservation equations (Reuss, 1976).

In addition to the equation for the liquid phase (according to Equ. 3.97), which needs only extension with one reaction term, namely $-(1-\varepsilon_G)r_o$, the equation for the gas phase in the steady case can be written

$$\frac{do}{dt} = 0 = k_L \cdot a(o^* - o_L) + \frac{d}{dz}(v_{SG} \cdot o_G) \tag{3.103}$$

v_{SG} is the superficial gas velocity [$m \cdot h^{-1}$] and is, along with o_L^*, dependent on the hydrostatic pressure in the column.

Some insight into possible types of fluid flow of the reaction phase was given in Fig. 3.10. Both the liquid and gas phases can exhibit all intermediate stages and combinations of the extremes of maximum mixing and total segregation. Close observation of the situation in a reactor to be simulated provides the first hint as to which model to use.

Performance models of biofilm reactors (fixed and fluidized bed) following the concepts of pseudohomogeneous and true heterogeneous modeling are outlined in Sect. 6.7.

3.6 "Perfect Bioreactors" in Bench and Pilot Scale for Process Kinetic Analyses

Bioreactors must fulfill a twofold purpose in bioprocessing—industrial production and process kinetic analysis. A standard research bioreactor has been recommended by the European Federation of Biotechnology (Dechema Monograph, 1982): It is a stirred tank in which all dimensions are standardized (cf. Table 3.7).

For modeling purposes, on the other hand, the reactor used should exhibit pseudohomogeneity (see Sect. 4.3); such a reactor is called "perfect bioreactor" (A. Moser, 1977b, 1983b). Several constructions are used. Atkinson (1974) pointed out that only the biological film fermenter (BFF) configurations are able to separate biological from physical parameters. Normally, in conventional stirred tank fermenters operating with flocs, external and internal transport limitations cannot be completely excluded.

Figure 3.42 shows several bioreactors suitable as perfect bioreactors for quantifying biological rates unaffected by physical transport: (a) the sloping plane biological film fermenter (BFF) of Atkinson (1974), (b) the thin-layer tubular biofilm fermenter (ThLTBFF) (Moser, 1977a), (c) the horizontal rotary BFF (Hoehn and Ray, 1973; Tomlinson and Snaddon, 1966) similar to the horizontal rotary fermenter of Phillips et al. (1961), (d) the vertically rotating drum of Kornegay and Andrews (1968) and subsequently used by La Motta (1976); and (e) the completely mixed microbial film fermenter (CMMFF) of Atkinson and Davies (1972). The so-called multipurpose bioreactor (MBR, 1982; Moser, 1983b) includes several concepts of floc and film bioprocessing: Both vertical and horizontal positions are possible using different stirring devices (impellers, rotating drums of different shape) Fig. 3.42f. Another type of floc bioreactor is the completely filled bioreactor, which realizes pseudohomogeneity to a high degree (Karrer, 1978; Puhar et al., 1978).

Biofilm thickness and density are important factors for quantification. Biological film thickness can be measured using an optical procedure (Kornegay and Andrews, 1968; Sanders, 1966; Trulear, 1980) or by means of electrical conductance (Hoehn and Ray, 1973; Norrman et al., 1977; see also Charaklis et al., 1982). The biofilm thickness can be controlled by mechanical means (Atkinson, 1974) or hydrodynamically in the CMMFF or in the ThLTBFF. The special property of a perfect reactors is illustrated in the case of G|L processing, where the OTR and $k_{L1} \cdot a$ are to be included in modeling. Both terms, k_{L1} and a, are affected by biological reactions in different ways, so that G|L reactors with constant value of a are to be preferred. Figure 3.43 shows three known configurations of G|L-model reactors with a = constant and exhibiting no (or a negligible) velocity profile at the G|L interface (penetration depth). This profile is of interest, because mass transfer, according to Equ. 2.3c, is affected by fluid velocity, and the simple mass transfer theories mentioned in Equ. 3.30b and c are only valid in the case of constant flow velocity with no or a known gradient. As indicated in Fig. 3.43a, the laminar film on a vertical

TABLE 3.7. Recommendations for bioreactor dimensioning.

Variable	*Recommendation**	*Bench scale*: $V_R = 42$ l	*Pilot scale*: $V_R = 300$ l	*Technical scale*: $V_R = 3000$ l
Material	r	no. 4541, 4571	Same	Same
Pressure [bar]	s	4	Same	Same
Temperature [°C]	s	143	Same	Same
Height/diameter	s	3:1	3:1 (2:1)	3:1 (2:1)
d_T [mm]	s	259	492	1072
H_T [mm]	s	769	1595	3390
Heating	r	Jacket	Double jacket	Double jacket or coils
Motor drive	r	Single, from bottom	Same	Same
Power [kW]	r	1.5	5	20
Shaft-bearing assemblies	r	Double-crane pack	Same	Same
Impeller (turbine):	s	Changeable		
Diameter (impeller) d_i		$d_i = 0.4 \cdot d_T$	Same	Same
Width		$0.3 \cdot d_T$	Same	Same
Height		$0.2 \cdot d_T$	Same	Same
Diameter (plate)		$0.6 \cdot d_T$	Same	Same
No plates		6	Same	Same
No impellers		3	Same	Same
Distance from bottom		$0.3 \cdot d_{R,i}$	Same	Same
Distance to air distributor [mm]		15	40	60
Rotational speed [rpm] (thyristorized motor)	r	0–2500	0–1200	0–500
Air distributor ($d = 0.7 \cdot d_i$)	s	3 jets each at 3 mm	$d_{dist} = 25$ mm with 10 holes each at 4 mm	$d_{dist} = 50$ mm with 100 holes each at 4 mm
Maximum air flow rate [$Nm^3 \cdot m^{-3} \cdot min^{-1}$]	r	4	2	2

* r, recommended; s, strictly recommended. From Dechema, 1982.

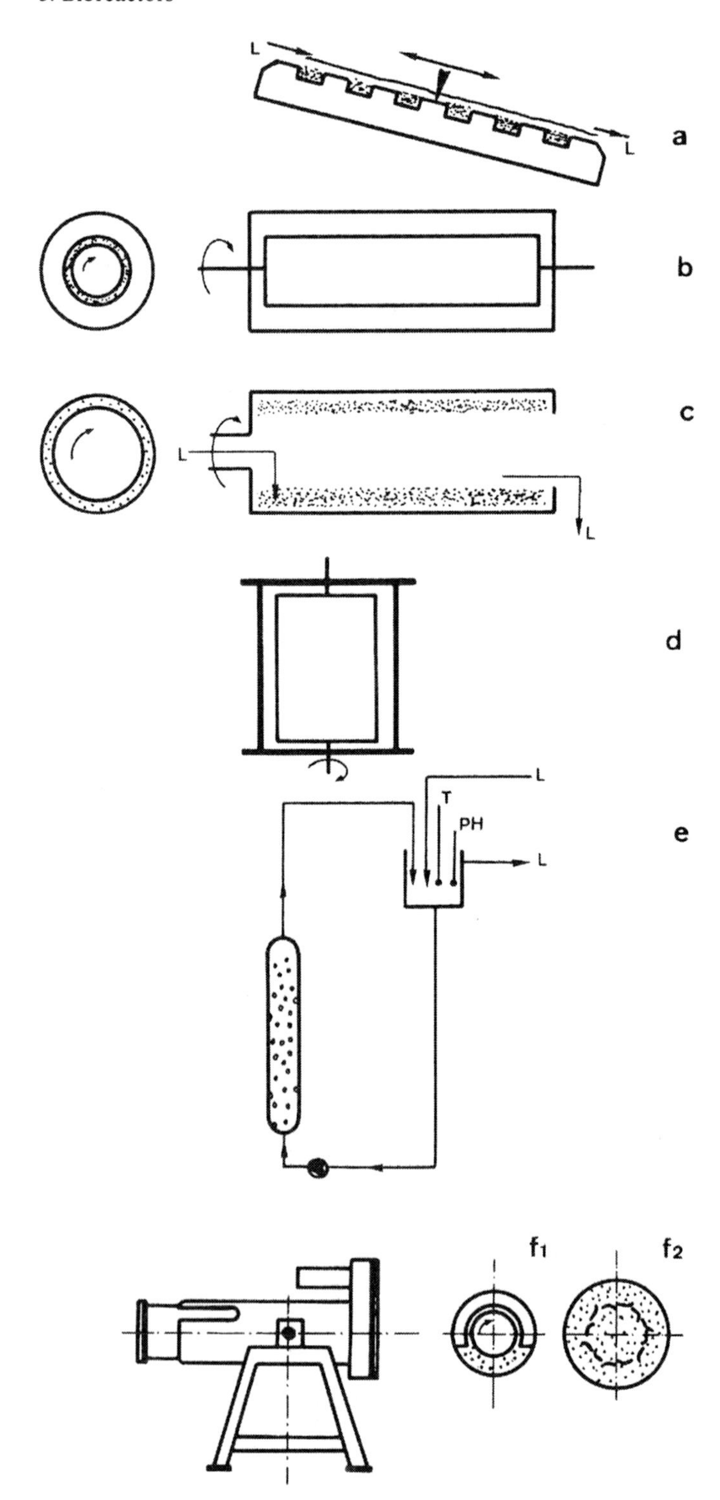

L
L
a
b
c
L
L
d
L
T
PH
e
L
f1
f2

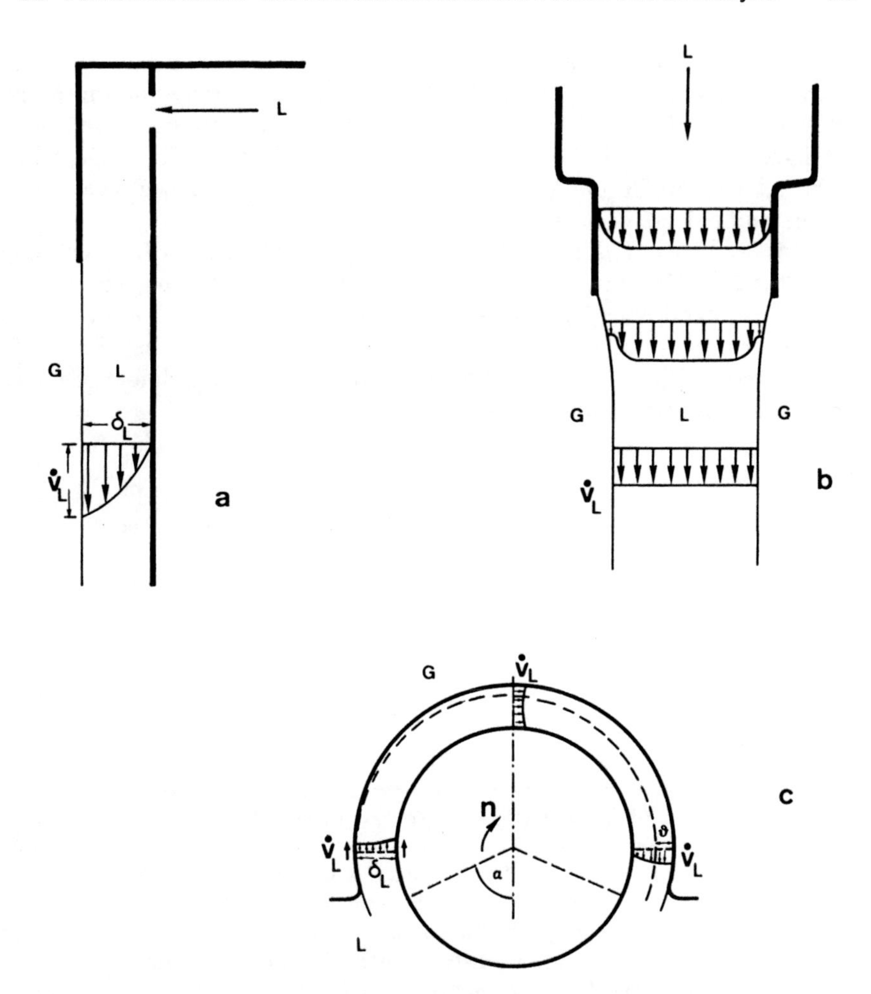

FIGURE 3.43. G|L model reactors for systematic process kinetic analyses. (a) Falling liquid film with thickness δ_L. (b) Falling liquid jet. (c) Thin-layer reactor with δ = penetration depth of gas, being always higher than δ_L (see A. Moser, 1973c). In all three cases the velocity profile of liquid flow is simple enough to allow the application of simple mass transfer theories.

◁ FIGURE 3.42. Biofilm-model reactors and so-called perfect reactors for a systematic process kinetic analysis. (a) Sloping plane biofilm reactor, according to Atkinson (1974). (b) Thin-layer fermenter, according to Gorbach (see Moser, 1977a). (c) Rotating cylinder biofilm reactor, according to Tomlinson and Snaddon (1966). (d) Rotating drum, according to Kornegay and Andrews (1968). (e) Completely mixed microbial film fermenter, according to Atkinson (1974). (f) Multipurpose bioreactor, according to Moser (1983b), including case b (f_1) and a horizontal stirred tank (f_2) (Bauer and Moser, 1985).

plate exhibits the well-known half-parabolic profile (Beek and Muttzal, 1975), while the laminar falling liquid jet (Fig. 3.43b) has a nearly uniform profile (Danckwerts, 1970). The situation in the horizontal rotating drum in the thin-layer tubular film fermenter (ThLTFF) is shown in Fig. 3.43c, where the velocity profile in the film can—by calculation—be neglected in the penetration depth of oxygen, δ_{O2} (Moser, 1973c).

The last type of ThLTFF is the most suitable one for aseptic bioprocessing. It offers favorable properties for systematic process kinetic analyses, because the G|L interface is given by geometry and k_{L1} follows strictly a theory of mass transfer (Equ. 3.30c), so that both terms need not be experimentally measured. The significant role of "model bioreactors" for physiological studies is recently stressed (Moser, 1988).

Bibliography

Aiba, S., Humphrey, A.E., and Millis, N. (1976). *Biochemical Engineering.* New York: Academic Press. 2nd edition.

Aiba, S., Nagai, S., and Nishizawa, Y. (1976). *Biotechnol. Bioeng.*, 18, 1001.

Aiba S., Huang S.Y. (1969) Chem. Engng. Sci., 24, 1149.

André, G., et al. (1981). *Biotechnol. Bioeng.*, 23, 1611.

Andrews, G.F. (1982). *Biotechnol. Bioeng.*, 24, 2013.

Atkinson, B. (1980). Presented at Conference on Biological Fluidized Bed Treatment of Water and Wastewater. University of Manchester, England, 14–17 April.

Atkinson, B. (1974). *Biochemical Reactors.* London: Pion.

Atkinson, B., and Davies, I.J. (1972). *Trans. Inst. Chem. Engrs.*, 50, 208.

Atkinson, B., and Knight, A.J. (1975). *Biotechnol. Bioeng.*, 17, 1245.

Atkinson, B., and Kossen, N.W.F. (1978). *In Proceedings of the* 1st European Congress on Biotechnology, Interlaken, Switzerland, Dechema Monograph., 82, 37.

Atkinson, B., and Ur-Rahman, F. (1979). *Biotechnol. Bioeng.*, 21, 221.

Baader, W., et al. (1978). *Biogas in Theorie und Praxis.* Hrsg. Kuratorium für Technik und Bauwesen in der Landwirtschaft, Münster-Hiltrup, West Germany.

Babivin, L.A., et al. (1981). *Eur. J. Appl. Microbiol. Biotechnol.*, 13, 15.

Bailey, J.E. (1973). *Chem. Eng. Commun.*, 1, 111.

Bajpaj, R.K., and Reuss, M. (1982). *Canad. J. Chem. Eng.*, 60, 384.

Baker, C.G.J., et al. (1980). In Moo-Young, M., et al. (eds.). *Advances in Biotechnology*, Vol. 1. Oxford: Pergamon Press, p. 635.

Bandyopadhyay, B., Humphrey, A.E., and Taguchi H. (1967). *Biotechnol. Bioeng.*, 9, 533.

Bauer, A., and Moser, A. (1985). In *Proceedings of the 5th European Mixing Conference* in Würzburg: BHRA Press Cranfield, Bedford, England, (J. Stanbury, ed.) p. 171

Bayer, K., and Fuehrer, F. (1982). *Proc. Biochem.*, July/August, 42.

Beek, W.J. (1969). "Stofftransport mit und ohne chemische Reaktion," *VSSD Skriptum.* Technical University, Delft, Netherlands.

Beek, W.J., and Muttzall, K.M.K. (1975). *Transport Phenomena.* New York and London: John Wiley.

Beyeler, W., et al. (1981). *Eur. J. Appl. Microbiol. Biotechnol.*, 13, 10.

Beyeler, W., et al. (1983). *Chem. Ing. Techn.*, 55, 869.

Biggs R.D. (1963). AIChE Journal 9, 636.

Blanch, H.W. (1979). *Annu. Rep. Ferm. Proc.*, 3, 47.
Blenke, H. (1979). *Adv. Biochem. Eng.*, 13, 121.
Bronn, W.K. (1971). *Chem. Ing. Techn.*, 43, 70.
Brown, D.E. (1967). *Proc. Biochem.*, 2, 27.
Brown, D.E. (1977). *Chem. Ind.*, 20 (August), 684.
Brown, D.E. (1984). In *Process Variables in Biotechnology.* Bioreactor Performance working party (chairman W. Crueger). Frankfurt: Dechema, Chap. 17, p. 94.
Brown, D.E., et al. (1979). *Biotechnol. Lett.*, 1, 159.
Bruxelmane M. (1983). Technique de l'Ingenieur 2, A5910.
Bryant, J. (1977). *Adv. Biochem. Eng.*, 5, 101.
Bryant, J., and Sadeghzadeh, N. (1979). Paper F3 presented at 3rd European Conference on Mixing, York, England.
Calderbank, P.H. (1958). *Trans. Inst. Chem. Eng.*, 36, 443.
Carberry, J. (1964). *Ind. Eng. Chem.*, 56, 39.
Cardosó-Duarte, J.M., et al. (1975). In (Dean, A.C.R., et al. (eds.). *Continuous Culture*, Vol. 6. London: SCI Chap. 3.
Charaklis, W.G., et al. (1982). *Water Res.*, 0, 1.
Charaklis, W.G., et al. (1981). *Biotechnol. Bioeng.*, 23, 1923.
Chibata, T., et al. (1972). In Terui, G. (ed.). *Fermentation Technology Today.* Tokyo: Soc. of Fermentation Technol., p. 383.
Cleland, N., et al. (1984). *Appl. Microbiol. Biotechnol.*, 20, 268.
CMEA Recommendation (1968). *Standard Methods of Water Analysis*, Vol. 1. Budapest: Comecon Vituki (Scientific Research Institute for Water Economy) Vol. 1.
Cooker B. et al. (1983). The Chem. Eng. 392, 81.
Cooney, Ch.L., Wang, D.I.C., and Mateles, R.I. (1968). *Biotechnol. Bioeng.*, 11, 269.
Cooper, C.M., Fernström, G.A., and Miller, S.A. (1944). *Ind. Eng. Chem.*, 36, 504.
Coughlin, R.W., Charles, M., Paruchuri, E.K., and Allen, B.R. (1975). *Chem. Ing. Techn.*, 47, 111.
Cow, J.S., Littlehailes, J.D., Smith, S.R.L., and Walter, R.B. (1975). In Tannenbaum, S.R., and Wang, D.I.C. (eds.). *Single Cell Protein, II.* Cambridge, Mass.: MIT Press, p. 424.
Danckwerts, P.V. (1958). *Chem. Eng. Sci.*, 8, 93.
Danckwerts, P.V. (1970). *Gas-Liquid Reactions.* New York: McGraw-Hill.
Dang, N.D.P., Karrer, D.A., and Dunn, I.J. (1977). *Biotechnol. Bioeng.*, 19, 853.
Dawson, P.S.S. (1974). In Sikyta, B., Prokop, A., and Novak, M. (eds.). *Advances in Microbial Engineering*, Part 2. New York Interscience Publications, p. 809.
Dechema (1982). *Arbeitsmethoden für die Biotechnologie.* Biotechnology working group (chairman H.J. Rehm). Frankfurt: Dechema.
Dechema (1984). *Process Variables in Biotechnology.* Bioreactor Performance, working party EFB (chairman W. Crueger). Frankfurt: Dechema.
Dechema (1987). *Physical Aspects of Bioreactor Performance.* Bioreactor Performance, working party EFB (W. Crueger et al., eds.). Frankfurt: Dechema.
Deckwer, W.D. (1977). *Chem. Ing. Techn.*, 49, 213.
Deckwer, W.D. (1979). *Fort. Verfahrenstechnik*, 17(D), 317.
Dohan, L.A., and Weinstein, H. (1973). *Ind. Eng. Chem. Fundam.*, 12, 64.
Dunn, I.J., and Mor, J.R. (1975). *Biotechnol. Bioeng.*, 17, 1805.
Dunn, I.J., and Einsele, A. (1975). *J. Appl. Chem. Biotechnol.*, 25, 707.
Driessen, F.M., Ubbels, J., and Stadhouders, J. (1977). *Biotechnol. Bioeng.*, 19, 841.
Dŭdŭkovic, M.P. (1977). *Ind. Eng. Chem. Fundam.*, 16, 3,385.

Ebert, K.H. (1971). *Chem. Ing. Tech.*, 43, 50.
Einsele, A. (1976a). Ph.D. Thesis, Technical University, Zurich.
Einsele, A. (1976b). *Chem. Rundschau*, 29, 53.
Einsele, A., and Fiechter, A. (1974). *Chem. Ing. Tech.*, 46, 701.
Einsele, A., Ristroph, D.L., and Humphrey, A.E. (1978). *Biotechnol. Bioeng.*, 20, 1487.
Einsele A. and Finn R.K. (1980). *Ind. Engng. Chem. Proc. Des. Dev.*, 19, 600.
Eriksson, R. (1971). In *Proceedings of the 1st European Biophysics Congress, Medical Academy*. Vienna: Verlag der Wiener medizinischen Akademie (Broda E. et al., eds.) IV, 319.
Eriksson, R., and Holme, T. (1973). *Biotechnol. Bioeng. Symp.*, 4, 581.
Falch, E. (1968). *Biotechnol. Bioeng.*, 10, 233.
Fan, L.T., Erickson, L.E., Shah P.S., and Tsai, B.I. (1970). *Biotechnol. Bioeng.*, 12, 1019.
Fiechter, A. (1974). In *Proceedings of the 4th International Symposium on Yeasts*, Part II. Vienna: University Bodenkultur Press (Klaushofer H., Sleytr U. eds.), 17.
Fiechter, A. (1978). In *Proceedings of the 1st European Congress on Biotechnology Interlaken, Switzerland. Dechema Monograph*, 82, 17.
Finn, R.K. (1969). *Proc. Biochem.*, 4, 17.
Ford, J.R., et al. (1972). *Biotechnol. Bioeng. Symp.*, 3, 267.
Forrest, W.W. (1972). In Norris, J.R., and Ribbons, D.W. (eds.). *Methods in Microbiology*, Vol. 6B. London-New York: Academic Press, p. 285.
Fu, B., Weinstein, H., Bernstein, B., and Shaffer, A.B. (1971). *Ind. Eng. Chem. Proc. Des. Dev.*, 10, 501.
Ghose, T.K., and Mukhopadhyay S.N. (1979). Paper presented at 32nd Annual Session of International Institutes of Chemical Engineering, Indian Institute of Technology-IIT, New Delhi, in Bombay.
Grieves, R.B., et al. (1964). *J. Appl. Chem.*, 14, 478.
Griot, M., et al. (1986). In *Proceedings Intern. Conf. on Bioreactor Fluid Dynamics*. (Cambridge, April): BHRA Press, Granfield, Bedford, England (J. Stanbury., ed.) p. 203
Grosse, H.H., et al. (1985). *Acta Biotechnolog.*, 5, 163, (esp. ref. 5).
Hartung, K.H., and Hiby, J.W. (1973). *Chem. Ing. Techn.*, 45, 522.
Heineken, F.G. (1970). *Biotechnol. Bioeng.*, 12, 145.
Heinzle, E. (1978). Ph.D. Thesis, Technical University, Graz, Austria.
Herbert, D., et al. (1956). *J. Gen. Microbiol.*, 14, 601.
Herzog, P., et al. (1981). *Chem. Ing. Techn.*, 55, 566.
Hesseltine, C.W. (1977a). *Proc. Biochem.*, 11 (July–August), 24.
Hesseltine, C.W. (1977b). *Proc. Biochem.*, 11 (November), 29.
Hiby, J.W. (1972). *Chem. Ing. Techn.*, 44, 907.
Hines, D.A., Bailey, M., Onsby, J.C., and Roesler, F.C. (1975). 1st Chem. Eng. Symp., Ser. 41, D1.
Hirner, W. (1974). Ph.D. Thesis, Technical University, Stuttgart.
Hirner, W., and Blenke, H. (1974). *Chem. Ing. Techn.*, 46, 353.
Hoehn, R.C., and Ray, A.D. (1973). *J. Water Poll. Contr. Fed.*, 45, 2302.
Holmes, D.B. (1964). Chem. Eng. Sci., 19, 201.
Howell, J.A., and Atkinson, B. (1976). *Biotechnol. Bioeng.*, 18, 15.
Imanaka, T., and Aiba, S. (1976). *J. Appl. Chem. Biotechnol.*, 26, 559.
Jones, P.H. (1973). *Proc. Biochem.*, September, 19.
Joshi J.B. (1980). *Trans. Inst. Chem. Engrs.*, 58, 155.
Joshi J.B., Pandit A.B., and Sharma M.M. (1982). *Chem. Eng. Sci.*, 37, 813

Jüttner, F. (1982). *Process Biochem.*, March/April, 2.

Käppel, M. (1976). *VDI Bericht*, 578.

Käppeli, O., and Fiechter, A. (1980). *Biotechnol. Bioeng.*, 22, 1509.

Käppeli, O., and Fiechter, A. (1981a). *Biotechnol. Bioeng.*, 23, 1897.

Käppeli, O., and Fiechter, A. (1981b). *Biotechnol. Lett.*, 3, 541.

Karrer, D. (1978). Ph. D. Thesis, Technical University, Zürich.

Katinger, H.W.D. (1976). Mitteil. d. Versuchsanstalt für das Gärungsgewerbe, Wien Nr. 7/8, 82

Katinger, H.W.D., Scheirer, W., and Krömer, E. (1979). *Ger. Chem. Eng.*, 2, 31.

Khang, S.J., and Levenspiel, O. (1979). *Chem. Eng. Sci.*, 31, 569.

King, C.J. (1966). *Ind. Eng. Chem. Fundam.*, 5, 1.

Kipke, K.D. (1984). In Dechema Monograph (1984). *Process Variables in Biotechnology.* Bioreactor Performance EFB-working party (W. Crueger chairman). Frankfurt: Dechema, Chap. 20.

Kishinevskii, M.Kh. (1951). *J. Appl. Chem.* (*USSR*), 24, 593.

Kono, H. (1980). *Hydrocarb. Proc.*, January, 123.

Korjagin, W.W., et al. (1978). In *Entwicklung von Laborfermentoren.* 6th Reinhardsbrunner Symposium, Academy of Science. Berlin: Akademie Verlag, Ringpfeil M., ed. p. 327.

Kornegay, B.H., and Andrews, J.F. (1968). *J. Water Poll. Contr. Fed.*, 40, 460.

Kramers, H., and Alberda, G. (1953). *Chem. Eng. Sci.*, 2, 1731.

Kramers, H., and Alberda, B. (1953). *Chem. Eng. Sci.*, 2, 35

Küng, W., and Moser, A. (1986). Bioproc. Eng., 1, 23.

Küng, W., and Moser, A. (1988). Biotechnol. Lett., in press.

Läderach, H., Widmer, F., and Einsele, A. (1978). *Proceedings of the* 1st *European Congress on Biotechnology*, Part I, Interlaken, Switzerland. *Dechema Monograph*, 82.

La Motta, E.J. (1976). *Biotechnol. Bioeng.*, 18, 1359.

Lee, Y.H., and Tsao, G.T. (1979). *Adv. Biochem. Eng.*, 13, 35.

Lehmann, J., et al. (1980). In Moo-Young, M., et al. (eds.). *Proceedings* 6*th International Fermentation Symposium* (*Advances in Biotechnology*), Vol. 1. Oxford: Pergamon Press, p. 453.

Lehmann, J., et al. (1982). In *Arbeitsmethoden für die Biotechnologie.* Working group on Biotechnology (H.J. Rehm, chairman). Frankfurt: Dechema.

Lehmann, J., et al. (1985). In Rehm, H.J., and Reed, G. (eds.). *Biotechnology.* Vol. 2. Deerfield Beach, Fla., and Basel: Verlag Chemie Weinheim, Chap. 25.

Lehnert, J. (1972). *Verfahrenstechnik*, 6, 58.

Leistner, G., Müller, G., Sell, G., and Bauer, A. (1979). *Chem. Ing. Techn.*, 51, 288.

Levenspiel, O. (1972). *Chemical Reaction Engineering.* New York: John Wiley.

Levenspiel, O. (1979). *The Chemical Reactor Omnibook.* Corvallis, Ore.: OSU Book Stores.

Levenspiel, O., and Smith, W.K. (1957). *Chem. Eng. Sci.*, 6, 227.

Liepe, et al. (1978). In *Proceedings of the* 1st *European Congress on Biotechnology*, Interlaken, Switzerland, Part 1, p. 78, *Dechema Monograph*, 82.

Linek, V., Sobotka, M., and Prokop, A. (1973). *Biotech. Bioeng.*, 15, 429.

Liu, M.S., et al. (1973). *Biotechnol. Bioeng.*, 15, 213.

Lücke, J., Oels, U., and Schügerl, K. (1977). *Chem. Ing. Techn.*, 49, 161.

Luft, G., and Herbertz, H.A. (1969). *Chem. Ing. Techn.*, 41, 667.

Luong, J.H.T., and Volesky, B. (1980). *Canad. J. Chem. Eng.*, 58, 497.

Luong, J.H.T., and Volesky, B. (1982). *Canad. J. Chem. Eng.*, 60, 163.

Marison, I.W., and von Stockar, U. (1983). In Proceedings BIOTECH 83, International Conference on Commercial Applications & Implications of Biotechnology, Online Publ. Ltd., Northwood/UK, p. 947.
Marison, I.W., and von Stockar, U. (1987). Enzyme and Microbial Technology, 9, 33.
Märkl, H., and Vortmeyer, D. (1973). In *Proceedings of 3rd Symposium on Technical Microbiology*, Berlin: Dellweg H., ed., Gärungsgewerbe und Biotechnologie p. 29.
Mathur, K.B., and Epstein, N. (1974). *Spouted Beds.* New York: Academic Press.
MBR (Microbial Bioreactor AG, in Wetzikon/Zürich CH) (1982). In *Neuentwicklungen*, ACHEMA 82 (Dechema, publ.) 1, 430.
Mersmann, A. et al. (1976). Intern. Chem. Engng., 16, 590.
Merz, A., Vogg, H. (1978). Chem. Ing. Techn. 50, 108.
Meyer, C., and Beyeler, W. (1984). *Biotechnol. Bioeng.*, 26, 916.
Meyer H.P. (1987) in *Physical aspects of Bioreactor Performance*, EFB-working party Bioreactor Performance, (Crueger W. et al., eds.) Dechema/Frankfurt D., Chap. 8.
Meyrath, J., and Bayer, K. (1973). In *Proceedings of the 3rd Symposium on Technical Microbiology.* Berlin: Dellweg, H. (ed.), Inst. für Gärungsgewerbe und Biotechnologie. p. 117.
Middleton, J.C. (1979). Paper A2 presented at 3rd Duropean Conference on Mixing, York, England.
Midler, M., and Finn, R.K. (1966). *Biotechnol. Bioeng.*, 8, 71.
Mihaltz, P., and Hollo, J. (1980). *Acta Biotechnolog.*, 0, 39.
Minkevich, I.G., and Eroshin, V.R. (1973). *Folia Microbiol.*, 18, 376.
Moes, J., et al. (1984). In *Proceedings of the 3rd European Congress on Biotechnology.* Munich: Vol. 2, p. 285. Verlag Chemie, Weinheim/D.
Monk, P.R. (1978). *Proc. Biochem.*, December, 4.
Moo-Young, M. et al. (1979). *Biotechnol. Bioeng.*, 21, 593.
Moo-Young, M. (1985). (ed.-in-chief). *Comprehensive Biotechnology*, Vol. 1–3. Oxford: Pergamon Press.
Moser, A. (1973a). *Biotechnol. Bioeng. Symp.*, 4, 399.
Moser, A. (1973b). In *Proceedings of the 3rd Symposium on Technical Microbiology.* Berlin: Dellweg, H. (ed.) Inst. für Gärungsgewerbe u. Biotechnologie. p. 62.
Moser, A. (1973c). *Chem. Ing. Techn.*, 45, 1313.
Moser, A. (1974). *Gas Wasserfach/Wasser-Abwasser*, 115, 411.
Moser, A. (1977). *Chem. Ing. Techn.*, 49, 612.
Moser, A. (1978). In preprints, *1st European Congress on Biotechnology*, Interlaken, Switzerland, Part 1, 88.
Moser, A. (1980a). In Ghose, T.K. (ed.). *Proceedings of Bioconversion and Biochemical Engineering Symposium*, Vol. 2, New Delhi: Indian Institute of Technology, New Delhi p. 253.
Moser, A. (1980b). In *Theoretical Basis of Kinetics of Growth, Metabolism and Product Formation of Microorganisms*, Part 2. UNEP/UNESCO/ICRO Training Course, Zentralinstitut für Mikrobiologie und Experimentelle Therapie, Academy of Science, Jena, (Knorre W., ed.) East Germany (publisher) p. 27.
Moser, A. (1981). *Bioprozesstechnik.* Vienna and New York: Springer-Verlag.
Moser, A. (1983a). *Biotechnol. Lett.*, 4, 73.
Moser, A. (1983b). *Proceeding Adv. in Ferm.*, 83 (Suppl. to *Process Biochemistry*) p. 202.
Moser, A. (1985a). In Moo-Young, M. (ed.-in-chief). *Comprehensive Biotechnology*, Vol. 2. Oxford: Pergamon Press, Chap. 4.

Moser, A. (1985b). In *Proc. Adv. Bioreactor Engng.*, Lødz, Poland, September 1985.
Moser, A. (1987). In *Physical Aspects of Bioreactor Performance.* EFB-working party Bioreactor Performance (Crueger W., et al., eds.). Dechema. Frankfurt; Chap. 4.
Moser, A. (1988). *Trends in Biotechnology*, Vol. 6, No. 11.
Moser, A., and Steiner, W. (1974). *Chem. Ing. Techn.*, 46, 695.
Moser, A., and Steiner, W. (1975a). *Chem. Ing. Techn.*, 47, 211.
Moser, A., and Steiner, W. (1975b). *VDI Bericht*, 232, 259
Moser, A., Kosaric N., and Margaritis A. (1980). Paper presented at 30th Canad. Chem. Eng. Conference, Edmonton, Canada, October.
Moser, A., Küng, W., Weiland, P. (1985). Paper presented at Achema/Frankfurt. In Dechema preprints Biotechnology ACHEMA, Intern. Meeting of Chem. Engng.
Moser, F. (1977). *Verfahrenstechnik*, 11, 670.
Mou, D.G., and Cooney, Ch.L. (1976). *Biotechnol. Bioeng.*, 18, 1371.
Mukataka, S. et al. (1980). *J. Ferm. Technol*, 58, 155.
Mukataka, S. et al. (1981). *J. Ferm. Technol.*, 59, 303.
Mulcahy, L.T., and La Motta, E.J. (1978). Rep. No. Env. E. 59-78-2, Dept. of Civil Engng., University of Massachusetts, Amherst.
Nagel, O., and Hegner, B. (1973). *Chem. Ing. Techn.*, 45, 913.
Nagel, O., Kürten, H., and Sinn, R. (1972). *Chem. Ing. Techn.*, 44, 367.
Nelböck, M., and Wandrey, C. (1978). In: *Biotechnologie.* Frankfurt: Umschau-Verlag. Bundesministerium f. Forschung & Technologie, Bonn, p. 81.
Ng, D.Y.C., and Rippin, D.W.T. (1964). *Proceedings of the Third European Symposium on Chemical Reaction Engineering Amsterdam.* Oxford: Pergamon Press, p. 161. (van Heerden C. et al., eds.)
Nikolaev, P.I., et al. (1976). *Theoret. Found. Chem. Eng.* (Transl.), 10, 13.
Norrman, G., et al. (1977). *Develop. Ind. Microbiol.*, 18, 581.
Norwood, K.W., and Metzner, A.B. (1960). AIChE Journal, 6, 432.
Ohyama, Y., and Endho, K. (1955). *Chem. Eng.* (*Jap.*), 19, 2.
Oosterhuis, N.M.G. (1984). Ph.D. Thesis, Technical University, Delft, Netherlands.
Paar H., et al. (1988). *Bioprocess Eng.*, 3, in press.
Parker, D.S., Kaufmann, W.J., and Jenkins, D. (1971). *J. Water Poll. Contr. Fed.*, 43, 1817.
Paspek, St. G., et al. (1980). *Chem. Eng. Educ.*, Spring, 78.
Pawlowski, J. (1962). *Chem. Ing. Techn.*, 34, 628.
Philipps, K.L. (1969). In Perlman, D. (ed.). *Fermentation Advances.* New York: Academic Press, p. 465.
Philipps, K.L., et al. (1961). *Ind. Eng. Chem.*, 53, 749.
Pickett, A.M., Topiwala, H.H., and Bazin, M.J. (1979). *Proc. Biochem.*, 13 (Nov., 10).
Pirt, S.J. (1974). *J. Appl. Chem. Biotechnol.*, 24, 415.
Pirt, S.J. (1980). Plenary Lecture, 6th Internation Fermentation Symposium London, Ontario. (Zajic J.E. et al., eds.) Publ. Sales & Distribution, National Research Council Canada (Ottawa).
Pirt, S.J., et al. (1983). *J. Chem. Techn. Biotechnol.*, 33B, 35.
Pitcher, W.H., Jr. (1978). *Adv. Biochem. Eng.*, 10, 1.
Popovic, M., Niebelschütz, H., and Reuss, M. (1979). *Eur. J. Appl. Microbiol. Biotechnol.*, 8, 1.
Powell, O., and Lowe, J.R. (1964). In Malek, I., (ed.). *Continuous Culture of Microorganisms.* Prague: Czechoslovakia Academy of Science, p. 45.

Prochazka J. and Landau J. (1961) Coll. Czech. Chem. Commun. 26, 2961.
Puhar, E., Karrer, D., Einsele, A., and Fiechter, A. (1978). In *1st European Congress on Biotechnology*, Interlaken, Switzerland, Part 2, 1/83. Dechema, Frankfurt.
Pye, E.K., and Wingard, L.B., Jr. (1974ff). *Enzyme Engineering*, Review Series. New York: Plenum Press.
Quicker, G., et al. (1981). *Biotechnol. Bioeng.*, 23, 635.
Rehm, H.J. (1980). *Technische Mikrobiologie*. Heidelberg–Berlin–New York: Springer-Verlag.
Rehm, H.J., and Reed, G. (eds.) (1982ff). *Biotechnology—A Comprehensive Treatise*, Vol. 1–9, esp. Vol. 2. Deerfield Beach, Fla., and Basel: Verlag Chemie Weinheim.
Reith, T. (1968). Ph.D. Dissertation, Technical University, Delft, Netherlands.
Reuss, M. (1976). In *Proceedings of the 5th International Fermentation Symposium*, Berlin. Dellweg H. (ed.), Inst. f. Gärungsgewerbe und Biotechnologie, Berlin.
Reuss, M. (1977). In *Proceedings of the Tutzing Symposium. Dechema Monograph*, 81, 45. Rehm H.J. (ed.) Verlag Chemie, Weinheim.
Reuss, M., and Bajpaj, R.K. (1988). In Ho, Ch., and Wang, D.I.C. (eds.). In *Biochemical Engineering: Bioreactor Design and Operation*. Butterworth. Stoneham MA., in press.
Reuss, M., and Wagner, F. (1972). In Dechema Monograph, 71, 9.
Richards, J.W. (1968). *Introduction to Industrial Sterilization*. London–New York: Academic Press.
Rippin, D. (1967). *Ind. Eng. Chem. Fundam.*, 6, 488.
Ruchti, G., et al. (1981). *Biotechnol. Bioeng.*, 23, 277.
Rudkin, J. (1967). *Br. Chem. Eng.*, 12, 1374.
Rushton, J.H., Costich, E.W., and Everett, H.J. (1950). *Chem. Eng. Progr.*, 46, 467.
Ryu, D., and Humphrey, A.E. (1972). *J. Ferm. Technol.*, 50, 424.
Sanders, W.M. (1966). *AirWater Poll. Int. J.*, 10, 253.
Sano, Y., Yamaguchi N., and Adachi, T. (1974). J. Chem. Eng. (Japan), 7, no. 4, 255
Schauerte, W.A. (1981). Ph.D. Thesis, Technical University, Munich.
Schmidt, K.G., and Blenke, H. (1983). *Verfahrenstechnik*, 17, 593.
Schneider G., Purgstaller A., Somitsch H. and Moser A. (1986). in Proc. Biochemical Engrg. Congress Stuttgart, (Chmiel H. et al., eds.) G. Fischer Verlag, p. 928.
Schneider H., and Moser, A. (1984). *Biotechnol. Lett.*, 6, 295.
Schneider, H. and Moser, A. (1985), ibid 7, 376.
Schreier, K. (1975). *Chemiker Zeit.*, 99, 328.
Schügerl, K. (1977). *Chem. Ing. Techn.*, 49, 605.
Schügerl, K. (1979). In Jottrand, R. (ed.). Ausbildungskurs über "Chemische Verfahrenstechnik in den biologischen Operationen" Brüssel, November. University Press Brussels.
Schügerl, K. (1980). *Chem. Ing. Techn.*, 52, 951.
Schugerl, K. (1982). *Adv. Biochem. Eng.*, 22, 94.
Schugerl, K., Lücke, J., Lehmann, J., and Wagner, F. (1978). *Adv. Biochem. Eng.*, 8, 63.
Schügerl, K., Oels, U., and Lücke, J. (1977). *Adv. Biochem. Eng.*, 7, 1.
Schumpe, A., et al. (1982). *Adv. Biochem. Eng.*, 24, 2.
Schumpe, A., et al. (1985). In Rehm, H.J., and Reed, G. (eds.). *Biotechnology*, Vol. 2, Deerfield Beach, Fla., and Basel: Verlag Chemie Weinheim, Chap. 10.
Sinclair, C.G. (1984). *Biotechnol. Lett.*, 6, 65.
Sinclair, C.G., and Brown, D.E. (1970). *Biotechnol. Bioeng.*, 12, 1001.
Sittig, W. (1977). *Fort. Verfahrenstechnik*, 15D, 354.

Sittig, W., and Heine, H. (1977). *Chem. Ing. Techn.*, 49, 595.
Sobotka, M., et al. (1982). In Tsao, G.T. (ed.). *Annual Report on Fermentation Processes.* New York: Academic Press.
Steiner, W., et al. (1977). In 4th FEMS Symp. (Abstr. B33). Federation of European Microbiological Societies, Bu'Lock, J., Meyrath, J. (eds.) University Bodenkultur Press.
Stenberg, O., (1984). Ph.D. Thesis, Technical University, Göteborg, Sweden. (Prof. N.H. Schöön).
Taguchi, H., and Humphrey, A.E. (1966). *J. Ferm. Technol.*, 44, 881.
Tanaka, H. et al. (1975a). *J. Ferm. Technol.*, 53, 18.
Tanaka, H. et al. (1975b). *J. Ferm. Technol.*, 53, 35.
Toda, K., and Dunn, I.J. (1982). *Biotechnol. Bioeng.*, 24, 651.
Tomlinson, T.G., and Snaddon, D.M. (1966). *Air Water Poll.*, 10, 865.
Trulear, M.G. (1980). MSc. Thesis, Rice University, Houston.
Tsai, B.I., Erickson, L.E., and Fan, L.T. (1969). *Biotechnol. Bioeng.*, 11, 181.
Tsai, B.I., Fan, L.T., Erickson, L.E., and Chen, M.S.K. (1971). *J. Appl. Chem. Biotechnol.*, 21, 307.
van de Vusse, J.G. (1955). *Chem. Eng. Sci.*, 4, 178.
van de Vusse, J.G. (1962). *Chem. Eng. Sci.*, 17, 507.
van Krevelen, D.W., and Hooftijzer, P.J. (1948). 21st Congress International de Chimie Industrielle Bruxelles, numero speciale, Chimie et Industrie. p. 168.
van Meyenburg, K., and Fiechter, A. (1968). *Biotechnol. Bioeng.*, 10, 535.
van Stockar, U., and Stravs, A.A. (1983). *Swiss Biotech.*, 6, 7.
van Uden, N. (1971). Z. *Allgem. Mikrobiol.*, 11, 6,541.
Venkatsubramanian, K. (ed.) (1979). *Immobilized Microbial Cells.* ACS Symposium Series 106. Washington, D.C.: American Chemical Society.
Volesky, B., et al. (1982). *J. Chem. Tech. Biotechnol.*, 32, 650.
Votruba, J., and Sobotka, M. (1976). *Biotechnol. Bioeng.*, 18, 1815.
Wandrey, Ch. (1983). In Proceeding "Biotech 83". International Conference on Commercial Applications & Implications of Biotechnology, Online Publ. Ltd., Northwood/UK, p. 577.
Wandrey, Ch., and Flaschel, E. (1979). *Adv. Biochem. Eng.*, 12, 148.
Wang, D.I.C., Cooney, Ch.L., Demain, A.L., Dunnill, P., Humphrey, A.E., and Lilly, M.D. (1979). *Fermentation and Enzyme Technology.* New York: John Wiley.
Wang, D.I.C., and Humphrey, A.E. (1969). *Chem. Eng.*, 76, 108.
Wang, H., et al. (1978). *Eur. J. Appl. Microbiol. Biotechnol.*, 5, 207.
Weinstein, H., and Adler, R.J. (1967). *Chem. Eng. Sci.*, 22, 65.
Wen, C.Y., and Fan, L.T. (1975). *Models for Flow Systems and Chemical Reactors.* New York: Marcel Dekker.
Werther, J. (1977). *Chem. Ing. Techn.*, 49, 777.
Werther, J. (1980). *Chem. Eng. Sci.*, 35, 372.
Zlokarnik, M. (1967). *Chem. Ing. Techn.*, 39, 539.
Zlokarnik, M. (1975). *Verfahrenstechnik*, 9, 442.
Zlokarnik, M. (1978). *Adv. Biochem. Eng.*, 8, 133.
Zlokarnik, M. (1979). *Adv. Biochem. Eng.*, 11, 157.
Ziegler, H., Meister, D., Dunn, I.J., Blanch, H.W., and Russel, T.W.F. (1977). *Biotechnol. Bioeng.*, 19, 507.
Ziegler, H. et al. (1980). *Biotechnol. Bioeng.*, 22, 1613.
Zwietering, Th.N. (1959). *Chem. Eng. Sci.*, 11, 1.

CHAPTER 4
Process Kinetic Analysis

4.1 Kinetic Analysis in Different Types of Reactors

A schematic diagram for a homogeneous biological process was shown in Fig. 2.3. For comparison, Fig. 4.1 shows a typical situation in a heterogeneous system; in it a rotating disk binds a microbiological film. In analogy to Fig. 2.3, all significant process variables in the three reaction phases (gas, liquid, and solid) are shown along with the limiting transport steps, in the order 1 to 4.

Mass transport between phases and the relative extent of each phase are important in deciding whether and if a homogeneous or heterogeneous model can be applied. In dealing both with microbiological particles and films, the variable characterizing the size has a distribution function, as discussed in Sect. 3.3.9: From the distribution of actual particle diameters one obtains $\bar{d}_P$, the mean diameter. Two extreme cases, of interest in process engineering, can be distinguished: *Uncontrolled thickness growth* will be found in all bioreactors in which shear forces vary greatly from place to place (stirred vessels). Uncontrolled growth is also normal in biofilm reactors, especially in trickling filters.

Controlled thickness growth of particles would be conceivable if a homogeneous shear force field could be established. Initiatives in this direction can be found in the literature (EPA, 1977; Moser, 1983; Schreier, 1975). The control of size can be better accomplished, however, in film-type bioreactors. This is done (a) using a mechanical scraping device of a particular height, as in the "biological film reactor" (Atkinson, 1974), (b) using mechanical methods involving surface friction to displace the biological film from the surface of the small particles in a fluidized bed reactor (completely mixed microbial film fermenter, Atkinson and Davies, 1972), or (c) using a method of controlling the hydrodynamic stress with the aid of peripherical speed of a rotating drum, which is the support for the growth of a particular biological strain (thin film fermenter, Moser, 1977b). The characteristics of these film reactors with controlled thickness makes them especially well suited for obtaining kinetic data for bioprocesses (Atkinson, 1974), as outlined in Sect. 3.6.

To illustrate the complexity of real fermentation, Fig. 4.2 shows a plot of

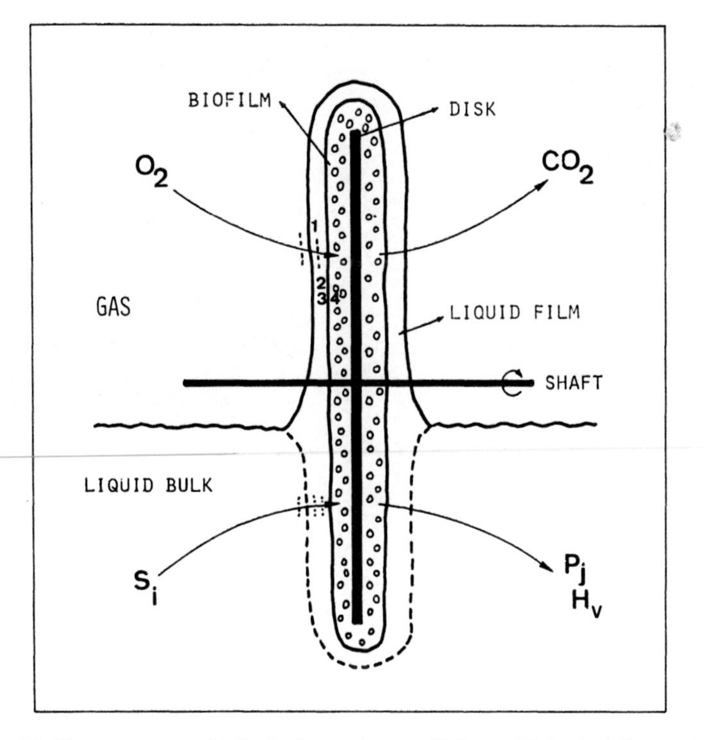

FIGURE 4.1. Heterogeneous biological processes utilizing a biological film (a biodisk reactor). A group of disks is fastened to a horizontal shaft. Each disk supports the growth of the desired culture. The disks are turned through the liquid and gas phases. Mass transport is subject to the same limitations as given in Fig. 2.3 for pseudo-homogeneous processes. With the biodisk, however, step 4 (limitation in the solid phase) becomes significant. (From Moser, 1981.)

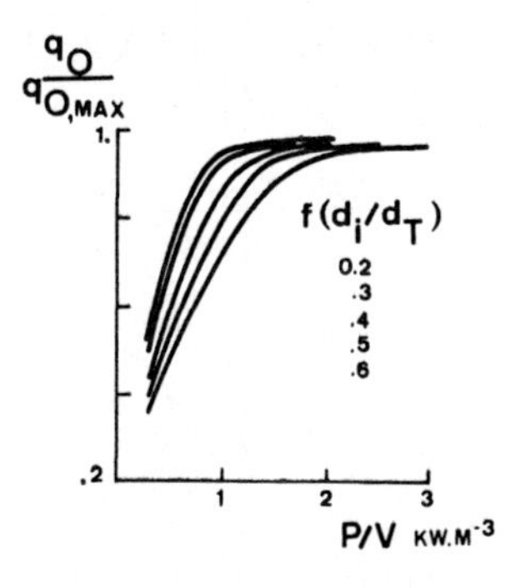

FIGURE 4.2. Simulations of oxygen consumption kinetics $q_O/q_{O,max}$ at different impeller/tank diameter ratios d_i/d_T as a function of power consumption P/V, as a typical example of macrokinetics in a 20 l vessel. (From Reuss et al., 1982.)

the respiration rate, q_O, versus the specific power input at varied impeller tank diameter ratio for a stirred rank reactor ($V_R = 20$ l) as observed for mycelial growth (Steel and Maxon, 1966) and later on simulated by Reuss et al. (1982). The appearance of these pseudokinetics is the consequence of interactions between physics (transport in a bioreactor influenced by geometry) and biol-

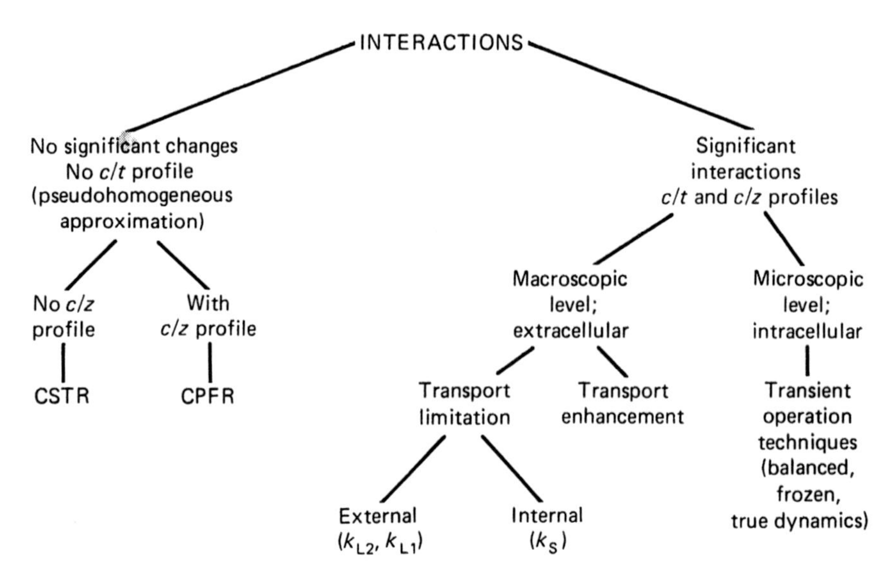

FIGURE 4.3. Classification of different types of interactions between microbial metabolism, respectively kinetics, and physical transport phenomena in the environment (bioreactor).

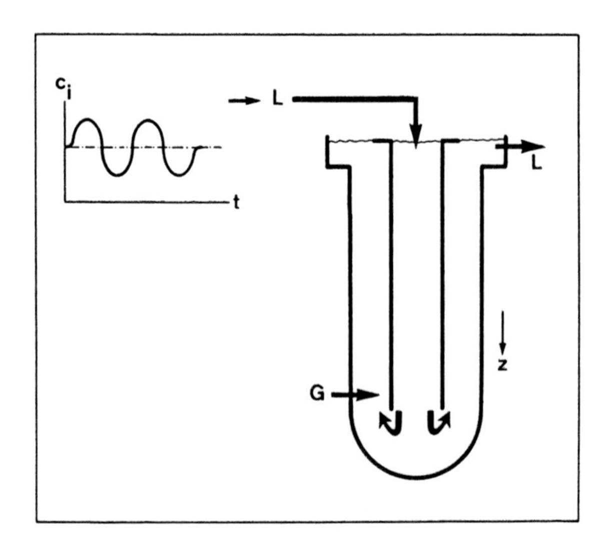

FIGURE 4.4. Schematic illustration of typical experimental situations in real technical scale bioprocessing: transient conditions due to environmental changes caused by gradients in concentrations, temperature, shear, and pressure in the reactor as well as in the inflowing liquid stream.

ogy. Figure 4.3 shows different types of interactions possible in bioprocessing. With the exception of conventional batch processes, we have thus far considered only bioreactors operating under steady-state conditions with no external perturbation. However, most industrial bioprocesses operate under nonstationary conditions, as illustrated in Fig. 4.4. Fluctuations in the inflowing stream (e.g., waste water pollution), gradients of concentrations of components (X, S, O, C, P, H_v), pH, shear, pressure, and so on over the length or depth of reactors are obvious, resulting in changes in the cell's environment and composition.

The strategy presented in the following section enables bioprocess engineers to approach such complex situations.

4.2 Regime Analysis—General Concept and Guidelines

By analogy to an interesting approach to the depiction of complex systems in thermodynamics (Prigogine and Defay, 1954), the concept of characteristic times of internal processes can be extended to the treatment of bioengineering systems (Bailey, 1973; Harder and Roels, 1982; Kossen, 1979; Moser, 1977b, 1981, 1982, 1984; Oosterhuis, 1984; Roels, 1982; 1983; Sweere et al. 1987). A systematic approach to problems involving interactions between the environment and cells seems to be possible for case 1, transport effects, including transport to and in measuring electrodes; and for case 2, transient effects of reactions, in which enzyme regulation phenomena such as induction and repression must be considered (cf. Fig. 4.9). A guideline for quantification and modeling can be summarized as follows (Moser, 1982):

1. Describe significant phenomena (see Table 4.1): case 1, transports: RTD, mixing, CTD, OTR, H_vTR, and so on, and/or case 2, metabolic rates: significant and compositional changes in cell content, for example, proteins, DNA, RNA, enzymes, macromolecules such as PHB, as internally balanced units together with macroscopic process variables.
2. Determine characteristic rate constants or characteristic times from experiments: case 1: mean residence time ($\bar{t}$), mixing time (t_m), cycle time (t_c), OTR time (t_{OTR}), and/or response time of measuring devices, for example, O_2 electrode (t_E); case 2: at macroscopic level, characteristic reaction times $t_{r,i}$, which are indirectly proportional to the specific rates ($\mu, q_S, q_O, q_P, q_C, q_H$); and at microscopic level, individual time responses for changes in DNA, RNA, protein enzymes, and so on (see Table 4.1).
3. Compare rate constants or, for significant phenomena, characteristic times of changes over the period of interest, and determine hierarchy of individual response time or rate constant ("regime"). Table 4.2 lists possible cases and results of regime analysis in bioreactor operation.

A conclusion from these considerations on mathematical modeling of biokinetics is that in both interaction cases (transport and metabolic changes),

TABLE 4.1. Characteristic values of k and t used in regime analyses of bioprocesses.

Macroscopic process variable	$k_{charact}$	$t_{charact}$	*Typical range* [s]
Kinetics (k_r resp. t_r)			
X	μ_{max}	$t_\mu = 1/\mu_{max}$	$1 \cdot 10^5$–$1 \cdot 10^4$
S	$q_{s,max}$	t_{qs}	$1 \cdot 10^4$–$5 \cdot 10^4$
O	$q_{o,max}$	t_{qo}	0·1–20
P	$q_{p,max}$	t_{qp}	—
C	$q_{c,max}$	t_{qc}	—
H_v	$q_{H_v,max}$	$t_{qH_v} = \dfrac{\bar{c}_P \cdot \rho \cdot \Delta T}{(r_{H_v})_r + (r_{H_v})_{TR}}$	—
Cell compartments (internal compounds)	q_{int}	$t_{q_{int}}$	—
Transports (k_{TR} resp. t_{TR})			
Micromixing time		t_m as $f(m)$	10–$3 \cdot 10^2$
Circulation time		t_c	1–30
Circulation time distribution		$\bar{t}_c s^2$	—
Mean residence time		$\bar{t}_L = V_L/F_L$	—
OTR global	$k_{L1}a$	$t_{OTR} = 1/k_{L1}a$	5–10
OTR gas bubble		$t_{OTR}^B = He/k_{L1}a$	$3 \cdot 10^2$–$6 \cdot 10^2$
gas residence time		$\bar{t}_G = (1 - \varepsilon_G)\dfrac{V_L}{F_G}$	10–20
Heat transfer	$k_{H_v}a_{H_v}$	$t_{H_v TR} = \dfrac{\bar{c}_P \rho}{k_{H_v} \cdot a_{H_v}}$	$3 \cdot 10^2$–$8 \cdot 10^2$
Electrode response	k_E	$t_E = 1/k_E$	
Substrate addition		$t_{S,A} = V \cdot S/F_S \cdot S_0$	10^1–10^2
Concept of effectivity factors	$\eta_r = f(k_r/k_{TR})$		
Transport limitation			
Transport enhancement	$\eta_{Tr} = f(k_{Tr}/k_R)$		

a drastic reduction of model complexity can be achieved with the aid of the rds, respectively qss, principle.

Figure 4.5 provides an overview of the range of characteristic times of significant phenomena in bioprocessing, including physical transport and biological reactions on the macroscopic and microscopic level and indicating regions where the organism and environment interact (Roels, 1983). Interesting papers on this subject can be found in the literature (Bailey, 1973; Bazine, 1982; Harder and Roels 1982; Kossen, 1979; Pickett et al. 1979; Roels, 1982; Romanovsky et al., 1974; Sweere et al., 1987).

In cases involving complex situations, simple unstructured models are often insufficient. Both bioreactor models and biological reaction models, therefore, must be structured to some degree. Figure 4.6 is a graphical representation of a two-compartment model for oxygen transfer in a stirred tank reactor, where the liquid phases are structured into mixed and segregated (bubble) zones (connected by a pumping capacity), while the gas phase is assumed to be well mixed. A quite simple model for OTR is shown in Fig. 4.7, where the liquid phase is separated into a stirrer zone with high hydrodynamic stress and a zone of bulk flow (Oosterhuis, 1984; see also Sect. 6.9.4). A

TABLE 4.2. Consequences of comparison of characteristic values in applying regime analysis to bioprocessing.

Comparison	*Consequence*
$\bar{t}_L \gg t_m$	Completely mixed reactor (L phase)
$t_m < t_{r,i}$	Pseudohomogeneous reactor model
$t_c < t_{r,i}$	Pseudohomogeneous reactor model
$t_{OTR} < t_{qo}$	No O_2 limitation
$t_{OTR} \sim t_{qo}$	O_2 limitation (O_2 gradients)
$t_{OTR}^B < \bar{t}_G$	Gas bubbles not depleted
$t_{H,TR} \sim t_{qH_v}$	Isotherm process
$t_c < t_{qH_v}$	No T gradient
$t_E < t_{OTR}$	No electrode effect on OTR
$\eta_r = 1\ (k_r < k_{TR})$	Kinetic regime
$k_r > k_{L1}$	External transport limitation (G|L)
$k_r > k_{L2}$	External transport limitation (L|S)
$k_r > k_s$	Internal transport limitation (S phase)
$\eta_{Tr} < 1\ (k_r > k_{TR})$	Diffusion transport regime
$\eta_{Tr} > 1$	Transport enhancement
$t_e \gg t_r$	"Balance" system (see Sect. 5.7)
$t_e \ll t_r$	"frozen" biosystem (see Sect. 5.7)
$t_e \sim t_r$	Resonance, thus true dynamics
$t_{qs} \sim t_{S,A}$	S-limited growth
$t_m \sim t_{S,A}$	S gradients in reactor
$t_m \sim t_{OTR}$	O_2 gradients in reactor

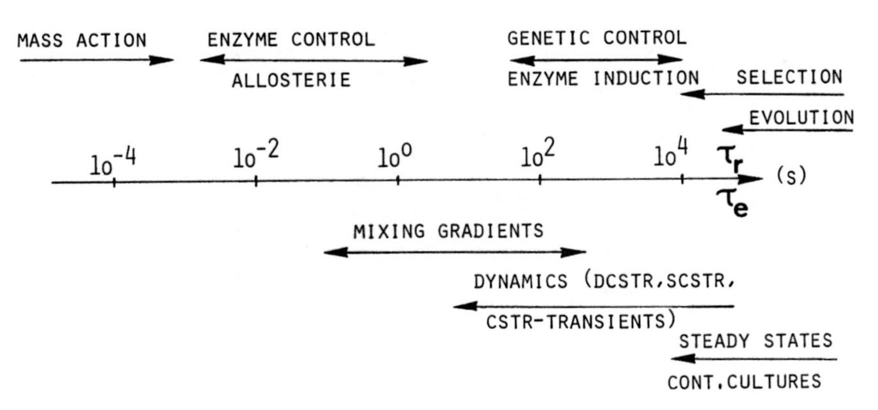

FIGURE 4.5. Comparison of the relaxation times (characteristic response times, see Table 4.2), of the reaction mechanisms inside organisms (τ_r) and those of characteristic times of environment in bioreactors (τ_e), according to Roels (1983).

more structured (two-compartment) biokinetic model is shown in Fig. 4.8 (Williams, 1967, 1975): a synthetic compartment of biomass x_K, which is mainly RNA, and a genetic–structural compartment, x_G, which consists mainly of proteins and DNA. Models with more than three compartments for cells are seldom needed (Harder and Roels, 1982; see also Sect. 5.7.2). Mathe-

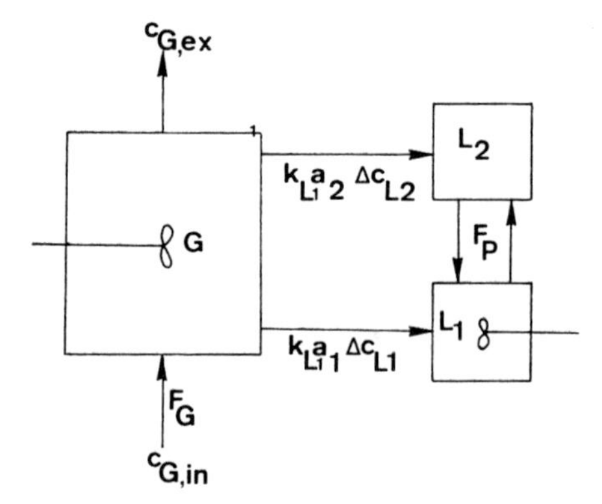

FIGURE 4.6. Scheme of a two-compartment model for oxygen transfer in a stirred tank. The liquid phase is structured into a mixed zone (L_1) and into a zone with freely rising bubbles in analogy to a bubble column (L_2), with a connecting exchange (F_P) due to impeller pumping. (From Oosterhuis, 1984.)

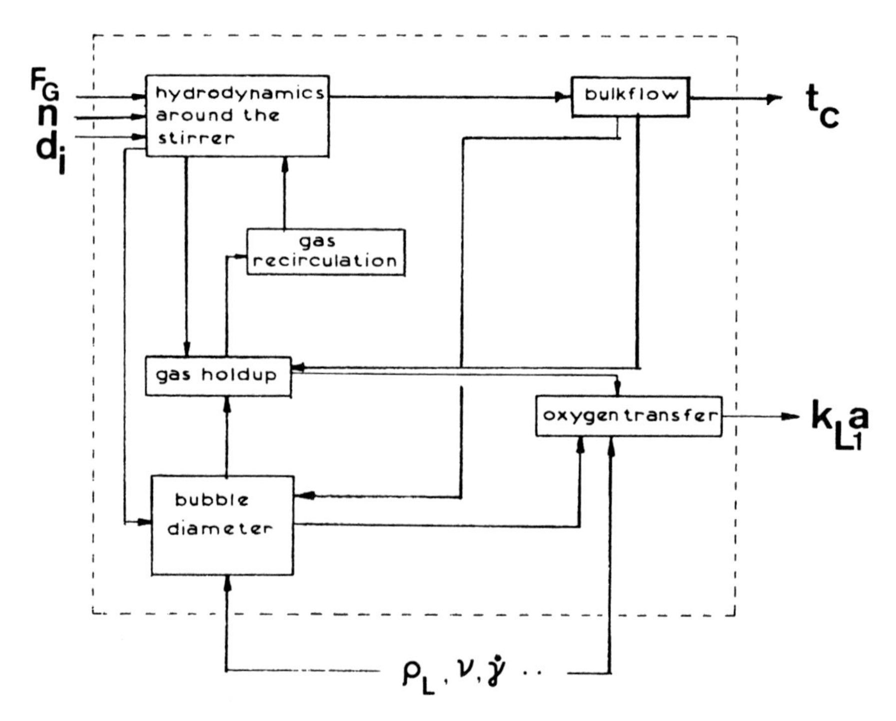

FIGURE 4.7. A somewhat structured model for oxygen transfer in a stirred-tank reactor in which only the hydrodynamics around the stirrer are taken into account. (See Oosterhuis, 1984.)

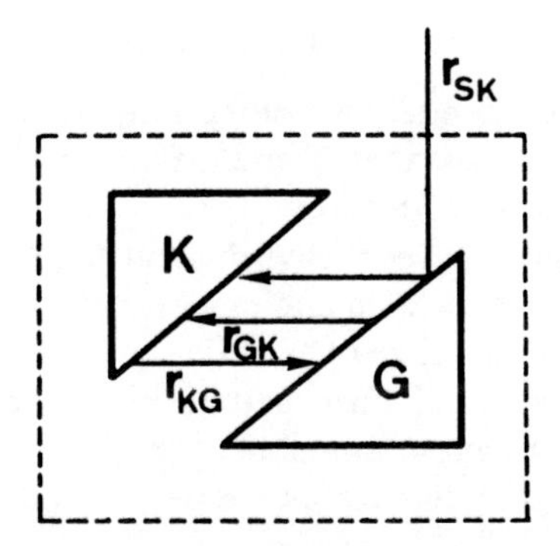

FIGURE 4.8. Block diagram of a two-compartment model (K and G) of cells following the concept of Williams (1967,1975). The rates are r_{SK} = rate of conversion of substrate to K compartment, r_{KG} = rate of conversion of K to G compartment, and r_{GK} = rate of depolymerization of G to K compartment. (From Williams (1967) and Harder and Roels (1982).)

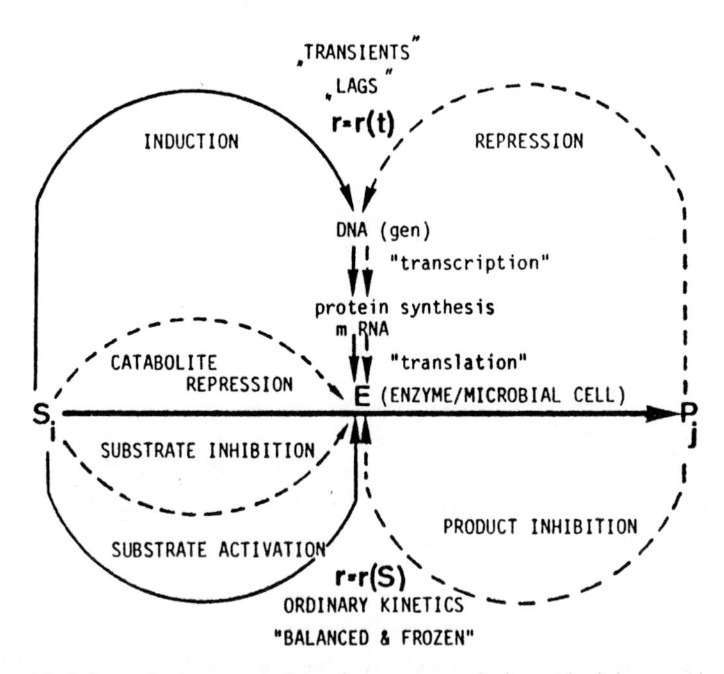

FIGURE 4.9. Schematic representation of enzyme regulation principles as a basis for biokinetic model building. While ordinary kinetics are analogous to chemical kinetics (heterogeneous catalysis), enzyme induction and repression are typical biological phenomena creating transients. (From Moser, 1984.)

matical models of biokinetics, thus, must include both ordinary kinetics (substrate activation and inhibitions, in analogy to chemical kinetics) and transients (induction and repression of enzymes), as shown in Fig. 4.9. (see also Sect. 5.2.2.2).

For process kinetic analysis, the following rules can be given:

1. The validity of experimentally measured kinetics on any level is restricted to the reactor type investigated. These macrokinetics should not be extrapolated to another scale or reactor.
2. Because interactions between environment and metabolism are often poorly understood, kinetic data and models should be transferred to other reactors only within the range of "kinetic similarity."
3. To some extent, the simple unstructured kinetic models can be used for describing complex bioprocessing if the parameters are adapted to process behavior by formulating them in a time-dependent way [e.g., μ_{max}, K_s, and $Y = f(t)$].
4. Simple unstructured models, e.g. Esener et al. (1983), can be successfully used to describe macroscopic behavior if the model is modified to include additional phenomena such as S inhibition, 2-S limitations, a lag phase, a stationary phase, or a dead phase. Adaptive modeling is essential (Moser, 1981). This applies in all cases of transients, in balanced, frozen, and real transient situations.
5. Structured modeling is needed only in cases of real transient processing with $\tau_e \sim \tau_r$, or in cases where the purpose of the investigation is not engineering calculations of conversion but rather a microscopic description of metabolic mechanisms (Toda, 1981). Such models should be of more interest for engineering when better process control can be realized through development of more specific sensors. The reactivity of different strains is waiting to be investigated.

4.3 Test of Pseudohomogeneity

A systematic analysis of bioprocesses must take into account the interaction of biology and physics. With so-called "perfect reactors," pseudohomogeneity is assumed, but this approximation should be carefully checked in advance.

Pseudohomogeneity involves quantitative tests to ensure that this condition is fulfilled. These tests, along with their quantitative criteria, are presented in Table 4.3 for transport steps 1 through 4.

As previously indicated, the criteria of mixing time, circulation time, and CTD should be connected in a logical way with a biological parameter to permit a biologically relevant conclusion. The comparison of a "reaction time," t_r, with a "mixing time" under conditions of a biologically determined degree of mixing presents one poor possibility. A reaction time for a biotechnological process is best defined on the basis of oxygen consumption (see Table 4.1): This is one of the most rapid and easily measurable reactions (see Fig. 3.16 and Bryant, 1977; Moser, 1977c).

A quantitative test to establish that k_{L2} at the liquid–solid interface is not a limiting factor is difficult: In general, one assumes that, with good mixing in the liquid phase, the turbulence also has an effect on the L film at the L|S

TABLE 4.3. Quantitative tests for Pseudohomogeneity of bioprocesses as a case of three-phase systems.

Problems	*In the case of*	*Criteria*
1. G|L (k_{L1})	High soluble gases	k_G
η_{Tr}	Aerobic processes	$k_{L1} \cdot a(o^* - o) \geq q_0 \cdot x$ (gas dynamics!)
	$d \ll \delta_{L1}$, flocs only	$0.3 \leq \text{Ha} \leq 1 \rightarrow \eta_{Tr} > 1$ with $\text{Ha} = \frac{1}{k_{L1}}\left(\frac{2}{n+1} \cdot k_r \cdot c^{*n-1} \cdot D\right)^{1/2}$
2. L phase	DCSTR and CSTR	$t_m \leq 0.1\ t_r$
Micromixing	Recycle reactors	$t_c \leq t_r$
Macromixing	CSTR	$N \rightarrow 1$
	CPFR	$\text{Bo} \rightarrow \infty$
3. L|S (k_{L2})	Flocs	Calculate (analogy to chemical processes) $\text{Sh}_{L2} = 2 + 0.4\ \text{Re}_\varepsilon{}^{1/4} \cdot \text{Sc}^{1/3}$
	Films	$k_{L2} \sim v_L{}^{0.7}$
4. S phase (D_s)	Flocs and films	$d \leq d_{crit}$
	1-S limited	Calculate $d_{crit} = \frac{c^*}{K_S} \cdot \frac{(1 + 2 \cdot c^*/K_S)^{1/2}}{1 + c^*/K_S} \cdot \left(\frac{Y \cdot D_S \cdot K_S}{\sigma_{max} \cdot \rho}\right)^{1/2}$

From Moser, 1980b.

interface. Despite this, calculations of k_{L2} can be made using either the theory of "local isotropic turbulence" (Kolmogoroff, 1941) or the theory of "relative velocities" between the L and S phases (Frössling, 1938). In essence, the Kolmogoroff theory gives equal values of k_{L2} for equal power used in mixing; the theory of relative velocities assumes that the k_{L2} value is that which reflects the proper terminal velocity. Sano, Yamaguchi, and Adachi (1974) presented a comparison of the most important literature on the L|S coefficient k_{L2}. They showed the apparent existence of a uniform correlation between Sh_{L2} and a Reynolds number Re_ε based on the energy dissipation ε (ε = mean energy distribution or power utilization per unit mass [cm^2/sec^3]). This is relevant to stirred vessels (ST) and also bubble columns (BC):

$$\text{Sh}_{L2} = (2 + 0.4 \cdot \text{Re}_\varepsilon^{1/4} \cdot \text{Sc}^{1/3}) \cdot \phi_c \tag{4.1a}$$

with

$$\text{Re}_\varepsilon = \frac{\varepsilon \cdot d_p^4}{v_L^3} \tag{4.1b}$$

and

$$\varepsilon_{ST} = \frac{P \cdot g_c}{V_L \cdot \rho_L} = \frac{N_p \cdot d_i^5 \cdot n^3}{V_L} \tag{4.1c}$$

or

$$\varepsilon_{BC} = \frac{g_c \cdot \Delta p \cdot F_G}{V_L \cdot \rho_L} = v_{SG} \cdot g \tag{4.1d}$$

The factor ϕ_c is, after Carman, the surface factor, which is calculated from the particle thickness, grain diameter, and specific particle surface area. With this, an approximate value for k_{L2} can be estimated using Equ. 4.1a. For biological films, the dependence of k_{L2} on the flow velocity has been found to be 0.7 (La Motta, 1976a). A preliminary estimate can be made of the relative importance of the individual transport steps in G|L|S process, using the concept of effectiveness factors η (cf. Sect. 4.5 and Equ. 2.50).

Sylvester, Kulkarni, and Carberry (1975) introduced a strategy in which a judicious estimate of certain physical parameters permits estimation of the intrinsic reaction rate coefficients in G|L|S processing. An overall effectiveness factor $\tilde{\eta}$ is defined as

$$\tilde{\eta} = \frac{1}{1 + \mathrm{Da}_0} \tag{4.2}$$

with

$$\mathrm{Da}_0 = \frac{\eta_{\mathrm{L|S}} \cdot k_r \cdot \eta_r \cdot c_S}{k_{L1} \cdot a} \tag{4.3}$$

where η_r is the intraparticle effectiveness factor (cf. Sect. 4.5.2) and $\eta_{\mathrm{L|S}}$ is the liquid particle effectiveness factor given by

$$\eta_{\mathrm{L|S}} = \frac{1}{1 + \mathrm{Da}_{\mathrm{L|S}}} \tag{4.4}$$

with

$$\mathrm{Da}_{\mathrm{L|S}} = \frac{k_r \cdot \eta_r \cdot c_S}{k_{L2} a_S} \tag{4.5}$$

Also a G|L effectiveness factor $\eta_{\mathrm{G|L}}$ can be defined

$$\eta_{\mathrm{G|L}} = \frac{1}{1 + \mathrm{Da}_{\mathrm{G|L}}} \tag{4.6}$$

with

$$\mathrm{Da}_{\mathrm{G|L}} = \frac{k_{L1}}{k_G} \tag{4.7}$$

Substituting these equations into the full equation for the effective rate in the S phase for a three-phase heterogeneous bioprocess at steady state

$$r^{s}_{\mathrm{eff}} = c^* \frac{1}{(1/k_{L1} a_L) + t_m + (1/k_{L2} a_S) + (1/k^{s}_{r,i})} \tag{4.8}$$

which results in the following equations for the total effective rate constant

$$k_{\mathrm{eff}} = \frac{\eta_{\mathrm{L|S}} \cdot \eta_r \cdot k_r \cdot c_S}{1 + \eta_{\mathrm{L|S}} \cdot \eta_r \cdot k_r \cdot (c_S / k_{L1} a)} \tag{4.9}$$

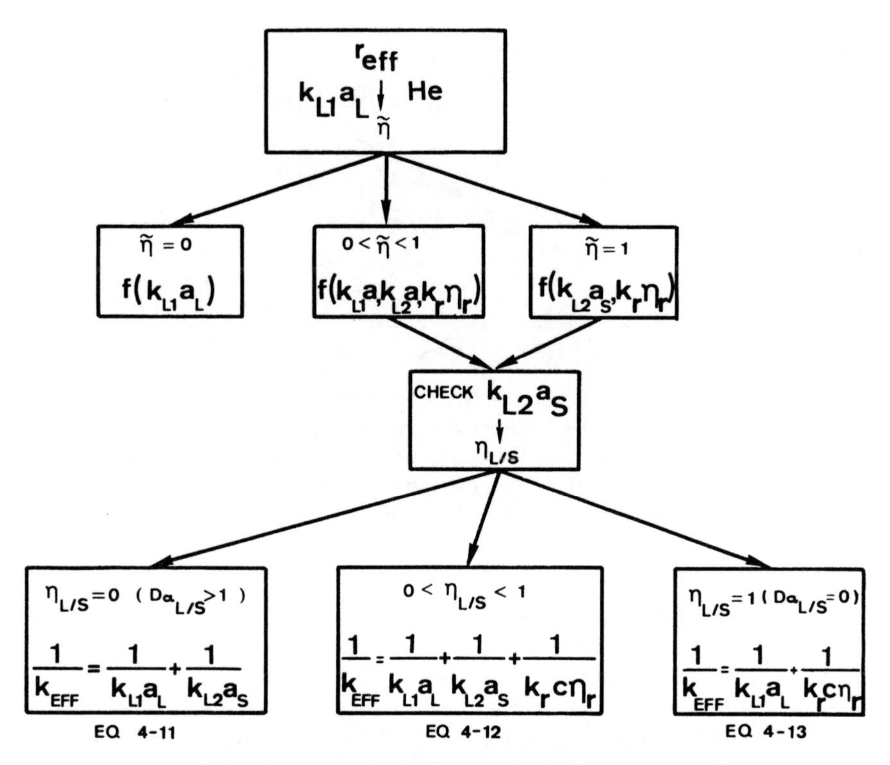

FIGURE 4.10. Flowchart of procedure for checking different transport limitations, according to a strategy presented by Sylvester et al. (1975). For explanation see text and Equs. 4.2 through 4.10.

and

$$k_{\mathrm{eff}} = \eta_{\mathrm{L|S}} \cdot \mathrm{Da}_{\mathrm{L|S}} \cdot \tilde{\eta} \cdot k_{\mathrm{L2}} \cdot a_{\mathrm{S}} \tag{4.10}$$

Figure 4.10 summarizes the procedure of intrinsic kinetic parameter estimation. The following steps are included:

1. $\tilde{\eta}$ can be checked using Equ. 4.2, estimating $k_{\mathrm{L1}}a$ and η_{GL} from an experimental value of r_{eff} and a known value of He.
2. $\eta_{\mathrm{L|S}}$ must then be checked on the basis of above data using known correlations for $k_{\mathrm{L2}} \cdot a_{\mathrm{S}}$ (cf. Equ. 4.1a).
3. With known $\eta_{\mathrm{L|S}}$, $\eta_{\mathrm{r}} \cdot k_{\mathrm{r}}$ can be calculated from $\mathrm{Da}_{\mathrm{L|S}}$ using Equ. 4.5, and then η_{r} from the plot η_{r} versus Thiele modulus ϕ (cf. Fig. 4.36).

As shown in Fig. 4.10, finally, the effective rate of individual experiments can be interpreted as a function of different types of interactions between the reaction and external or internal transport (cf. Equ. 4.11–4.13 in Fig. 4.10).

In principle, internal transport limitation, can be checked by measuring the characteristic dimension of the biocatalytic mass and comparing it with

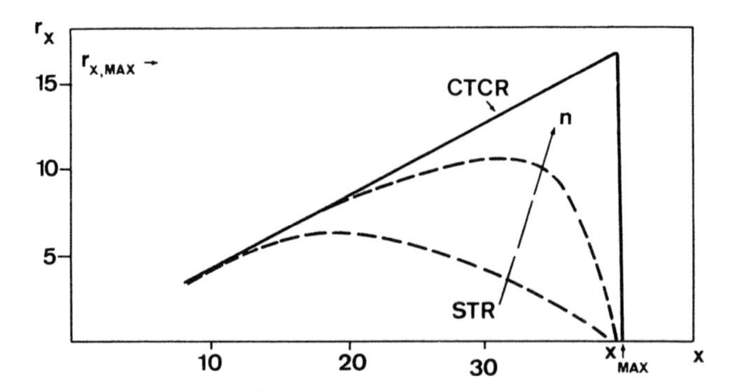

FIGURE 4.11. Schematic illustration of the influence of mixing phenomena on microbial growth behavior creating macrokinetics: Operation conditions, for example, stirrer speed n in stirred tank reactors (STR), or reactor construction, as in a cycle tube cyclone reactor (CTCR) (see Ringpfeil, 1980), affect the plot of growth rate r_x versus biomass concentration x drastically, especially when x values are high.

the critical diameter, d_{crit}, which can be found in the literature (for example, Atkinson and Daould, 1970; Kornegay and Andrews, 1968). The values vary, but they are typically about 0.1 mm. The application of this concept of biofilm kinetics to biological flocs is described in the literature (Atkinson and Ur-Rahman, 1979). Experimental data are lacking for various types of particles and films (Wuhrmann, 1963). Atkinson and Knight (1975) gave an estimate for the ideal or critical film thickness for the 1-S limitation; the value of K_S and the effective diffusion coefficient in the solid phase are particularly important factors in this calculation. The same group (Howell and Atkinson, 1976) also proposed an extension of the method to deal with cases of 2-S limitation.

No doubt additional effort is needed in the kinetic analysis of a process to optimize a laboratory-scale bioreactor using the criteria listed in Table 4.3. The results, however, should justify the effort. Figure 4.11 illustrates this problem showing that biokinetics seems to be dependent from the type of bioreactor. Productivity is plotted for a stirred tank with increasing rotational speed n and is compared with that in the cycle tube cyclone reactor (Ringpfeil, 1980).

Finally, it should be mentioned that experimental measurements of process variables are made almost exclusively in the liquid phase. The assumption is that there is at least equilibrium, if not identity, between concentrations in the L and the S state. It has been shown that the correlation between external and internal fluxes is poor (Barford and Hall, 1979).

An overview of the principal mathematical techniques for process kinetics is given in Fig. 4.12. With real processes, the concentration terms of type 1 and type 2 are coupled with transport terms in or between fluid (G|L) and solid phases, leading to types 4 and 5. This scheme is equally applicable to chemical and biological reactions: Analogies to chemical kinetics often

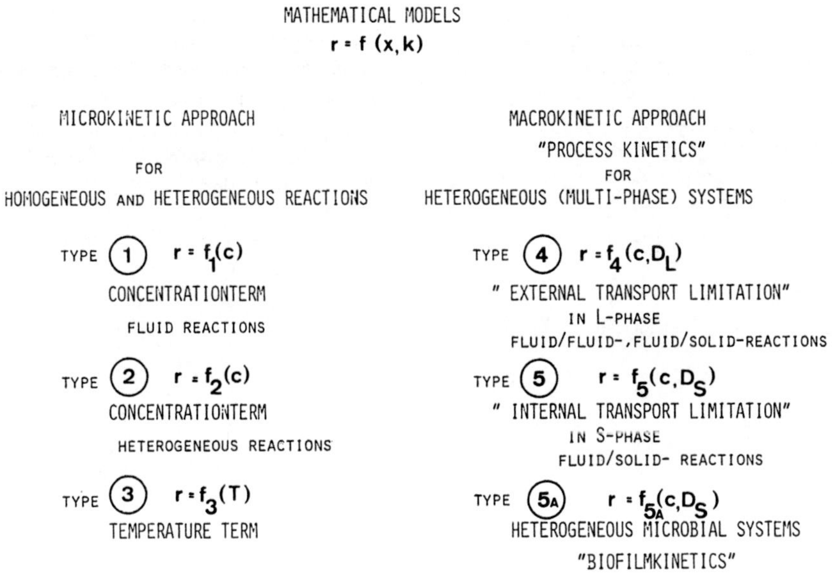

MATHEMATICAL MODELS
$r = f(x, k)$

MICROKINETIC APPROACH
FOR
HOMOGENEOUS AND HETEROGENEOUS REACTIONS

TYPE (1) $r = f_1(c)$
CONCENTRATIONTERM
FLUID REACTIONS

TYPE (2) $r = f_2(c)$
CONCENTRATIONTERM
HETEROGENEOUS REACTIONS

TYPE (3) $r = f_3(T)$
TEMPERATURE TERM

MACROKINETIC APPROACH
"PROCESS KINETICS"
FOR
HETEROGENEOUS (MULTI-PHASE) SYSTEMS

TYPE (4) $r = f_4(c, D_L)$
" EXTERNAL TRANSPORT LIMITATION"
IN L-PHASE
FLUID/FLUID-,FLUID/SOLID-REACTIONS

TYPE (5) $r = f_5(c, D_S)$
" INTERNAL TRANSPORT LIMITATION"
IN S-PHASE
FLUID/SOLID- REACTIONS

TYPE (5A) $r = f_{5A}(c, D_S)$
HETEROGENEOUS MICROBIAL SYSTEMS
"BIOFILMKINETICS"

FIGURE 4.12. Classification of mathematical models of kinetics (types 1–5) and their areas of application. (From Moser, 1981.)

provide useful problem-solving methods. The description in Fig. 4.12 results from the use of the expression "homogeneous reaction" for reactions in the G or L phases, and "heterogeneous reaction" for reactions at a solid surface (internal or external).

4.4 Parameter Estimation of Kinetic Models with Bioreactors

Under the assumption of pseudohomogeneity (cf. Table 4.3) and under constant temperature conditions, the kinetic analysis of a process is reduced to the search for functions $f_1(c)$ and/or $f_2(c)$, as specified under types 1 and 2 in Fig. 4.12. The distinctions between reactors with or without a concentration profile is also a decisive factor; that is, the distinction between so-called integral and differential reactors is a necessary one.

4.4.1 Integral and Differential Reactors

As an aid in the precise definition of integral and differential reactors, Fig. 4.13 shows a conversion versus time diagram, and the transfer of these data to a continuous tubular reactor (Moser and Lafferty, 1976).

A *differential reactor* works with small concentration differences, for exam-

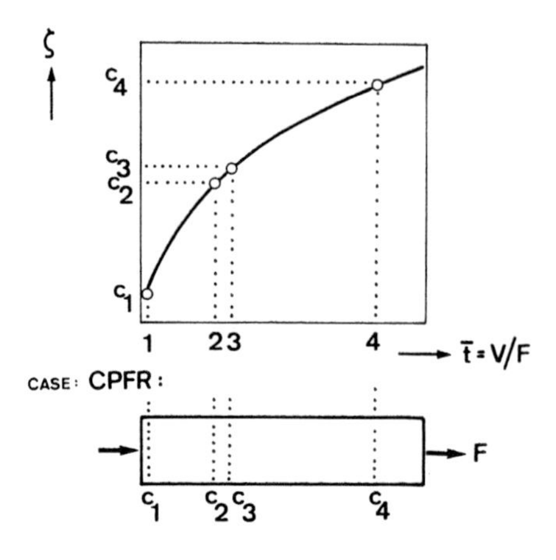

FIGURE 4.13. The principles of integral and differential reactors as shown by a conversion/time plot (ζ/t), and the transfer of the corresponding concentration data to its position along the length of a continuous tubular reactor. (From Moser and Lafferty, 1976.)

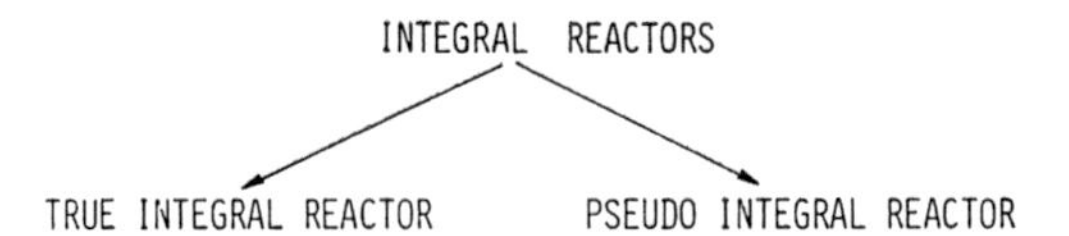

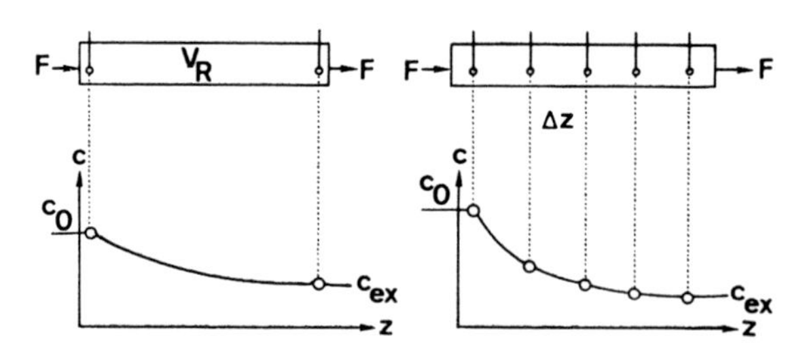

FIGURE 4.14. Integral reactors: classification and concentration profile for a continuous tubular reactor.

ple $c_3 - c_2$, and is usually defined such that the conversion ζ is small (rule: $\zeta < 10\%$). An *integral reactor* is one in which the conversion is large (rule: $\zeta > 50\%$) and in which large concentration differences are found, for example, $c_4 - c_1$. A closer look at the behavior of these two reactor types is given in Figs. 4.14 and 4.15.

For the integral reactor

$$\bar{t} = -\int_{c_0}^{c} \frac{1}{r} dc \equiv c_0 \int_{0}^{\zeta_{ex}} \frac{1}{r} d\zeta \tag{4.14}$$

follows from Equs. 2.3d and 3.94. The conversion data in an integral reactor represent the integral value of the reaction rate operating in each volume element of the reactor with the hypothetical concentration profile shown in

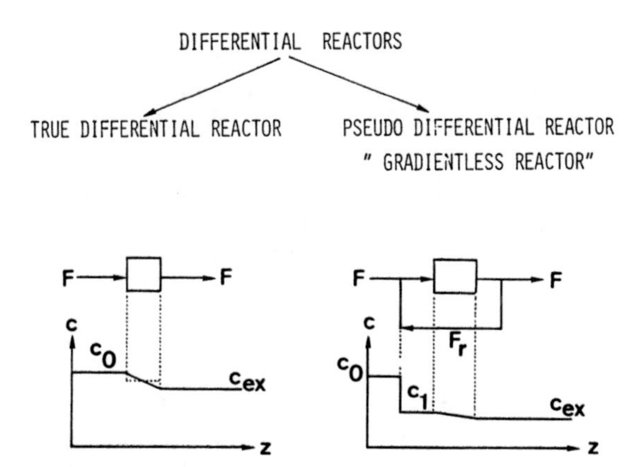

FIGURE 4.15. Differential reactors: concentration profiles for a differential reactor and for a gradient-free (recycle) reactor.

Fig. 4.14. A special type of integral reactor (pseudo-integral reactor) is one constructed with taps at various distances along the length so that samples may be removed and the actual concentration profile measured. A disadvantage of integral reactors is that the balance equations are a system of coupled differential equations. The measured conversion often is due to a complex interaction of transport and reaction processes. For quick, empirical, and pragmatic process development, the integral reactor may be well suited, especially now that fast digital computers and effective integration algorithms facilitate parameterization.

Precise kinetic measurements are, however, better (made) using a differentially operated reactor. A small measured change in conversion may be directly related to a reaction rate

$$r \approx \frac{c_0 - \bar{c}}{\bar{t}} = \frac{\zeta_{ex} - \zeta_0}{\bar{t}} \tag{4.15}$$

and transport influences or temperature gradients are of negligible consequence. This approximation is permissible when the reactor volume is small enough or when the amount of material flowing through a continuous tubular reactor is sufficiently large. Under such a circumstance, the reactor may be described by a system of algebraic equations, and this facilitates evaluating the experimental measurements. For a complete series of measurements, the reaction rate must be considered dependent on the amount of product already present (see Fig. 4.13). A differential reactor must be primed with different starting mixtures; this is most conveniently done by inserting an integral reactor in front of the differential reactor. The high degree of precision required of the analytical methods used to measure small differences in conversion is another problem with differential reactors.

The analytical problems with differential reactors can be overcome by using so-called gradient-free reactors; for example, loop reactors (Fig. 4.15). In this method, after conversion in a differential reactor, a portion of the reaction mixture is recirculated (F_r) and mixed with fresh input (F) before being reintroduced into the reactor (V_R).

When $F_r \gg F$, the concentration difference ($c_1 - \bar{c}$) between the actual reactor input and output is small, although the concentration difference between the fresh input and the output ($c_0 - \bar{c}$) is large. This minimizes the analytical difficulties ($c_{in} = c_0$ and $c_{ex} = \bar{c}$), and

$$r = \frac{(F + F_r)(c_1 - \bar{c})}{V_R} \equiv \frac{F(c_0 - \bar{c})}{V_R} = \frac{c_0 - \bar{c}}{\bar{t}} \tag{4.16}$$

The conversion per reaction cycle is a differential amount, but it is at the high level of conversion typical of an integral reactor. The loop reactor is therefore referred to as a "pseudo-differential reactor" or "gradient-free reactor." The identity indicated in Equ. 4.16 can be more easily recognized when the two parts that result from the mass equivalence of a recycling reactor (left side) and set equal to that of an ideal stirred vessel (right side), and the resulting equation is then solved for c_1:

$$c_1 = \frac{F_r \cdot c_{ex} - F \cdot c_{in}}{F + F_r} \tag{4.17}$$

Only for $F_r \gg F$ does $c_1 \to \bar{c}$ (Hoffmann, 1975). At $F_r \geq 100 \cdot F$, the recycling reactor is kinetically equivalent to the CSTR (Luft and Herbertz, 1969), which is the type most often used for the kinetic analysis of homogeneous, liquid phase reactions. The recycling reactor is most often used for kinetic measurements involving a solid phase (heterogeneous catalysis); due to the recycled stream there is high fluid velocity, which results in good transport properties at the L|S interface (see Equ. 4.1a). The recycling reactor combines the advantages of differential reactors (specifically, that the reaction rate can be obtained directly from experimental data and related directly to conditions and concentrations) with the advantages of integral reactors (large differences in input and output concentrations that are easily measured). The different reactor behaviors require individually adapted methods for evaluating the kinetic data, and these are now to be discussed.

4.4.2. Integral and Differential Reactor Data Evaluation Methods

The connection between integral and differential reactors and that between integral and differential methods of data evaluations are shown in Fig. 4.16 (after Froment, 1975). Data from integral reactors can be evaluated in the same way as data from differential reactors if the data are first numerically differentiated, or differentiated analytically or, more often, graphically. In cases where integral data are to be evaluated differentially, the following steps should be followed:

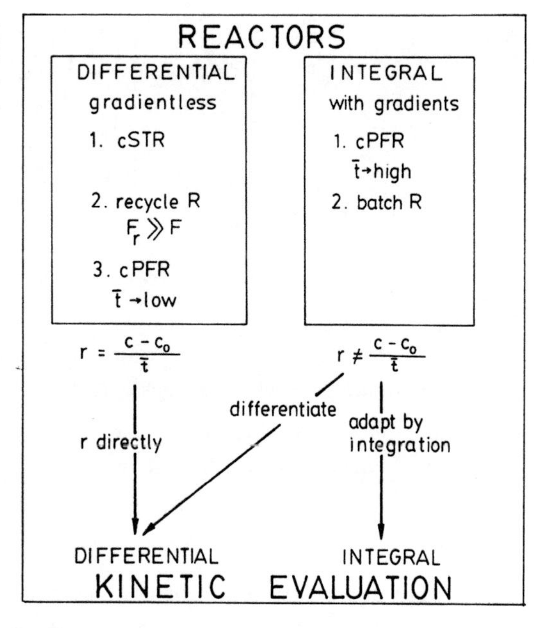

FIGURE 4.16. Relationship between differential and integral reactors, and between differential and integral methods for evaluating kinetic parameters.

TABLE 4.4. General definition of rates in different reactor operation modes.

Reactor operation (T = *constant*)	*Profile*	*Rate*, r_i	*Equ. No.*
DCSTR (V = constant)	$f(t)$	$r_i = \frac{dc_i}{dt}$	2.3d (2.4a–c)
DCSTR (V = variable), SCSTR	$f(t)$	$r_i = \frac{dN}{V(t) \cdot dt}$	$V(t) = V_0 + F(t) \cdot t$ 3.86
DCSTR + recycle	$f(t)$	$r_i = \frac{(V_R + V_r)dc_i}{V_R \cdot dt}$	3.100; if $V_r \ll V_R$ then 2.3d
CPFR	$f(z)$	$r_i = \bar{v}_z \frac{dc_i}{dz}$	3.94
CSTR (V = constant)	—	$r_i = \frac{F}{V}(c_{ex} - c_{in})$	3.87
CSTR + recycle	—	$r_i = \frac{(F + F_r)}{V}(c_{ex} - c_{in})$	3.101; if $F_r \gg F$ then 3.87

1. Draw the concentration versus time curve.
2. Draw a smooth curve through the measured data points.
3. Determine the slope of the curve using numerical, analytical, or graphical methods: e.g., $(dc_i/dt) = r_i$ for DCSTR. Table 4.4 lists the definitions of "reaction rates" in different reactor operations.

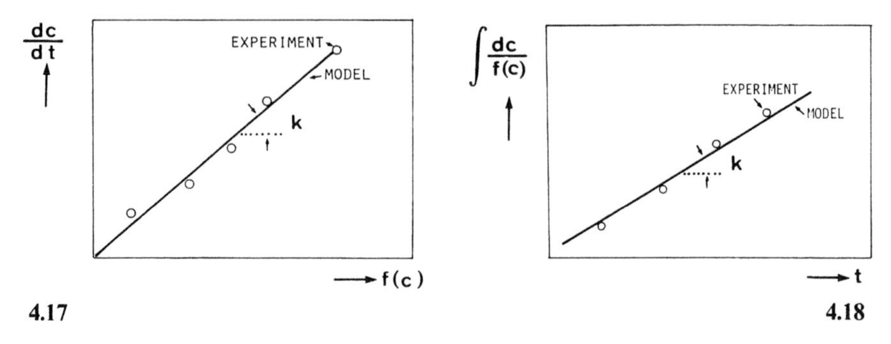

4.17 **4.18**

FIGURE 4.17. General plot showing the differential method of evaluating experimental data to obtain the parameter k of the kinetic equation $r = k \cdot f(c)$.

FIGURE 4.18. General plot showing the integral method of evaluating experimental data to obtain the kinetic parameter k.

With measurements made in a differential reactor, the values of r_i are obtained directly. The important step of a differential data evaluation procedure consists of a comparison between the experimental data and the hypothesis of Equ. 2.53. Usually, this is done graphically using a linearization procedure (see Sect. 2.4), as is shown in Fig. 4.17 for the general case. The function $f(c)$ represents the mathematical function that was postulated in constructing the model. With the computer, nonlinear regression methods are also a realistic alternative.

In an integral data evaluation scheme using data from integral reactors, the following steps are included:

1. Fit the function representing the hypothesis (Equ. 2.53) to the experiment using analytical or graphical techniques so that

$$\int_{c_0}^{c_{ex}} \frac{dc}{f(c)} = k \cdot \int_0^t dt \tag{4.18}$$

2. Compare the experimental data with the prediction that follows from the hypothesis; this may often be done using the linearization shown in Fig. 4.18.

If the agreement between the hypothesis and the experimental data is satisfactory (by statistical tests), the kinetic parameters can be taken from the slope of the straight line and, in some cases, from the intercepts with the axes. The correct model must be identified before parameterization. This procedure for the analysis of process kinetic data is a general one and has been successfully used for chemical processes also (Levenspiel, 1972).

The application of the procedure to bioprocesses should be considered an analogy. Limitations on the validity are to be expected due to the adaptability

TABLE 4.5. Bioreactor operation techniques and biological behavior.

Reactor operation	*Biological growth*	*"Balanced"*	*"Unbalanced"*
Gradient-free	Steady state	CSTR	
	Quasi-steady state	"Extended culture"	
	Unsteady state	←--- CSTR, SCSTR ---→	
	"Periodic"	←-- "Transient operation techniques" --→	
With gradients	Steady state	←--- CPFR, NCSTR ---→	
	Unsteady state	←-- DCSTR, CPFR, NCSTR --→	
	"Periodic"	←-- "Transient operation techniques" --→	

From Moser, 1980b.

of biological materials: Organisms are able to adapt themselves to changes in their environment by altering their metabolism (cf. Sect. 4.2). Therefore, a criterion should exist for partitioning a process into reactor operations and biological factors. The concept of "balanced growth" is used in this way, and the organism lives in completely compensated conditions. Table 4.5 opposes bioreactor operation and biological behavior (Moser, 1980b). A narrow classification system is not yet possible due to lack of relevant data. A factor that may be significant is the relative behavior of the time constants (kinetic rate constants) between a disturbance of the external environment and a disturbance of the inner environment of the biological organism (as outlined in Sect. 4.2 as a case of regime analysis).

There is a difference in the use for process development of the different kinetic parameters obtained from the two calculation methods, each referring to a different process operating technique. The kinetic parameters for the various process operation techniques are primarily bound to the special type of reactor system investigated through the coupling indicated in Table 4.4 (Moser, 1978). The values can be extrapolated to other reactor systems only to a first approximation. However, Esener et al. (1983) have recently pointed out that under certain circumstances no significant difference exists between the parameter values obtained in different modes of cultivation.

4.4.3 Results of Differential and Integral Analysis: Linearization Diagrams

Since kinetic data from a series of DCSTR experiments not only serve as the basis for process development in the DCSTR but also serve to a first approximation for the design of continuous reactors, integral and differential data evaluation methods for the case of a DCSTR will be considered in more detail.

Figure 4.19 shows the contrasting concentration/time curves for a discontinuous fermentation (Moser and Lafferty, 1976). For integral analysis, the typical procedures are represented as if carried out in a microbiological laboratory using shaking flasks. A series of i experiments is carried out, each

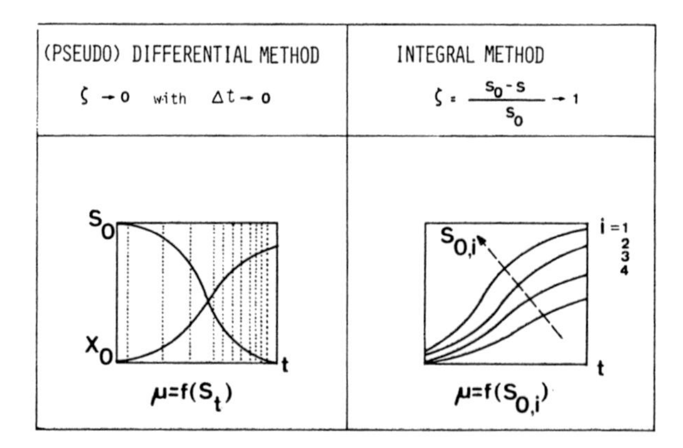

FIGURE 4.19. Differential and integral methods for the kinetic analysis of a process in a discontinuous reactor with microbial flocs. (From Moser and Lafferty, 1976.)

with a different starting concentration of substrate, s_0. In each trial only the cell concentration, x, requires measurement. The data evaluation follows from Equ. 4.18. The integration of the function representing the hypothesis, Equ. 2.5a

$$\int_{x_0}^{x} \frac{dx}{x} = \mu \int_{t_0}^{t} dt \tag{4.19a}$$

results in

$$\ln\left(\frac{x}{x_0}\right) = \mu(t - t_0) \tag{4.19b}$$

By means of a semilogarithmic plot of the values over the range t_2 to t_1, μ may be obtained

$$\mu = \frac{\ln x_2 - \ln x_1}{t_2 - t_1} \tag{4.19c}$$

As a result of the integral evaluation one finds that, for example in the model identification of Monod kinetics (Equ. 2.54), μ can be given as a function of the initial concentrations $s_{0,\mathrm{i}}$

$$\mu = f(s_{0,\mathrm{i}}) \tag{4.20}$$

In contrast, with the differential method of analysis, only a single experiment need be carried out. However, in this experiment $x(t)$ and $s(t)$ are simultaneously measured. As suggested in Fig. 4.19, measurements made at about 2-hour intervals are sufficient from the beginning of the experiment to the beginning of the exponential growth phase. In the range where Monod kinetics are to be characterized in effect, measurements should be repeated every few minutes (this is also the reason why, for example, semicontinuous addition methods make possible more satisfactory kinetic measurements; Reuss, 1976; Webster, 1981). As previously indicated, calculation of the rate of growth, μ, is done directly from the slope of the x/t curve by graphical differentiation over a particular range of Δt:

$$\mu = \frac{1}{\bar{x}} \cdot \frac{\Delta x}{\Delta t} \tag{4.21}$$

where $\bar{x}$ is the average value of x in the measured interval Δt.

The result of a differential analysis of a DCSTR process is a relationship giving μ as a function of s, which is itself a variable changing over the course of the reaction process

$$\mu = f(s(t)) \tag{4.22}$$

In this connection it might be mentioned that Monod (1942) originally regarded the relationship that he proposed for microbial growth kinetics (based on an analogy to enzyme kinetics) as equally valid for continuous and discontinuous processes. One was here dealing with formal kinetics and a case of parameter fitting (K_s instead of K_m).

Finally, briefly presented here are the results of a differential or integral analysis of the various kinetic formulas shown in Fig. 4.12. The solutions are also useful as analogies to many processes in chemical, biological, and food technology.

Figure 4.20 gives a comparison of the results for type 1 kinetics: the so-called "power law equation," which operates with the concept of the reaction order, n. The diagram allows determination of n and the rate constant k. A number of variants are to be found in the literature (Levenspiel, 1972). In this case, the

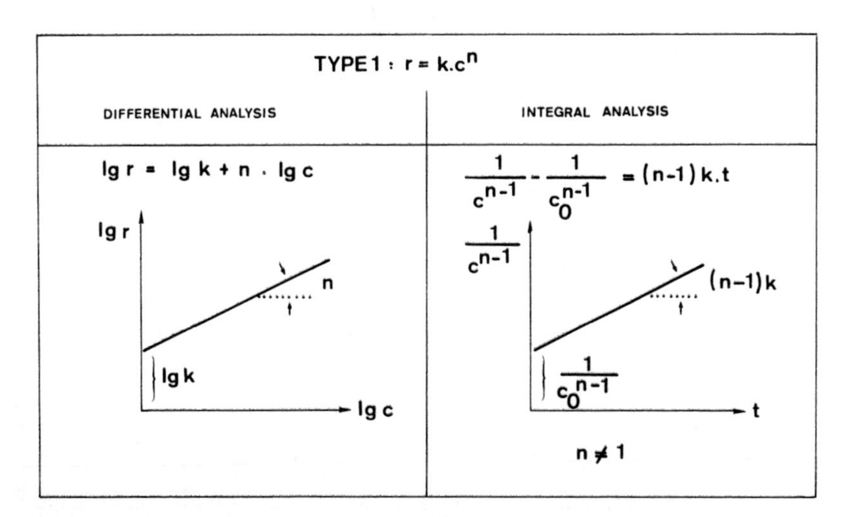

FIGURE 4.20. Determination of the parameters of a type 1 kinetic model (Fig. 4.12) using differential or integral methods.

integral evaluation requires a "trial and error" method where n and k should be obtained simultaneously:

$$\frac{1}{c^{n-1}} - \frac{1}{c_0^{n-1}} = (n-1)\cdot k \cdot t \tag{4.23}$$

This method may be simplified to a graphical presentation when one simply assumes in turn that $n = 0, 1, 2$, and so on.

In contrast, the differential evaluation of the hypothesis $r = k \cdot c^n$ gives

$$\log r = \log k + n \cdot \log c \tag{4.24}$$

with the appropriate graphical presentation, Fig. 4.20.

The type 2 kinetics of Fig. 4.12, with the general form

$$r_i = \frac{k_1 \cdot c}{1 + k_2 \cdot c} \tag{4.25}$$

is applicable to several different cases:

1. Homogeneous chemical reactions in which the order of the reaction changes
2. Heterogeneous chemical catalysis reactions following the Langmuir adsorption isotherm relationship with $k_1 = k \cdot K$, $k_2 = K$, and K = adsorption equilibrium constant, k = rate constant for adsorption which, when multiplied by the occupancy factor (i.e., fraction of surface covered by adsorbed (molecules), gives the rate of adsorption (cf. Equ. 2.56)
3. Enzymatic reactions, where $r_{max} = k_1/k_2$ and $K_m = 1/k_2$ (cf. Equ. 2.54)
4. Microbiological reactions, where $r_i \sim \mu$ (or q_i), $c = s$, and $k_1/k_2 \sim \mu_{max}$ (or $q_{i,max}$), $K_S = 1/k_2$

Figure 4.21 shows the result of the integral analysis of type 2 kinetics in graphical form. For part (a), after the integration of Equ. 4.25, one obtains

$$\frac{\ln c_0/c}{c_0 - c} = -k_2 + \frac{k_1 \cdot t}{c_0 - c} \tag{4.26}$$

For part (b) of the figure, after integrating Equ. 2.54 one obtains the so-called "Henri equation," that is, utilizing the nomenclature for enzyme reactions

$$r_{max} \cdot t = K_m \cdot \ln\frac{s_0}{s} + (s_0 - s) \tag{4.27}$$

Multiplying the Henri equation by $1/t$ makes possible a linear, graphical representation:

$$\frac{s_0 - s}{t} = -K_s \cdot \frac{1}{t} \ln\frac{s_0}{s} + r_{max} \tag{4.28}$$

This method is known as the "Walker plot" method (Walker and Schmidt, 1944), and it is often used in the area of biological waste water treatment (Wilderer, 1976). The Walker plot has the advantage that, for various types of kinetic models (power law equations with different reaction orders as well

TYPE 2 : $r = \frac{k_1 \cdot c}{1 + k_2 \cdot c}$

$$\frac{\lg(c_0/c)}{c_0 - c} = -k_2 + \frac{k_1 \cdot t}{c_0 - c}$$

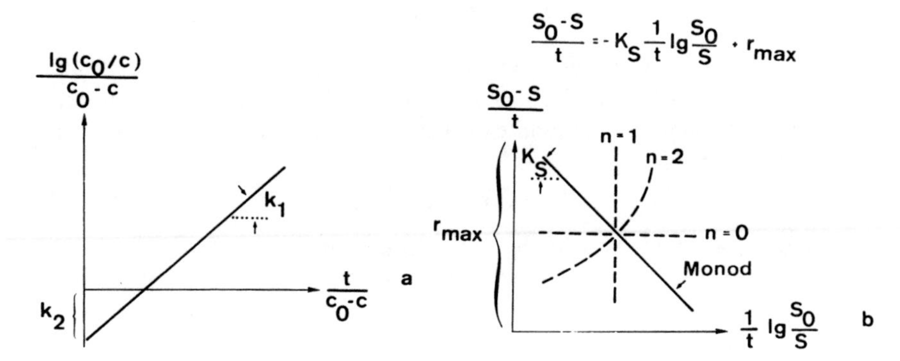

FIGURE 4.21. Determination of the parameters of a type 2 kinetic model (Fig. 4.12) using integral data evaluation methods. In (b), the "Walker diagram" shows a corresponding integral evaluation method for various kinetic equations. (Adapted from Wilderer, 1976.)

as type 2 kinetics), straight lines of differing slopes are obtained, as indicated in Fig. 4.21b.

Strictly considered, the Walker plot applies to reactions in which only substrate is consumed; the growth of cells is not anticipated and there is no other reaction. This linearization starts from the enzyme kinetic equation

$$-\frac{ds}{dt} = r_S = r_{S,\max}\frac{s}{K_m + s} \tag{2.54}$$

which differs from the equation used in microbial kinetics for the rate of use of substrate

$$-\frac{ds}{dt} = r_S = q_{S,\max}\frac{s}{K_s + s}\cdot x \tag{4.29}$$

The connection between enzyme and microbial growth kinetics is

$$r_{S,\max} = q_{S,\max}\cdot x \tag{4.30}$$

which is also reflected in the dimensions of $r_{S,\max}$ [mg/l · min] and $q_{S,\max}$ [h^{-1}]. The Walker plot is, therefore, a special case with $dx/dt = 0$, that is, with $Y_{X|S} = 0$.

A complete description of microbial reactions includes at least two differential equations: one for substrate (S and/or O_2) and the other for growth. The system of equations for the reaction scheme of Equ. 2.1 is, in the simplest case

$$\frac{dx}{dt} = \mu_{max} \frac{s}{K_S + s} \cdot x \tag{4.31}$$

$$-\frac{ds}{dt} = q_{S,max} \frac{s}{K_S + s} \cdot x \tag{4.32}$$

$$-\frac{do}{dt} = q_{O,max} \frac{o}{K_O + o} \cdot x \tag{4.33}$$

The solution of this system of coupled differential equations for the variables X and S (or O_2) is complicated, and it is done primarily using the concept of the yield coefficient, Y, to eliminate the variable S or O_2. Equation 2.13, written in another form, is

$$(X - X_0) = Y_{X|S} \cdot (S_0 - S) \tag{2.13}$$

The complete solution with determination of the kinetic parameters in this case is, however, still a graphical trial and error procedure, linearizations being used only in secondary plots using intercepts and slopes of the primary plot of, for example, $1/t \cdot \ln(x/x_0)$ versus $1/t \cdot \ln(x_{max} - x_0/x_{max} - x)$ for growth data (Gates and Marlar, 1968). Instead of the Henri equation, the following equation is found after integrating Equ. 4.32 (using Equ. 2.13 for substrate consumption)

$$q_{S,max} \cdot x_{max} \cdot t = K_s \ln \frac{s_0}{s} + \left(\frac{1}{Y_{X|S}} \cdot x_{max} + K_S \right) \ln \frac{x_{max} - Y_{X|S} \cdot s}{x_0} \tag{4.34}$$

Equation 4.27 is a special case of this equation: It can be obtained as the limiting case when $Y_{X|S} \rightarrow 0$ in the second term of the right side of Equ. 4.34. Linearization of Equ. 4.34 is possible by multiplying by $1/t$, resulting in

$$\frac{1}{t} \ln \frac{x_{max} - Y_{X|S} \cdot s}{x_0} = \frac{q_{S,max} \cdot Y_{X|S} \cdot x_{max}}{x_{max} + Y_{X|S} \cdot K_S} - \frac{Y_{X|S} \cdot K_S}{x_{max} + Y \cdot K_S} \cdot \frac{1}{t} \ln \frac{s_0}{s} \tag{4.35}$$

In contrast to the Walker plot method, in the Gates linearization method the curves for various conditions (s_0, x_{max}, and x_0) are not superimposed; they vary depending on the conditions. An alternative Gates diagram can be drawn using a graphical trial and error method and Equ. 4.36

$$\frac{1}{t} \ln \frac{s}{s_0} = c \left[\frac{\ln(1 + ad)}{t} \right] - b \tag{4.36a–e}$$

where $a = Y_{X|S}/x_0$

$$b = \frac{\mu_{max}}{Y_{x|s} \cdot K_S} (x_0 + Y_{x|s} \cdot s_0)$$

$$c = 1 + \frac{x_0 + Y_{x|s} \cdot s_0}{Y_{x|s} \cdot K_S}$$

$$d = s_0 - s$$

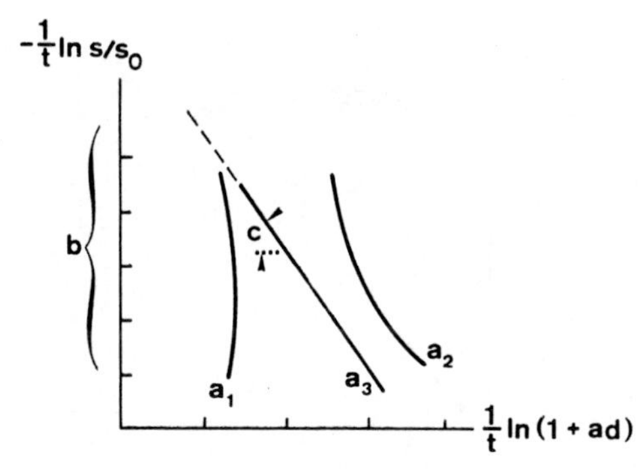

FIGURE 4.22. Batch process evaluation using the integrated form of biokinetics (see Equs. 4.36a–e) with the aid of a graphical trial and error method.

Plotting $1/t \ln(s/s_0)$ versus $[\ln(1 + ad)/t]$ by assuming values for a, and iterating this procedure until linearization appears, leads to the estimation of kinetic parameters (see Figs. 4.22 and 4.23). A variant of this procedure, which also starts from the integrated form of the Monod equation, is used for determining K_S for a discontinuous process from the position of the inflection point of the growth curve (Meyrath and Bayer, 1973).

This short presentation of integral solution techniques for biokinetic equations shows that integration of the function incorporating the hypothesis works only in the simplest cases. In addition, the integrated form lacks much of the flexibility desired for fitting data; there is an immediate tendency to resort to descriptive polynomial functions. One also tries to use simplified forms of enzyme kinetics such as Monod kinetics, and this involves "power law"-type equations (cf. type 1 in Fig. 4.12 or Equ. 2.2a).

For the case of low substrate concentration, such as in biological waste water treatment, one can ultilize a formal first-order equation. From Equ. 4.29

$$-\frac{ds}{dt} = \frac{\mu_{max}}{Y} \frac{s}{K_S} \cdot x \tag{4.37}$$

and with $x = x_0 =$ constant, the integration can easily be carried out to give

$$\ln \frac{s_0}{s} = \left(\frac{\mu_{max}}{Y \cdot K_S}\right) x_0 t \tag{4.38}$$

so that using a semilogarithmic plot, the factor in parentheses may be obtained from the slope.

The missing value for Y must be calculated for the region $s_0 \gg K_S$ using an equation that is formally zero order with respect to S. With $\mu = \mu_{max}$, integrating Equ. 2.5a one obtains

$$x = x_0 \cdot e^{\mu_{max} \cdot t} \tag{4.39}$$

and substituting this in a kinetic equation of zero order in S

$$-\frac{ds}{dt} = \frac{\mu_{max}}{Y} \cdot x \tag{4.40}$$

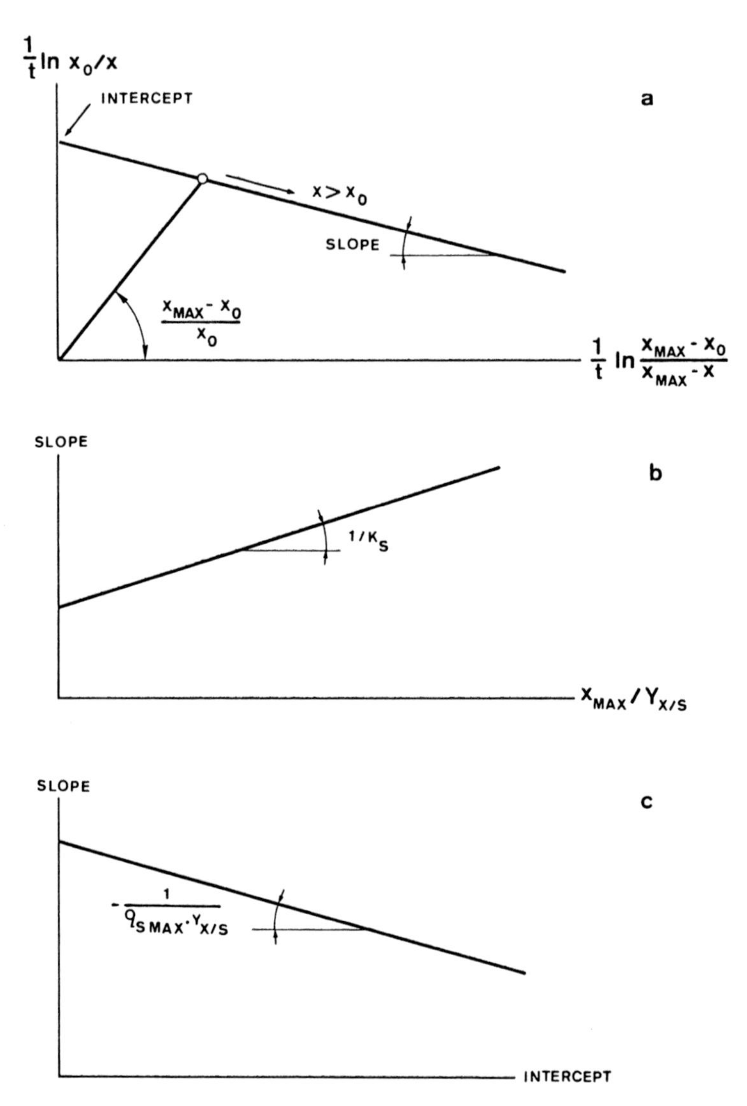

FIGURE 4.23. Alternative integral method following linearization according to Gates and Marlar (1968) on the basis of a primary plot (a) and secondary plots using the slopes and the intercepts of the primary plot (b and c).

one obtains on substitution and integration

$$\frac{s_0 - s}{x_0} = \frac{1}{Y}(e^{\mu_{max} \cdot t} - 1) \tag{4.41}$$

The corresponding plot gives a straight line with slope $1/Y$.

In general, it is apparent that one can use integral data evaluation methods only with the simplest types of kinetic equations. When additional kinetic effects appear (see Chap. 5), as they always do in real processes, the integration step is more difficult or is impossible.

For this reason, in complex cases the differential data evaluation method is used. The differential analysis of DCSTR is of the greatest significance: The DCSTR is the type most often used in practical situations. Also, for developing processes for unconventional bioreactors (for example, bubble columns, towers, and tubular reactors), it is significant that these all have integral reactor behavior.

There are several alternatives to a graphical solution for the type 2 kinetic equations (Fig. 4.12) that form the basic type of kinetic model. In Fig. 4.24c, the commonly used Lineweaver–Burk (1934) linearization is presented by Equ. 4.42:

$$\frac{1}{\mu} = \frac{K_S}{\mu_{max}} \cdot \frac{1}{s} + \frac{1}{\mu_{max}} \tag{4.42}$$

The full procedure is shown from experimental data manipulation in Fig. 4.24a, with experimental errors to the setup of a Monod diagram in Fig. 4.24b. Significant distortions result from the double reciprocal plot and the concentration of data in a certain region (points 2–7).

Due to the distortion in the treatment of errors associated with the double reciprocal plot, it is useful to look for alternatives. A simple reciprocal plot due to Eadie (1942) and Hofstee (1952) and corresponding to Equ. 4.43

$$\frac{\mu}{s} = \frac{\mu_{max}}{K_S} - \frac{\mu}{K_S} \tag{4.43}$$

is shown in Fig. 4.25, along with the same error range of $\pm 5\%$ (that is, the 95% confidence interval). The most satisfactory treatment is, however, the one proposed by Langmuir (1918) for heterogeneously catalyzed chemical reactions. The general form of this linearization is

$$\frac{c}{r} = \frac{1}{k_1} + \frac{k_2}{k_1} \cdot c \tag{4.44}$$

which may be written for Monod kinetics as

$$\frac{s}{\mu} = \frac{K_S}{\mu_{max}} + \frac{s}{\mu_{max}} \tag{4.45}$$

This linearization is shown in Fig. 4.26, again with a $\pm 5\%$ error range. Here the error distortion is minimal.

A completely new method of determining enzyme kinetic parameters is one that may be called "direct linearization." It is due to Eisenthal and Cornish–Bowden (1974) and is based on the direct dependence of μ on S, leading to the dependence of μ_{max} on K_S in the form

$$\mu_{max} = \mu + \frac{\mu}{s} K_S \tag{4.46}$$

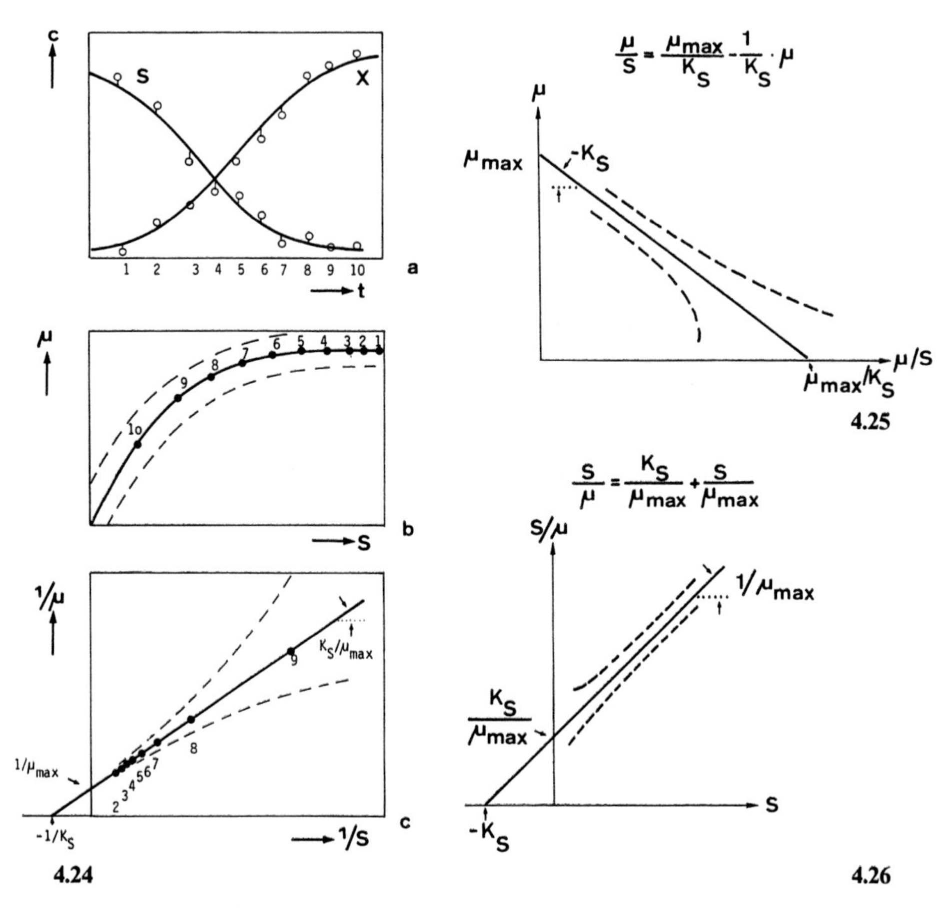

FIGURE 4.24. Growth and substrate utilization curves versus time for a discontinuous process with measured data 1–10. (a) Measured data in a c/t diagram. (b) Data shown in form of Monod plot, (---) indicates 95% confidence interval. (c) Data shown in form of Lineweaver–Burk (1934) double reciprocal plot.

FIGURE 4.25. Model identification and parameter estimation of the Monod kinetic equation using differential methods and a simple reciprocal plot (Eadie, 1942; Hofstee, 1952). (---), 95% confidence interval.

FIGURE 4.26. Model identification and parameter estimation of the Monod kinetic equation using differential methods following the Langmuir (1918) approach. (---), 95% confidence interval. The deviation is minimum.

These authors showed that if the experimental S values are plotted on a negative horizontal axis and the observed r values are plotted on a vertical axis, then straight lines drawn through the corresponding $-s$ and r points intersect at $s = K_m$ and $r = r_{max}$.

While Fig. 4.27b represents the conventional $r = f(s)$ plot (see also Fig. 4.24b), Fig. 4.27a demonstrates the simple graphical procedure for estimat-

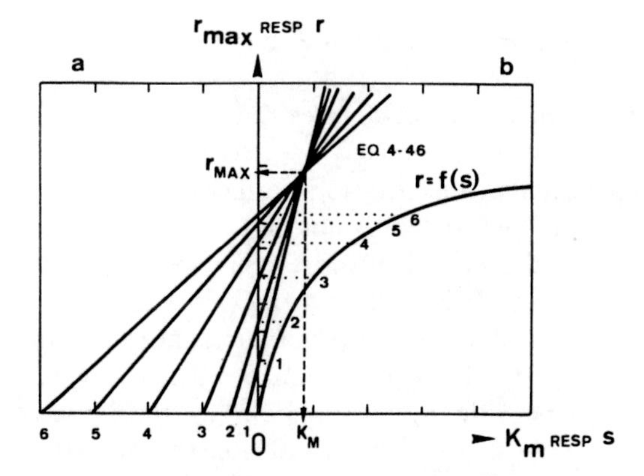

FIGURE 4.27. Parameter estimation without model identification of saturation-type kinetics (enzyme and Monod kinetics) using the "direct linear plot" (Eisenthal and Cornish-Bowden, 1974) of r versus s or r_{max} versus K_m (a) according to Equ. 4.46. The right-side graph (b) shows the conventional plot of saturation-type kinetics, for better comprehensiveness. (Reprinted by permission from Biochem. Journal vol. 139, p. 715, copyright (c) 1974. The Biochemical Society, London.)

ing enzyme kinetic constants, which is in agreement with the rearranged Michaelis–Menten equation as given in Equ. 4.46.

For any values of s and r it is possible to plot r_{max} against K_m as a straight line with slope r/s, intercept $-s$ on the K_m axis, and intercept r on the r_{max} axis. The most obvious advantage of this "direct linear plot" is that it requires no calculation.

Finally, the values of kinetic model parameters can also be estimated using nonlinear regression methods or following a direct method of applying a computer for the simulation and comparison with experimental data (e.g., Aiba, 1978). Because analytical solutions of differential equations are rarely possible, numerical solutions are preferred. The Runge–Kutta method is used: Normally it is available from computer program libraries and is used together with the maximum principle of Pontryagin for optimization (e.g., Bojarinow and Kafarow, 1972; Fan and Wang, 1968).

Logically, the accuracy of parameter estimation using this simulation method depends on the quality of the model function chosen. An inadequate model results in pseudokinetic parameters. The fitting of simulation to experimental data is carried out with the aid of statistical criteria such as the F-test, minimizing deviations σ_i^2:

$$\sum \sigma_i^2 \approx \sum (c_{exp,i} - c_{sim,i})^2 \rightarrow \min \tag{4.47a}$$

Multiresponse analysis or joint analysis uses the concept

$$\sum (x_{exp,i} - x_{sim,i})^2 + (s_{exp,i} - s_{sim,i})^2 \rightarrow \min \tag{4.47b}$$

With real biotechnical processes, almost without exception one uses differential data evaluation methods (see Chap. 5).

4.5 Modeling Heterogeneous Processes

In this section the modeling procedures that can be used for a true heterogeneous system are briefly outlined.

Figure 4.28 presents a concentration profile of, as an example, oxygen in a three-phase process. The transport steps that could be rate limiting are shown. These steps involve the mass transport coefficients k_G and k_{L1} at the G|L interface and k_{L2} at the L|S interface, or k_S as a measure of a critical film thickness $\delta_{S,crit}$ above which the oxgyen concentration sinks below a critical value, o_{crit}. These types of concentration profiles are typical of multiphase processes, and they provide the starting point for the derivation of the model equations. Apparently, the relative proportions of the various phases play a major role in deciding whether one can work with a homogeneous model or must work with a heterogeneous model system. Changes in the relative velocities between the liquid and solid phase (v_{rel}) result in changes in the concentration profile in the solid phase (curves a–c, Fig. 4.28).

It is possible to apply the concept of pseudohomogeneity in cases where the diameter of small particles d_p is slightly less than the film thickness δ, which, in the two-film theory, corresponds to the transport resistance. This situation is illustrated in Fig. 4.29. In this example of a three-phase process, the S phase particles are suspended in the L phase. Five different concentration profiles are shown at the G|L interface (cf. curves a–e) for various ratios of the reaction rate coefficient k_r to the transport coefficient k_{TR}. In the case of a L|S reaction with a large d_p, only profiles a–c apply (cf. Fig. 4.28).

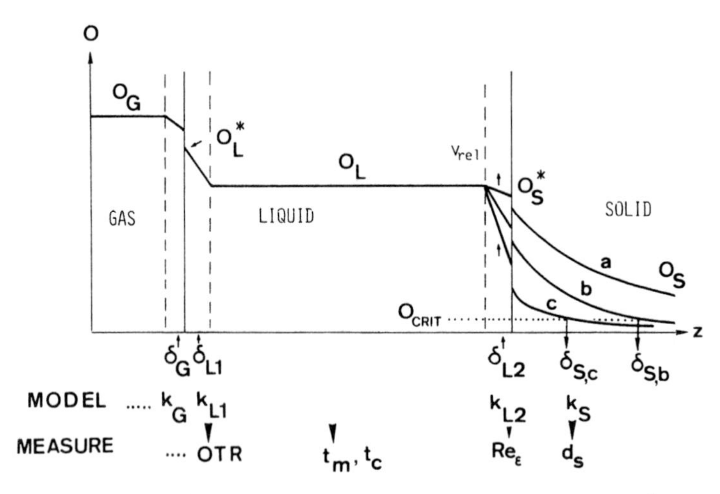

FIGURE 4.28. Heterogeneous three-phase model for a gas–liquid–solid process with the oxygen concentration profile across the different films of thickness δ ($\delta_G, \delta_{L1}, \delta_{L2}, \delta_S$), which are, according to two-film theory, linked to the mass transport coefficient k (k_G, k_{L1}, k_{L2}, k_S).

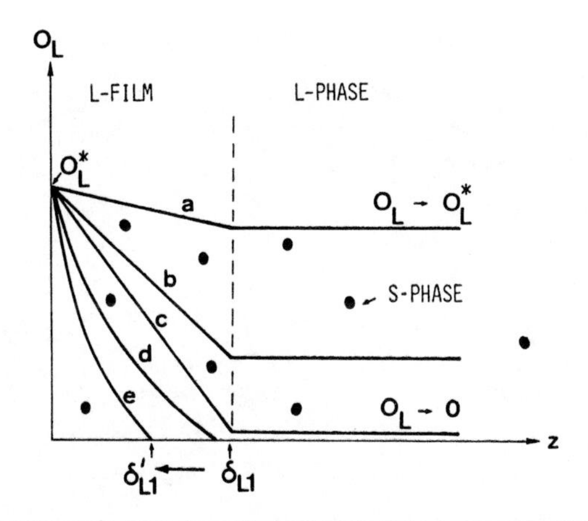

FIGURE 4.29. Pseudohomogeneous model approach for a heterogeneous G|L|S process. The solid phase, S, is homogeneously suspended in the liquid phase as small particles (flocs with $d_S < \delta_{L1}$). It may affect mass transport through the G|L interface, as shown in curve a–e in case of, for example, a reaction occurring in the liquid phase (or inside the flocs) consuming O_2.

All of these cases of coupling between kinetics (k_r) and transport (k_{TR}) phenomena can be handled and solved with a uniform scheme, which is shown in Fig. 4.30 with Equs. 4.48 through 4.59. One uses the concept of an effectiveness factor, η, which is regarded as a function of the ratio k_r/k_{TR}. In individual regions of applying the η concept, individual, traditional descriptions of the relationships between k_r and k_{TR} have established themselves and will be discussed hereinafter (Thiele moduli, ϕ_{ext} and ϕ_{int}; Damkoehler number, Da_{II}, Hatta number, Ha).

For model building when the problem is one of "mass transport with simultaneous reaction," the conservation of mass equation (Equ. 2.3a–c) is again utilized. In the one-dimensional stationary case it has the form

$$D\frac{d^2c}{dz^2} - r = 0 \tag{4.60}$$

The boundary conditions for fluid (gas or liquid)–solid processes are

$$c = c^* \qquad \text{for } z = 0 \tag{4.61a}$$

and

$$\frac{dc}{dz} = 0 \qquad \text{for } z = \delta \tag{4.61b}$$

In the case of fluid–fluid processes

$$c = c^* \qquad \text{for } z = 0 \tag{4.61c}$$

and equilibrium exists between on transport and off reactions for $z = \delta$.

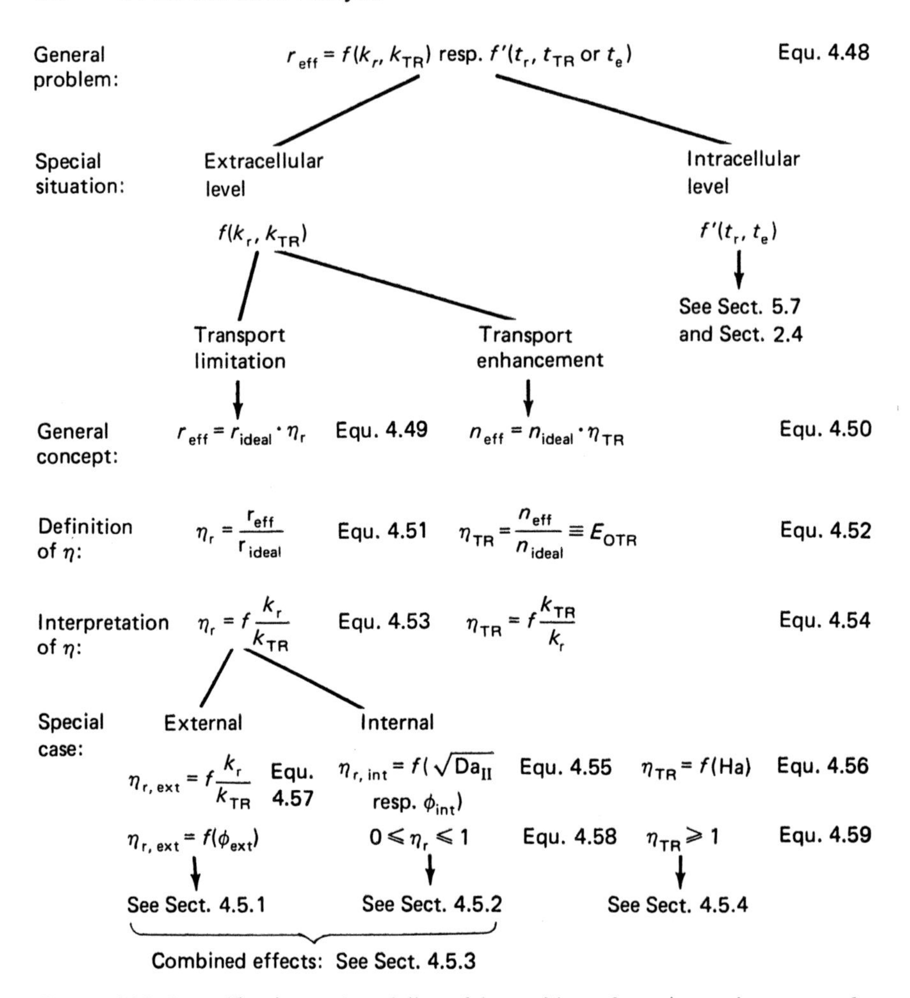

FIGURE 4.30. Quantification and modeling of the problem of reaction and mass transfer interaction in microbial systems: general concept for the solution.

The following general catagories characterize the interaction between kinetic and transport considerations:

1. Transport limitations between fluid–fluid and fluid–solid processes
   a. "External" transport limitations at the G|L and L|S interface (cf. Sect. 4.5.1)
   b. "Internal" transport limitations within the S phase (heterogeneous chemical catalysis and heterogeneous biological catalysis, cf. Sect. 4.5.2)
2. Transport enhancement only at the G|L and L|L interfaces of (pseudo) homogeneous systems (cf. Sect. 4.5.4)

The concentration profiles a through c in Figs. 4.28 and 4.29 reflect transport-limiting processes; the profiles d and e in Fig. 4.29 reflect the enhancement of transport due to a reaction.

4.5.1 External Transport Limitations

With curve a of Fig. 4.29, the reaction is unimpeded ($c \rightarrow c^*$); here the reaction is under kinetic control and is in the "kinetic regime." With curve c, transport is so slow in relation to the rate of reaction that the concentration in the L phase (or S phase) decreases to zero; the reaction here is in the regime of "diffusion control" and transport limitation is paramount.

Curve b is an intermediate case in which both phenomena are at work. The equation valid for a so-called "effective" reaction rate, r_{eff}, is (for the case of a first-order reaction, Fitzer and Fritz, 1975) at steady state

$$r_{eff} = \frac{c^*}{(1/k_L \cdot A) + (1/k_r \cdot V)} \tag{4.62a}$$

The reciprocal rate constants are additive in this case in which transport and reaction steps operate in series.

One may take the ratio of the kinetic (k_r) to transport (k_{TR}, Equ. 2.2b) rate constants as a criterion for determining the boundary between the different types reflected in curves a and c. If $(k_r/k_{Tr}) \ll 1$, then the reaction is "very slow" in relationship to transport, and Equ. 4.61 may be reduced for the kinetic regime to

$$r = k_r \cdot V \cdot c^* \tag{4.62b}$$

which is the ideal rate of reaction. If, on the other hand, $(k_r/k_{TR}) \gg 1$, then, although the rate of the reaction may be slow, the transport process is even slower. One thus has transport limitation, and Equ. 4.62a is reduced to the already well-known equation for physical adsorption. Due to the reaction in the liquid phase, $c_L \rightarrow 0$.

To obtained undistorted kinetic parameters in cases such as these, transport processes must be arranged well enough so that Equ. 4.50 applies. This is accomplished mainly by good mixing in the fluid phase: With good turbulence, a high k_L value is attained. If Equ. 4.62a is recast in the form, for example,

$$r_{eff} = k_r \cdot V \cdot c^* \frac{1}{1 + k_r/k_{TR}} \tag{4.62c}$$

clearly this equation is of the same type as that found in all other cases of the interaction of kinetic and transport factors (cf. Fig. 4.30). The solution is Equ. 4.57 resp. Equ. 4.49. A comparison of Equs. 4.57 and 4.62c shows that Equ. 4.62c reflects r_{ideal} (the ideal case, without transport limitation). The effectiveness factor η_r, defined in Equ. 4.51, is often found in the range

$$0 \leq \eta_r \leq 1 \tag{4.58}$$

in case of external and/or internal transport limitation, and it is only greater than unity in the case of complex kinetics or nonisothermal conditions.

The definition of η_r in the case of a purely external limitation is based on a substrate modulus (Thiele modulus), ϕ_{ext}, given by Horvath and Engasser (1974) as

$$\phi_{ext} = \frac{r_{max}}{K_S \cdot k_{L2}} \tag{4.63}$$

in dimensionless form.

For simple saturation-type kinetics, Equ. 2.54, at steady state one has

$$r_{max} \cdot \frac{s}{K_S + s} = k_{L2}(s_L - s_L^*) \tag{4.64}$$

where s_L and s_L^* are the liquid bulk and surface concentration of substrate, which can be calculated using Henry distribution coefficient $s_L^* = s_S^*/He$.

A graphical representation of Equ. 4.64 can easily be given using dimensionless quantities like ϕ_{ext} (see Equ. 4.63) and is shown in Fig. 4.31 for different values of ϕ_{ext}.

Fig. 4.32 represents the dependence of the effectiveness factor η_r on the modulus ϕ_{ext} at different values of s/K_S. One observes the kinetic regime with $\eta_r \to 1$ and the diffusion regime, which approaches a limiting value of ϕ_{limit}

$$\phi_{limit} = \frac{1}{1 + \phi_{ext}} \tag{4.65}$$

for reactions that are first order in substrate. Under these conditions, the apparent first-order rate constant $k_{1,app}$ is given by

$$k_{1,app} = \frac{r_{max}}{K_S} \cdot \phi_{limit} = K_S(1 + \phi_{ext}) \tag{4.66}$$

Similarly results have been reported by Toda (1975) examining the dependence of pseudokinetic parameters $r_{max,app}$ and $K_{S,app}$ on fluid flow rate. The following equation was found to be valid for describing the dependence of K_S on $k_{L2}a$

$$\frac{K_{S,app}}{K_S} = 1 + \frac{\varepsilon_P}{a_S} \cdot \phi_{ext} \tag{4.67}$$

where ε_p is the volume fraction of enzymic agar particles in a packed column of immobilized cells.

The conventional method of estimation of kinetic parameters is a plot of $s_0\zeta_s$ versus $\ln(1 - \zeta_s)$ according to

$$s_0 \cdot \zeta = K_{S,app} \ln(1 - \zeta_S) + r_{max,app} \cdot \bar{t} \tag{4.68}$$

This is valid for plug flow reactors such as packed beds (Kobayashi and Moo-Young, 1973; Lilly, et al., 1966). Varying the liquid flow rate should give

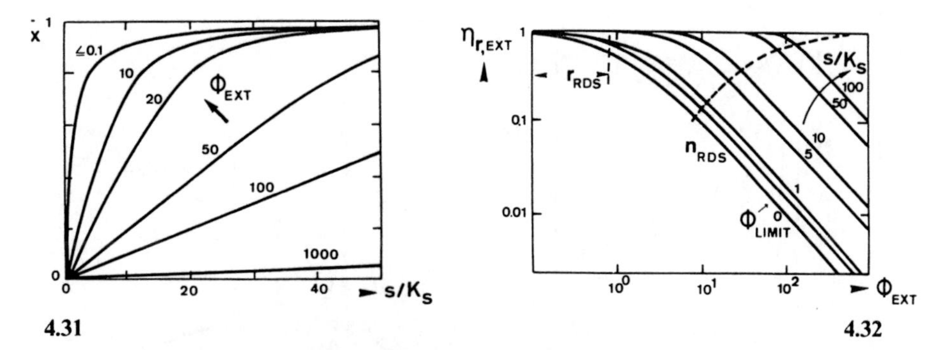

4.31 **4.32**

FIGURE 4.31. Plot of normalized effective reaction rate r_{eff}/r_{max} against dimensionless substrate bulk concentration s/K_S for different values of modulus for external transport limitation ϕ_{ext} in the case of Monod-type kinetics. (From Horvath and Engasser, 1974.)

FIGURE 4.32. Plots of effectiveness factor of reaction $\eta_{r,ext}$ versus modulus ϕ_{ext} in case of external transport limitation as a function of substrate concentration s/K_S. The rate-determining steps (rds) are indicated with their range of validity by dotted lines, together with the limiting first-order effectiveness factor ϕ_{limit} at low s values. (From Horvath and Engasser, 1974.)

the value of K_S unaffected by external transport. A more accurate method was presented by Lee and Ryu (1979), by dividing Equ. 4.68 by $\bar{t}$. This yields the Walker plot (cf. Fig. 4.21b), which allows a more accurate estimation by means of isoconcentration lines.

A new graphical method of the estimation of K_m values (K_S) free from external transport was proposed by Seong et al. (1981). The intercept at the ordinate obtained by the straight-line extrapolation of data points in the plot of $K_{m,app}$ values versus the reciprocal of superficial velocity in column permits an easy and accurate calculation of K_m.

The influence of external transport on kinetic parameter estimation can also be illustrated in an Eadie–Hofstee plot, as shown in Fig. 4.33 (Hartmeier, 1972; Horvath and Engasser, 1974). Significant departures from linearity, however, are observed with increasing external transport limitation ($\phi_{ext} > 0.1$), particularly when a wide range of s is examined.

Atkinson and Ur-Rahman (1979) used an approach based on Equ. 4.62a, replacing the rate constant k_r by an equation for $k_{r,app}$, which also takes into account internal transport limitation ($\eta_{r,int}$; see next section).

With

$$k'_{L2} = \frac{k_{L2}}{\rho_s a_S} \tag{4.69}$$

they concluded that if

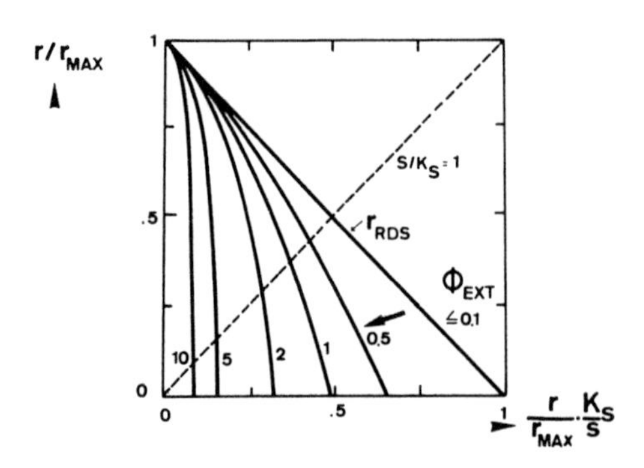

FIGURE 4.33. Eadie–Hofstee-type plots in dimensionless form for different values of the modulus ϕ_{ext} in case of external transport limitation with Monod-type kinetics. (From Horvath and Engasser, 1974.)

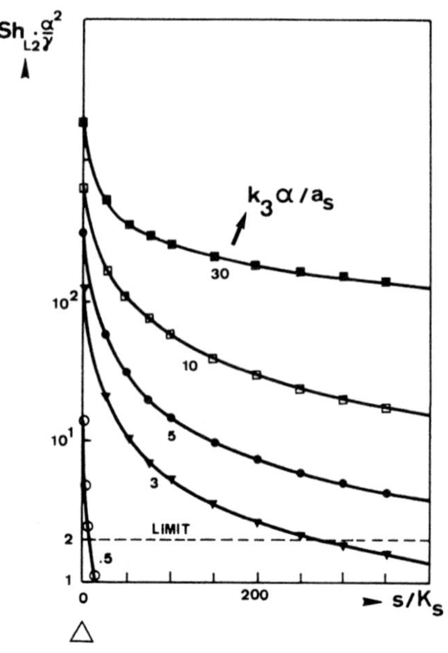

FIGURE 4.34. Sherwood number Sh_{L2} required for a minimal liquid-phase transport limitation at L|S interface (δ_{L2}) as a function of substrate concentration s/K_S, with varied values of solid-phase transport limitation represented by k_3, which is proportional to particle diameter. The limiting Sh_{L2} number ($Sh_{L2} \cdot \alpha^2/\gamma = 2$) is indicated. (From Atkinson and Ur-Rahman, 1979.)

$$\frac{k'_{L2}}{k_{r,app}} \geq 10 \tag{4.70}$$

the liquid-phase diffusion limitation will have only a small effect on the overall rate of substrate uptake. Finally, these authors derived an equation using Equ. 4.69 and 4.70 together with $k_{r,app}$, leading to the following expression

$$Sh_{L2}\frac{\alpha^2}{\gamma} = \frac{60 \cdot \eta_r [k_3 \cdot \alpha \cdot (1/a_S)]^2}{1 + s/K_S} \tag{4.71}$$

which is plotted in Fig. 4.34 in terms of $Sh_{L2} \cdot (\alpha^2/\gamma)$ versus s/K_S for various values of $k_3 \cdot \alpha \cdot a_S$, k_3 being a coefficient of the biological rate equation representing solid-phase diffusion limitation (cf. Equ. 5.248). The curves represent the value of Sh_{L2} required to achieve a minimum L-phase diffusional limitation for bioflocs of given size ($k_3 \cdot a_S$) exposed to a given concentration. Clearly the larger the particle and the smaller the concentration, the larger the k_{L2} value required. Since γ, relating diffusivity within biomass D_{eff} to that of the free solution D, is probably only slightly less than unity, and α, which is a shape parameter, is only slightly greater than unity (1.06 for cylindrical and 1.16 for spherical flocs), it follows that the smallest value for Sh_{L2} (α^2/γ) is approximately 2, sufficient only for the smallest flocs, as shown in Fig. 4.34. Substituting this value into Equ. 4.71 for the conditions most sensitive to a k_{L2} limitation, that is $c \rightarrow 0$, on the assumption that (with $d_S = 1/a_S$) $k_3 \cdot \alpha \cdot d_S < 0.5$ ($\eta_r \sim 1$), leads to the criterion that there will be no k_{L2} limitation even under quiescent conditions (Atkinson and Ur-Rahman, 1979):

$$k_3 \cdot \alpha \cdot d_S \leq 0.182 \tag{4.72}$$

In the case of a vertical rotating drum biofilm reactor, La Motta (1976a) measured the k_{L2} value and correlated it with rotational speed as a measure of the relative velocity v_{rel}

$$k_{L2} = 5.3 \cdot 10^{-4} \cdot v_{rel}{}^{0.7}\ [\mathrm{cm} \cdot \mathrm{s}^{-1}] \tag{4.73}$$

This result is in agreement with similar situations under turbulent conditions. External transport limitation in this biofilm reactor was eliminated when the linear fluid velocity exceeded $0.8\ \mathrm{m} \cdot \mathrm{s}^{-1}$.

Turbulence in the reactor is thought to be the parameter of greatest concern, because liquid film resistances (k_{L2} and k_{L1}), active microbial volume, and mixing (t_c, t_m) are all affected by velocity. Active microbial volume is impacted because at higher velocities greater depth of substrate penetration increases the active film volume available for reaction (e.g., Miura, 1976). An apparent abrupt change in the physical properties of a biofilm reactor occurs at a critical magnitude of Re. Data from Castaldi and Malina (1982) indicate that $\mathrm{Re}_{crit} \sim 1000$, which is less than the transition value of Re for the flow of water in a tube. Below Re_{crit}, the apparent uptake rate is primarily a surface reaction regardless of v_{rel}.

In conclusion, it is obvious that the influence of external transport limitation significantly falsifies the estimation of kinetic parameters. Special safeguards are needed in using high turbulence STRs. The use of unconventional (perfect) bioreactors for process kinetic analyses is preferable to use of STRs. As discussed in Sect. 4.3, liquid–solid mass transfer coefficients, k_{L2}, are correlated by means of the empirical expressions referred to in Table 4.3 (similar equations have been reported by Deckwer, 1980, and by Moo-Young and Blanch, 1981). The relative importance of $k_{L2} \cdot a_S$ depends mainly on the size of the biomass particle, as illustrated in Fig. 4.34. In case of cell pellets of, for example, 0.5 to 2 mm diameter, k_{L2} should be regarded as the major resistance ($k_{L2} a_S < k_{L1} a$), while with flocs with $d_p \sim 10\ \mu\mathrm{m}$, $k_{L2} a_S > k_{L1} a$, and OTR limitation at the G|L interface will dominate.

4.5.2 Internal Transport Limitations

The concept of Equs. 4.49 resp. 4.51 is also applicable in the case of internal transport limitation; here the ratio of rate constants is treated as a dimensionless number, the so-called "Thiele modulus" ϕ (Thiele, 1939) or the Damkoehler number of the second degree, Da_{II}.

The solutions in all cases of internal transport have the form represented by Equ. 4.55 (Emig and Hofmann, 1975) as a consequence of Equ. 4.60 with boundary conditions 4.61a and 4.61b, representing the case of complete S penetration within the film. Thus the relationship between ϕ and $\mathrm{Da_{II}}$ is, in general form,

$$\phi_{\mathrm{int}} = \sqrt{\mathrm{Da_{II}}} = d_p \sqrt{\frac{k_r \cdot c^{n-1}}{D_{\mathrm{eff}}}} \tag{4.74}$$

In this equation, d_p is the diameter of the particle and D_{eff} is the effective diffusion coefficient, which is related by two-film theory to transport over a layer of thickness d_p with a transport coefficient in the solid phase of k_S. The physical meaning is the ratio between reaction rate at surface conditions and diffusivity rate through the outer surface of the particle with diameter d_p. If ϕ is small, the reaction rate becomes the rate-determining step, while diffusion is the rate-determining step if ϕ is high.

The results of Equ. 4.55 are often given in the form of a graph; Fig. 4.35 shows the effect of internal transport on overall effective rate. Compared with Fig. 4.31, obviously the internal effect is much more pronounced than the external. Another typical plot is shown in Fig. 4.36. In the range of kinetic control, $\eta_r = 1$ (i.e., $r_{\mathrm{eff}} = r_{\mathrm{ideal}}$). With increasing $\mathrm{Da_{II}}$ (smaller mass transport coefficient, faster reactions), the effective rate of the reaction is throttled by the transport limitation.

For true kinetic measurements of reaction rates in the S phase, the particle diameter must be kept so small that $\bar{d}_p \leq d_{\mathrm{crit}}$. In Fig. 4.35b, the particle diameter is the ordinate; this part of the figure is drawn in correspondence to part a of the same figure. Measurements of the effectiveness factor, η_r, are carried out experimentally by varying the particle size. If the average diameter of the S-phase particles in an engineering process is $\bar{d}_p$ (cf. Fig. 4.2), then a plot such as Fig. 4.36b can be used in determining the effective rate of the process. Other methods for determining η_r are described in the literature (cf. Emig and Hofmann, 1975).

The slope of the curves in Fig. 4.36 is only moderately influenced by the reaction order and/or the geometry of the S-phase particles (plates, spheres, or cylinders). However, the form of the curve is strongly affected by the material transport coefficients and/or by D_{eff} in the S phase. Exceptions are possible where $\eta > 1$ due to nonisothermal behavior or complex kinetic equations (Aris, 1975).

The theoretical approach must be modified in cases of incomplete S penetration. When substrate is depleted within a biofilm layer δ_S, which is less than δ, boundary conditions must be modified for this fact of no mass transport beyond $z = \delta_S$. Thus, instead of Equ. 4.61b,

$$\frac{dc_i}{dz} = 0 \qquad \text{at } z = \delta_S \tag{4.61d}$$

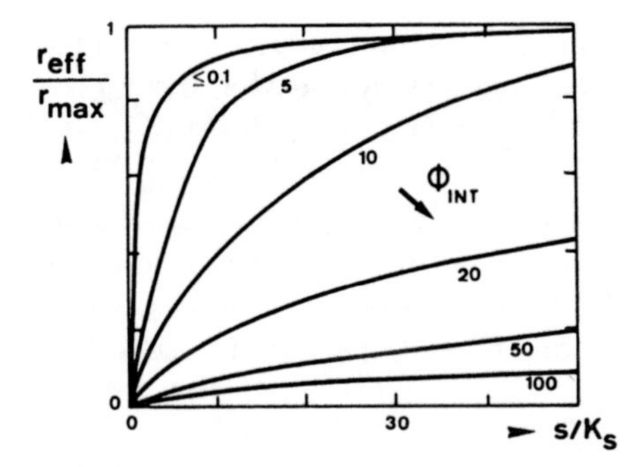

FIGURE 4.35. Plots of normalized overall reaction rate r_{eff}/r_{max} versus dimensionless substrate concentration s/K_S for different values of the modulus ϕ_{int} in case of internal transport limitation with saturation-type kinetics. (From Horvath and Engasser, 1974.)

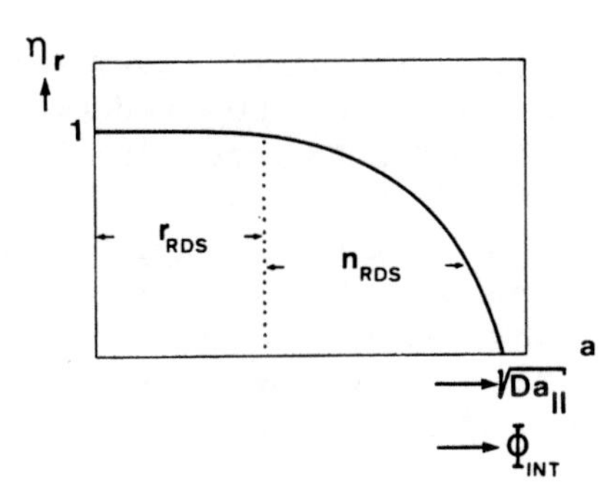

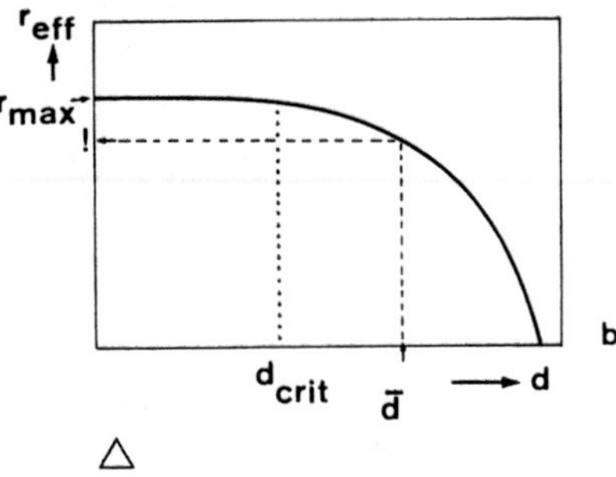

FIGURE 4.36. Graphical representation of the concept of effectiveness factors: (a) The effectiveness factor of reaction, η_r, as a function of the Damkoehler number of the second kind, Da_{II}, or of the Thiele modulus, ϕ (cf. Equ. 4.74). (b) The effective reaction rate, r_{eff}, as a function of the diameter of the particle, d. Part (b) can be used to obtain the value of r_{eff} for a given average diameter, $\bar{d}$, of a population of flocs with a distribution d. The range of validity of kinetic control (r_{rds}) and of diffusion control (n_{rds}) is indicated in part a.

The mass balance differential equation in the case of a spherical shell of a pellet yields, instead of Equ. 4.60

$$\frac{\partial c_i}{\partial t} = D_{eff}\left(\frac{\partial^2 c_i}{\partial R^2} + \frac{\partial c_i}{\partial R}\right) - q_i(c_i) \tag{4.75}$$

At steady state with a Monod-type reaction, Equ. 4.75 (with $r_{max} = q_{S,max} \cdot x$) becomes

$$D_{eff}\frac{d^2 s}{dz^2} = a_f \cdot r_{max}\frac{s}{K_S + s} \tag{4.76}$$

where a_f is the biofilm surface and s is identical to s_f, the surface concentration.

Integration of this equation with appropriate boundary conditions would give the concentration gradient ds/dz in the biofilm. Because it is nonlinear, Equ. 4.76 cannot be solved explicitly. Numerous workers have presented analytical solutions for cases when the Monod relation reduces to the first- and zero-order approximations (Harremoes, 1977; Haug and McCarty, 1971). Although not solving Equ. 4.76 for s_f, Atkinson and others (Atkinson and Daould, 1968; Atkinson and Davies, 1974; Rittmann and McCarty, 1978; Williamson and Chung, 1975) have presented solutions for the flux of substrate into the biofilm for the general case of saturation-type kinetics. The most convenient form of solution of this set of equations is obtained when dimensionless terms are introduced, reducing the effective number of parameters from 6 to 3. Thus

$$\frac{d^2\hat{s}_f}{d\hat{z}^2} - \phi^2 \frac{\hat{s}_f}{1 + (s_L/K_S)\hat{s}_f} = 0 \tag{4.77}$$

$$\hat{s}_f = \left[\frac{s_f}{s_L}\right] = 1 \qquad \text{at } \hat{z} = \left[\frac{\delta - z}{\delta}\right] = 1 \tag{4.78}$$

$$\frac{d\hat{s}_f}{d\hat{z}} = 0 \qquad \text{at } \hat{z} \Rightarrow 0 \tag{4.79}$$

Thereby, the parameter ϕ, representing the dimensionless biofilm thickness δ, is defined as modulus (cf. Equ. 4.74)

$$\phi = \delta \sqrt{\frac{a_f \cdot r_{max}}{K_S \cdot D_{eff}}} \tag{4.80}$$

The solution to Equ. 4.77 with 4.78 has been given in numerical form by Schneider and Mitschka (1965) as the "biological rate equation," (cf. Equ. 5.239).

The functional representation of η_r developed for biofilms can then be expressed as

$$\eta_r = 1 - \frac{\tanh k_3 \cdot d}{k_3 \cdot d}\left(\frac{\phi_p}{\tanh \phi_p}\right) \qquad \text{for } \phi_p \leq 1 \tag{4.81}$$

and

$$\eta_r = \frac{1}{\phi_p} - \frac{\tanh k_3 \cdot d}{k_3 \cdot d}\left(\frac{1}{\tanh \phi_p} - 1\right) \qquad \text{for } \phi_p \geq 1 \tag{4.82}$$

where ϕ_p is the generalized Thiele modulus as a function of ϕ and K_S/s. A graphical representation of this biological rate equation is demonstrated in Sect. 5.8.

In the case of the first-order reaction $r_f = k_{1f} \cdot s_f$, the modulus becomes

$$\phi_1 = \delta \sqrt{\frac{k_{1f}}{D_{eff}}} \tag{4.83}$$

with a solution, well known in literature (Levenspiel, 1972)

$$r'_S = k_{1f} \cdot \delta \cdot s_L \cdot \frac{\tanh \phi_1}{\phi_1} \tag{4.84}$$

with $k_{1,f}$ [s^{-1}].

In the case of a zero-order reaction ($r_f = k_{0f}$), the Thiele modulus β has been modified to suit the interpretation for biofilm kinetics. It is then the reciprocal of the normal Thiele modulus for a zero-order reaction,

$$\beta \equiv \frac{1}{\phi_0} = \sqrt{\frac{2 \cdot D_{eff} \cdot s_f}{k_{0f} \cdot \delta^2}} \tag{4.85}$$

In the case of fully penetrated biofilm, β is greater than 1, and thus

$$r'_S = k_{0f} \cdot \delta \equiv k'_0 \tag{4.86}$$

with k_{0f} [$g \cdot cm^{-3} \cdot s^{-1}$]. In the case of incomplete substrate penetration (partly penetrated biofilm), β is less than 1, and thus (Harremoes, 1977; La Motta, 1976b)

$$r'_S = (2D_{eff} \cdot k_{0f})^{1/2} \cdot s_f^{\ 1/2} \ [kg \cdot m^{-2} \cdot s^{-1}] \tag{4.87}$$

One concludes that a zero-order reaction in a biofilm either becomes a bulk zero-order reaction or a half-order reaction—depending on the characteristics of the biofilm (k_{0f}, δ, D_{eff}) and on the S concentration at the L|S interface.

The half-order reaction (see Equ. 4.87) with

$$k_{1/2} = 2D_{eff} \cdot k_{0f} \ [mg^{1/2} \cdot dm^{-1/2} \cdot s^{-1}] \tag{4.88}$$

(as recognized by Atkinson and Fowler, 1974) can easily be used as a formal kinetic approach quantifying biofilm-processing behavior (see Sect. 5.8) (Harremoes, 1978).

The depth of penetration of substrate within the biofilm is given by

$$\delta_{crit} = \left(\frac{2 \cdot D_{eff} \cdot s_S^*}{k_{0f}} \right)^{1/2} \tag{4.89}$$

Combining Equ. 4.89 with Equ. 4.87 results in

$$r_{eff} = k_{0f} \cdot A \cdot \delta_{crit} \ [mg \cdot s^{-1}] \tag{4.90}$$

This equation illustrates that when there is incomplete S penetration in thick films, the observed rate depends only on the magnitude of the depth of penetration and not of the total thickness (La Motta, 1976b). This so-called active depth was specified by Kornegay and Andrews (1969), and it is often used for design purposes. The significance of internal mass transfer resistance to the interpretation of the kinetic data of suspended growth systems will be explained quantitatively. The biological rate equation, written as

$$r_{i,eff} = r_{i,max} \frac{s}{K_S + s} x \cdot \eta_r \tag{4.91}$$

usually is transformed for parameter estimation into one of the following equations:

Lineweaver–Burk plot (cf. Fig. 4.24c)

$$\frac{r_{\max}}{r_{\text{eff}}} = \frac{1}{\eta_{\text{r,V}}} + \frac{K_{\text{S}}}{\eta_{\text{r,V}}} \cdot \frac{1}{s} \tag{4.92}$$

Eadie–Hofstee plot (cf. Fig. 4.25)

$$\frac{r_{\text{eff}}}{r_{\max}} = \eta_{\text{r,V}} - \frac{r_{\text{eff}} \cdot K_{\text{S}}}{s \cdot r_{\max}} \tag{4.93}$$

Langmuir plot (cf. Fig. 4.26)

$$\frac{r_{\max}}{r_{\text{eff}}} = \frac{K_{\text{S}}}{s} \cdot \frac{1}{\eta_{\text{r,V}}} + \frac{1}{\eta_{\text{r,V}}} \tag{4.94}$$

The effect of internal mass transfer resistances can be analyzed using these plots, as illustrated in Figs. 4.37 through 4.39 and Equs. 4.92 through 4.94 (Shieh, 1981). As can be seen, the linearity of all three linearized plots is significantly distorted by internal mass transfer (quantified by ϕ_{int}^2), especially with large diameter particles. Larger K_{S} values are obtained, while $r_{\max}$ ($= q_{\text{S,max}} \cdot x$) is not affected provided the S concentration is reasonably high. If the S concentration is not sufficiently high, then both K_{S} and $r_{\max}$ would be affected. Similar plots have been presented by different authors for comparable situations: Ngian et al. (1977) for small and large flocs, Shieh (1979), and Atkinson and Ur-Rahman (1979). Clearly, approximate linearity of experi-

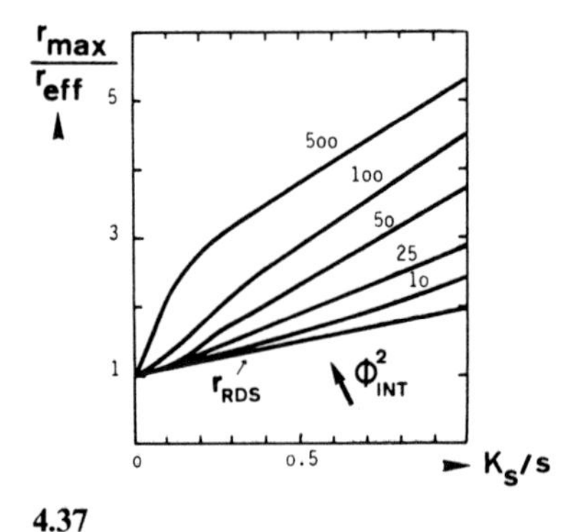

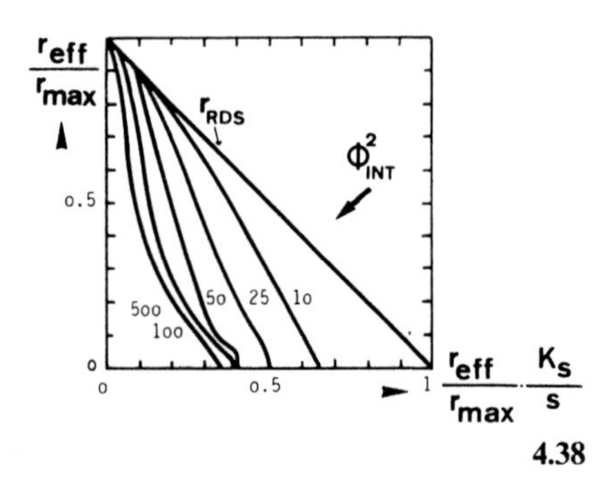

4.37 **4.38**

FIGURE 4.37. Effect of internal transport limitation on Monod-type kinetics demonstrated on a Lineweaver–Burk plot (see Fig. 4.24c) with varied values of a modulus ϕ^2_{int} according to Shieh (1980a). The limiting case of no transport limitation is indicated (r_{rds}).

FIGURE 4.38. Effect of internal mass transport limitation on Monod-type kinetics demonstrated in an Eadie–Hofstee plot (see Fig. 4.25) with variation of modulus ϕ^2_{int} according to Shieh (1980a).

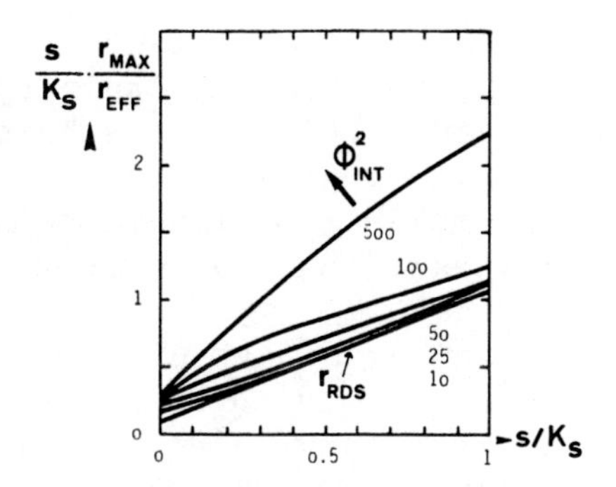

FIGURE 4.39. Effect of internal mass transport limitation on Monod-type kinetics demonstrated in a Langmuir plot (see Fig. 4.26) with varied values of the modulus ϕ^2_{int} according to Shieh (1980a).

mental data in a certain region of conditions does not establish the absence of a solid-phase diffusion limitation. Only by application of different floc sizes can it be proved that $\eta_r \rightarrow 1$.

The mass principles of the theory of mass transfer with simultaneous reaction have been used to characterize the diffusional limitation of suspensions of filamentous microorganisms (Reuss et al., 1980, 1982). If mycelial broth can be considered as consisting of small spheres or filaments, the net specific uptake rate of oxygen in the broth can be calculated using Equ. 4.91 with appropriate values for η_r (Aktinson, 1974) using the Thiele modulus ϕ (see Equs. 4.80–4.82). If the assumptions concerning the hypothetical geometry of mycelium structure are reasonable, then ϕ should be constant at given fluid dynamics, and this has been experimentally verified. The effect of agitation speed can be satisfactorily accounted for by ϕ. Applying the statistical theory of turbulence (Kolmogoroff, 1941), it may be concluded that the mean size of particles is controlled by turbulent shearing action proportional to the energy dissipation ε:

$$\bar{d}_p = c \cdot \varepsilon_{max}{}^{-0.25} \tag{4.95}$$

where, according to Liepe et al. (1971),

$$\varepsilon_{max} = 0.5 \cdot \bar{\varepsilon} \left(\frac{d_T}{d_i}\right)^3 \tag{4.96}$$

where d_T = tank diameter
d_i = impeller diameter
$\bar{\varepsilon}$ = average energy dissipation rate per unit mass

Equation 4.95 means that for elements large enough to correspond to eddies of inertial subrange, the maximum stable diameter is only a function of ε and is independent of broth viscosity. Combining Equs. 4.95 and 4.96 and substituting into Equ. 4.80, it becomes evident that ϕ should be correlated with the power input raised to exponent of (-0.25); this has also been verified by experiments (Reuss et al., 1982). Equation 4.95 also provides an opportunity to verify the theory for those experiments carried out with mycelium, in which macrokinetic O_2 uptake rates varied with the impeller/tank diameter ratio (Steel and Maxon, 1966). This experimental fact can be explained by Equs. 4.91, 4.80, and 4.95: As already shown in Fig. 4.2, the effective reaction rate of O_2 uptake is indeed a function of power input and tank impeller diameter ratio, indicating that smaller impellers are always more effective than larger ones at a given power input per unit volume in this case of mycelium broth aeration.

Finally, the practicability of the η_r concept will be illustrated for STRs. Here it is known that due to the gradients in shear rates, floc size distribution will be of consequence. Therefore, estimation of the biological rate equation coefficient k_3 is not possible in STRs—it is possible only in biofilm reactors. Nevertheless, the η_r concept remains a useful approach because it has been shown that a single floc size, $\bar{d}_p$ (i.e., the mean value), is sufficient to characterize a given distribution function (Atkinson and Ur-Rahman, 1979). The mean floc size closely corresponds to the "surface" mean floc size ("Sauter diameter")

$$\bar{d}_p = \frac{\sum_i n_i \cdot d_i^3}{\sum_i n_i \cdot d_i^2} \tag{4.97}$$

where n_i is the number of flocs with diameter d_i. This mean diameter can then be used in the graphical plot of the biological rate equation shown schematically in Fig. 4.36 to determine the effective rate to be expected for the experimental situation. Clearly, Fig. 4.36 is analogous to the plots of $\eta_r = f(\phi, s/K_S)$ from Equs. 4.81 and 4.82 shown, for example, in Fig. 5.70.

The use of a surface-based effectiveness factor $\eta_{r,A}$ instead of the volume-base factors previously described ($\eta_r = \eta_{r,V}$) was suggested by Fujie et al. (1979) and defined as

$$\eta_{r,A} = \frac{\int_0^\delta (-r_S)dz}{\int_0^\infty (-r_S)dz} = \frac{-D_{eff}\left(\frac{ds}{dz}\right)_{z=0}}{n'_{max}} \tag{4.98}$$

Here n'_{max} [$g \cdot m^{-2} \cdot h^{-1}$] is the maximum mass flux through biofilm surface. The physical model used by these authors was the same as that proposed by Atkinson, extended to double S limitation (glucose and O_2). Their authors' "case A" corresponds to no S or O_2 limitation, "case B" involves an inactive aerobic film (S limitation), and "case C" involves O_2 limitation (partly anaerobic film). With a modified Thiele modulus ϕ_m and a pseudo-first-order approximation for the enzyme-catalyzed rate under the given surface conditions, Fig. 4.40 shows the result of the calculation of $\eta_{r,A}$ versus ϕ_m as a function of s_L/K_S under various conditions. For biofilm reactor operation, $\eta_{r,A}$ is more convenient in the form

$$-r'_S = \eta_{r,A} \cdot n'_{max} \tag{4.99}$$

and is simpler than

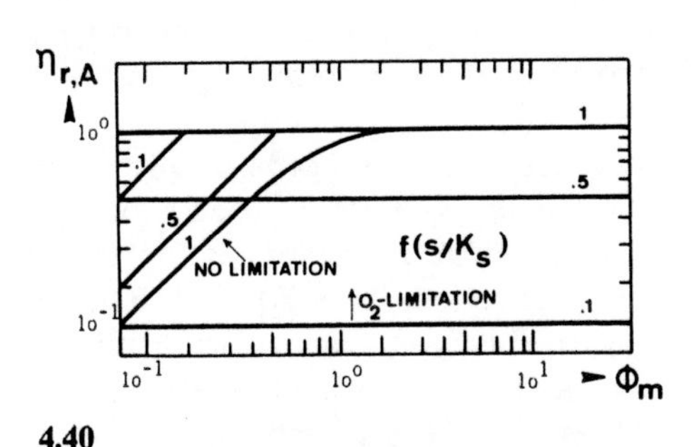

4.40

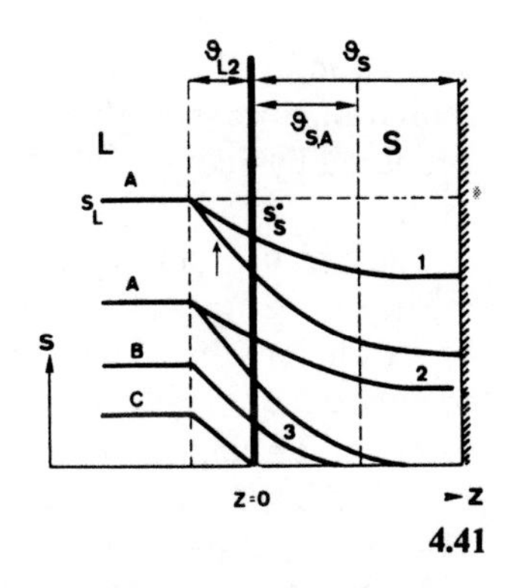

4.41

FIGURE 4.40. Graphic presentation of the surface-base effectiveness factor $\eta_{r,A}$ (see Equ. 4.98) with Monod-type kinetics according to Fujie et al. (1979) and using a modified modulus ϕ_m. The cases of no limitation (in S or O_2) and O_2 limitation are indicated in the graph with varied values of s/K_S.

FIGURE 4.41. Possible cases of concentration profiles s versus z in L|S bioprocessing with partially aerobic biofilms: (1) fully penetrated biofilm, (2) shallow biofilm, (3) deep biofilm with incomplete penetration. Profiles A–C represent the situations of combined external and internal mass transport limitations. The thickness of individual films of mass transfer resistance (δ_{L2}, δ_S) are indicated; $\delta_{S,A}$ represents the active layer of biomass (aerobic biofilm). Increasing turbulence at the L|S interface increases the concentration profile, as shown in δ_{L2}.

$$-r'_S = \eta_{r,V} \cdot (-r_{S*}) \cdot \delta \qquad (4.100)$$

The conventional form of a plot of $\eta_{r,V}$ versus ϕ is shown in Sect. 5.8.

More complex cases, such as cases with S- and P-inhibition kinetics, have been solved numerically by Moo-Young and Kobayashi (1972). Recently criteria have been developed specifically for Monod-, S-inhibited-, Teissier-, and maintenance-type kinetics to quantify and predict diffusional control within whole cells and cell flocs (Webster, 1981). Further details concerning the use of the effectiveness factor concept for the quantification of biological processes will be given in Sect. 5.8, presenting simple formal kinetics in the case of internal transport limitation.

4.5.3 Combined Internal and External Transport Limitations

Normally the phenomena of external and internal mass transport cannot be treated as completely separate. Theoretically, separation of these two effects may be difficult, but experimentally it is quite possible to create conditions in which (a) the effect of external transport is negligible (by increasing the relative fluid velocity) or (b) the effect of internal limitation is minimal (by reducing particle size). Generally, however, both phenomena must be taken into consideration by checking the experimental conditions. If, for example, internal limitation does strongly influence reaction rate, the plot of $K_{m,app}$ versus $1/v_{rel}$ proposed by Seong et al. (1981) will not give a straight line even at high flow rates.

The typical experimental situation in which both phenomena are significant is shown in Fig. 4.41. In most aerobic biofilm systems, O_2 does not fully penetrate into the whole biofilm (Fujie et al., 1979). Therefore it is convenient to divide the overall biofilm system into four regions: bulk liquid, a diffusion layer conforming to two-film mass transfer theory, δ_{L2}, an active or aerobic biofilm δ_S, and an anaerobic or inactive biofilm. Increasing the relative fluid velocity v_{vel} at the L|S interface would result in a deeper active biofilm, thus indicating the combined action of both phenomena. Different concentration profiles are obtained for different ratios of k_{L2} and k_S (or D_{eff}), and they are expressed by dimensionless numbers $k_{L2} \cdot \delta / D_{eff}$ (see Fink et al., 1973). In the steady state, the combined effects can be formulated using the concentration change at the distance $z = \delta$ as

$$n' = k_{L2}\left(c_{L,B} - \frac{c_{S*}}{\text{He}}\right) = D_{eff}\left(\frac{dc}{dz}\right)_{z=\delta} \tag{4.101}$$

where n' is the molar flux per unit surface area of biofilm [$\text{mol} \cdot \text{m}^{-2} \cdot \text{s}^{-1}$] and $c_{L,B}$ and c_{S*} are the respective concentrations in the bulk liquid and at the surface of the biofilm.

An analytical solution to this problem of combined external and internal transport, calculating c_{S*} and n' with Equ. 4.101, is possible only for simple reaction rate equations. A numerical solution using dimensionless forms can be achieved in more difficult cases of enzyme kinetics (Lee and Tsao, 1972) or by assumed first-order approximations (Rony, 1971). It is also possible to solve the problem without numerically solving the differential equation by using the results of computations that consider only internal transport. Horvath and Engasser (1974) showed that external and internal transport can best be identified using Eadie–Hofstee plots (cf. Figs. 4.33 and 4.38). In doing so, one finds the value of c_{S*} where the transport of the reactant from the L bulk to the catalyst equals the conversion in the particle. Thus

$$k_{L2}\left(c_{L,B} - \frac{c_{S*}}{\text{He}}\right) = \delta \cdot \eta_r \cdot r(c_{S*}) \tag{4.102}$$

After calculating the modulus ϕ by assuming a value for c_{S*}, η_r may be found from existing η_r charts (Atkinson, 1974; Moo-Young and Kobayashi, 1972). As the true value of c_{S*} must be found by an iterative procedure, Equ. 4.102 holds only for one particular value of c_{S*}.

A more straightforward method for determining c_{S*} is a graphical method that uses the existing η_r charts (Frouws et al., 1976). Defining a normalized mass transport rate, η'_r, in contrast to η_r, which is a dimensionless reaction rate

$$\eta'_r = \frac{k_{L2}(c_{L,B} - c_{S*}/\text{He})}{\delta \cdot r\,(c_{S*})} \tag{4.103}$$

this dimensionless factor η_r' can then be plotted against the Thiele modulus ϕ, which includes c_{S*}, on the same graph as η_r versus ϕ. The intersection of η_r and η_r' on both plots yields the correct relative values η_r and c_{S*} needed. An advantage of the graphical method is that it clearly shows to what extent the diffusion resistance is inside or outside the particle. It often seems to be the case that both effects are about equally important. In Equ. 4.104, the combined effects of external and internal transport limitations are quantified, showing the type of interaction

$$-r_{eff} = \frac{1}{1/(k_r \cdot \eta_r) + 1/k_{L2}} \cdot s \cdot x = k_{app} \cdot s \cdot x \tag{4.104}$$

where $s \equiv c_{L,B}$.

The behavior of this type of equation can be shown graphically by plotting k_{app} against $s_{L,B}$ in a log–log diagram, Fig. 4.42 (Watanabe et al., 1980).

In the case of external limitation (see profile c in Fig. 4.41), c_S is equal to zero and Equ. 4.102 reduces to (cf. Equ. 2.2b)

$$-r_S = k_{L2} \cdot c_{L,B} \tag{4.105}$$

which states that a bulk first-order reaction occurs at extremely low values of $c_{L,B}$, which is in accordance with enzyme kinetics. If L|S mass transfer is negligible, that is, if

$$c_{L,B} \cong c_S \qquad \text{and} \qquad k_{app} \cong k_r \tag{4.106}$$

then Equ. 4.104 is simplified to obtain Equs. 4.86 and 4.88 for the concentration profiles a and b in Fig. 4.41. Case A is characterized as a zero-

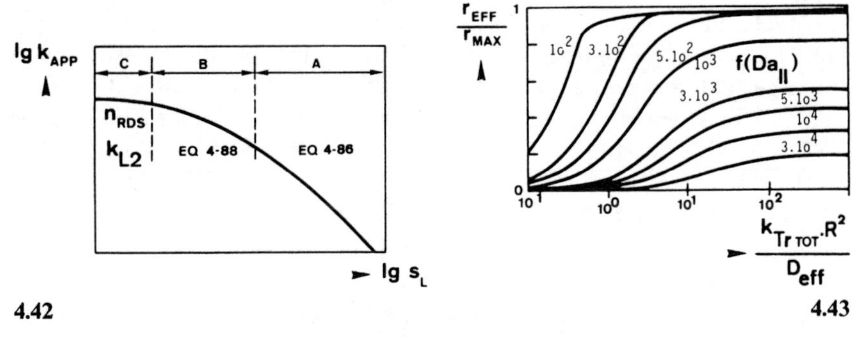

4.42

4.43

FIGURE 4.42. Graphical plot of the relationship between apparent overall rate coefficient k_{app} and bulk substrate concentration s_L according to Equ. 4.104. Three distinct regions (A–C) correspond to the three cases in Fig. 4.41. (Reprinted with permission from *Prog. Wat. Tech.*, vol. 12, Watanabe et al., copyright 1980, Pergamon Journals Ltd.)

FIGURE 4.43. Effect of external O_2 transfer rate $k_{Tr,tot} \cdot R^2/D_{eff}$ on effective respiration rate r_{eff}/r_{max} as a function of Damkoehler number Da_{II} as a measure for particle diameter d_p (see Equ. 4.74 with $n = 1$ and $k_r = q_{0,max}/K_S$). (From Reuss, 1976.)

order reaction, where δ_0 is the oxygen penetration thickness, and case B is characterized by a half-order reaction rate (cf. Equ. 4.88). Nitrification experiments have been carried out, and they adequately verify the proposed model (Watanabe et al., 1980).

The case of coupled external and internal transports has also been dealt with by Reuss (1976), who proposed a graphical solution to the problem, assuming saturation kinetics for the respiration of pellets and defining an overall transfer coefficient

$$\frac{1}{k_{\mathrm{TR,tot}}} = \frac{1}{k_{\mathrm{L1}} a_{\mathrm{L}}} + \frac{1}{k_{\mathrm{L2}} a_{\mathrm{S}}} \tag{4.107}$$

which takes into account the external transports $k_{\mathrm{L1}} a_{\mathrm{L}}$ and $k_{\mathrm{L2}} a_{\mathrm{S}}$. With the proper boundary condition (see Equ. 4.61), Equ. 4.75 was solved numerically and plotted as in Fig. 4.43. Calculations demonstrate the extent to which r_{eff} of O_2 utilization of pellets increases when external transport is increased. O_2 limitation still remains as a function of pellet size, maximum respiration rate, and the diffusion coefficient ($\mathrm{Da_{II}}$; see Equ. 4.74), and it can only be avoided by increasing the external O_2 concentration (cf. profile A_1 in Fig. 4.41). On the basis of these calculations, the potential influence of such process variables as agitation or aeration can be predicted (cf. Fig. 4.2). In the case of the immobilized glucose oxidase and catalase system, such considerations explain the fact that the external mass transfer rate may cause a shift from glucose limitation to O_2 limitation (Reuss and Buchholz, 1979). This concept was also applied to complex kinetics with S and P inhibition (Wadiak and Carbonell, 1975), in which multiple steady states are possible and $\eta_{\mathrm{r}} > 1$.

Another concept, similar to the approach just described, has been proposed by many workers (Aris, 1975; Chen et al., 1980; Fink et al., 1973; Gondo et al., 1974; Hamilton et al., 1974). It designates an overall effectiveness factor $\hat{\eta}_{\mathrm{r}}$, and it is a valuable approach when L-film resistance of mass transfer cannot be neglected. Since $\hat{\eta}_{\mathrm{r}}$ approaches $\eta_{\mathrm{r,V}}$ when L|S mass transport becomes infinite, the latter can be regarded as a limiting case of the former (Yamane, 1981). As an extension of the work of Kobayashi and Moo-Young (1973), Kobayashi et al. (1976) proposed the following approximate expression for the solution of Equ. 4.77 with Equs. 4.78 and 4.79, containing terms for reaction in the zero- and first-order regions of enzyme kinetics

$$\eta_{\mathrm{r}}^{*} = \frac{\eta_{\mathrm{r,0}} + \left(\dfrac{K_{\mathrm{S}}}{s}\right) \cdot \eta_{\mathrm{r,1}}}{1 + K_{\mathrm{S}}/s} \tag{4.108}$$

where for biofilms

$$\eta_{\mathrm{r,0}} = \begin{cases} 1 & \text{for } \phi_{\mathrm{p}} \leq 1 \\ \dfrac{1}{\phi_{\mathrm{p}}} & \text{for } \phi_{\mathrm{p}} \geq 1 \end{cases} \tag{4.109a}$$

and

$$\eta_{r,1} = \frac{\tanh \phi_p}{\phi_p} \tag{4.109b}$$

Using Equ. 4.108, Yamane (1981) attempted to estimate $\hat{\eta}_r$, giving different expressions to both $\hat{\eta}_{r,0}$ and $\hat{\eta}_{r,1}$ by incorporating the Biot number, Bi, as a measure of external film resistance. For a biofilm configuration, thus

$$\hat{\eta}_{r,0} = \begin{cases} 1 & \text{for } \phi_p \le \sqrt{\dfrac{2}{(1 + (2/\text{Bi}))}} \\ -\dfrac{1}{\text{Bi}} + \sqrt{\dfrac{1}{\text{Bi}^2} + \dfrac{2}{\phi_p^2}} & \text{for } \phi_p \ge \sqrt{\dfrac{2}{(1 + (2/\text{Bi}))}} \end{cases} \tag{4.110a}$$

and

$$\hat{\eta}_{r,1} = \left(\frac{\phi_p}{\tanh \phi_p} + \frac{\phi_p^2}{\text{Bi}} \right)^{-1} \tag{4.110b}$$

When $\text{Bi} \to \infty$, the external transport limitation becomes negligible (cf. Equ. 4.110) and hence $\hat{\eta}_r$ approaches η_r. Comparing the magnitude of relative deviation from the corresponding true numerical values for film and floc geometries, Yamane showed that Equs. 4.81 and 4.82 can be recommended for a slab but never for a sphere. The variation of the relative error with the Thiele modulus is shown in Fig. 4.44. Equations 4.108 and 4.110 give quite similar degrees of accuracy.

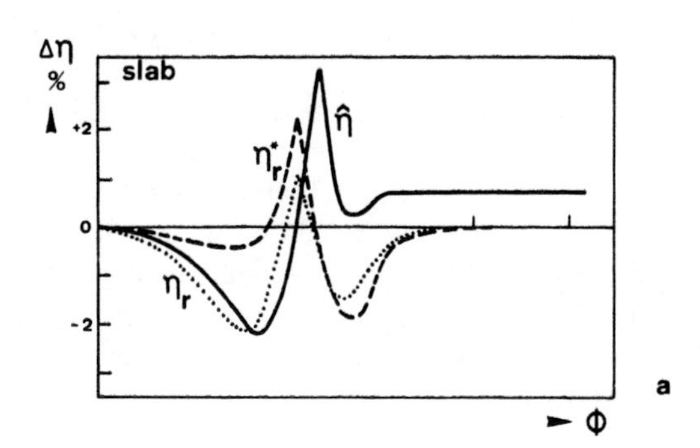

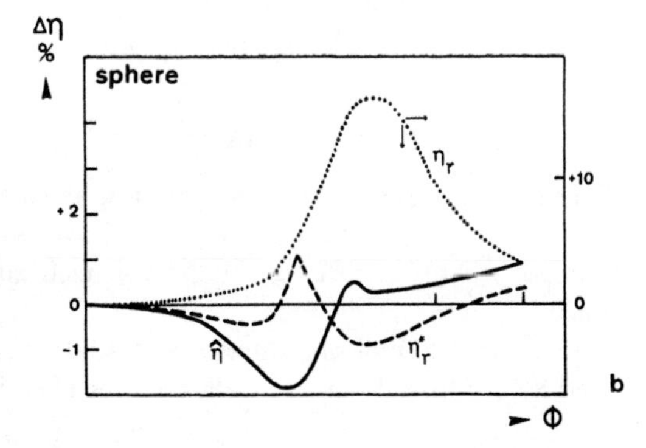

FIGURE 4.44. Comparison of different effectiveness factor concepts (η_r, η_r^* and $\hat{\eta}$) by demonstrating the variations in the relative error $\Delta\eta(\%)$ with the Thiele modulus ϕ for different geometry: slab (a) and sphere (b). Atkinson concept: η_r, see Equs. 4.81 and 4.82; Kobayashi et al.: η_r^*, see Equ. 4.108; Yamane: $\hat{\eta}$, see Equ. 4.110. (From Yamane, 1981.)

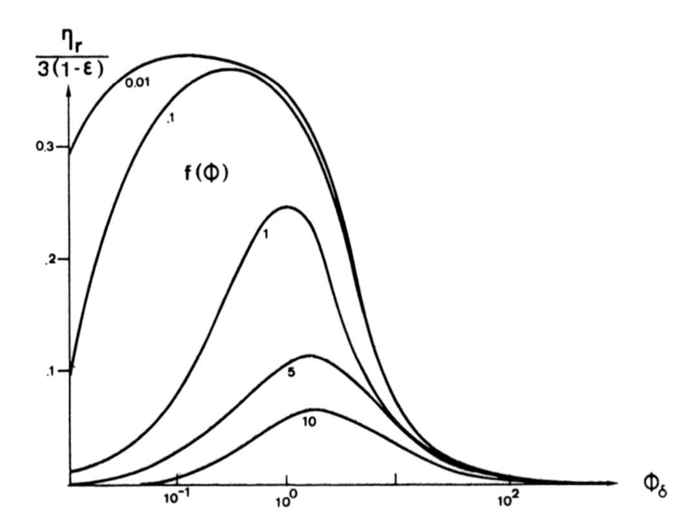

FIGURE 4.45. Effectiveness factor of reaction η_r based on biofilm thickness δ in dependence of Thiele modulus ϕ_δ (see Equ. 4.111) as a function of the Thiele modulus ϕ for support particle size. (Kargi and Park, 1982.)

A Thiele modulus ϕ_δ based on biofilm thickness $(R_p - R_m) = \delta$

$$\phi_\delta = (R_p - R_m)\sqrt{\frac{k_r}{D_{eff}}} \equiv \phi\left(\frac{R_p}{R_m} - 1\right) \tag{4.111}$$

was introduced by Kargi and Park (1982) for dealing with fluidized bed bioreactors. The effectiveness factor was shown to vary only with biofilm thickness (ϕ_δ) and support particle size (i.e. modulus ϕ) when the bed porosity ε was constant. This behavior is depicted in Fig. 4.45. It can be seen that effectiveness increases as ϕ decreases for a given value of ϕ_δ. There is an optimum value of ϕ_δ resulting in maximum effectiveness factor for any given value of ϕ. This optimum value of ϕ_δ decreases with decreasing ϕ. In contrast to results of other work on fluidized bed biofilm reactors (cf. Sect. 6.7), this theoretical analysis reveals that there is no $d_{p,opt}$ value, but rather there exists an optimal biofilm thickness for a given support particle size, that is, an optimal ratio of biofilm thickness/support particle diameter.

4.5.4 TRANSPORT ENHANCEMENT

Thus far only cases in which interaction between kinetic and transport factors has led to a limitation of the reaction have been presented. We now take up the case where the rate of transport is itself increased by the reaction taking place.

The same method presented in Fig. 4.30 for solving the problem can also be used here. The only difference is that the effectiveness factor is referred to the transport rate and is called the enhancement factor, E ($E \equiv \eta_{TR}$). By analogy to Equ. 4.51, E is defined in Equ. 4.52 and the solution equation, Equ. 4.56, with boundary conditions Equ. 4.61c also applies.

Two extreme cases exist, as shown in Fig. 4.29 by concentration profiles c (for $E = 1$) and e (for $E > 1$). As in all other cases, the relative magnitudes of the rate constants define the kinetic and diffusion-controlled regimes. It is, however, customary to describe the relative k values with a dimensionless number, Ha. For G|L reactions with k_L (k_{L1})

$$\mathrm{Ha} = \frac{1}{k_L}\sqrt{\frac{2}{n+1}\cdot D\cdot c^{*n-1}\cdot k_r} \tag{4.112}$$

after the numerical, approximate solution of the differential Equs. 4.60 and 4.61a, c due to Reith (1968). The general accurate form of the solution is due to Hirner (1974):

$$\frac{dc}{dt} = a(c^* - c)\left(\sqrt{\frac{2}{n+1}\cdot D\cdot c^{*n-1}\cdot k_r}\right)(\sqrt{1+C}) \tag{4.113}$$

where the integration constant, C, is actually a function of several variables

$$C = f\left(\mathrm{Ha}, \frac{c^*}{c}, \mathrm{Hl}\right) \tag{4.114a}$$

although the Reith approximation uses

$$C = \frac{1}{\mathrm{Ha}^2} \tag{4.114b}$$

The approximate solution is in error relative to the exact solution by only +6%. The general solution is therefore also given in the form of a function of Ha

$$E = f(\mathrm{Ha}) \tag{4.115a}$$

for example, that due to Reith (1968)

$$E = \sqrt{1 + \mathrm{Ha}^2} \tag{4.115b}$$

The graphical representation of this solution is shown in Fig. 4.46. The region $\mathrm{Ha} \leq 0.3$ is the transport-limited regime, where the effective process rate is determined by physical adsorption. The reaction is slow here relative to the rate of transport: It occurs completely in the liquid phase (bulk), and the reaction therefore has no influence on the transport rate. In the same regime, deviations due to different Hl values of G|L reactors (cf. Equ. 3.59) may occur, giving the curves shown in Fig. 4.56 (for smaller Hl values).

In the region $\mathrm{Ha} \geq 3$, pure kinetic control is in effect, so that here the rate of the process is determined by the rate of the reaction. The reaction is very fast, and it occurs almost entirely in the L film at the G|L interface. In cases where one is dealing with a suspension of catalytically active particles, a fall

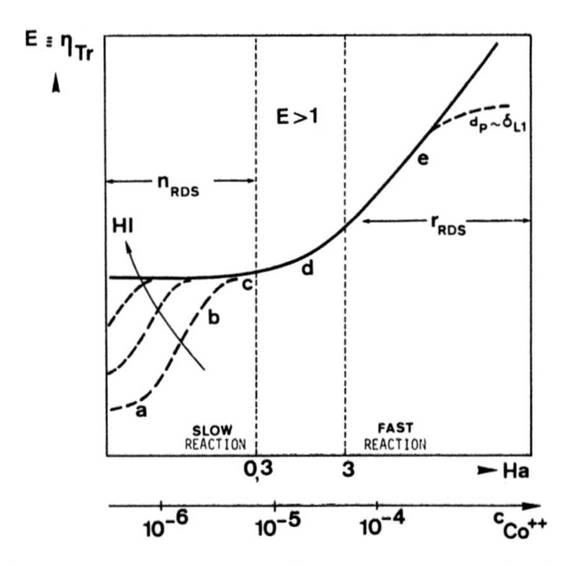

FIGURE 4.46. Mass transport enhancement factor, E, or η_{TR}, as a function of the Hatta number, Ha, or the catalyst concentration, $c_{Co^{++}}$, in the case of sulphite oxidation with Ha $= f(c_{Co^{++}})$. Curves a–e correspond to those in Fig. 4.29. (Adapted from Hirner, 1974; Reith, 1968.)

off in rate is possible if the particle diameter is comparable to the film thickness δ (cf. Fig. 4.29; Alper, Wichtendahl, and Deckwer, 1980; Deckwer and Alper, 1980).

Concentration profile 3 in Fig. 4.41 corresponds to the region Ha ≥ 3. In the intermediate region $0.3 <$ Ha < 3, represented by profile d in Fig. 4.29, the reaction is fast enough so that it also runs partly at the G|L interface, and it therefore also increases the rate of transport.

The distinctions made according to $0.3 \leq$ Ha ≤ 3.0 are not arbitrary ones; rather they are based on the realization that when one process takes place at a rate 1/10th that of another process, one of the processes may be ignored (cf. Equ. 4.115b).

There are many examples of enhanced transport among homogeneous chemical processes (Dankwerts, 1970). There are several procedures for obtaining $k_L \cdot a$ values: Co^{++}-catalyzed sulphite oxidation is a valuable method (Reith, 1968); Cu^{++}-catalyzed sulphite oxidation has some kinetic problems (cf. Sect. 3.3 and Moser, 1973); the Dankwerts CO_2/NaOH method which also exhibits some problems (Dankwerts and Roberts, 1963); the glucose oxidase method (Tsao and Kempe, 1960; Hsieh, Silver, and Mateles, 1969), and more recent methods using anhydrase (Alper et al., 1980b). They all function according to the theory of simultaneous reaction and mass transport.

In the special case of Co^{++} sulphite oxidation, the reaction rate is directly dependent on the Co^{++} concentration, and consequently Ha $= f(Co^{++})$.

This is taken into consideration in Fig. 4.46 in that the catalyst concentration is shown as the second ordinate of the plot. The special importance of this method (and also of the CO_2/NaOH method) is that the values for k_{L1} and a can be obtained separately in a G|L reactor. This will be briefly described. The solution in this case, relying on Eq. 4.113, is in case of oxygen consumption (o):

$$\left(\frac{do}{dt}\right)_{\text{eff}} = a(o^* - o)\sqrt{\frac{2}{n+1} \cdot D \cdot o^{*^{n-1}} \cdot k_r + k_L^2} \qquad (4.116a)$$

It may be simplified for the region Ha ≥ 3 to

$$\left(\frac{do}{dt}\right)_{\text{eff}} = a \cdot o^* \sqrt{\frac{2}{n+1} \cdot D \cdot o^{*^{n-1}} \cdot k_r} \equiv k_L \cdot a \cdot o^* \cdot \text{Ha} \qquad (4.117)$$

or, for the case Ha ≤ 0.3, it may be simplified to

$$\frac{do}{dt} = k_L \cdot a(o^* - o) \qquad (4.118)$$

In the intermediary region $0.3 < \text{Ha} < 3$, Equ. 4.116a is fully valid, and it may be rewritten in another form which $o \to 0$ to give

$$\left(\frac{do}{dt}\right)_{\text{eff}} = k_L \cdot a \cdot o^* \cdot E \qquad (4.116b)$$

The experimental measurement of values for k_{L1} and a, thus, is possible using the following procedure:

1. Select a fast reaction (for example, $Co^{++} > 10^{-4}$ gmol/l) in which the process rate constant is dependent on k_r and a, according to Equ. 4.117. The reaction rate constant k_r may then be obtained as a function of temperature, pH, and catalyst concentration, assuming that one is working with a so-called "perfect" or "model" reactor for G|L reactions in which the interfacial area, a, is known and the hydrodynamics are simple and also known. Examples of such reactors are the falling film reactor, the liquid jet reactor (Astarita, 1967), and the thin film reactor with rotating drum (Dankwerts and Kennedy, 1954; Moser, 1973) as shown in Fig. 3.43.
2. Run the fast reaction in the reactor whose value is to be determined. After k_r is known, use Equ. 4.117 to calculate the exchange surface area a.
3. Working then with the slow reaction (for example, $Co^{++} \leq 10^{-5}$ gmol/1) in the reactor whose value is to be determined, calculate k_L (a is known from the previous experiment). Note, however, that physical factors such as viscosity, ionic strength, surface tension, and the presence of small particles all can have an additional influence on k_L and a (Alper et al., 1980; Linek and Benes, 1977).

The possible rate enhancement of G|L oxygen transfer by viable respirating microbial cells was first studied by Tsao and his group (Tsao, 1968, 1969; Lee and Tsao, 1972; Tsao et al., 1972) using a surface-agitated stirred vessel.

The importance of OTR enhancement occurring in bioprocessing was later challenged by two other groups: Yagi and Yoshida (1975), in their experiments in a sparged stirred tank fermenter, found on significant enhancement. Linek et al. (1974) employed an equation based upon Dankwerts' model, calculated very low enhancement factors (1.025), and also challenged Tsao's work. Furthermore, Tsao et al. (1972) found an even more pronounced effect of OTR enhancement due to respiration of viable cells or in a glucose oxidase system and explained the difference between theory and experiment by the accumulation of microorganisms in the surface region (so-called "two-zone model"). Yagi and Yoshida (1975) explained the discrepancy with Tsao's data by the fact that with "young" bubbles, in which most of the mass transfer takes place, the degree of accumulation of cells at the surface is likely to be low. As a first conclusion, it is evident that two different problems are involved here. The first question is whether OTR enhancement occurs at all in bioprocessing, and the second question is why experimentally determined enhancement factors are substantially greater than the theoretical values calculated from the analogy to chemical enhancement. The situation, and thus experimental verification, is complex, and the reader is referred to the literature: Küng and Moser (in press); Lohse et al. (1980); Merchuk (1977); Moser (1977c, 1980c); Sada et al. (1981); Tsao (1978a); Tsao (1978b). Elucidation seems possible by including cell distribution profiles in the G|L film and experiments with fixed kinetics (e.g., enzymes) rather than microbial cells, which rapidly adapt to environmental changes.

4.5.5 Concluding Remarks

Finally, we recall that the relative extent of each reaction phase and the relative magnitudes of the rate constants determine the extent to which a pseudohomogeneous model approximation may be applied to a three-phase process. If there are no transport limitations in or between phases (cf. Fig. 4.28) (if, for example, $o_S^* \rightarrow o_L^*$), one may ignore the differential equations pertaining to mass transport, and the system of equations is reduced to that appropriate for a pseudohomogeneous model.

In this connection, a special circumstance should be mentioned. Due to the similarity in density between cells ($\rho \sim 1.05$) and water ($\rho = 1$), one cannot rely on any great difference in relative velocity between L and S phases (Fig. 4.1). However, with film reactors an elevated metabolism has been found, corresponding to the higher relative velocity that is possible in the case of carrier-bound films of microbial cells (La Motta, 1976a; Moser, 1977b). The k_{L2} value shows a corresponding dependence (power index 0.7) on flow velocity (or rotation velocity). Relying on the effectiveness factor, one comes to the conclusion that with elevated k_{L2} value, the Thiele module must be smaller. There will thus be a concentration difference $o_S^* \rightarrow o_L^*$ at the L|S interface, and with this, d_{crit} in the S phase will be larger ($\delta_{S,b}$ in Fig. 4.28). The effect of accelerated metabolism is, at least in part, interpreted as attributable to the elimination of a transport-limiting factor, which apparently is not completely eliminated with biofloc reactors ($o_S^* \neq o_L^*$). Similar effects have been observed with pellets (Miura, 1976).

Bibliography

Aiba, S. (1978). *Biotechnol. Bioeng. Symp.*, 9, 269.

Alper, E., Wichtendahl, B., and Deckwer, W.D. (1980a). *Chem. Eng. Sci.*, 35, 217, and 1264.

Alper, E., et al., (1980b). In Moo-Young, M., et al. (eds.). *Advances in Biotechnology*, Vol. 1. Oxford: Pergamon Press, p. 511.

Aris, R. (1975). *The Mathematical Theory of Diffusion and Reaction in Permeable Catalysts*, Vol. 1. Oxford: Clarendon Press.

Astarita, G. (1967). *Mass Transfer with Chemical Reaction*. Amsterdam: Elsevier.

Atkinson, B. (1974). *Biochemical Reactors*. London: Pion.

Atkinson, B., and Daould, I.S. (1968). *Trans. Inst. Chem. Eng.*, 46, 19.

Atkinson, B., and Daould, I.S. (1970). *Trans. Inst. Chem. Eng. (Lond.)*, 48, T245.

Atkinson, B., and Davies, I.J. (1972). *Trans. Inst. Chem. Eng.*, 50, 208.

Aktinson, B., and Davies, I.J. (1974). *Trans. Inst. Chem. Eng.*, 52, 248.

Atkinson, B., and Fowler, H.W. (1974). *Adv. Biochem. Eng.*, 3, 221.

Atkinson, B., and Knight, A.J. (1975). *Biotechnol. Bioeng.*, 17, 1245.

Aktinson, B., and Ur-Rahman, F. (1979). *Biotechnol. Bioeng.*, 21, 221.

Bailey, J.E. (1973). *Chem. Eng. Commun.*, 1, 111.

Barford, J.P., and Hall, R.J. (1979). *7th Australian Conf. Chem. Eng.*, 21.

Bazine, M.J. (ed.) (1982). *Microbial Population Dynamics* CRC press, Boca Raton, Florida.

Bojarinow, A.I., and Kafarow, W.W. (1972). *Optimierungsmethoden in der Chemischen Technologie* (trans. from Russian). Berlin: Akademie Verlag.

Bryant, J. (1977). *Adv. Biochem. Eng.*, 5, 101.

Cassano, A.E. (1980). *Chem. Eng. Educ.*, Winter, 14.

Castaldi, F.J., and Malina, J.F. (1982). *J. Water Poll. Contr. Fed.*, 54, 261.

Chen, K.-C., et al (1980). *J. Ferment. Technol.*, 58, 439.

Danckwerts, P.V. (1970). *Gas Liquid Reactions*. New York: McGraw-Hill.

Danckwerts, P.V., and Kennedy, A.M. (1954). *Trans. Inst. Chem. Eng.*, 32, 51.

Danckwerts, P.V., and Roberts, D. (1963). *Chem. Eng. Sci.*, 18, 63.

Deckwer, W.D. (1980). *Adv. Biotechnol.*, 1, 203.

Deckwer, W.D., and Alper, E. (1980). *Chem. Ing. Techn.*, 52, 219.

Eadie, G.S. (1942). *J. Biol. Chem.*, 146, 85.

Eisenthal, R., and Cornish-Bowden, A. (1974). *Biochem. J.*, 139, 715.

Emig, G., and Hofmann H. (1975). *Chem. Ing. Techn.*, 47, 889.

Environmental Protection Agency (1977). Report No. 6254-77-003a. Washington, D.C.: U.S. Government Printing Office.

Esener, A.A., et al. (1983). *Biotechnol. Bioeng.*, 25, 2803.

Fan, L.-T., and Wang, Ch.-S. (1968). *The Discrete Maximum Principle—A Study of Multistage Systems Optimization*. New York: John Wiley.

Fink, D.J., et al. (1973). *Biotechnol. Bioeng.*, 15, 879.

Fitzer, E., and Fritz, W. (1975). *Technische Chemie*. Berlin–Heidelberg–New York: Springer-Verlag.

Froment, G.F. (1975). *AIChEJ*, 21, 1041. Americ. Inst. Chem. Eng. Journal
Frössling, N. (1938). *Beitr. Geophys.*, 52, 170.
Frouws, M.J., et al. (1976). *Biotechnol. Bioeng.*, 18, 53.
Fujie, K., et al. (1979). *J. Ferment. Technol.*, 57, 99.
Gates, W.E., and Marlar, J.T. (1968). *J. Water Poll. Contr. Fed.*, 40, R469.
Gondo, S., et al. (1974). *J. Ferment. Technol.*, 17, 423.
Hamilton, B.K., et al. (1974). *AIChEJ*, 20, 503. Americ. Inst. Chem. Eng. Journal
Harder, A., and Roels, J.A. (1982). *Adv. Biochem. Eng.*, 22, 56.
Harremoes, P. (1977). *Vatten*, 2, 122.
Harremoes, P. (1978). *Water Poll. Microbiol.*, 2, 71.
Hartmeier, W. (1972). Ph.D. Thesis, Technical University, West Berlin.
Haug, R.T., and McCarty, P.L. (February 1971). *Techn. Rep. No.* 149, Civil Engineering Department, Stanford University, Stanford, Calif.
Hirner, H. (1974). Ph.D. Thesis, Technical University, Stuttgart.
Hofmann, H. (1975). *Chimia*, 29, 159.
Hofstee, B.H.J. (1952). *J. Biol. Chem.*, 199, 357.
Horvath, C., and Engasser, J.-C. (1974). *Biotechnol. Bioeng.*, 16, 909.
Howell, J.A. and Atkinson, B. (1976). *Biotechnol. Bioeng.*, 18, 15.
Hsieh, D.P.H., Silver, R.S., and Mateles, R.I. (1969). *Biotechnol. Bioeng.*, 11, 1.
Kargi, F., and Park, J.K. (1982). *J. Chem. Tech. Biotechnol.*, 32, 744.
Kobayashi, T., and Moo-Young, M. (1973). *Biotechnol. Bioeng.*, 15, 47.
Kobayashi, T., et al. (1976). *J. Ferment. Technol.*, 54, 260.
Kolmogoroff, A.N. (1941). *Compt. Rend. Acad. Sci. USSR*, 30, 301; and 32, 16.
Kornegay, B.H., and Andrews, J.F. (1968). *J. Water Poll. Contr. Fed.*, 40, 460.
Kornegay, B.H., and Andrews, J.F. (1969). *Proceedings of the 24th Industrial Waste Conference*, Purdue, Univ. Press p. 1398. Lafayette, Ind.
Kossen, N.W.F. (1979). In *Symp. Soc. Gen. Microbiol.*, 20, 327.
Küng, W., and Moser, A. *Biotechnol. Lett.* (in press). 8.
La Motta, E.J. (1976a). *Biotechnol. Bioeng.*, 18, 1359.
La Motta, E.J. (1976b). *Environ. Sci. Technol.*, 10, 765.
Langmuir, I. (1918). *J. Am. Chem. Soc.*, 40, 1361.
Lee, S.B., and Ryu, D.D.K. (1979). *Biotechnol. Bioeng.*, 21, 1499.
Lee, Y.H., and Tsao, G.T. (1972). *Chem. Eng. Sci.*, 27, 2601.
Levenspiel, O. (1972). *Chemical Reaction Engineering.* New York: John Wiley.
Liepe, F. et al. (1971). *Chem. Tech.*, 23, 23. Chemische Technik
Lilly, M.D., et al. (1966). *Biochem. J.*, 100, 718.
Linek, V., and Benes, P. (1977). *Biotechnol. Bioeng.*, 19, 565.
Linek, V., and Turdik, J. (1971). *Biotechnol. Bioeng.*, 13, 353.
Linek, V., et al. (1974). *Chem. Eng. Sci.*, 29, 637.
Lineweaver, H., and Burk, D. (1934). *J. Am. Chem. Soc.*, 56, 658.
Lohse, M., et al. (1980). *Chem. Ing. Techn.*, 52, 536.
Luft, G., and Herbertz, H.A. (1969). *Chem. Ing. Techn.*, 41, 667.
Merchuk, J.C. (1977). *Biotechnol. Bioeng.*, 19, 1885.
Meyrath, J., and Bayer, K. (1973). *In* Proc. 3rd Symposium Technische Microbiologie in Berlin, Dellweg, H., ed., Inst. f. Gärungsgewerbe und Biotechnolgie, p. 117.
Miura, Y. (1976). *Adv. Biochem. Eng.*, 4, 3.
Monod, J. (1942). *Recherches sur la Croissance des Cultures Bacteriennes.* Paris: Hermann.

Moo-Young, M., and Blanch, H.W. (1981). *Adv. Biochem. Eng.*, 19, 1.
Moo-Young, M., and Kobayashi, T. (1972). *Canad. J. Chem. Eng.*, 50, 162.
Moser, A. (1973). *Chem. Ing. Techn.*, 45, 1313.
Moser, A. (1977a). Thesis, Habilitation, Technical University, Graz, Austria.
Moser, A. (1977b). *Chimia*, 31, 22.
Moser, A. (1977c). *Chem. Ing. Techn.*, 49, 612.
Moser, A. (1978). 1st European Congress on Biotechnology, Interlaken, Switzerland, Part 1 p. 88. Dechema, Frankfurt.
Moser, A. (1979). Paper presented at Training course "Genie chimique dans les Operations Biologiques". Sociéte de Chimie Industrielle, Bruxelles, November 26–28. (Prof. R. Jottrand, ed.) Univ. Libre de Bruxelles.
Moser, A. (1980a). In Proc. UNEP/UNESCO/ICRO Course, *Theoretical Basis of Kinetics of Growth, Metabolism and Product Formation of Microorganisms* Jena, East Germany: Academy of Sciences, Zentralinstitut für Mikrobiologie und Experimentelle Therapie, Part II, p. 27. (Knorre W., ed.)
Moser, A. (1980b). In Moo-Young, M., and Robinson, C.M. (eds.). *Waste Treatment and Utilization*, Vol. 2, p. 177.
Moser, A. (1980c). In Ghose, T.K. (ed.) *Proceedings of the 2nd International Symposium on Bioconversion and Biochemical Engineering*. Vol. 2. New Delhi: IIT, p. 253.
Moser, A. (1981). *Bioprozesstechnik*. Vienna-New York: Springer-Verlag.
Moser, A. (1982). In *Proceedings of the 3rd Austrian-Italian-Yugoslavian Chemical Engineering Conference*, Graz, (Moser F., ed.) T.U. Graz, Austria, September 14–16, Vol. 2, p. 620.
Moser, A. (1983). *Proc. Adv. Ferm.* 83 (Suppl. Process Biochemistry), 221.
Moser, A. (1984). *Acta Biotechnolog.*, 4, 3.
Moser, A., and Lafferty, R.M. (1976). In Dellweg H. (ed.) Proc. 5th International Fermentation Symposium, Berlin, Abstract No. 5.21.
Ngian, K.F., et al. (1977). *Biotechnol. Bioeng.*, 19, 1773.
Oosterhuis, N.M.G. (1984). Ph.D. Thesis, Technical University, Delft, Netherlands.
Pickett, A.M., and Topiwala H.H., and Bazin, M.J. (1979). *Proc. Biochem.*, 14, 10.
Pitcher, W.H. (1978). *Adv. Biochem. Eng.*, 10, 1.
Prigogine, I., and Defay, R. (1954). *Chemical Thermodynamics*. London: Longman Green.
Reith, T. (1968). Ph.D. Thesis, Technical University, Delft, Netherlands.
Reuss, M. (1976). *Fort. Verfahrenstechnik*, 551.
Reuss, M., and Buchholz, K. (1979). *Biotechnol. Bioeng*, 21, 2061.
Reuss, M., and Wagner, F. (1973). In Dellweg, H. (ed.). *Proceedings of the 3rd Symposium on Technical Microbiology*. Inst. für Gärungsgewerbe und Biotechnologie Berlin:, p. 89.
Reuss, M., et al. (1980). Paper presented at 6th International Fermentation Symposium, London, Ontario.
Reuss, M., et al. (1982). *J. Chem. Tech. Biotechnol.*, 32, 81.
Ringpfeil, M. (1980). Paper presented at 2nd Symposium on Socialist Countries on Biotechnology. Leipzig, East Germany 2–5 December.
Rittmann, B.E., and McCarty, P.L. (1978). *J. Environ. Eng. Div., ASCE*, 104 (EE 5), 889.
Rittmann, B.E., and McCarty, P.L. (1980). *Biotechnol. Bioeng.*, 22, 2343, 2359.
Rittmann, B.E. (1982a). *Biotechnol. Bioeng.*, 24, 501.
Rittmann, B.E. (1982b). *Biotechnol. Bioeng.*, 24, 1341.

Roels, J.A. (1982). *J. Chem. Tech. Biotechnol.*, 32, 59.

Roels, J.A. (1983). *Energetics and Kinetics in Biotechnology.* Amsterdam: Elsevier Biomedical Press.

Romanovsky, J.M., Stepanova, N.V., and Chernavsky, D.S. (1974). *Kinetische Modelle in der Biophysik*, Jena: VEB G. Fischer.

Rony, P.R. (1971). *J. Chem. Tech. Biotechnol.*, 13, 431.

Sada, E., et al. (1981). *Biotechnol. Bioeng.*, 23, 1037.

Sano, Y., Yamaguchi, N., and Adachi, T. (1974). *J. Chem. Eng.* (*Jap.*), 7(4), 225.

Schreier, K. (1975). *Chemiker Zeit.*, 99, 328.

Schneider, P., and Mitschka, P. (1965). *Colln. Czech. Chem. Commun.*, 30, 146.

Seong, B.L., et al. (1981). *Biotechnol. Lett.*, 3, 607.

Shah, Y.T. (1979). *Gas-Liquid-Solid Reactor Design.* New York: McGraw-Hill.

Shieh, W.K. (1979). *Biotechnol. Bioeng.*, 21, 503.

Shieh, W.K. (1980a). *Water Res.*, 14, 695.

Shieh, W.K. (1980b). *Biotechnol. Bioeng.*, 22, 667.

Shieh, W.K., et al. (1981). *J. Water Poll. Contr. Fed.*, 53, 1574.

Steel, R., and Maxon, W.D. (1966). *Biotechnol. Bioeng.*, 8, 97.

Sweere, A., et al (1987). *Enzyme and Microbial Technology*, 9, 386.

Sylvester, N.D., Kulkarni, A., and Carberry, J.J. (1975). *Canad. J. Chem. Eng.*, 53, 313.

Thiele, E.W. (1939). *Ind. Eng. Chem.*, 31, 916.

Toda, K. (1975). *Biotechnol. Bioeng.*, 17, 1729.

Toda, K. (1981). *J. Chem. Tech. Biotechnol.*, 31, 775.

Tsao, G.T., and Kempe, L.L. (1960). *Biotechnol. Bioeng.*, 2, 129.

Tsao, G.T. (1968). *Biotechnol. Bioeng.*, 10, 765.

Tsao, G.T. (1969). *Biotechnol. Bioeng.*, 11, 1071.

Tsao, G.T. (1978a). *Biotechnol. Bioeng.*, 20, 157.

Tsao, G.T. (1978b). *Chem. Eng. Sci.*, 33, 627.

Tsao, G.T., et al. (1972). In Proceedings of the 4th International Fermentation Symposium Kyoto, Japan, p. 65.

Tsao, G.T., et al. (1977). *Biotechnol. Bioeng.*, 19, 557.

Wadiak, D.T., and Carbonell, R.G. (1975). *Biotechnol. Bioeng.*, 17, 1761.

Walker, A.C., and Schmidt, C.L.A. (1944). *Arch. Biochem.*, 5, 445.

Watanabe, Y., et al. (1980). *Prog. Water Tech.*, 12, 233.

Webster, I.A. (1981). *J. Chem. Tech. Biotechnol.*, 31, 178.

Wilderer, P. (1976). In Hartmann, L. (ed.). *Karlsruher Berichte zur Ingenieurbiologie*, Vol. 8, Karlsruhe, West Germany: University of Karlsruhe.

Williams, F.M. (1967). *J. Theoret. Biol.*, 15, 190.

Williams, F.M. (1975). In Pattern, B.V. (ed.). *System Analysis and Simulation in Ecology*, Vol. 1. New York: Academic Press, Chap. 3, p. 197.

Williamson, K.J., and Chung, T.H. (1975). Paper presented at 49th National Meeting of the American Institute of Chemical Engineers, Houston.

Wuhrmann, K. (1963). In Eckenfelder, W.W., and MacCabe, J. (eds.). *Advances in Biological Waste Treatment.* New York: Pergamon Press, p. 27.

Yagi, H., and Yoshida, F. (1975). *Biotechnol. Bioeng.*, 17, 1083.

Yamane, T. (1981). *J. Ferment. Technol.*, 59, 375.

Yoshida, F., and Yagi, H. (1977). *Biotechnol. Bioeng.*, 19, 561.

CHAPTER 5
Bioprocess Kinetics

Model building is considered here as an adaptive process (cf. Fig. 2.18): It involves stepwise fitting of the parameters, k, and discrimination of the function, f, itself. All of the models in this chapter should be considered as working hypotheses. The nature of formal kinetic descriptions means that other mathematical functions can always be found that serve the same descriptive function to within the precision of the measurements (see Esener et al., 1983). The lack of basic content in formal kinetic models in comparison with structured models (Harder and Roels, 1981; Roels and Kossen, 1978) can to some extent be compensated for by subsequent analysis (Esener et al., 1983; A. Moser, 1978b, 1984a).

The basic kinetic scheme for a typical bioprocess is presented in Fig. 5.1 as the starting point for the analysis (A. Moser, 1980a). A chosen rate, somehow connected with the biological growth process, is plotted as a function of a chosen, relevant quantity—for example, a concentration, the pH, or the temperature. One observes a typical pattern of four regions: maintenance level endogenous metabolism, stimulation, inhibition, and toxicity. The typical biological optimum, with r_{max} intermediate between two critical concentration values, is a consequence of opposing stimulatory and inhibitory factors. The influence of a transport limitation is schematically illustrated by the quantity representing an effective diffusion coefficient D_{eff} in the liquid and/or solid phase. The goal of a formal kinetic analysis is description of the various factors influencing this course of events.

To a large extent, the formal kinetic analysis techniques presented in this chapter relate to discontinuous batch operations. Even if the goal is a continuous operation, the batch process kinetic model serves as a start-up. The most significant element of a kinetic analysis is the time dependence of the macroscopic process variables mentioned in Chap. 2. Bacteria, molds, viruses, and yeasts all have different reproduction mechanisms, and formulating a structured kinetic model more closely related to the actual mechanism is a desirable goal. More structured models are desirable not only to deal with active cells but also to extend kinetic analysis to more complex situations involving inactive cells, mixed populations of cells, multiple substrates, and

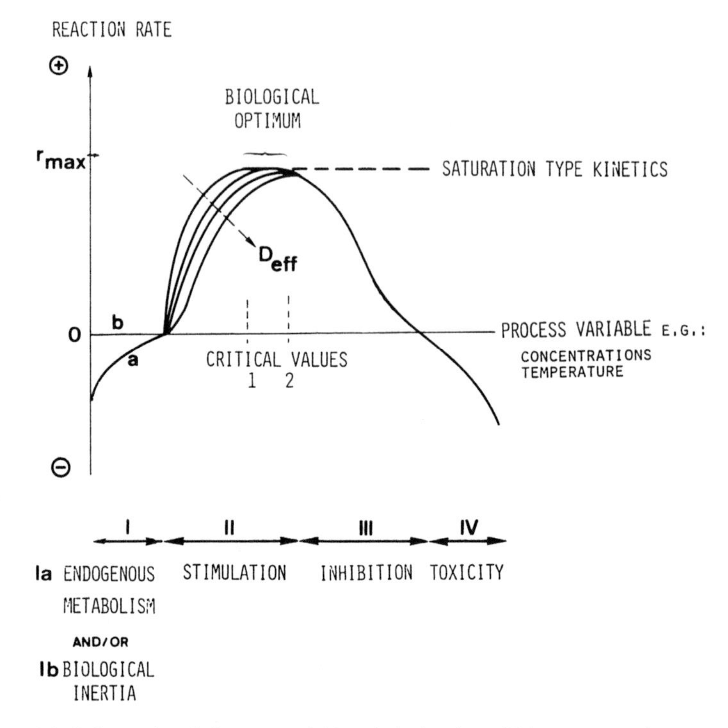

FIGURE 5.1. Schematic of the general kinetic behavior of bioprocesses in terms of the various regions of the effect of reaction (growth or substrate utilization) rate influencing factors. (From A. Moser, 1980a.)

so on. However, with a formal approach dealing mainly with experimental observations, such structuring is undertaken not for its own sake: Rather, structuring primarily serves the objective of improving the accuracy or extending the range of the model. Computer simulations also play an important role in formal kinetic analysis (cf. App. II).

5.1 Temperature Dependence, $k(T)$, Water Activity, a_W, and Enthalpy/Entropy Compensation

In Equ. 2.53, the very general model equation $r = f(x, k)$ is shown separated into concentration- and temperature-dependent parts. The reaction rate constants are therefore regarded as temperature dependent, as illustrated with type 3 in Fig. 4.12.

For a general equation, especially one for use in food technology (e.g., drying processes), the water concentration or water activity, a_W, should be incorporated. The water activity is defined as

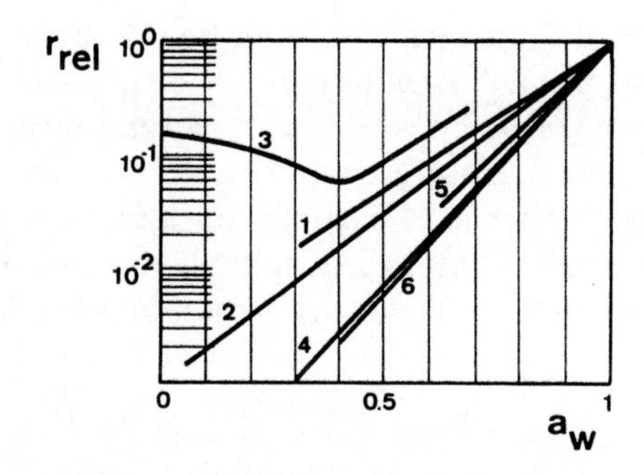

FIGURE 5.2. Dependence of the relative rates of various processes on water activity at constant temperature. Curve 1: relative destruction rate of chlorophyll in spinach at 37°C. Curve 2: relative decay rate of ascorbic acid. Curve 3: relative oxidation rate of potato chips stored at 37°C. Curve 4: relative rate of inactivation of phosphatase in skim milk. Curve 5: relative rate of inactivation of *Clostridium botulinum*. Curve 6: drying rate of a slab of glucose at 30°C slab temperature. (From Thijssen, 1979.)

$$a_W = \frac{(p_{H_2O})_{T,\text{material}}}{(p_{H_2O})_T} \tag{5.1}$$

and can be experimentally determined using hygrometers or the freezing point depression method (Bol et al., 1980).

The rate of drying and of loss of aromatic components, and also the rates of enzymatic, chemical, and physical processes including the loss of microbial cells, are all dependent on a_W (Thijssen, 1979), as shown in Fig. 5.2. Thus

$$k = k(T, a_W) \tag{5.2}$$

The T dependence of k generally follows the Arrhenius equation from chemical reactions

$$k = k_\infty(a_W)\cdot\exp\left[-\frac{E_a(a_W)}{RT}\right] \tag{5.3a}$$

where k_∞ and E_a are determined from the experimental dependence of a_W. The value k_∞ is the maximum value of the rate constant, which, according to collision theory or the theory of activated complexes, continues to contain a T dependence. Deviations are thus found at high temperatures. E_a is the activation energy in [kJ mole^{-1}]. A linearization of Equ. 5.3a is possible using Equ. 5.4, which represents the Arrhenius plot

$$d\ln k = -\left(\frac{E_a}{R}\right)\cdot d\left(\frac{1}{T}\right) \tag{5.4a}$$

or

$$\ln k = \ln k_\infty - \frac{E_a}{R}\cdot\frac{1}{T} \tag{5.4b}$$

when E_a is considered T independent. With Equ. 5.4b the parameters E_a and k_∞ can be estimated using Arrhenius plots.

For microbial processing one must expect a certain complexity in the law of temperature variation, since growth depends on a sequence of reactions. In formal analogy to chemical reactions, an approximate agreement with the Arrhenius law can be considered a suitable working hypothesis, because the effective growth rate is the rate of synthesis minus the rate of degeneration. Thus

$$k_{bio} = k_{\infty,1} e^{-E_{a,1}/RT} - k_{\infty,2} e^{-E_{a,2}/RT} \tag{5.5}$$

where $E_{a,2}$ will be considerably greater than $E_{a,1}$ (Hinshelwood, 1946).

At low values of T, the second term is negligible, and the rate will formally follow the simple Arrhenius law with $E_{a,1}$. Over a certain narrow range of T, the two terms will be of the same order of magnitude. After that the negative second term far outweighs the first; the rate will fall rapidly to zero with a formal activation energy of microbial death $E_{a,2}$.

This plot is shown in Fig. 5.3 for a microbiological process (Humphrey, 1978). Corresponding to the dual character of T, which first stimulates growth and then, when high, destroys the microorganisms, the Arrhenius plot has two straight lines. The slopes of these lines correspond to the activation energies of growth, E_{gr}, identical with $E_{a,1}$, and of the killing process, E_d, identical with $E_{a,2}$.

The rate constants in the case of biological processes are, for example, μ_{max} and k_d. The parameter K_S is also T dependent, as may be seen from the definition in App. III.

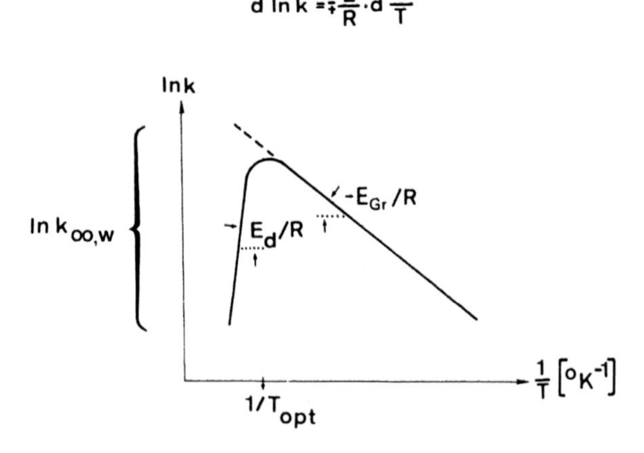

FIGURE 5.3. Kinetic effect of temperature on a biocatalytic process: Arrhenius plot of the growth and death of microbial cells. The apparent activation energies E_{a1} for growth (E_{gr}) and E_{a2} for death (E_d) can be estimated from the slopes, and the factor k_∞ from the intercept.

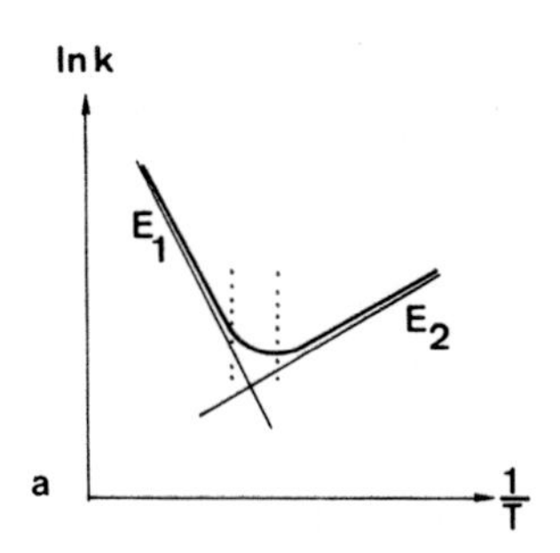

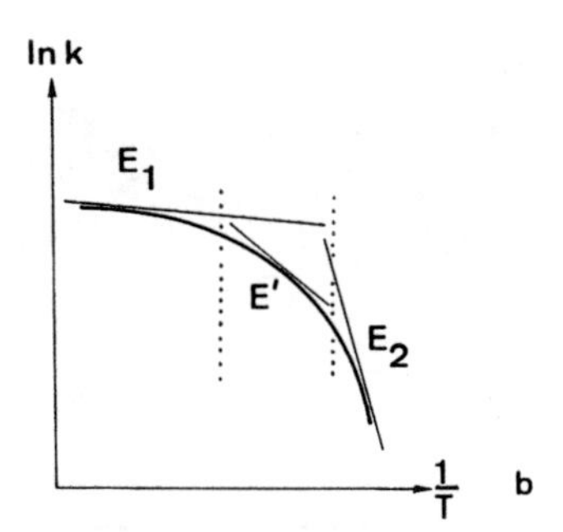

FIGURE 5.4. Deviations from simple Arrhenius equation kinetics of bioprocesses. (a) Sudden change in activation energy ($E_1 \rightarrow E_2$) due to a second enzyme that becomes rate limiting. (b) Continuous change of activation energy to another value (after Talsky, 1971). Quantification is possible by separating the problem into segments in which separate linear approximations (E_1, E_2, E') are valid.

It is an experimental fact that the Arrhenius equation is applicable to bioprocesses. A theoretical prerequisite for the Arrhenius equation is well known—that the velocities of, for example, gas particles follow a Boltzmann distribution. The distribution becomes narrower with increasing molecular weight, so that in the case of enzymes and cells it barely exists. Without the theoretical background, application of the Arrhenius equation to a microbiological processes must be considered a formal kinetic procedure with appropriate fitting parameters.

Deviations from the simple Arrhenius relationship are known: They may, in principle, correspond to Fig. 5.4. Part (a) of this figure represents the sudden change from one rate-determining step to another; for example, from enzyme E_1, with its activation energy, to enzyme E_2, and its own activation energy. In part (b) the transition is more gradual (Talsky, 1971).

The following functions are useful for linear, exponential, and hyperbolic cases as alternatives (see also Equ. 2.57):

$$f(T) = \alpha + \beta \cdot T \tag{5.6a}$$

$$f(T) = \alpha \cdot T^{\beta} \tag{5.6b}$$

$$f(T) = \frac{\alpha}{\beta - T} \tag{5.6c}$$

where α and β are empirical coefficients (Saguy and Karel, 1980). In situations where the Arrhenius equation is not applicable, the best correlation is usually found with hyperbolic equations. Another numerical fit to the typical optimum curve, as schematically represented in Fig. 5.1, is achieved with Equ. 5.7

$$f(T) = \alpha(\mathrm{k} - T)^{\beta} \cdot e^{\gamma(\mathrm{k} - T)} \tag{5.7}$$

used by Setzermann et al. (1982).

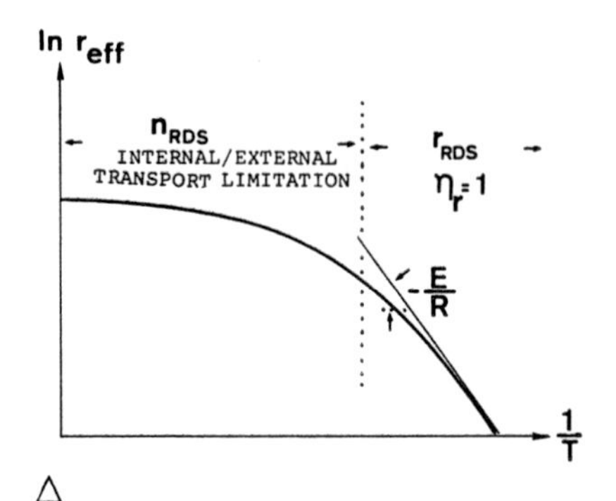

△

FIGURE 5.5. Alteration of the observed activation energy due to the influence of mass transport (after Dialer and Löwe, 1975).

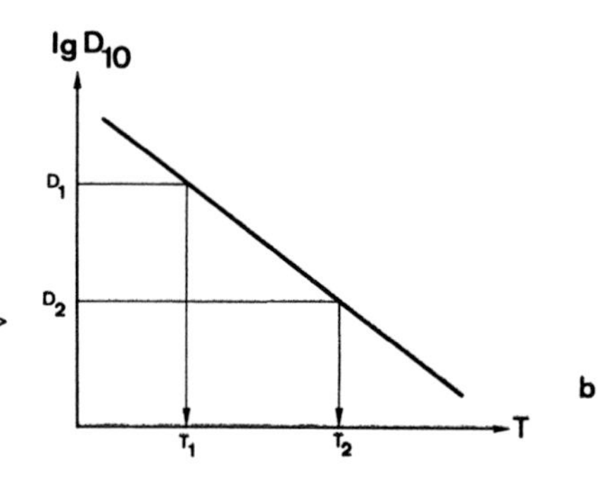

FIGURE 5.6. Alternative concepts for ▷ quantification of temperature effects: (a) "decimal reduction time," D_{10} (see Equ. 5.8), and (b) "Z value" (see Equ. 5.9).

Further deviations are the result of transport influences represented in Fig. 5.5 by an Arrhenius-type plot for a case with strong internal and external diffusion control (Dialer and Löwe, 1975). One can see that only for the lowest temperatures is the value of E calculated from the linear slope a true value.

An additional alternative to the Arrhenius law for correlation of T dependencies is given by the concept of "decimal reduction time," or the "D_{10} value," which is used in the food-processing industry. D_{10} is defined as the time required to reduce a microbial population by one $\log_{10}$ cycle, as illustrated in Fig. 5.6a. The T dependence of the D_{10} value is characterized by the "Z value," as shown in Fig. 5.6b. The following relations apply:

$$D_{10} = \frac{2.3}{k_d} \tag{5.8}$$

$$Z = \frac{T_2 - T_1}{\log D_2 - \log D_1} \tag{5.9}$$

The D_{10} concept is adequate only over very narrow T intervals, because according to the Arrhenius equation, Fig. 5.6b is linear only in a narrow range.

Difficulties in analyzing the T dependence are given by the facts that

- The effects of T and pH are, in general, not independent. Only pH-corrected parameters should be used.

- Very little significance can be attached to studies on the T dependence of enzyme kinetic parameters, unless the mechanistic meaning of these parameters is known.
- Comparison of the activation parameters ΔH and ΔS is often much more informative than comparison of simple rate constants.

Despite difficulties of interpretation, the application of theoretical concepts permits conclusions, or at least hints at conclusions, concerning the mechanism. As a result of the application of the theory of activated complexes, Equ. 5.3a can be recast as

$$k = \left(\frac{k_B \cdot T}{h} \cdot e^{-\Delta S^{\neq}/R}\right) \cdot e^{\Delta H^{\neq}/RT} \qquad (5.3b)$$

where $\Delta H^{\neq}$ = enthalpy of formation of activated complex [$J \cdot mole^{-1}$]

$\Delta S^{\neq}$ = entropy [$J \cdot K^{-1} \cdot mole^{-1}$]

k_B = Boltzmann constant [1.38×10^{-16} erg/°C]

h = Planck constant [6.6×10^{-27} erg · sec]

The expression within the brackets of Equ. 5.3b is identical with the factor k_∞. Taking the derivative of Equ. 5.3b one obtains (Johnson, Eyring, and Polissar, 1954):

$$\ln\frac{k}{T} = \ln\frac{k_B}{h} + \frac{\Delta S^{\neq}}{R} - \frac{\Delta H^{\neq}}{R} \cdot \frac{1}{T} \qquad (5.10)$$

From the slope of the straight line in the Arrhenius plot $\left(\ln\frac{k}{T} \text{ vs. } \frac{1}{T}\right)$ one obtains $\Delta H^{\neq}$, and from the intercept one may calculate $\Delta S^{\neq}$. For a particular reaction in an aqueous medium, if one then plots $\Delta H^{\neq}$ against $\Delta S^{\neq}$ for various conditions, one often finds a linear relationship, as illustrated in Fig. 5.7.

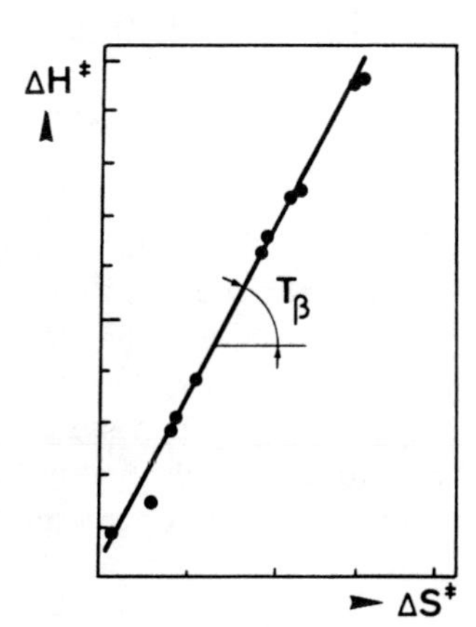

FIGURE 5.7. Enthalpy/entropy plot ($\Delta H^{\neq}/\Delta S^{\neq}$) according to Equ. 5.11 exhibiting a linear relationship ("linear enthalpy/entropy compensation law"). From the slope of the line the isokinetic temperature T_β can be estimated.

The slope of this straight line is referred to as the "isokinetic temperature, T_β" (Barnes, Vogel, and Gordon, 1969; Leffler, 1966). This relationship with a linear thermodynamic compensation of rates is called "enthalpy/entropy compensation"; it is important for the physiological stability of proteins (Lumry and Eyring, 1954). The value T_β is generally between 270°K and 320°K; for the thermal destruction of cells T_β is between 320°K and 350°K. The importance of this fact is the possible implication of a uniform mechanism for cell death through protein denaturation.

In considering enthalpy/entropy compensation, using the basic laws of thermodynamics

$$\Delta H^{\neq} = \Delta G_\beta^{\neq} + T_\beta \cdot \Delta S^{\neq} \tag{5.11}$$

Equ. 5.10 may be rewritten. Microbial cell death, for example, can be related to the optimal growth temperature of yeast cells, T_{max}. A slightly modified form of Equ. 5.11

$$\Delta H^{\neq} = \Delta G^{\neq}_{T_{max+n}} + (T_{max} + n)\Delta S^{\neq} \tag{5.12}$$

with

$$\Delta G^{\neq}_{T_{max+n}} = c(T_{max} + n) \tag{5.13}$$

can successfully be used for this purpose where c and n are empirical coefficients (van Uden, Abranches, and Cabeca-Silva, 1968).

Even though, in many cases, compensation may be considered artifactual (van Uden and Vidal-Leiria, 1976), this method is nevertheless a useful instrument for predicting rate constants and activation energies under various conditions.

The variation of E_a with respect to a_W can be explained in this way, without resorting to a particular change in mechanism (Labuza, 1980).

5.2 Microkinetic Equations Derived from the Kinetics of Chemical and Enzymatic Reactions

5.2.1 The Dynamic Flow Equilibrium Approach to Life Processes

The so-called "dynamic flow equilibrium" is a basic consideration in the kinetics of biological systems following the concept of von Bertalanffy (1942) (see also Netter, 1969, and von Bertalanffy et al., 1977). The dynamic flow equilibrium takes into account once again the conservation of mass (cf. Equ. 2.3a). In open systems, with continuous input and output, reversible reactions take place, but the entire process may be irreversible because of transport. Even for the simplest stationary-state case, the reaction equation, which is

$$A_{ex} \xrightarrow[D_A]{} A \underset{k-1}{\overset{k+1}{\rightleftarrows}} B \xrightarrow[D_B]{} B_{ex} \tag{5.14}$$

with the transport of external materials A and B (diffusion constants D_A and D_B and transport coefficient k_{TR}), the typical dynamic flow equilibrium behaviors are manifest, such as:

1. The steady-state concentrations, $\bar{c}_i$, are constant at constant flux.
2. A time-dependent or permanent change in the dynamic equilibrium can come about only through a change in the kinetic parameters $k_{\pm 1}$ (that is, through enzyme concentration changes). From a second simple example

$$\begin{array}{l} A_{ex} \xrightarrow[D_A]{} A \underset{k-1}{\overset{k+1}{\rightleftharpoons}} B \\ \qquad\qquad \downarrow k_2 \\ \qquad\qquad C \xrightarrow[D_C]{} C_{ex} \end{array} \tag{5.15}$$

further characteristics of the dynamic equilibrium of life processes can be deduced.
3. The conversion ratio of the reactants is constant and independent of external concentrations, which means that the reaction is self-directing (autoregulation, or "equifinality"). This fact becomes clear from the following type of equation, which is the steady-state solution of the system in Equ. 5.15:

$$\bar{a} : \bar{b} : \bar{c} = 1 : \frac{k_{+1}}{k_{-1}} : \frac{k_2}{k_{TR,C}} \tag{5.16}$$

with

$$\bar{a} = a_{ex} \frac{1}{1 + k_2/k_{TR,A}} \tag{5.17}$$

Equation 5.17 shows the typical result of all open systems (living cells, continuous flow reactors), that the ratio of rate coefficients of reaction to transport determines the system behavior (cf. Equ. 4.62c).

All open systems, not just living ones, seem to operate as if they had a goal of maintaining their dynamic equilibrium. This has the effect of making the system independent of the environment. For example, in a negative feedback system, a temporary disturbance in equilibrium constants will cause the system to oscillate until the old steady state is reestablished. If the disturbance is permanent, a new steady state will be found.

A well-known case is the glycolytic pathway, where oscillations in glycolysis and respiration are the result of feedback activation and inhibition of the allosteric oscillophores phosphofructokinase (pfk) and pyruvate kinase (pyk), which are stoichiometrically coupled to each other through ATP/ADP and AMP, or through NADH/NAD in case of glyceraldehyde 3-phosphate dehydrogenase (gapdh) and alcohol dehydrogenase (adh). This system is schematically represented in Fig. 5.8, a block diagram of glycolysis that has been used in computer simulations (e.g., Garfinkel, 1969; Heinzle et al., 1982; Hess and Boiteux, 1967; Higgins, 1967; Mochan and Pye, 1973). In the network sequences of metabolic pathways regulatory enzymes underlie the allosteric effect (Hill kinetics), especially when situated at the beginning and/or end of reaction sequences.

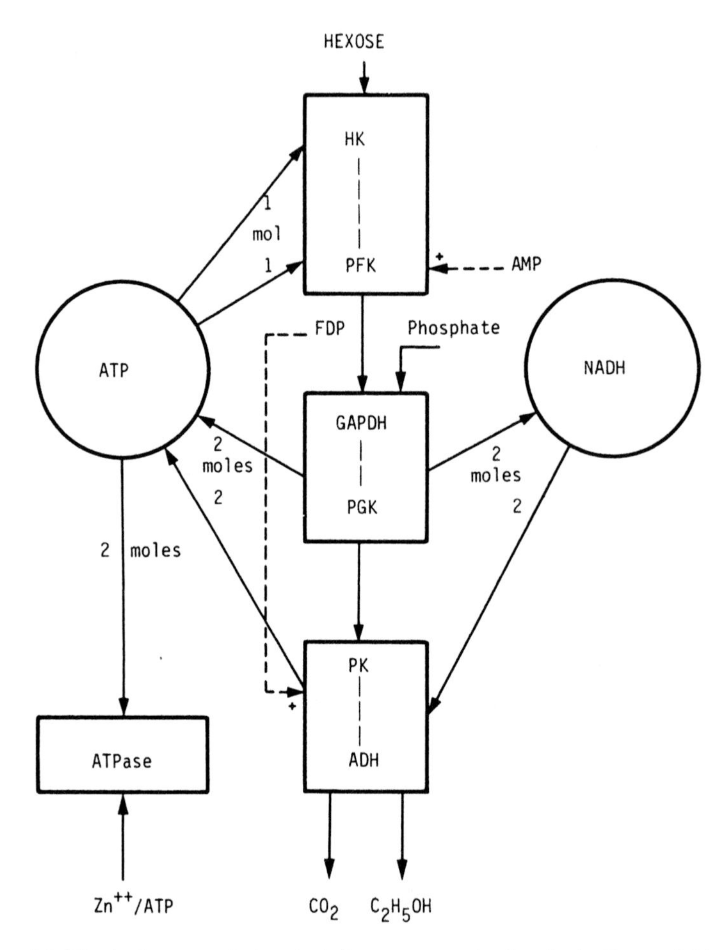

FIGURE 5.8. Block diagram of glycolysis in yeast metabolism including stoichiometric (——→) and informative (- - -→) interactions. (From Hess and Boiteux, 1973.)

5.2.2 Contribution of Enzyme Mechanism to Bioprocess Kinetic Models

5.2.2.1 Simple Enzyme Reactions

Microbial activities like growth and product formations can be regarded as a sequence of enzymatic reactions. On this basis Perret (1960) constructed a kinetic model for a growing bacterial cell population. The main pathways for major nutrients are considered together with pathways for minor nutrients and trace elements linked to each other. This metabolic network can be simplified with the aid of the concept of the rate-determining step (rds), resulting in a "master reaction" or bottleneck that limits the total flux and the rate of the process.

This picture can be used to demonstrate that enzymatic reactions are the center of kinetic considerations for microbial systems, even if it is hard to verify directly such a mechanistic approach in real fermentations.

The differences between steady state, stationary state, and transient state may be clarified with the aid of a consecutive reaction of order $n = 1$:

$$A \xrightarrow{k_{r,1}} B \xrightarrow{k_{r,2}} C \tag{5.18}$$

with the differential equations

$$\frac{da}{dt} = -k_{r,1} \cdot a \tag{5.19a}$$

$$\frac{db}{dt} = k_{r,1} \cdot a - k_{r,2} \cdot b \tag{5.19b}$$

$$\frac{dc}{dt} = k_{r,2} \cdot b \tag{5.19c}$$

The concentration time curve is presented in Fig. 5.9, which shows the ranges of the different states mentioned.

The following equation summarizes all possible reaction mechanisms containing the reversible steps of adsorption, desorption, and reaction:

$$E + S \underset{k_{-1}}{\overset{k_{+1}}{\rightleftharpoons}} ES \underset{k_{-2}}{\overset{k_{+2}}{\rightleftharpoons}} EP \underset{k_{-3}}{\overset{k_{+3}}{\rightleftharpoons}} E + P \tag{5.20}$$

From this generalized mechanism, all singular kinetic equations known in the literature can be derived (see App. III), including the following five types of kinetics:

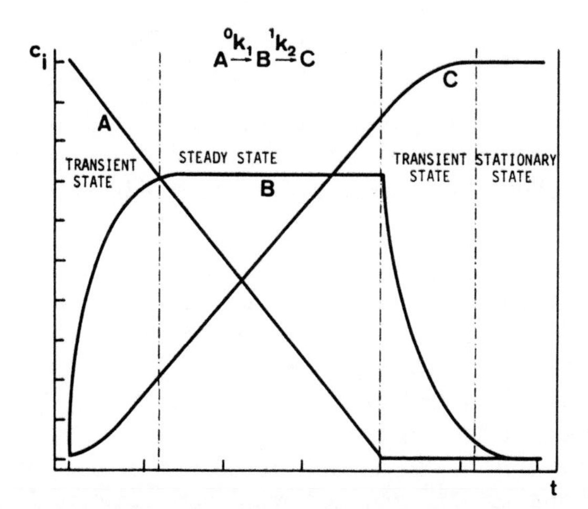

FIGURE 5.9. Illustration of different states in reaction kinetics in case of a simple consecutive chemical reaction with zero order in the first step ($^0k_1$) and first order in the second ($^1k_2$): transient states, steady state, and stationary state.

1. Michaelis–Menten equation (cf. Equ. 2.54) in case of irreversible reaction ($k_{-2} = 0$) and of no distinction between ES and EP ($k_{+3} = 0 = k_{-3}$).
2. Briggs–Haldane equation as an alternative with a different interpretation of the K_m value.
3. Langmuir–Hinshelwood approach, with irreversible reaction as further alternative applied to chemical and enzymatic reactions ($k_{-2} = 0$).
4. The reversible case with $k_{-2} > 0$ for Michaelis–Menten kinetics ($k_{+3} = 0 = k_{-3}$).

   The resulting final kinetic equation (e.g., Mahler and Cordes, 1966)

$$r_{tot} = r_+ - r_- = \frac{r_{+max} \cdot r_{-max} \cdot s - r_{+max} \cdot r_{-max} \cdot p/K_{eq}}{K_S \cdot r_{-max} + r_{-max} \cdot s + r_{+max} \cdot p/K_{eq}} \tag{5.21}$$

   is valid not only for the so-called "initial rate kinetics in the forward direction" (cf. Equ. 2.54), as it is utilized in biochemistry, but over the entire range, including the final phase where the reverse reaction (r_-) is noticeable, until the equilibrium between r_+ and r_- is established. The equilibrium constant K_{eq} also called the "Haldane relationship," shows the limitation of reversible enzymatic reactions due to the thermodynamic equilibrium:

$$K_{eq} = \frac{k_{+1} \cdot k_{+2}}{k_{-1} \cdot k_{-2}} \tag{5.22}$$

   The individual reaction rates (r_+ at $p = 0$ and r_- at $s = 0$) govern simple enzyme kinetics. The kinetic parameters, among others, are taken from Equ. 5.22 using special methods for experimental verification.
5. The reversible Langmuir–Hinshelwood kinetic equation, utilizing the full mechanism in Equ. 5.20, results in the following equation:

$$r_{tot} = r_- - r_- = \frac{r_{+max} \cdot K_p \cdot s - r_{-max} \cdot K_S \cdot p}{K_S \cdot K_p + K_p \cdot s + K_S \cdot p} \tag{5.23}$$

   and doubtless it presents a realistic representation useful for engineering calculations in situations in which the accumulation of product inhibits further production due to the reversible nature of the reaction (cf. App. III).

The evaluation of the parameters of the model is complex in all cases of reversible reactions. For various P concentrations, one draws, for example, a double reciprocal plot (similar to Fig. 4.24c) of $1/r$ versus $1/s$; all of the straight lines pass through the point $1/r_{max}$. The K_S value is greater with greater P concentration. The kinetic parameters can be calculated with the help of the initial rate parameters r_+ and r_-.

In the same spirit as the simple reaction mechanisms already discussed (cf. Equ. 5.20 and App. III), and their final kinetic equations (Equs. 2.54, 2.55, 5.21, 5.23), mechanisms and kinetic equations can also be derived for the inhibition of various types of enzymatic reactions. A special case has already been considered in Equ. 5.23. The most important class of inhibitors reducing the rate of enzyme-catalyzed reactions is that of reversible inhibitors forming dynamic complexes with an enzyme whose catalytic properties differ from those of the free enzyme. A general scheme proposed by Botts and Morales (1953) is useful in discussing inhibitors that are not reaction products:

$$\begin{array}{ccc} E & \underset{P}{\overset{S}{\rightleftharpoons}} & ES \\ I \updownarrow & P & \updownarrow I \\ EI & \underset{S}{\overset{P}{\rightleftharpoons}} & EIS \end{array} \tag{5.24}$$

where I is the inhibitor. This scheme includes most of the simple types of inhibition:

1. Competitive inhibition if EIS and the reactions involving it are missing
2. Uncompetitive inhibition if EI is missing and EIS does not break down
3. Noncompetitive inhibition if EIS complex is inactive and no effect on substrate binding occurs (E and EI have equal affinities for S)

The kinetic behavior of various inhibition types is represented in the graphs of Fig. 5.10a–c. An additional type of inhibition is the so-called "substrate inhibition," which can occur at high S concentrátions, graphically shown in Fig. 5.10d (cf. Equ. 5.88). The resulting kinetic equations are:

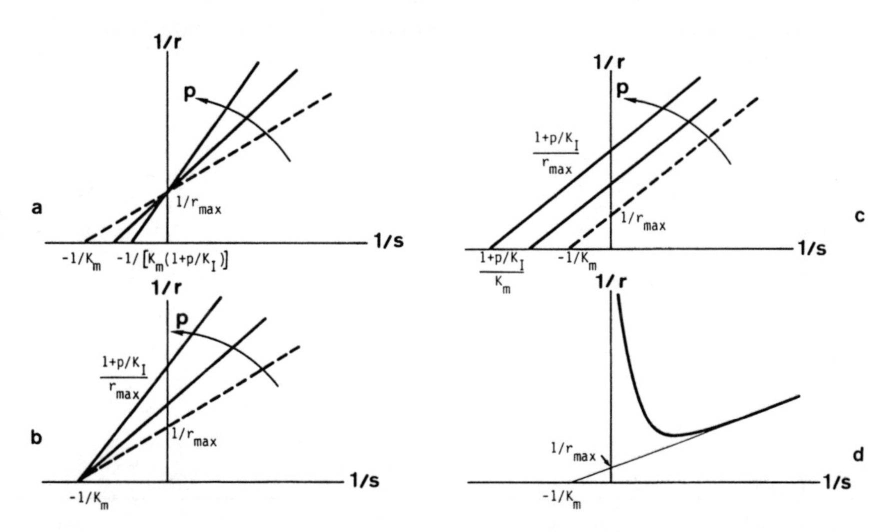

FIGURE 5.10. Demonstration of the four basic types of inhibitions of enzyme kinetics in Lineweaver–Burk plots (see Fig. 4.24c): (a) competitive, (b) noncompetitive, (c) uncompetitive, and (d) substrate inhibition. The parameters r_{max} and K_m can be estimated from the intercepts and slope of the line with $p = 0$, where p = inhibitor concentration.

For noncompetitive inhibition

$$r = \frac{r_{max}}{1 + i/K_I} \cdot \frac{s}{K_S + s} \tag{5.25}$$

For competitive inhibition

$$r = r_{max} \frac{s}{[K_S/(1 + i/K_I)] + s} \tag{5.26}$$

For uncompetitive inhibition

$$r = \frac{r_{max}}{1 + i/K_I} \cdot \frac{s}{[K_S/(1 + i/K_I)] + s} \tag{5.27}$$

A comparison of these equations shows that the various types of inhibitors (with concentrations i and inhibition constants K_I) result in a change of r_{max} or K_S or both of these parameters. An approximate, graphical presentation can show this (cf. Figs. 5.10a–c and 5.11) with $p = i$.

One can use a plot of $1/r$ versus $1/i$ (the Dixon plot) to obtain the parameter K_I, as demonstrated in Fig. 5.12a–c. The Walker plot can also be used for this purpose (Wilderer, 1976) as an integral evaluation method (cf. Fig. 4.20).

Although there are basic types of inhibition (see preceding section), in practice inhibition has mixed effects on multisubstrate enzymes and multiple inhibitors, and further classification is necessary:

1. Linear, if the replot of slope and/or intercept versus i is a straight line
2. Parabolic, if the slope and/or intercept replot is a parabola
3. Hyperbolic, if the replot is a hyperbola (Segel, 1975).

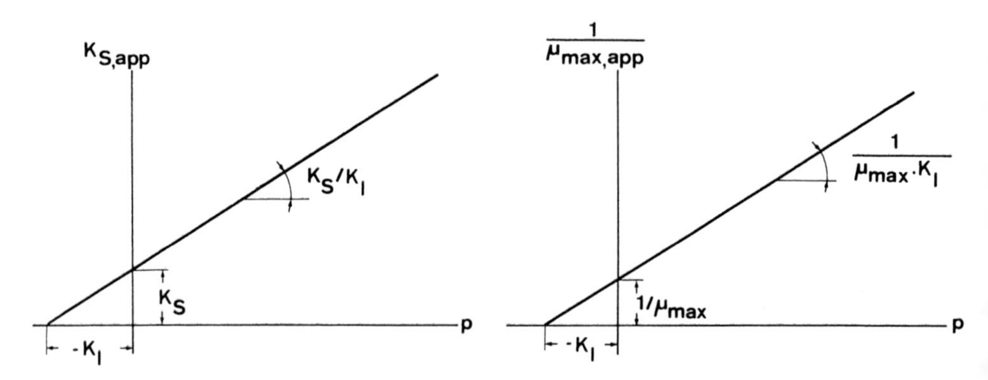

FIGURE 5.11. Estimation of true parameters K_S (a) and μ_{max} (b) from apparent values in replots of data taken from previous plots in case of competitive (a) and noncompetitive (b) inhibition.

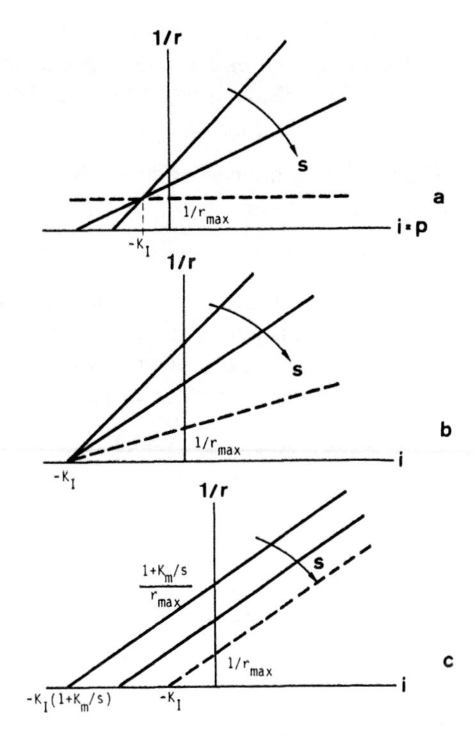

FIGURE 5.12. Dixon plots for the estimation of inhibition parameter K_I, respectively r_{max} and K_m, for different cases of enzyme inhibitions: (a) competitive, (b) noncompetitive, and (c) uncompetitive inhibition.

5.2.2.2 Enzyme Regulation and Control

In addition to enzymatic reactions, microbial growth phenomena involve the pronounced ability of living cells to adapt to environmental changes. A microorganism adapts its activities to changes in the environment by the following types of mechanisms (Fig. 4.9):

1. Simple regulations by direct mass-action law. These changes are generally covered by ordinary kinetics (unstructured models). Rapid response is characteristic in this case, and the time constants of these changes are small.
2. Regulation of enzyme activity by small molecules. These effectors cause changes in conformation and activity that can be positive or negative. These so-called "allosteric controls" are vital to the integration of metabolism in organisms (Monod, Changeux, and Jacob, 1963). The key enzymes in metabolic networks are especially affected by allosteric phenomena, and change in kinetic behavior is rapid. The time constant is normally small.
3. Regulation of enzyme amount. Enzyme synthesis is governed by such regulatory mechanisms as induction, repression, and catabolite repression. These mechanisms, which may have drastic effects on cellular composition, have time constants on the order of minutes to hours, and hence result in a much slower adaption than the mechanisms treated before. Drastic

changes in the kinetic behavior of the cell are the result. These changes may severely lag behind the changes in environmental conditions. Such "lag times" in transient modes of operations constitute a "biological inertia" that must be taken into account in mathematical modeling (Roels, 1978).

If an enzyme molecule binds more than one molecule of S, so that the overall binding can be represented as

$$E + nS \rightleftharpoons \{ES_n\} \rightarrow E + nP \tag{5.28}$$

the response of the enzyme activity to S may be sigmoidal rather than hyperbolic. This may also be interpreted on the basis of an allosteric change in conformation of the enzyme (Monod, Changeux, and Jacob, 1963).

A mathematical function describing allosteric inhibition (Hill, 1910)

$$r = r_{max} \frac{s^{n_H}}{K_H + s^{n_H}} \tag{5.29}$$

is known as the Hill equation, where n_H is the number of substrate-binding sites per molecule of enzyme, normally in the range of 1 to 3.2 (Hill coefficient), and K_H is a constant comprising the intrinsic dissociation constant K_m and some interaction factors of substrate binding. The definition is analogous to that for K_m (see Fig. 5.13). The estimation of the model parameters n_H and K_H can be carried out with the logarithmic form of the Hill equation

$$\log \frac{r}{r_{max} - r} = n_H \cdot \log s - \log K_H \tag{5.30}$$

A plot of $\log r/(r_{max} - r)$ versus $\log s$ (cf. Fig. 5.14) is a straight line with a slope of n_H and an intercept of $-\log K_H$. When $\log r/(r_{max} - r) = 0$, $r/(r_{max} - v) = 1$ and the corresponding position on the $\log s$ axis gives $\log s_{0,5}$:

$$s_{0,5} = \sqrt[n]{K_H} \tag{5.31}$$

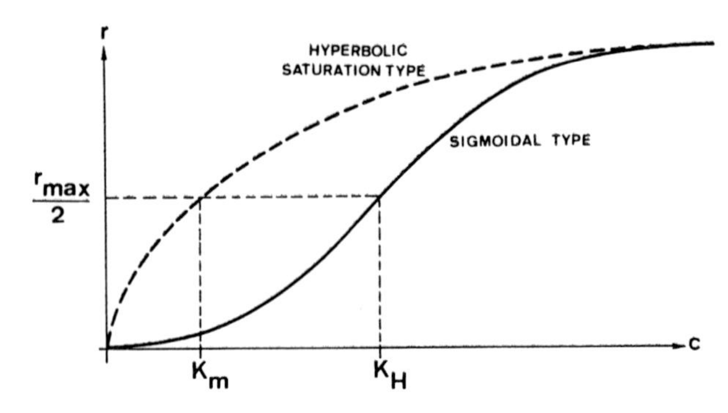

FIGURE 5.13. Basic biokinetic models including simple hyperbolic saturation type (e.g., Monod with K_S) and sigmoidal type (e.g., allosteric Hill with K_H).

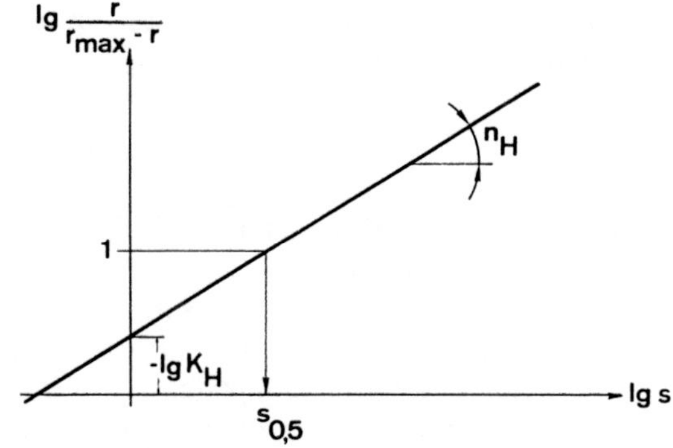

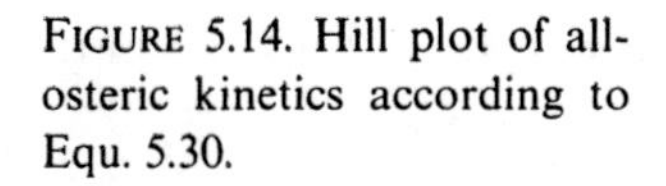
FIGURE 5.14. Hill plot of allosteric kinetics according to Equ. 5.30.

The influence of experimental conditions on Hill parameter estimation is described by Glende and Reich (1972). According to Reich et al. (1972), a nonlinear regression with a minimum of 15 measuring points should be used for an unbiased determination of parameters. Also, the Hill model has been extended by Adair (1925) to allow for the existence of stable, partly liganded intermediates.

It is interesting to note that this sigmoidal character of enzyme kinetics plays an essential role in regulation. Most of the regulatory enzymes in the metabolic pathways of amino acid biosyntheses and in carbohydrate metabolism exhibit such behavior, with the result that the kinetics can change in the presence of effectors (intermediary metabolities) from sigmoid to hyperbolic (see glycolysis oscillations, Sect. 5.2.1).

The nature of some simplifications is outlined in detail in the articles of Frederickson et al. (1970), Roels (1978), and Roels and Kossen (1978). However, the construction of such a model requires the structuring of the biomass (see Fig. 4.8).

A number of publications have appeared on the dynamics of enzyme synthesis in a variety of situations. Most of the models are based on more or less sophisticated versions of the operon model of Jacob and Monod. The role of m-RNA and its stability were modeled by Terui (1972). Repressor and inducer control was treated by Knorre (1968), Imanaka et al. (1972; 1973), van Dedem and Moo-Young (1973), and Suga et al. (1975). Allowance for dual control and catabolite repression was made by Toda (1976). [See also the kinetic treatment by Yagil and Yagil (1971), Imanaka and Aiba (1977), and Bajpai and Ghose (1978)]. A simple structured model was developed by Roels (1978) showing a combination of the features of the models published. More recently Toda (1981) reviewed the effects of induction and repression of enzymes in microbial cultures and their modeling.

The basic concept in the modeling of enzyme synthesis in transient situations involving a change in environment such as an upward shift in substrate level is that the level of enzyme concentration in the cells must increase. For this reason, the structural units responsible for enzyme synthesis (m-RNA in the ribosomes R) must first be produced.

$$\frac{dr}{dt} = k_E \cdot r \cdot x \tag{5.32}$$

where k_E is the kinetic constant and r is the concentration of ribosomes; x is indirectly a measure of e.

The concentration of ribosomes in a state of balanced growth is a function of the growth rate μ (Maaloee and Kjeldgaard, 1965; Tanner, 1970). Thus, $r \approx \mu$. In this way, the changing number of ribosomes represents a measure of biological inertia (lag time), which will also affect the growth rate (Romanovsky et al., 1974). This is expressed in the following differential equation:

$$\frac{dr}{dt} = \frac{1}{t_L}(\bar{r} - r) \tag{5.33}$$

where $\bar{r}$ is the stationary concentration of ribosomes, the value that will be reached with a time constant t_L (lag time, $t_L \sim 1/\mu$). This simple but adequate approach uses the concept of a characteristic time (see Sect. 4.2, Fig. 4.5, and Table 4.2).

5.2.3 Contribution of Chemical Kinetic Laws to Bioprocess Kinetic Modeling

In addition to the laws of enzyme microkinetics, the kinetic equations from chemical reactions based on the type 1 situation shown in Fig. 4.12 also provide a suitable approach. The power law equations with various reaction orders, n, differ in the c/t relationship of their reaction components. In Fig. 5.15, the time course of substrate concentration is compared for $n = 0$, 1/2, 1, and 2 and for Michaelis–Menten enzyme kinetics. The substrate disappearance and the oxygen utilization in biological waste water treatment may be cited as realistic examples. For the simple case, integration is possible (Levenspiel, 1972). The integrated solutions for various reaction orders are

$$n = 0 \qquad c_0 - c = k_0 \cdot t \tag{5.34}$$

$$n = 1/2 \qquad 2(\sqrt{c_0} - \sqrt{c}) = k_{1/2} \cdot t \tag{5.35}$$

$$n = 1 \qquad \frac{\ln c_0}{c} = k_1 \cdot t \tag{5.36}$$

$$n = 2 \qquad \frac{1}{c} - \frac{1}{c_0} = k_2 \cdot t \tag{5.37}$$

A graphical representation of concentration, the square root of the concentration, the logarithm, or the reciprocal concentration versus time leads to a linear form from which the slope, k_r, may be determined.

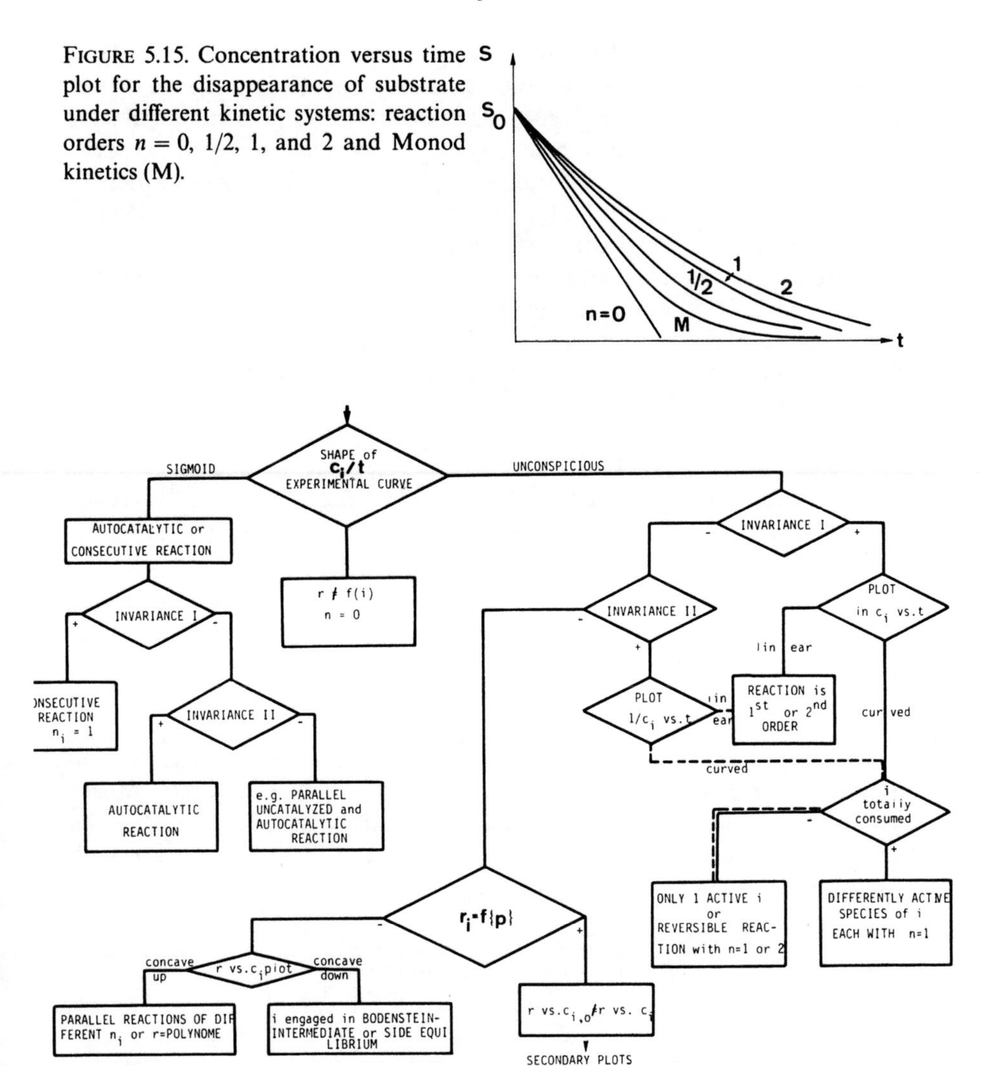

FIGURE 5.15. Concentration versus time plot for the disappearance of substrate under different kinetic systems: reaction orders $n = 0$, 1/2, 1, and 2 and Monod kinetics (M).

FIGURE 5.16. Flowchart of systematic analysis of formal kinetics using the shape of experiments as decision criteria together with the property of invariance I and II, according to Schmid and Sapunov (1982).

In the application of chemical kinetics, a formal kinetic evaluation method has been proposed (Schmid and Sapunov, 1982). An operation scheme is illustrated in Fig. 5.16; it uses two properties of c/t curves as decision criteria, called invariance I and invariance II. These properties concern the linear transformation capability of first- and second-order reactions. Kinetic curves with various initial concentrations $c_{i,0}$ can be superimposed over arbitrary standard curves $(c_{i,0})_s$ by multiplying ordinates by ratios $(c_{i,0})_s/c_{i,0}$ in the case of first-order kinetics, and by multiplying ordinates in the same way plus dividing the abscissa by the same ratio in the case of second-order reactions. As recently outlined, the arbitrary reaction-order concept is still a valuable method for the formal kinetic modeling of bioprocesses in which the experiments are distorted by physical transport phenomena. These simple pseudo-kinetic types give interesting hints at the mechanism (A. Moser, 1983a,b). While a zero-order reaction can be the result of a $k_{L1}a$ limitation in the presence of a first-order reaction or of multi-S kinetics (overlapping S utilization), pseudo-first-order kinetics are often the consequence of catalyst poisoning in side reactions of a G|L process or of internal transport limitations such as in the case of biofilm processes in which the half-order concept is also adequate (see Sects. 4.5.2 and 5.8).

5.3 Basic Unstructured Kinetic Models for Growth and Substrate Utilization (Homogeneous Rate Equations)

In general, unstructured models can be considered a good approximation when the cell composition is time independent (i.e., in balanced growth) or when the composition of cells is irrelevant at the industrial scale. In applying such simplified models, one must be cautious in using extrapolations (see Sect. 2.3).

Fig. 5.17 shows a summary of growth and S utilization phenomena as a function of substrate concentration under various circumstances (A. Moser, 1978b). The simple Monod diagram $\mu = \mu(s)$ is selected as the form for graphical presentation. The cases 1 through 7 can be classified as shown (A. Moser, 1981; Roels and Kossen, 1978).

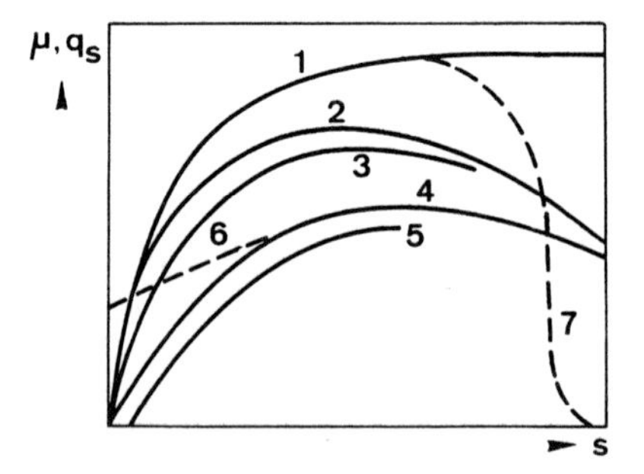

FIGURE 5.17. Diagram of the mathematical model describing the relationship between the specific growth rate μ (or specific rate of substrate utilization, q_S) and the limiting substrate concentration s (A. Moser, 1978b): 1, Monod kinetics; 2, Monod with S inhibition; 3, Monod with S inhibition at high X; 4, as 3 with P inhibition; 5, as 4 with endogenous metabolism; 6, as 4 with sequential metabolism of two substrates, that is, biosorption; and 7, transients such as lag phase.

5.3.1 $\mu = \mu(s)$: Simple Model Functions of Inhibition-free Substrate Limitation (Saturation-type Kinetics)

The present-day theory of microbial growth kinetics stems from, and is still dominated by, Monod's formulation (1942, 1949) of the function $\mu = \mu(s)$, given in Equ. 2.54. Also, this relation is a homologue of the Michaelis–Menten equation: Monod derived it empirically, and thus this is a formal kinetic equation. The consequence is a different interpretation of the parameters μ_{max} and K_S. The microbial growth rate is

$$\mu = \mu_{max}\frac{s}{K_S + s} \tag{5.38}$$

and the S consumption rate is

$$q_S = q_{S,max}\frac{s}{K_S + s} \tag{5.39}$$

These equations are related to each other with the aid of the Y concept ($Y_{X|S}$; see Equ. 2.13).

Several formulas have been proposed as alternatives to the Monod equation, due to the fact that some failures in applying the Monod equation have come to light:

- Teissier (1936) in an analogy to the organic growth in length

$$\mu = \mu_{max}(1 - e^{-s/K_S}) \tag{5.40}$$

- H. Moser (1958) in an analogy to allosteric Hill kinetics (see Equ. 5.29, and also Kargi, 1977)

$$\mu = \mu_{max}\frac{s^n}{K_S + s^n} \tag{5.41}$$

- Contois (1959) and Fujimoto (1963)

$$\mu = \mu_{max}\frac{s}{K_S x + s} \tag{5.42}$$

Powell (1967) compared these functions and elaborated another equation, taking into account cell wall permeability, substrate diffusion, and cell size by a factor K_D (see Fig. 5.74):

$$\mu = \mu_{max}\frac{s}{(K_S + K_D) + s} \tag{5.43}$$

It can be shown that the value of μ according to Monod kinetics approaches its asymptote too slowly to be a good representation of experimental data even in simple cases (cf. Fig. 5.18). The experimental model of Teissier (1936) and of Blackman kinetics (1905)

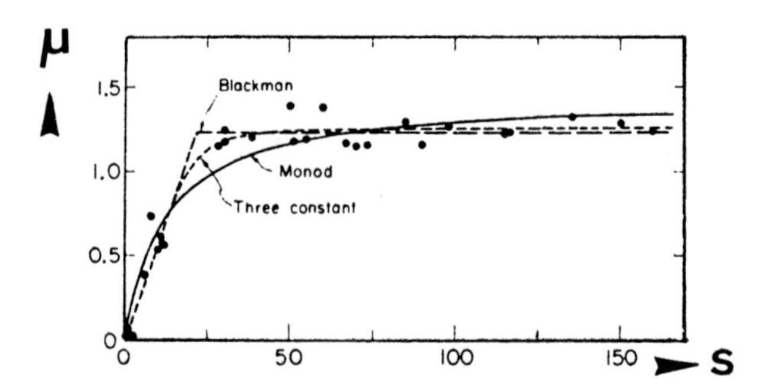

FIGURE 5.18. Comparison of three basic biokinetic models for substrate-limited growth in a conventional plot of μ versus s: Monod equation (cf. Equ. 5.38), Blackman kinetics (cf. Equ. 5.44), and the "three-constant (parameter) equation" (cf. Equ. 5.47). (From Dabes et al., 1973.)

$$\frac{\mu}{\mu_{max}} = \frac{1}{2} \cdot \frac{s}{K} \qquad \text{for } s < 2K \tag{5.44a}$$

and

$$\frac{\mu}{\mu_{max}} = 1 \qquad \text{for } s > 2K \tag{5.44b}$$

is generally thought to give a better fit because saturation is faster than with Monod kinetics.

Dabes, Finn, and Wilke (1973) in a more general approach suggested that (a) only the upper limit of growth rate is fixed by a single enzymatic step (rds concept proposed by Blackman in 1905), and (b) at low substrate concentrations, more than one step in a series of enzymatic reactions influences growth rate. These authors analyzed the linear sequence of enzymatic reactions

$$P_{n-1} + E_n \rightleftharpoons \{P_{n-1} \cdot E_n\} \rightleftharpoons P_n + E_n \tag{5.45}$$

and, using the concept of qss, they derived a general equation suggesting that reaction rates in a linear sequence cannot exceed the velocity of the forward reaction rate r_t of its slowest step, because this causes a term [the product $\Pi(r_i - \mu)$] to go to zero (rds concept).

$$s = \sum_n \frac{K_n \prod_i (r_{-,i-1} + \mu)}{\prod_i \frac{K'_{i-1}}{K_{i-1}} (r_{+,i} - \mu)} + \frac{P_n \prod_n (r_{-,n} + \mu)}{\prod_n \frac{K'_n}{K_n} (r_{+,n} - \mu)} \tag{5.46}$$

This equation simplifies to Monod's equation if one of the forward reaction rates is much slower than the others. A more complex situation obtains if two slow reactions are assumed. In the case that the second is slower than the first, a "three-parameter equation" results

$$s = \mu \cdot K + \frac{\mu - K_S}{\mu_{max} - \mu} \tag{5.47}$$

If the constant K is very small we simply have the Monod equation, but if K_S is very small the discontinuous function of Blackman kinetics is achieved. Model discriminations have shown that in a given confidence region of experimental error, the rival models cannot be distinguished (e.g., Boyle and Berthouex, 1974). Model functions with a minimum number of parameters are therefore preferred.

From the standpoint of chemical reaction kinetics, Kono (1968) and Kono and Asai (1968, 1969a,b) derived equations for growth and production rate that include a so-called "consumption activity coefficient," ϕ. The equation is more flexible than the simple Monod relation, and the growth rate is given by

$$r_x = \mu \cdot x \cdot \phi \tag{5.48}$$

The consumption coefficient ϕ can be numerically estimated from experimental data on growth, as shown in Fig. 5.19. The concept of a critical cell mass concentration x_{crit} is used, and it is determined by numerical methods from the plot of Fig. 5.19. The basic concept of the Kono approach is that reaction order changes at x_{crit} following the general growth equation

$$\frac{dx}{dt} = k_1^i \cdot k_2^j \cdot c_1^i \cdot c_2^j - k_3 c_3 \tag{5.49a}$$

where c_1, c_2, and c_3 are the concentrations of limiting substrate, co-substrate,

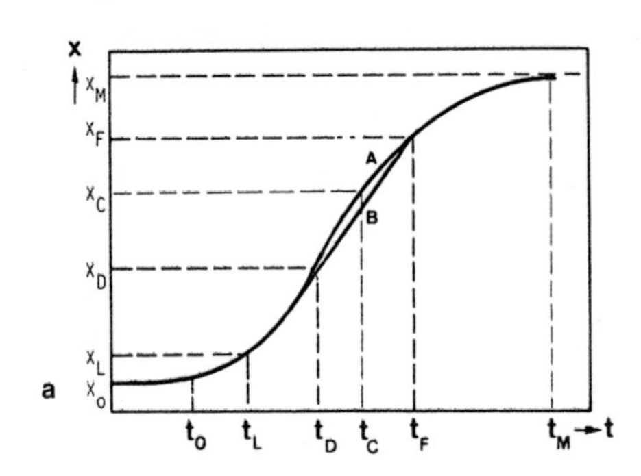

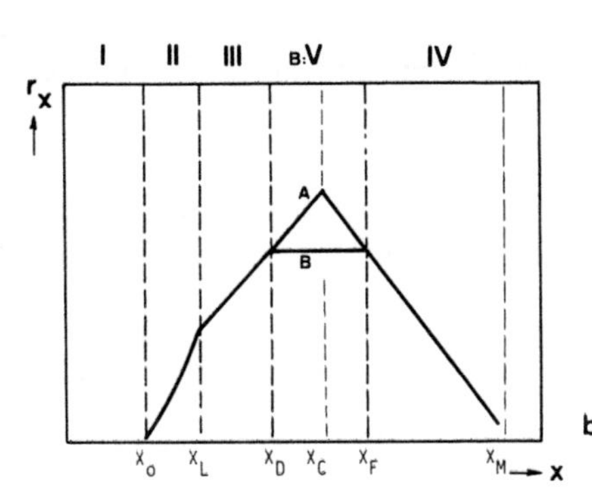

FIGURE 5.19. Representation of the numerical Kono approach in a c/t diagram of growth (a) and a kinetic plot r_x versus x (b). In agreement with different well-known growth phases I–IV, (cf. Fig. 5.23, lag phase, I; transition, II; exponential, III; and S-limitation decline phase, IV), and incorporating a linear growth phase in case of transport limitations (V in case B instead of exponential case A), numerical values can be taken from the plots to quantify the growth behavior with concentrations at certain times: x_0 = inoculum at t_0, t_L = lag-time, x_c = critical concentration at critical time when exponential growth is finished and reaction order changes from zero to one, x_m = maximal cell mass concentration. (Adapted from Kono, 1968, and from Kono and Asai, 1968, 1969.)

and product. Thus

$$i = 1 \text{ and } j = 0 \qquad \text{if } x < x_{crit} \tag{5.49b}$$

$$i = 0 \text{ and } j = 1 \qquad \text{if } x \geq x_{crit} \tag{5.49c}$$

By this numerical approach, the appearance of linear growth phases (characteristic for transport limitation) can also be handled, as shown in Fig. 5.19.

Another approach was derived by Mason and Millis (1976) and is given by

$$\mu = \mu_{asym} \frac{s}{K_S + s} + r_S^+ \cdot s^n \tag{5.50}$$

where μ_{asym} is the value of μ at which the asymptote of the curved function in a plot of μ versus s cuts the line $s = 0$ (Heidel and Schultz, 1978). The constant r_S^+ is the rate of simple diffusion of the substrate into the cell, given in $[m^3 \cdot kg^{-1} \cdot h^{-1}]$. Equation 5.50 applies when transport of substrate across the cell membrane is permease mediated and when this is the rate-determining step. Thus, this model describes deviations from Monod growth when the substrate is at a high concentration, and hence the term $r_S^+ \cdot s$ is significant. Schultz et al. (1974) showed that carrier-mediated diffusion through membranes also obeys a kinetic equation such as Equ. 5.50 with a variant interpretation of parameters ($\mu_{asym} = (D_c \cdot c)/d_m$ and $r_S^+ = D_S/d_m$, with D_S and D_c the diffusivities of substrate and carrier complex in the membrane of thickness d_m, and c the carrier concentration).

Alternatives to Monod-type kinetics for growth and S-utilization models postulate more structured mechanisms. Verhoff et al. (1972) divided cell growth into two steps, assimilation and ingestion (including cell division):

$$x_{ass} + \nu_2 S \underset{k_2}{\overset{k_1}{\rightleftharpoons}} x_{ingest} \xrightarrow{k_3} (1 + \nu_1) x_{ass} + \nu_3(\nu_2 - \nu_1) S \tag{5.51}$$

Assuming that assimilation is the rds, the following equation has been developed for $\mu = \mu(s)$:

$$\frac{\mu}{k_3 \cdot \nu_1} = -\left(\frac{k_1 \cdot \Delta s}{2k_3 \cdot \nu_1} + \frac{k_2 + k_3}{2k_3 \cdot \nu_1}\right) + \left[\left(\frac{k_1 \cdot \Delta s}{2k_3 \cdot \nu_1} + \frac{k_2 + k_3}{2k_3 \cdot \nu_1}\right)^2 + \frac{k_1 \Delta s}{k_3 \cdot \nu_1}\right] \tag{5.52}$$

where

k_i = rate constants

ν_i = stochiometric coefficients

Δs = concentration difference of limiting substrate outside and inside cell

Equation 5.52 reduces to simple Monod kinetics under certain conditions and also to the Lotka–Volterra relationship (cf. Sect. 5.6).

Another approach, quite different from that previously proposed, also operates with a dual S-utilization pattern, including uptake (assimilation or elimination from the liquid phase) and growth (degradation or utilization of S for growth):

$$S \xrightarrow{k_{\text{elim}}} S_e \xrightarrow{k_{\text{degrad}}} S_a \tag{5.53}$$

where S_a is the structural substrate not available for biomass synthesis and S_e is the functional substrate used for growth (Nyholm, 1976). The growth rate is expressed here as a function of intracellular concentration of the limiting substrate (s_{int}/x) and of "conserved" substrates such as inorganic ions (phosphate, nitrogen) or vitamins that are not broken down after uptake into the cells but rather are stored. Thus

$$\mu = \mu \frac{s_{\text{int}}}{(x)} \tag{5.54a}$$

and

$$\frac{d(s_{\text{int}})}{dt} = r_{\text{S,elim}} - r_{\text{S,degrad}} \tag{5.54b}$$

Obviously, this model is of practical interest in case of biological waste water treatment (see Sect. 5.9), because it accounts for the transition between limiting and nonlimiting conditions. In these situations, Monod-type kinetics are not applicable to conserved substrates, because μ is controlled by the amount of intracellular substrate, which may deviate significantly from extracellular concentration.

Another S-utilization mechanism has been proposed consisting of a consequence of two phenomena (adsorption and penetration of substrate molecules on specific sites followed by transportation through the cell membrane). In this approach, the following generalized kinetic model equation was derived (Borzani and Hiss, 1979; Hiss and Borzani, 1983):

$$\Psi \cdot \mu + \phi \cdot \dot{\mu} + 1 = 0 \tag{5.55}$$

where Ψ and ϕ are complex parameters (described in the original paper), depending on several fermentation parameters. Both parameters can be calculated from experimental data, even though at present no reasonable interpretion of individual variations can be given.

The objectives in developing a generalized model are (a) to aid in evaluating experimental growth data and (b) to aid in determining the conceptual basis and in suggesting other useful forms to describe microbial growth. From the general behavior of growth as a function of substrate concentration, a "driving force" for change in μ with respect to s may be thought of in terms of $(\mu_{\max} - \mu)$. Thus

$$\frac{d\mu}{ds} = k(\mu_{\max} - \mu)^{\text{p}} \tag{5.56}$$

where k is the kinetic constant and p is the reaction order (Konak, 1974). This equation is analogous to the well-known power law concept in chemical kinetics. It is not implicitly derived from a mechanism, but rather is a purely formal procedure. Using the concept of "relative" growth rate r_{rel} in a deeper biological sense (Tempest, 1976)

$$r_{rel} = \frac{\mu}{\mu_{max}} \tag{5.57}$$

Equ. 5.56 can be rewritten

$$\frac{d(\mu/\mu_{max})}{ds} = k \cdot \mu_{max}^{p-1} \cdot \left(1 - \frac{\mu}{\mu_{max}}\right)^{p} \tag{5.58}$$

Konak demonstrated that this equation reduces to the simple Monod relationship for $p = 2$ and to the Teissier form for $p = 1$. At the same time a quite interesting relationship between μ_{max} and K_S becomes apparent

$$K_S = \frac{1}{\mu_{max} \cdot k} \tag{5.59}$$

Similar generalizations of the differential form of the specific growth rate equation have been derived by Kargi and Shuler (1979), combining the differential forms of the four aforementioned specific growth rate equations. The following expression was the result of this work:

$$\frac{d(\mu/\mu_{max})}{ds} = K\left(\frac{\mu}{\mu_{max}}\right)^{m} \cdot \left(1 - \frac{\mu}{\mu_{max}}\right)^{p} \tag{5.60}$$

where K, m, and p are constants. This equation is of the same mathematical form as the "logistic equation" (Kendall, 1949). Comparing Equs. 5.60 and 5.57 shows, with $K = k \cdot \mu_{max}$, the full identity of both approaches. Table 5.1 gives the values of the constants of the generalized rate equation. The generalized rate equation for microbial growth, therefore, includes most widely used models as special cases with different constants. Equation 5.60 can be written as

$$\ln v = \ln K + m \cdot \ln v + p \cdot \ln(1 - v) \tag{5.61}$$

where $v = dr_{rel}/ds$; this is a more convenient form for parameter evaluation.

TABLE 5.1. Values of Constants in Equ. 5.60.

Model	*K*	*m*	*p*
Monod	$1/K_S$	0	2
Teissier	$1/K_S$	0	1
H. Moser	$n/K_S^{1/n}$	$1 - 1/n$	$1 + 1/n$
Contois	$1/K_S \cdot x$	0	2
Konak	$k \cdot \mu_{max}$	p	p

It is interesting to note that Vavilin (1982) developed a generalized model function independent of reactor type used for aerobic treatment of waste water (aeration tank, trickling filter, rotating disk) with a minimum number of coefficients:

$$q_S = q_{S,\max} \frac{s^n}{K_S^{n-p} \cdot s_0^p + s^n} \tag{5.62}$$

Another generalization of microbial kinetics was recently introduced (Levenspiel, 1980; Han and Levenspiel, 1987).

Finally, the influence of saline environment on the growth of mono and mixed culture is to be reported and summarized (Esener et al., 1982a). In Fig. 5.20, typical results are shown for the dependence of μ resp. q_o on NaCl concentration. Mixed culture data where taken from Imai et al. (1979a,b). A similar decrease was found also in the case of influence of salinity on yield factors, thermodynamic efficiency and the observed ratio of COD to BOD in activated sludge processes (Esener et al. 1981). A consequent decrease in $Y_{O|X}$ will than call for a higher aeration capacity. Thus the engineer will be faced with an optimization problem. In Fig. 5.21 $Y_{S|X}$ shows the amount of sludge

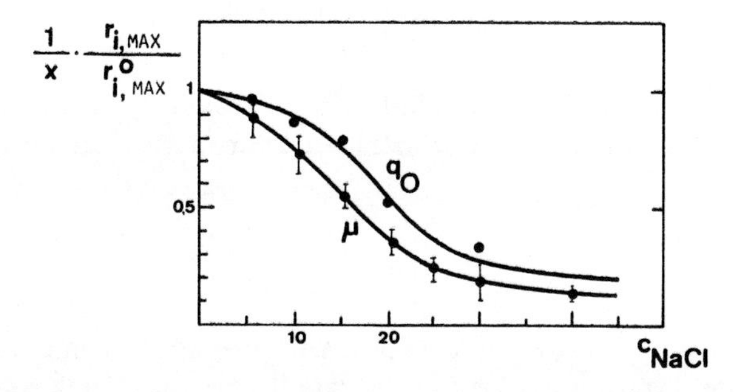

FIGURE 5.20. Effect of salinity quantified with salt-concentration c_{NaCl} on relative rates of maximum growth μ_{max} resp. respiration $q_{0,max}$ in pure microbial cultures. (Reprinted with permission from Esenér et al., *Proc. Second IWTU*, Vol. 2, copyright 1982, Pergamon Books Ltd.)

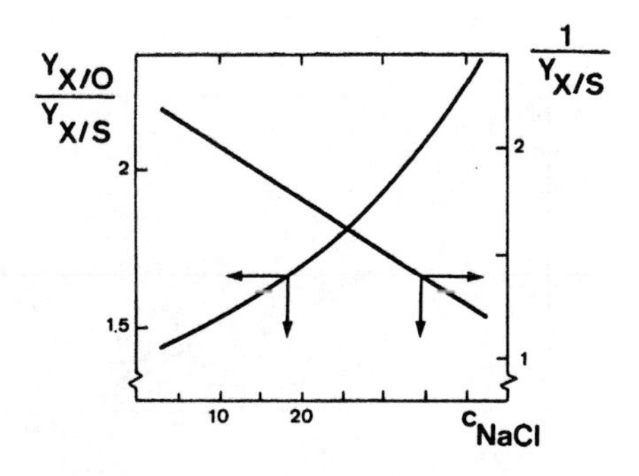

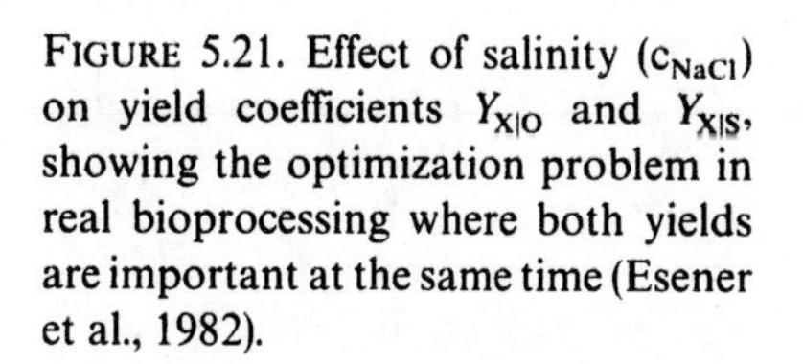
FIGURE 5.21. Effect of salinity (c_{NaCl}) on yield coefficients $Y_{X|O}$ and $Y_{X|S}$, showing the optimization problem in real bioprocessing where both yields are important at the same time (Esener et al., 1982).

formed per mole substrate consumed and $Y_{S|X}/Y_{O|X}$ is the amount of oxygen taken up per mole substrate consumed. Therefore it is desirable to minimize both. Efficient waste water operation e.g. must be carried out to minimize sludge production while maximizing yield on oxygen.

5.3.2 $\mu = \mu(x)$: Influence of Biomass Concentration on Specific Growth Rate

Normally it is assumed that microbial growth follows first-order kinetics with respect to biomass concentration (cf. Equ. 2.7). In exponential growth in batch processes and in continuous operation, $\mu \approx \mu_{max}$ can be realized. However, Monod (1949) himself presented different hypotheses concerning the function $\mu = \mu(s, x)$, illustrated in Fig. 5.22.

While Verhulst (1845) assumed a linear decrease according to

$$\mu = \mu_{max}^{0} - k_{x} \cdot x \tag{5.63}$$

represented by curve 3, a more realistic function is given in curve 2, quantified by

$$\mu(s, x) = \mu_{max} \frac{s_0 - (x/Y)}{K_S + s_0 - (x/Y)} \tag{5.64}$$

as a result of substrate limitation following a Monod-type kinetics (Meyrath, 1973). Curve 4 represents the improbable case where μ is inversely proportional to x

$$\mu = \frac{k_x}{x} \tag{5.65}$$

Obviously, the growth rate constant will formally become dependent on biomass concentration in case of substrate limitation at high cell concentrations. This fact is of great significance in the area of population dynamics. Using Equ. 5.63 for the calculation of cell mass as a function of time, a kinetic equation can be derived

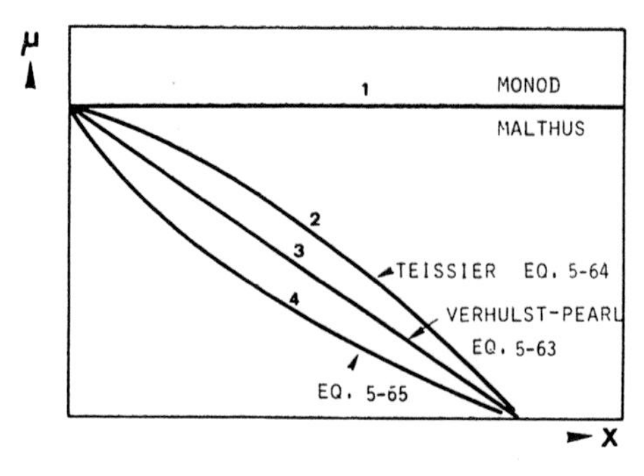

FIGURE 5.22. Graphical representation of the formal dependence of specific growth rate μ on biomass concentration x with several approaches from the literature taken as hypotheses.

$$N = \frac{N_0 \cdot \mu^0_{\max} \cdot e^{\mu^0_{\max} \cdot t}}{\mu^0_{\max} + m_x \cdot N_0 (e^{\mu^0_{\max} \cdot t} - 1)} \tag{5.66}$$

This is known as Verhulst–Pearl's equation, and it can be used in place of the normal non-substrate-limited growth of a population

$$N = N_0 \cdot e^{\mu_{\max} \cdot t} \tag{5.67}$$

5.3.3 $\mu = \mu(t)$: Extensions of Monod-Type Kinetics to Stationary and Lag Phase

While the simple Monod-type function is valid only in the exponential and decelerating growth phases, extensions can be introduced to extend the range of validity to the lag, stationary, and death phases.

Figure 5.23 gives the complete time course of a discontinuous growth processes; the growth curve as a function of time is reflected in Fig. 5.1. Whenever a bioprocess shows an exceptional lag phase, the simple growth kinetics $\mu(s)$ should be augmented with a time-dependent element $\mu = \mu(s, t)$.

5.3.3.1 Kinetic Modeling of Lag Phases

The duration of a lag phase before the specific growth rate reaches its maximum value is conveniently defined by the method of Hinshelwood (1946) and Lodge and Hinshelwood (1943):

$$t_L = t - \left(\frac{2.3}{\mu_{\max}}\right) \log \frac{x}{x_0} \tag{5.68}$$

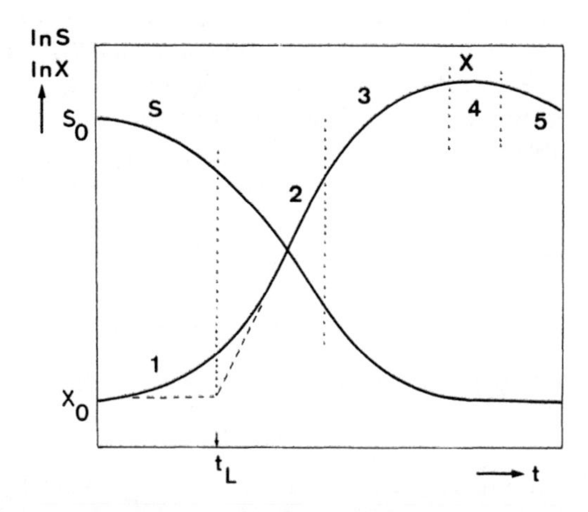

FIGURE 5.23. Concentration/time plot representing the growth phase of a discontinuous culture of microorganisms: (1) lag phase with t_L = lag time and $\mu = 0$, (2) exponential phase with $\mu = \mu_{\max}$, (3) decay phase with $\mu = f(s)$, (4) stationary phase with $\mu = -k_d$, and (5) lysis phase (endogenous metabolism phase) with $k_d > 0$.

A plot of $\log x$ versus t may be successful, and it follows from Equ. 2.6 (Pirt, 1975)

$$x = x_0 \cdot e^{\mu(t - t_L)} \tag{5.69}$$

It is known that the lag time t_L (cf. App. II) may depend on the size of inoculum x_0, quantified by Hinshelwood (1946) as

$$t_L = \frac{\text{const}}{N_0 - \text{const}} \tag{5.70}$$

and may be correlated with substrate concentration (Edwards, 1969)

$$t_L = 1.16(\bar{s} - 0.4) \tag{5.71}$$

Other formal kinetic equations for the quantification of lag phases in microbial growth are found in the literature. A simple extension of Monod-type kinetics using the lag time t_L as model parameter is given by Bergter and Knorre (1972):

$$\mu(s, t) = \mu_{max} \frac{s}{k_S + s}(1 - e^{-t/t_L}) \tag{5.72}$$

The plot of Fig. 5.23 may be used for the determination of μ_{max} and t_L. Other formal kinetic approaches are described by Kawashima (1973) and Edwards (1969). From a physiological point of view, the lag gives information about the mechanism of RNA and enzyme synthesis (see also Fig. 4.9). Indeed, the study of lag times was the basis of Jacob's and Monod's discovery of the regulation of enzyme synthesis (see Sect. 5.2.2.1). A lag phase is found not only after inoculation but also in cases of sequential utilization of substrates; this is called "diauxie." This phenomenon can be quantified by the formal macro-approach (A. Moser, 1983a) as shown in Appendix II. A mechanistic model of lag phases has been introduced by Pamment et al. (1978) and discussed by Barford et al. (1982). Starting from the observation that t_L was not inoculum-size dependent, Pamment and colleagues found the major factor involved to be the availability of the enzymes for glycolysis and respiration. This can be described by a structured model (Pamment et al., 1978) with two different kinds of biomass, but the number of parameters is high, and Equ. 5.72 is more practical in bioengineering calculations (A. Moser, 1984a).

5.3.3.2 Kinetic Modeling of Stationary Phase of Growth

The logarithmic or exponential law, given by the equation

$$r_x = \mu_{max} \cdot x \tag{5.73}$$

can be modified to incorporate a stationary growth phase. This revised form is known as the "logistic law" (Kendall, 1949; see also Equ. 5.58)

$$r_x = \alpha \cdot x \left(1 - \frac{x}{\beta}\right) \tag{5.74}$$

where α and β are empirical constants.

With $\alpha = \mu_{max}$ and $\beta = x_{max}$, La Motta (1976) used the logistic equation for the quantification of continuous growth cultures. Even though this approach has been successfully applied in fitting growth curves (e.g., Constantinides et al., 1970a and b), the drawback is that there is no clear relationship between μ and s in Equ. 5.74. However, the need for this equation is apparent in situations where product formation in fermentations occurs when microbial growth has ceased, for example, in antibiotic fermentation and also in biological waste treatment processes with $\bar{t} \rightarrow$ high.

5.3.4 Negative Biokinetic Rates—The Case of Microbial Death and Endogenous Metabolism

5.3.4.1 Kinetic Modeling of Microbial Death Phase

The microbial death phase must also be considered, especially in bioprocesses with long residence times, such as in biological waste treatment where one might expect an increasing chance of death with an increasing time of exposure. Various hypotheses exist (Fig. 5.24), including transitions between a strictly exponential form (curve I) and an abrupt death (curve IV). The Monod equation, therefore, must be enlarged by including a specific death rate $k_d[h^{-1}]$ in the balance equation for x, defined as follows:

$$k_d = -\frac{1}{x}\cdot\frac{dx}{dt} \tag{5.75}$$

Thus, a first model approach to a reaction $S \rightarrow X$ is

$$r_x = (\mu - k_d)x \tag{5.76}$$

together with

$$-r_s = \frac{1}{Y_{X|S}}\cdot r_x \tag{5.77}$$

This model was successfully used for activated sludge processing (Chiu et al., 1972).

In App. II a computer simulation of this model equation in batch processes (A. Moser and Steiner, 1975) with varying values for k_d is presented. The influence of a neglected k_d value on the evaluation of Monod parameters in linearization diagrams is shown in Fig. 5.25, whereby false or even negative values for K_S can be the consequence (Grady et al., 1972). With the known value of k_d, it is apparent that the undisturbed values for μ_{max} and K_S can be estimated (Moser and Steiner, 1975) only from a plot corresponding to Equ. 5.78

$$\frac{1}{\mu + k_d} = \frac{K_S}{\mu_{max}}\cdot\frac{1}{s} + \frac{1}{\mu_{max}} \tag{5.78}$$

However, physiological considerations lead to the conclusion that two independent factors are responsible for the decline of the overall mass of the

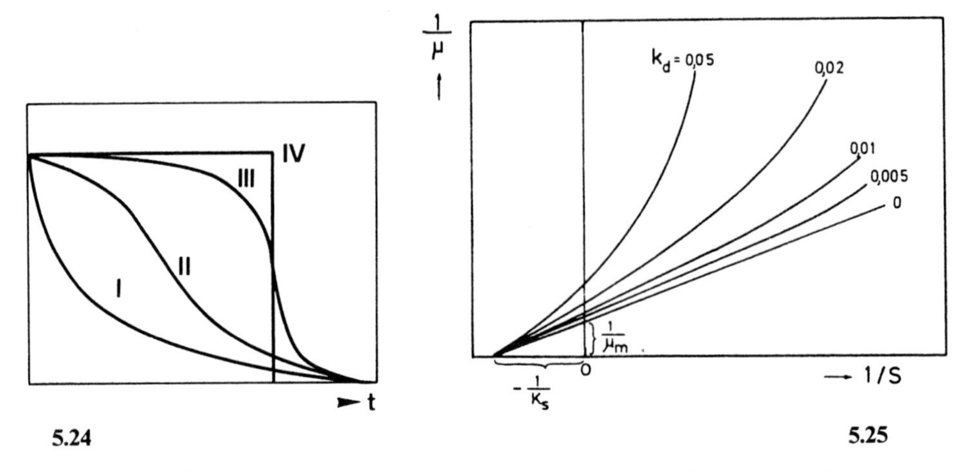

5.24 **5.25**

FIGURE 5.24. Various hypotheses for microbial death rate: strictly exponential form, curve I; sudden death after a certain time of exposure, curve IV; or curves II and III due to inequalities among cells in a population (Hinshelwood, *The Chem. Kinetics of the Bacteria Cell*, 1946, Oxford University Press.)

FIGURE 5.25. Representation of the influence of endogenous metabolism (as per Equ. 5.76 with k_d) on obtaining model parameters for microbial growth with Monod kinetics using a double reciprocal plot. (From Moser and Steiner, 1975.)

biomass. Viable cells are loosing mass due to endogenous metabolism, and they are also being converted to dead cells. A structured modeling of these phenomena, therefore, is necessary. Sinclair and Topiwala (1970) have done so by assuming individual rates of death of viable cells k_d^0 and of endogenous metabolism k_e. Thus, as illustrated in the case of a CSTR in Fig. 5.26 in dependence of dilution rate D

$$k_d = k_d^0 + k_e \tag{5.79}$$

The effect of the true rate of death k_d^0 on the steady-state yields becomes significant only at very low dilution rates, as shown in Fig. 5.26. Also, an apparent lag can be explained in batch cultures if the inoculum is of very low viability.

5.3.4.2 Kinetic Modeling of Endogenous Metabolism

Living cells that are thermodynamically characterized by a state far from equilibrium must take in high-energy substances and transform chemical energy to heat. This energy is necessary, for instance, for maintaining osmotic pressure and for repair processes on DNA, RNA, and other macromolecules. Thus, a source of energy must be consumed not only for macroscopic reactions like growth but also for maintenance of cellular structure. Therefore, the substrate balance equation must include a term m_s for maintenance (Pirt, 1965)

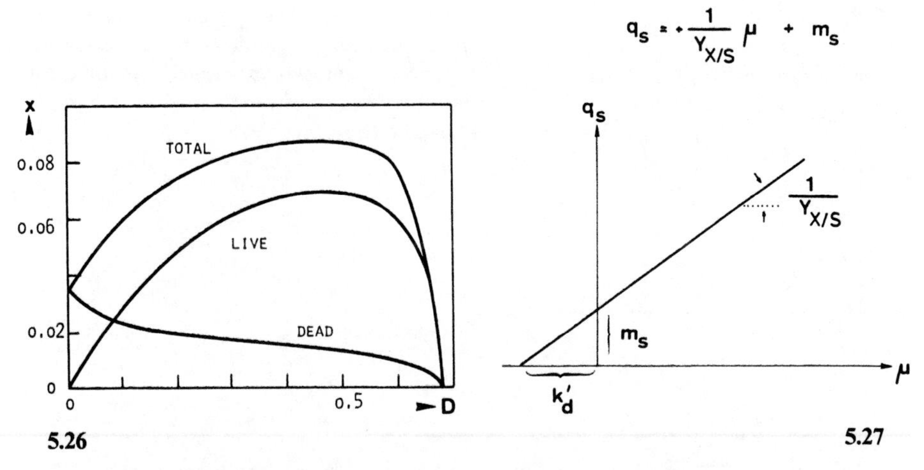

5.26

5.27

FIGURE 5.26. Plots of steady-state kinetics in a CSTR by assuming individual rates of death of viable cells and endogenous metabolism, showing the differences in biomass concentration x versus dilution rate D in this structured approach. (From Sinclair and Topiwala, 1970.)

FIGURE 5.27. Diagram of the estimation method for yield coefficient, $Y_{X|S}$, and m_S or k_d' of the kinetic model for endogenous metabolism, Equ. 5.80. (Adapted from Pirt, 1975.)

$$-r_S = \frac{1}{Y_{X|S}} \mu \cdot x + m_S \cdot x \tag{5.80}$$

together with

$$r_X = \mu \cdot x \tag{5.81}$$

where

$$m_S = -\left(\frac{1}{x} \cdot \frac{ds}{dt}\right)_e \tag{5.82}$$

The balance for energy-source utilization given in Equ. 5.80 can be used for the estimation of the maintenance coefficient m_S, plotting r_S versus r_x, as shown in Fig. 5.27. It is evident from this graph that

$$m_S = \frac{k_e}{Y_{X|S}} \tag{5.83}$$

which means that the specific maintenance rate k_e may be regarded as a turnover rate of biomass and is useful for comparing the turnover rates of biomass components with maintenance energy requirements (Herbert, 1958; Marr et al., 1963). An alternative plot for a CSTR is represented in Fig. 5.28. It is interesting to compare the different approaches given with Equs. 5.76 and 5.77 (model 1) and Equs. 5.80 and 5.81 (model 2) to show the model approaches

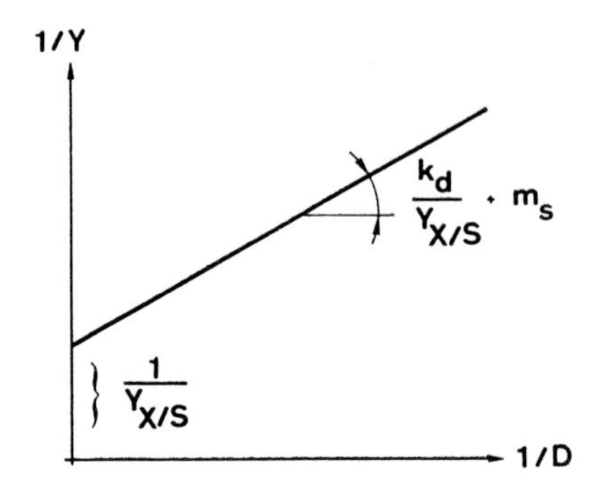

FIGURE 5.28. Alternative graphical method for the estimation of maintenance coefficient m_S by using chemostat experiments at different dilution rates D, as represented in the graph. (Adapted from Pirt, 1975.)

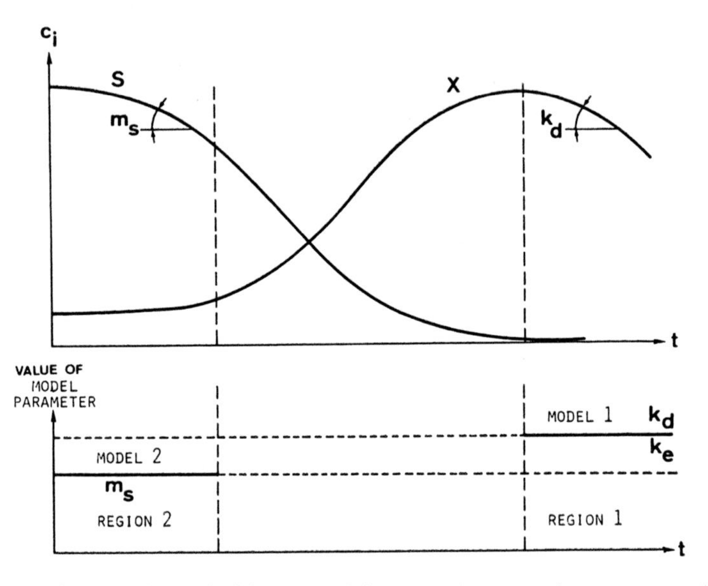

FIGURE 5.29. Comparison of different model approaches to endogenous metabolism and maintenance by showing the validity of the different model parameters k_d, k_e, and m_S.

used to quantify endogenous metabolism. In Fig. 5.29 the concentration time curves of a batch process are illustrated.

Even when the parameter k_d used in model 1 is, strictly speaking, valid only in the region where k_d can be estimated from the decline in the X curve (region 1), this model has been successfully applied to real bioprocessing (e.g., Chiu et al., 1972). The reason for this is that the reaction scheme behind model 1

$$S \xrightarrow{\mu} X_v \xrightarrow{k_d} X_d \tag{5.84}$$

where x_v is the viable cell mass and X_d is the death biomass, indirectly quantifies at the same time endogenous metabolism (see Equ. 5.83).

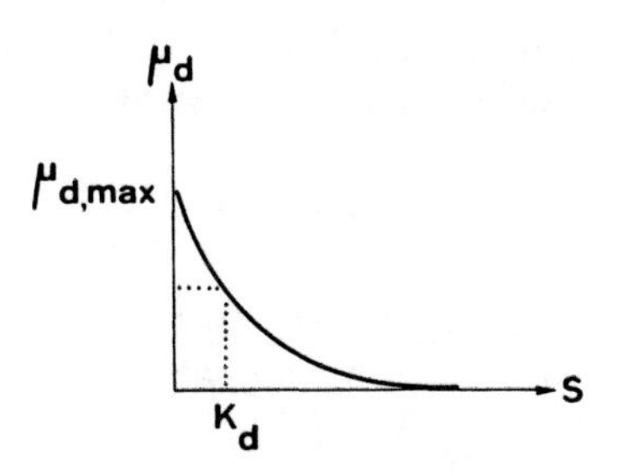

FIGURE 5.30. Kinetics of cell death due to the absence of substrate, and the method for obtaining the parameters $\mu_{d,max}$ and K_d of the model Equ. 5.85. (Reprinted with permission from *Chem. React. Eng. Adv. Chem. Ser.*, Vol. 109, p. 603, Humphrey. Copyright 1972 American Chemical Society.)

On the other hand, model 2 operating with the parameter m_S, which can be estimated directly from the decline of the S curve in the beginning of the batch run with $r_x = 0$ (region 2), represents a scheme of parallel reactions:

$$S \begin{cases} \xrightarrow{\mu} X \\ \xrightarrow{m_S} \text{endogenous metabolism} \end{cases} \tag{5.85}$$

where m_S is assumed to be constant over the whole range of the c/t curve. Other modifications of the basic models for endogenous metabolism have been proposed by Humphrey (1978) and by Gutke (1980, 1982). These show an S-concentration dependence of the parameters μ_d and m_S as given in Equs. 5.86 and 5.87 and Fig. 5.30:

$$\mu_d = \mu_{d,max}\left(1 + \frac{s}{K_d + s}\right) \tag{5.86}$$

$$m_S = m_{S,max}\left(\frac{s}{K_e + s}\right) \tag{5.87}$$

The most probable model will include both terms k_d and m_S, thus combining Equs. 5.76 and 5.80.

A comparison of unstructured growth models of microorganisms is given by Takamatsu et al. (1981), and Esener et al. (1983) discuss the applicability.

5.3.5 KINETIC MODEL EQUATIONS FOR INHIBITION BY SUBSTRATES AND PRODUCTS

In most cases of inhibition, the formal kinetic model equations are, like Monod's relationship, derived from theories of the inhibition of single enzymes. The equations are, however, only hypotheses; they may be replaced by any other adequate model.

5.3.5.1 Substrate Inhibition Kinetics

There are a large number of formal kinetic equations in the literature:

1. Andrews (1968, 1969, 1971) and Noack (1968) analyzed substrate inhibition in chemostat culture using

$$\mu = \mu_{max} \frac{1}{1 + K_S/s + s/K_{I,S}} \approx \mu_{max} \cdot \frac{s}{K_S + s} \cdot \frac{1}{1 + s/K_{I,S}} \quad (5.88)$$

   where $K_{I,S}$ is the substrate inhibition constant.
2. Allosteric substrate inhibition (Webb, 1963)

$$\mu = \mu_{max} \frac{s(1 + s/K_S')}{s + K_S + s^2/K_S'} \quad (5.89)$$

3. Multiple inactive enzyme–substrate complexes (Yano et al., 1966)

$$\mu = \mu_{max} \frac{1}{1 + K_S/s + \sum_j (s/K_{I,S})^j} \quad (5.90)$$

4. Empirical function (Aiba et al., 1968)

$$\mu = \mu_{max} \frac{s}{K_S + s} \cdot e^{-s/K_{I,S}} \quad (5.91)$$

5. A semiempirical form of the Teissier equation

$$\mu = \mu_{max} \left(\exp - \frac{s}{K_{I,S}} - \exp - \frac{s}{K_S} \right) \quad (5.92)$$

6. An equation derived by Webb (1963) assuming the Debye–Hückel theory for the effect of ionic strength, σ, on the rate

$$\mu = \mu_{max} \frac{s}{s + K_S(1 + \sigma/K_{I,S})} \cdot e^{1.17\sigma} \quad (5.93)$$

Edwards (1970) compared Equs. 5.89 through 5.93 with experimental results from the literature and found that discriminating among them was not possible. He therefore recommended the first form, Equ. 5.88, as the simplest of these equivalent equations. The behavior of Equ. 5.88 is shown in Fig. 5.31 and as a computer simulation in App. II.

The estimation of parameters is indicated in Figs. 5.31 and 5.32. $K_{I,S}$ is the value of s when half the maximum rate μ is reached. While Fig. 5.32 is adequate for an initial estimate, the three equations in Fig. 5.31 provide more accurate estimates (Humphrey, 1978). Other methods include a rearrangement of Equ. 5.88 for $s \gg K_S$ leading to

$$\frac{1}{\mu} = \frac{1}{\mu_{max}} + \frac{1}{\mu_{max} \cdot K_{I,S}} \cdot s \quad (5.94)$$

which can be used to determine $K_{I,S}$.

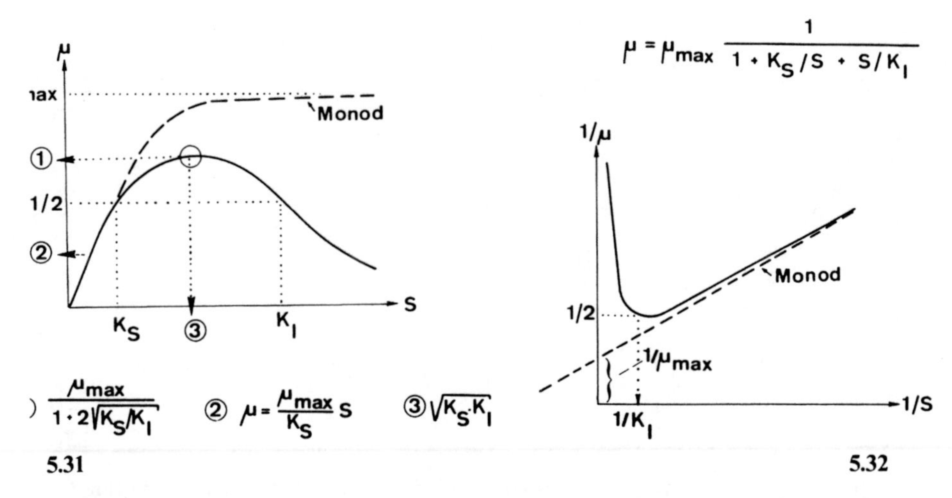

5.31 5.32

FIGURE 5.31. Kinetics of substrate inhibition and method of parameter estimation by precise method using computer simulation of the kinetic equation (cf. Equ. 5.88). (Reprinted with permission from *Chem. React. Eng. Adv. Chem. Ser.*, Vol. 109, p. 603, Humphrey. Copyright 1972 American Chemical Society.)

FIGURE 5.32. Kinetics of substrate inhibition and method of parameter estimation by approximation using a double reciprocal plot based on Equ. 5.88.

A quite different equation for the quantification of substrate inhibition kinetics has been proposed by Tseng and Waymann (1975) (Waymann and Tseng, 1976). Substrate concentrations higher than a characteristic threshold substrate concentration s_c inhibited growth linearly in accordance with

$$\mu = \mu_{max} \frac{s}{K_S + s} - K_{I,S}(s - s_c) \tag{5.95}$$

where $K_{I,S}$ is the inhibition constant. This approach should be used in cases where the inhibition pattern does not have the shape of the curve shown in Fig. 5.31 but rather shows a linear decrease, as schematically shown in Fig. 5.33 (Kosaric et al., 1984).

A generalized concept of a rate equation for one-substrate enzymatic reactions was proposed by Siimer (1978) for situations where the substrate S and products P_i can inhibit the reaction. The following general basic reaction mechanism was suggested as a basis for the derivation of a rate equation:

$$\begin{array}{ccccccc}
 & & SE & \xrightleftharpoons{(1/\beta)K_S} & SES & \xrightarrow{\delta \cdot k_{cat}} & SE + P_1 + P_2 \\
 & & \upharpoonleft\downharpoonright \beta \cdot K_S' & & \upharpoonleft\downharpoonright & & \\
EP_1 & \xrightleftharpoons{K_{P_1}} & E & \xrightleftharpoons{K_S} & \{ES\} & \xrightarrow{k_{cat}} & E + P_1 + P_2 \\
\upharpoonleft\downharpoonright & & \upharpoonleft\downharpoonright \alpha \cdot K_{P_2} & & \upharpoonleft\downharpoonright K_{P_2} & & \\
EP_1EP_2 & \xrightleftharpoons[(\gamma/\alpha)K_{P_1}]{} & EP_2 & \xrightleftharpoons[(1/\alpha)K_S]{} & EP_2S & \xrightarrow{\varepsilon \cdot k_{cat}} & EP_2 + P_1 + P_2
\end{array} \tag{5.96}$$

According to this scheme, SES and EP_2S complexes are supposed to be

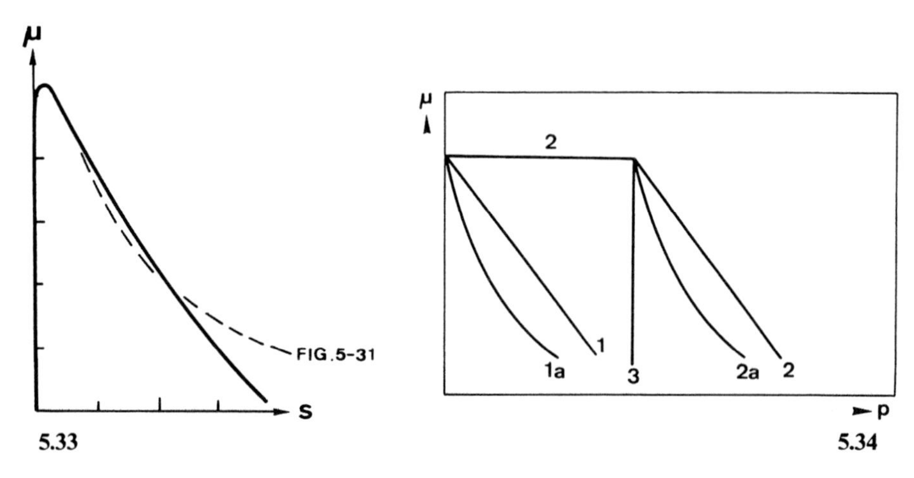

FIGURE 5.33. Comparison of two essential substrate inhibition models according to Equ. 5.88 and Equ. 5.95 (from Kosaric et al., 1984) in a diagram of specific growth rate μ versus substrate concentration s.

FIGURE 5.34. Possible hypotheses for the effect of product concentration p on microbial growth rate μ: linear decrease (1), first-order decrease (1a), sudden stop (3), and decrease after a period of no effect (2,2a). (Hinshelwood, *The Chem. Kinetics of the Bacteria Cell*, 1946, Oxford University Press.)

productive. P_1 inhibits competitively and P_2 noncompetitively. Using the concept of qss with $P_1 = P_2 = P$, the rate equation as a function of substrate conversion ζ_S (see Equ. 2.45) with parameters used in the scheme of Equ. 5.96 has the following form:

$$r = \frac{k_{cat} \cdot e_0(1 - \zeta_S)[1 + \delta(1 - \zeta_S)/K'_S + \varepsilon\zeta_S/K_{P_2}]}{a + b\zeta_S + c\zeta_S^2} \tag{5.97}$$

where a, b, and c are functions of s_0 and the dissociation constants K_S, K_{P_1}, and K_{P_2}.

5.3.5.2 Product Inhibition Kinetics

Hinshelwood (1946) distinguished different types of concentration–action curves (Fig. 5.34): linear decrease, exponential decrease, or a stepwise function. When there is no threshold concentration, the effect is often given by a simple linear relation (Dagley and Hishelwood, 1938)

$$\mu(s, p) = \mu_{max} \frac{s}{K_S + s}(1 - kp) \tag{5.98}$$

where k is the kinetic constant.

Beyond this numerical fit to product inhibition, which was modified by Holzberg et al. (1967)

$$\mu = \mu_{max} - k_1(p - k_2) \tag{5.99}$$

and by Ghose and Tyagi (1979) in the form

$$\mu = \mu_{max}\left(1 - \frac{p}{p_{max}}\right) \tag{5.100}$$

for quantifying ethanol inhibiton of yeast growth, the function $\mu(s, p)$ can be modeled by analogy to enzyme kinetics. Jerusalimsky and Neronova (1965) represented the dependence of μ on p by hyperbolic or sigmoidal curves, and Jerusalimsky (1967) recommended the following function

$$\mu(s, p) = \mu_{max}\frac{s}{K_S + s}\cdot\frac{K_{I,P}}{K_{I,P} + p} \tag{5.101}$$

where $K_{I,P}$ is the product inhibition constant. This model was used for a computer simulation of growth in discontinuous culture (Bergter, 1972), and it represents the most commonly used equation (Aiba et al., 1968, 1969a,b; Fukuda et al., 1978; Peringer et al., 1974). At the same time, Aiba and Shoda (1969) developed an empirical approach for the product inhibition pattern of yeast

$$\mu(s, p) = \mu_{max}\frac{s}{K_S + s}\cdot e^{-kp} \tag{5.102}$$

with k the empirical kinetic constant.

A quite flexible class of models has been described by Ramkrishna et al. (1966, 1967) using a sequence of reactions in which intermediary products inactivate viable cells. Oscillations of growth can be quantified using this model (Knorre, 1976), and a model of this type was used by Fishman and Biryukov (1974) for growth in penicillin fermentation. A similar approach is used in general analyses of growth patterns subject to the influence of intermediates (Knorre, 1980; Petrova et al., 1977).

Finally, Bazua and Wilke (1977) compared different approaches to ethanol inhibition of continuous yeast cultures and concluded that observed differences in the results are caused not only by different strains but also by altering experimental conditions. They proposed a three-parameter equation

$$\mu_{max} = \mu_0 - k_1\cdot\bar{p}(k_2 - p) \tag{5.103}$$

where $\bar{p}$ is the average value of p in continuous culture and k_1, k_2 are empirical constants. This equation shows that there is a limiting concentration of p, beyond which the cells grow or produce. These authors also used a two-parameter of an equation similar to Equ. 5.100 to fit experimental data

$$\mu_{max} = \mu_0\left(1 + \frac{\bar{p}}{p_{max}}\right)^{1/2} \tag{5.104}$$

with p_{max} the concentration of P at which cells are still viable. This equation shows some similarity to the generalized Monod equation developed by

Levenspiel (1980), with a Monod-type equation replacing μ_{max} by

$$k_{obs} = k\left(1 - \frac{p}{p_{crit}}\right)^n \tag{5.105}$$

where

$k = \mu_{max}$

p_{crit} = critical product concentration at which fermentation ends

n = toxic power number

To unravel the interacting effects of the four rate constants of the complete rate equation with Equ. 5.105 (k, K_S, p_{crit}, and n), it is more convenient to use a CSTR than a CPFR or a batch reactor. Making a series of runs at different p values, a plot can be made as shown in Fig. 5.10b, resulting in the parameter values of K_S and $k\ (1 - p/p_{crit})$. Then, with the value of p_{crit} easily estimated in advance, k and n can be found from a plot, as given in Fig. 5.35. The specific nature of the full rate equation with Equ. 5.105 says that there is a definite upper p limit for inhibitory effects (p_{crit}) above which fermentation ceases.

The empirical constant n accounts for the fact that inhibitory effects may work in different ways: linearly ($n = 1$), rapid initial drop followed by a slowing rate to p_{crit} ($n > 1$), or vice versa ($n < 1$), as shown schematically in Fig. 5.36 (Levenspiel, 1980). Experiments often show these types of inhibition effects (e.g., Hoppe and Hansford, 1982). Recently, Han and Levenspiel (1987) proposed a basic approach to biokinetic modelling by extending the Monod-equation to inhibition by substrate, product and biomass.

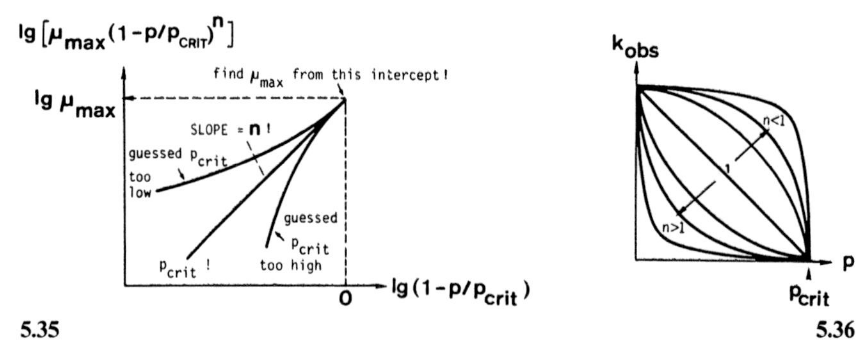

5.35

5.36

FIGURE 5.35. Evaluation of the toxic power coefficient n in the generalized microbial kinetic equation according to Levenspiel (1980). See Equ. 5.105.

FIGURE 5.36. Decrease of the observed rate constant in the generalized Monod equation modified with Equ. 5.105 with increase in the inhibiting product concentration p, showing the effect of toxic power number n. (From Levenspiel, 1980.)

5.3.6 Kinetic Model Equations for Repression

Catabolite repression plays a central role in many industrial fermentations such as in yeast technology (diauxic growth) and secondary metabolite productions. Even though the exact nature of the biochemical mechanisms involved in these regulations is often unknown (see Demain et al., 1979), the following type of formal kinetic equation

$$r_i \frac{1}{x} = r_{i,\max} \frac{s}{K_S + s(1 + s/K_R)} \cdot \frac{1}{x} \tag{5.106}$$

can be used as adequate approach. This equation is a formal analogy to the case of S inhibition (cf. Equ. 5.88 and Fig. 5.30) and was used for modeling yeast diauxie (A. Moser, 1978c; 1983a) and antibiotic production (Bajpaj and Reuss, 1981; Schneider and Moser, 1986; Moser and Schneider, 1988). Kinetic equations of structured models for induction and repression have been reviewed by Toda (1981).

5.3.7 $\mu = \mu(\text{pH})$

The influence of pH of a bioprocess can be dealt with in combination with inhibition kinetics. Andreyeva and Biryukov (1973) gave a number of different models for dealing with combined inhibition (see also Brown and Halsted, 1975).

For example, using Equ. 5.88 a formal kinetic equation can be constructed wherein the H^+ ion concentration h^+ is treated as if it were a substrate concentration. The value of the constant K is found by a method analogous to that in Fig. 5.29 using the concentration at the half-maximum value of μ from the two parts of the curve representing stimulation K_1 and inhibition K_2 (Humphrey, 1977b, 1978). Thus, a mixed inhibition function is obtained

$$r_i(s, \text{pH}) = r_{i,\max} \frac{s}{(K_S + s)(1 + K_1/h^+ + h^+/K_2)} \tag{5.107}$$

Andreyeva and Biryukov also gave a numerical curve-fitting procedure similar to Equ. 2.58 for evaluating the pH dependence of a rate

$$r_i = r_{i,\max}(\pm\alpha_0 \pm \alpha_1 \cdot \text{pH} \pm \alpha_2 \cdot \text{pH}^2 \ldots) \tag{5.108}$$

with α_i the coefficients of a polynomial.

5.3.8 Kinetic Pseudohomogeneous Modeling of Mycelial Filamentous Growth Including Photosynthesis

Active growth of molds occurs only at the tips of the hyphae. Bergter (1978) presented a kinetic model for calculating the relative rates of apical growth and branching of mycelial microorganisms in submerged cultures by measuring

μ and the quotient L/N with L the total length of hyphae per unit of culture volume and N the number of hyphal tips per unit of volume. On the basis of the simplest case of Hinshelwood's network theory (Dean and Hinshelwood, 1966), the following equation may be used to quantify the growth of small mycelial trees of *Streptomyces hygroscopicus* on solid surfaces

$$\frac{dL}{dt} = k_1 N \qquad \text{and} \qquad \frac{dN}{dt} = k_2 L \tag{5.109}$$

where k_1 is the mean relative rate of apical growth ($\mu m \cdot h^{-1}$), and k_2 is the mean relative rate of branching ($\mu m^{-1} \cdot h^{-1}$).

It can be concluded that

$$\mu = \sqrt{k_1 \cdot k_2} \tag{5.110}$$

The different dependence of k_1 and k_2 on s is shown in Fig. 5.37. With increasing s, the density of the colonies, which depends on the branching rate k_2, decreases faster than the radial growth rate; this can be interpreted in connection with the regulation of transport of metabolites within the hyphae. Cellular differentiation and the connection to product formation in molds was elaborated from Megee et al. (1970).

The growth of fungi has been reviewed by Prosser (1982), and growth of photosynthetic microorganisms has been summarized by Aiba (1982) and by Pirt et al. (1983) together with production in tubular photobioreactors. An unstructured model of algal growth has been developed by ten Hoopen et al. (1980) based on the Monod equation.

5.3.9 Kinetic Modeling of Biosorption

The phenomenon of "biosorption" occurs in biological waste water treatment ("sludge adsorption," Jones, 1971; Walters, 1966), in which substrates are eliminated rapidly from the liquid phase due to "adsorption" and "storage" in the sludge, but degradation itself lags behind the initial storage. Other examples of biosorption can be found in the literature pertaining to conventional fermentations (Bayer and Meister, 1982; A. Moser, 1974). The phenomenon can be detected by measuring the heat of combustion of sludge (differential thermal analysis). Simultaneous utilization of both light energy and substrate is another type of biosorption (Follmann, Märkl, and Vortmeyer, 1977).

Typical experimental findings are shown in Fig. 5.38 (Theophilou, Wolfbauer, and Moser 1978). Elimination is quantified by COD measurements, while degradation is given by the BOD values, occurring with a time delay. The difference is the effect of biosorption.

Quantification and modeling of biosorption must account for the biphasic nature just mentioned. Pseudohomogeneous treatments of biosorption fail because pseudohomogeneous macrokinetics are represented by a zero-order equation

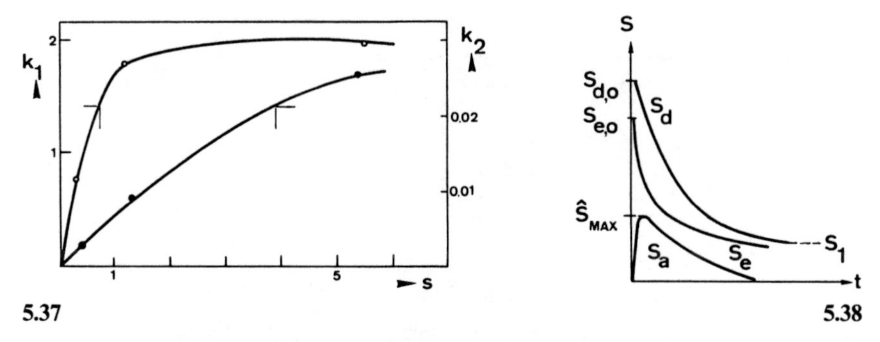

FIGURE 5.37. Dependence of kinetic parameters k_1 and k_2 in the model of filamentous growth (see Equ. 5.110) on glucose concentration s in case of chemostat culture of *Streptomyces hydroscopicus*. (From Bergter, 1978.)

FIGURE 5.38. Kinetics of adsorption (biosorption) in biological waste water purification. Time course of chemical and biological oxygen demand expressed as eliminated substrate S_e and degraded substrate S_d following the reaction scheme of substrate degradation and substrate elimination, and evaluation of the rate of adsorption S_a from the difference in chemical and biological rates (Theophilou et al., 1978). The final level S_1 represents the undegradable substrate. The value of S_{max} represents the maximal capacity of cells (sludge) to "adsorb" substrate (phenomenon of "biosorption").

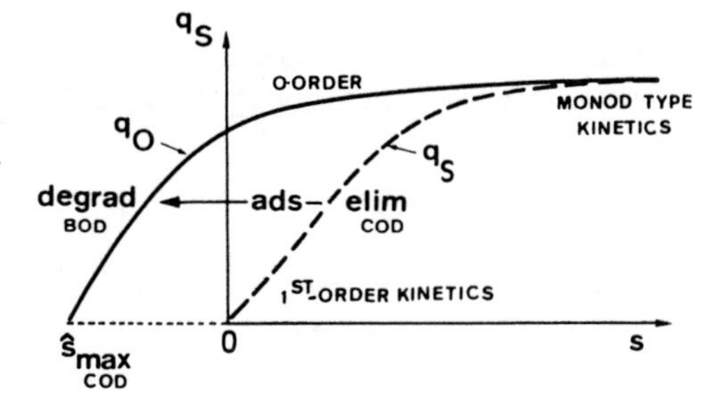

FIGURE 5.39. Consequences of "biosorption" effect for kinetics of biological waste water treatment shown in a plot of specific consumption rate q_S versus substrate concentration s with the resulting zero-order kinetics.

$$r_{S,\mathrm{eff}} = k_{0,\mathrm{app}} \cdot x \tag{5.111}$$

The appearance of this zero-order rate (BOD with respect to S) is seen in Fig. 5.39 (A. Moser, 1974). Several authors have developed a formal kinetic approach to biosorption effects (Busby and Andrews, 1975; A. Moser, 1977; Theophilou et al., 1978). Figure 5.40 illustrates the reaction scheme thought to be adequate as a starting point for the following set of model equations:

$$r_{\mathrm{elim}} = k_{\mathrm{elim}}(s_{\mathrm{elim}} - s_1) \tag{5.112a}$$

$$r_{\mathrm{degrad}} = k_{\mathrm{degrad}}(s_{\mathrm{degrad}} - s_1) \tag{5.112b}$$

$$r_{\mathrm{ads}} = r_{\mathrm{elim}} - r_{\mathrm{degrad}} \tag{5.112c}$$

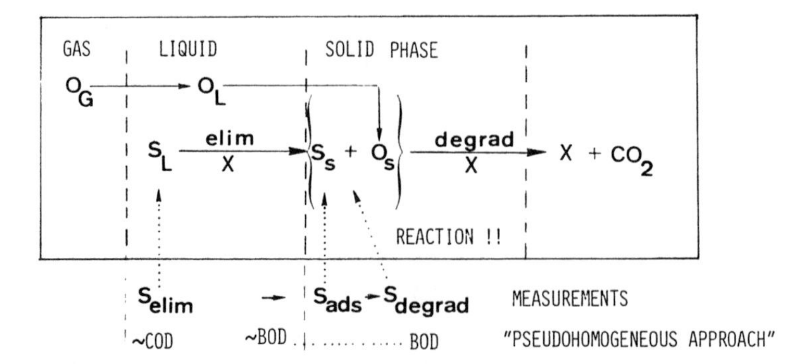

FIGURE 5.40. Elucidation of the "biosorption" phenomenon in a reaction scheme including substrate elimination (elim) quantified by COD removal, and degradation (degrad) accompanied by O_2 utilization (BOD) and "adsorption" (ads) following this pseudohomogeneous approach to the L|S process.

Adsorption rate often follows the approach given in Equ. 5.50. Thereby

$$s_{\mathrm{elim},0} = s_{\mathrm{degrad},0} - s_1 - s_{\mathrm{ads}} \tag{5.113}$$

and

$$\hat{s} \equiv s_{\mathrm{ads}} = \hat{s}_{\max} \frac{k(x \cdot s_{\mathrm{degrad},0} \cdot \xi_{\mathrm{ads}})}{1 + k(x \cdot s_{\mathrm{degrad},0} \cdot \xi_{\mathrm{ads}})} \tag{5.114}$$

Equation 5.114 is formulated by analogy to the adsorption isotherm (Langmuir kinetics, see Equ. 2.55) using the sorption capacity ξ_{ads} as measurement for sludge activity

$$\xi_{\mathrm{ads}} = 1 - \frac{\hat{s}}{\hat{s}_{\mathrm{mas}}} \tag{5.115}$$

($\hat{s}$ = adsorbed concentration, s_1 = undegradable S level). The term inside the parentheses in Equ. 5.114 can be regarded as the sorption potential. The value of $\hat{s}_{\max}$ is typical for a given waste water and is about 200 kg^{-3} at $x = 5$ kg^{-3}. It is interesting to note that Equs. 5.112a and 5.112b represent simple first-order reaction rates; this fact will be explained later as a formal kinetic approach to multisubstrate reactions (cf. Sect. 5.5). The most essential feature of such biphasic biokinetics is that they are able to explain the observed experimental discrepancies by incorporating nonidentical kinetics in the liquid and solid phase.

5.4 Kinetic Models for Microbial Product Formation

5.4.1 Metabolites and End Products

According to Gaden (1955, 1959), four types of product accumulation can be distinguished on the formal level based on the quantitative relationship

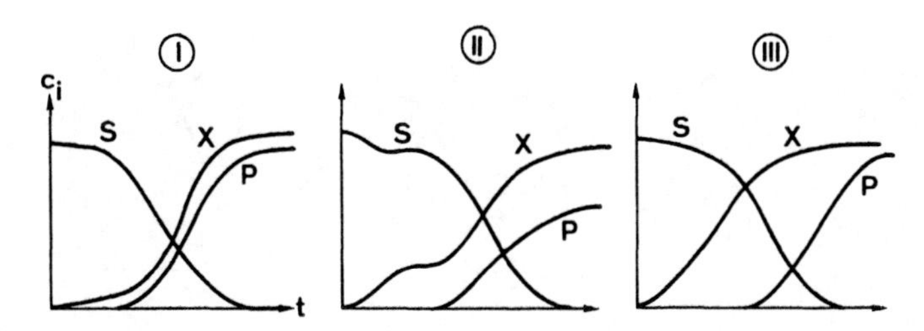

FIGURE 5.41. Concentration/time plot of the three basic types of microbial product formation: (I) growth association, (II) mixed growth association, and (III) nongrowth association (Gaden, 1955).

between the amount of product and the growth of cells. These types can be observed in Fig. 5.41, which is a c/t plot.

Type 0. Type-0 production as a supplementary case occurs even in resting cells that use only a little substrate for their own metabolism. The microbial cells function only as enzyme carriers. Steroid transformation and vitamin E synthesis by *Saccharomyces cerevisiae* are examples.

Type 1. Type-1 situations include processes in which product accumulation is directly associated with growth; this is the case for primary metabolites, in which the formation of the product is linked to the energy metabolism. Examples include fermentation to produce alcohol and gluconic acid (Koga et al., 1967), and situations in biological waste water treatment.

Type 2. Type-2 processes include fermentations in which there is no direct connection between growth and product formation and also no direct or indirect link to primary metabolism (secondary metabolites), for example, penicillin and streptomycin.

Type 3. Type-3 processes include those having a partial association with growth and thus an indirect link to energy metabolism. Examples include citric acid and amino acid production.

Recognition of the type of production is done with the aid of plots of r_i/t or q_i/t, as shown in Fig. 5.42 for types I through III. The most significant plot is given with specific rate q_i.

The diagram of the time dependence of the specific rate of a bioprocess is called the "quantification diagram": It gives the best insight into the process and is basic for designing mathematical models (cf. Sect. 2.4.2.1).

Product formation linked to microbial growth can be described by

$$r_p = Y_{P|X} \cdot r_x \tag{5.116a}$$

$$q_p = Y_{P|X} \cdot \mu \tag{5.116b}$$

where $Y_{P|X}$ is the product yield referred to biomass formed. If product yield is expressed in terms of substrate used we have (e.g., Constantinides, 1970a,b

$$r_p = Y_{P|S} \cdot r_S \tag{5.117}$$

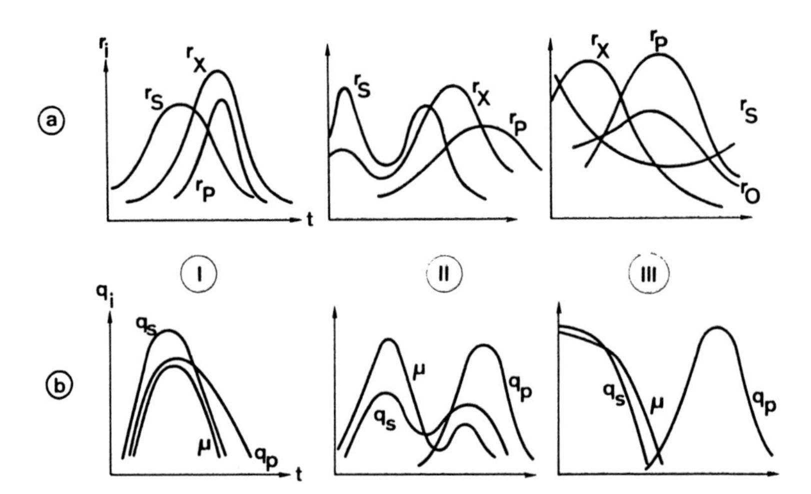

FIGURE 5.42. Formal kinetic diagrams of the three basic types of microbial product formation (I–III) expressed as volumetric rates r_X, r_S, r_P (a) or as specific rates μ, q_S, q_P (b), according to Luedeking and Piret (1959).

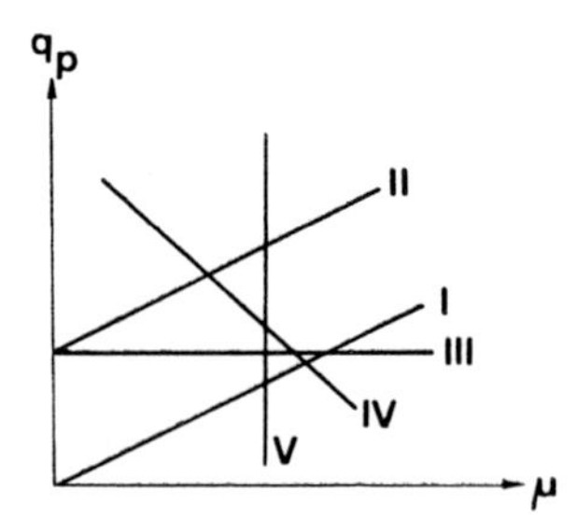

FIGURE 5.43. Formal kinetic linear relationships between specific rates of production q_P and growth μ exhibiting a different form: (I) growth association, (II) mixed growth association, (III) nongrowth association, (IV) negative correlation, and (V) no correlation. For equations see text.

Hence the following relationship exists between yield factors

$$\frac{Y_{P|S}}{Y_{X|S}} = Y_{P|X} \tag{5.118}$$

Substituting a Monod-type equation into Equ. 5.117 results in a hyperbolic function for production in the case of growth association

$$r_p = q_{P,\max} \frac{s}{K_S + s} \cdot x \tag{5.119}$$

Non-growth-linked product formation (curve III in Fig. 5.43) is more difficult to quantify because no direct relationship to growth exists. As an alternative in this case, the dependence of r_p on biomass concentration is often successfully used

$$r_p = k_p \cdot x \tag{5.120}$$

Product formation of this type can also be quantified by the dependence of substrate utilization (see Equ. 5.117). Similarly, the O_2 concentration o_L is useful in, for example, the quantification of penicillin synthesis (Giona et al., 1976)

$$q_p = k_1 \cdot o_L + k_2 \tag{5.121}$$

When product formation is partly growth linked and partly independent of growth, a combination of Equs. 5.116 and 5.120 is valid

$$q_p = Y_{P|X} \cdot \mu + k_P \tag{5.122}$$

as proposed by Luedeking and Piret (1959).

In addition to product formation types 0 through III, product formation types IV (and V) of Fig. 5.43 can also occur when there is a negative (or no) correlation between q_p and μ. This occurs, for example, in melanine production from *Aspergillus niger* (Rowley and Pirt, 1972). This type of production can be modeled by

$$q_p = q_{p,max} - Y_{P|X} \cdot \mu \tag{5.123}$$

Terui (1972) also proposed a kinetic model for enzyme formation in the nongrowing phase

$$q_P = q_{P,max} \cdot \exp[-k_2(t - t_{max})] + K_1\{\exp[-k_1(t - t_{max})] - \exp[-k_2(t - t_{max})]\} \tag{5.124}$$

where

k_1, k_2 = rate constant

K_1 = empirical constant

t_{max} = time of maximum production rate $q_{P,max}$

The general form of Equ. 5.122, with Equs. 5.120 and 5.116 as boundary cases, suggests a logistic equation (Luedeking and Piret, 1959). This has been similarly included in the generalized concept proposed by Kono and Asai (1968a–c, 1969a–c, 1971a,b) using the consumption coefficient ϕ as an apparent coefficient of growth activity. The value of ϕ in each phase of fermentation is (cf. Equ. 5.48):

Induction phase $\quad \phi = 0 \quad$ (5.125a)

Transient phase $\quad \phi = \phi \quad$ (5.125b)

Exponential growth phase $\quad \phi = 1 \quad$ (5.125c)

Declining growth phase $\quad \phi = \left(\frac{x_c}{x_m - x_c}\right)\left(\frac{x_m - x}{x}\right) \quad$ (5.125d)

TABLE 5.2. Types of fermentation processes classified according to values of production rate constants, k_{P_1} and k_{P_2}, in general formula of production rate (Equ. 5.126).

k_{P_1}	k_{P_2}	*Description*	*Case*
+	+	Product formation associated with growth and nongrowth	1a and 1b
+	0	Product formation associated with growth	2
0	+	Product formation associated with nongrowth	3
+	−	Product formation associated with growth and decreased with nongrowth	4

In a declining growth phase, the value of ϕ decreases from unity to zero. When a constant growth phase is included, the value of ϕ in this phase is expressed as follows:

$$\text{Constant growth phase} \qquad \phi = \frac{x_d}{x} \tag{5.125e}$$

x_d is the cell concentration at the boundary of an exponential growth phase and a constant growth phase. The general equation for product formation following the Kono concept, thus, is

$$r_P = k_{P_1} \cdot x \cdot \phi + k_{P_2} \cdot x(1 - \phi) \tag{5.126}$$

These authors described the various instances of growth-associated, or growth-independent, product formation using the parameter $k_{Pi} \gtrless 0$ as given in Table 5.2. The value of parameters k_{Pi} may be obtained in a purely numerical way directly from the experimental curves, as shown in Fig. 5.44. Case 1b (see Table 5.2) is included in Fig. 5.44a: It represents a linear growth phase and shows the similarity with the fermentation pattern described previously. The relationship between the Kono equation and the formal concept given in Fig. 5.41 becomes obvious in considering exponential phase with $\phi = 1$. For this situation the following equation results (see Equ. 5.48):

$$r_P = k_{P1} \cdot \frac{1}{\mu_{max}} \cdot r_x \tag{5.127}$$

Recently, Asai and Kono (1983) presented a modified equation and applied it to industrial fermentation processes.

When the product accumulation rate is influenced by the product decomposition rate, a term r_P must be added to general Equ. 5.122 to account for this event:

$$r_P = Y_{P|X} \cdot \mu \cdot x + k_P \cdot x - k_{P,d} \cdot p \tag{5.128}$$

Decomposition rate was used, for example, for penicillin production modeling (Constantinides et al., 1970a).

In bioprocesses in which r_P is directly coupled to energy metabolism, a generalized treatment also seems to be possible. On the basis of ATP pro-

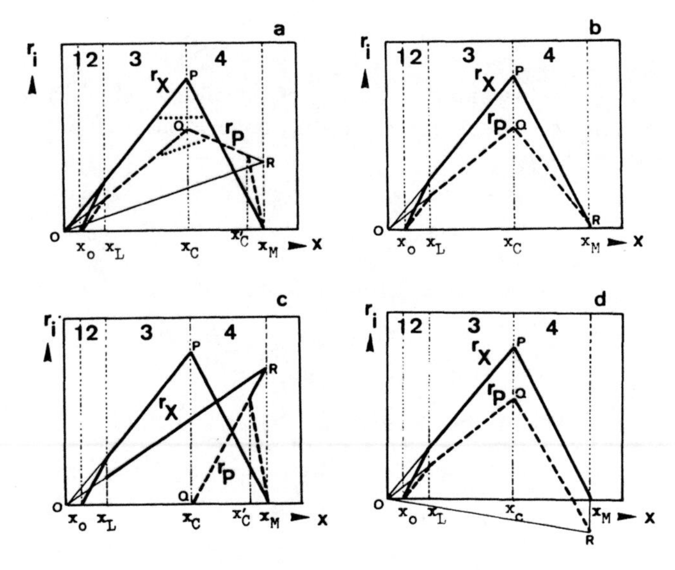

FIGURE 5.44. Schematic representation of the numerical Kono approach to microbial product formation expressed as the general formulas of the rates of growth r_X and production r_P, including different growth phases (1, induction; 2, transient; 3, exponential; and 4, declining), according to Equ. 5.126 and Table 5.2. (Adapted from Kono and Asai, 1968a–c, 1969a–c): (a) both k_{P1} and k_{P2} have a positive value. The dotted lines take into account a linear growth phase, as shown in Fig. 5.19. (b) $k_{P1} > 0$, $k_{P2} = 0$. (c) $k_{P1} = 0$, $k_{P2} > 0$. (d) $k_{P1} > 0$, $k_{P2} < 0$.

duction per amount of S converted by energy-producing pathways

$$r_{ATP} = Y^*_{ATP|S} \cdot r_{S,\text{energy}} \tag{5.129}$$

and by using a known concept for the relationship between the rate of substrate consumption due to material requirements for biomass precursor synthesis, the following equation results:

$$r_{S,\text{synth}} = Y^*_{S|X} \cdot r_x \tag{5.130}$$

This result is a type of equation for microbial production already given in Equ. 5.122 (cf. Roels and Kossen, 1978):

$$r_P = \frac{1}{a_P \cdot Y_{X|S}} - \frac{Y^*_{S|X}}{a_P} r_x + \frac{m_S}{a_P} x \tag{5.131}$$

with a_P the stoichiometric coefficient for product formation. Equation 5.131 states that, even in this case, a partly growth-associated term and a biomass-associated term both exist.

The difficult modeling of non-growth-associated product formation has been attempted by introducing time dependence. Shu (1961) proposed an empirical approach

$$r_P = \sum_i k_{1,i} \cdot e^{-k_{2,i} \cdot \Lambda} \tag{5.132}$$

on the basis of the assumption that r_P of individual cells is a genetically determined function of cell age Λ. Product concentration at arbitrary time, therefore, is given by

$$p = \int_0^t x(\Lambda) \int_0^\Lambda \sum_i k_{1,i} \cdot e^{-k_{2,i} \cdot \Lambda} \cdot d\Lambda \cdot d\Lambda \tag{5.133}$$

A mean cell age $\bar{\Lambda}(t_i)$ was later defined by Aiba and Hara (1965) as

$$\bar{\Lambda}(t_i) = \frac{x_0 \cdot \Lambda_0 + \int_{t_0}^{t_i} x \cdot d\tau}{x(t_i)} \tag{5.134}$$

with $t_0 \leq \tau \leq t_i$. The product formation rate can be quantified by

$$r_P = q_P(\bar{\Lambda}) \cdot x \tag{5.135}$$

where the specific production rate as a function of $\bar{\Lambda}$ can be evaluated directly from experiments. Fishman and Biryukov (1973) applied this concept for modeling the production of secondary metabolism. The value of q_P $(\bar{\Lambda})$ can be determined from the slope of a plot of q_P against $\bar{\Lambda}$, resulting in a model as given with Equ. 5.108.

Non-growth-associated production formation is the most difficult type to model. Another simple and yet plausible equation for such modeling involves the concept of the "maturation time" (Brown and Vass, 1973). The maturation time is taken as a kind of time delay between growth and production formation; this is shown in Fig. 5.45a. The equation is either

$$\left(\frac{dp}{dt}\right)_t = Y_{P|X} \cdot \left(\frac{dx}{dt}\right)_{t-t_M} \tag{5.136a}$$

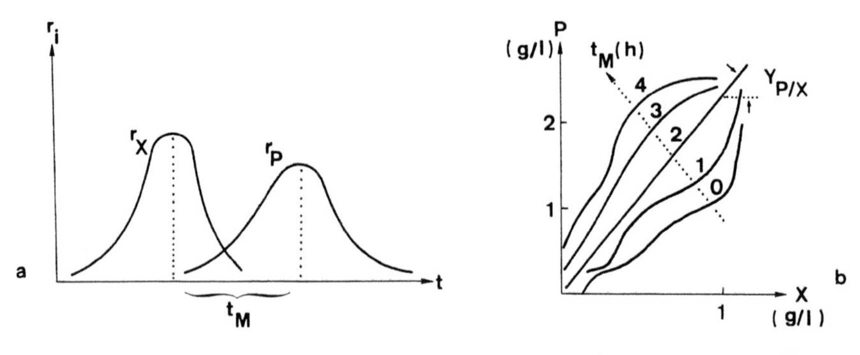

FIGURE 5.45(a). Graphic representation of the maturation time concept according to Brown and Vass (1973) in a plot of rates r_i versus time, with the possibility of a gross evaluation of the parameter t_M. (b) Evaluation of the model parameters t_M and $Y_{P|X}$ in the maturation time concept with the help of a graphical "trial-and-error" method (Brown and Vass, 1973).

or

$$(p)_t = Y_{P|X} \cdot (x)_{t-t_M} \tag{5.136b}$$

In a graph of the product concentration versus the concentration of cell mass, x, the line is straight when one has made the correct choice of t_M; the slope of the straight line represents the carrying coefficient. A graphical "trial and error" method of solution is given in Fig. 5.45b.

A systematic approach to unstructured modeling of microbial production has been presented by Ryu and Humphrey (1972) using a mechanism similar to that used for the derivation of Equ. 5.131. Here supplementary branching at a common intermediate I_i was considered. Assuming that the conversion of I_i to both cells and product is limited by the rate of one single enzyme in the chain, and assuming that both processes follow Monod-type kinetics, the following relationship holds:

$$q_P = q_{P,max} \frac{\varepsilon \cdot \mu/\mu_{max}}{1 + (\varepsilon - 1)\mu/\mu_{max}} \tag{5.137}$$

In this equation ε is the ratio of the Michaelis–Menten constants of the enzymatic reactions leading to growth and to product formation ($\varepsilon = K_{I,i}/K'_{I,i}$).

The implications of Equ. 5.137 are graphically presented in Fig. 5.46 (Roels and Kossen, 1978). These authors indicated that for $\varepsilon > 1$, a Monod-type equation similar to Equ. 5.119 applies; for $\varepsilon = 1$, an equation similar to Equ. 5.44 applies; and for $\varepsilon < 1$, one has a parabolic relationship.

Another generalization of product-formation kinetics based on mechanistic background but still using the formal kinetic approach has been presented by Bajpaj and Reuss (1980a, 1981). This model, (cf. App. II) along with rate equations for μ, q_S, and q_O, was successfully used to simulate experimental data. The concept of this model approach is to eliminate S between the equations for μ (see Equ. 5.38) and for q_P (see Equ. 5.106). The result is an equation relating q_P with μ in a more structured manner:

$$q_P = q_{P,max} \frac{[1 - (\mu/\mu_{max})](\mu/\mu_{max})}{[(K_P/K_S)\cdot x][1 - (\mu/\mu_{max})]^2 + [1 - (\mu/\mu_{max})](\mu/\mu_{max}) + (\mu/\mu_{max})^2/[(K_I/K_S)\cdot x]} \tag{5.138}$$

Plots of Equ. 5.138 are illustrated in Fig. 5.47 for different sets of values of the terms $[(K_P/K_s)\cdot x]$ and $[(K_I/K_s)\cdot x]$. The dotted curves (A) represent a modification due to inhibitions of growth (Moser and Schneider, 1988; Schneider and Moser, 1986).

5.4.2 Heat Production in Fermentation Processes

Heat production represents one of the significant macroscopic process variables (see Fig. 2.3). Experimentally, heat evolution can be determined with the aid of measurements of temperature changes in the reactor and the volumetric heat H_v [J/l] estimated according to the Uhlich approximation (see Equ. 3.70). An interpretation of the kinetic term $(dh_v/dt)_r$ in Equ. 3.69 is directly related

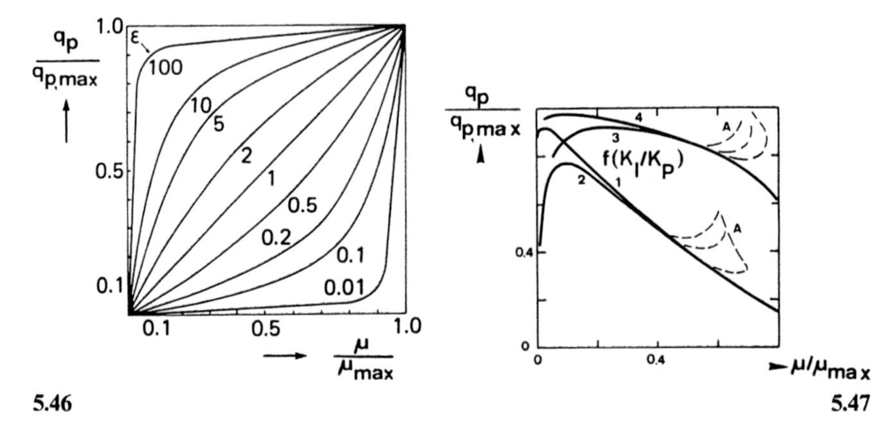

FIGURE 5.46. Graphical representation of the theoretical non-linear relationship between product formation and microbial growth according to Equ. 5.137 in a q_P versus μ plot. (Roels and Kossen, 1978, after Ryu and Humphrey, 1972.)

FIGURE 5.47. Kinetic pattern of secondary metabolite productions in a normalized q_P versus μ plot based on repression function (see Equ. 5.106) according to Equ. 5.138, with variation of the values K_I and K_P (Bajpaj and Reuss, 1982) in curves 1 through 4, by inclusion of other terms for inhibition of growth due to substrate and product K_{ISX} and K_{IPX} (Moser and Schneider, 1988) in curves A (---):

	$K_P/K_S x$	$K_I/K_S x$	K_{ISX}	K_{IPX}
Curve 1	0.0013	0.6667	—	—
2	0.013	0.6667	—	—
3	0.013	6.667	—	—
4	0.0013	6.667	—	—
A	Same values as in		500	10
	curves 1–4		200	10
			100	10

to fermentation heat analysis (cf. Sect. 3.3.8). Basically, the production of heat in a microbiological process can be dealt with quantitatively by the same equations as product formation. The validity of the following special equation has, however, been established: It includes a term for growth, one for product formation, and one for endogenous metabolism (Cooney et al., 1969; Mou and Cooney, 1976) using the specific rate q_{H_v}

$$q_{H_v} = \frac{1}{x} \cdot \frac{dh_v}{dt} = \frac{1}{Y_{X|H_v}} \cdot \mu + \frac{1}{Y_{P|H_v}} \cdot q_P + m_{H_v} \tag{5.139}$$

where $Y_{X|H_v}$ and $Y_{P|H_v}$ are the thermal yields related to biomass and product formation [g/J] and m_{H_v} is the maintenance coefficient [J/g · h]. The evaluation of Y and q_{H_v} is done in a way parallel to that shown in Fig. 5.27.

A simplified equation is possible for the rate of heat formation for aerobic processes. This equation utilizes the rate of oxygen utilization as a basis (Cooney et al., 1969):

$$q_{H_v} = \frac{1}{Y_{O|H_v}} \cdot q_O \tag{5.140}$$

This procedure for estimating heat evolution is simpler than the previous method. The proportionality constant ($1/Y_{O|H_v}$) for this correlation is independent of μ, slightly dependent on the substrate, and possibly dependent on the type of organism growth. This property of the constant ($1/Y_{O|H_v}$), which is thought to be identical with the reaction enthalpy of fermentation, $\Delta H_R^{(O)}$ [MJ/mol O_2] related to oxygen consumption,

$$\frac{1}{Y_{O|H_v}} = \Delta H_R^{(O)} \tag{5.141}$$

makes it extremely valuable when growth is to be carried out on complex substrates. According to Minkevich and Eroshin (1973), $\Delta H_R^{(O)}$ varies from 0.385 to 0.494 MJ/mole for filamentous fungi and from 0.385 to 0.565 MJ/mole for bacteria (Luong and Volesky, 1980), following the descending order bacteria, yeast, mold.

As mentioned previously (Sect. 3.3) in connection with the rate of heat transport, the reaction enthalpy $\Delta H_R^{(X)}$ or $\Delta H_R^{(S)}$ is given in units of [kcal/gX] or [kcal/gS] for practical reasons. Table 5.3 summarizes ΔH_R for fermentations (Bronn, 1971). A systematic treatment is possible using classifications of aerobic and anaerobic processing versus substrate; the numerical values are independent of the strain of microorganism.

For calculating heat balances, one should note that ΔH_R values are multiplied by a process rate and must therefore be in the correct dimensions

$$r_{H_v} = \frac{dH_v}{dt} = \Delta H_R \cdot r_x \qquad ([\text{kJ/l} \cdot \text{h}] = [\text{kJ/g}] \cdot [\text{g/l} \cdot \text{h}]) \tag{5.142}$$

Fermentations are principally free energy-yielding processes ($\Delta G < 0$) and/or are exothermic, $\Delta H < 0$. At present but few data are available in the literature.

A quite different approach in evaluating the heat evolution in bioprocessing uses the heat of combustions of the components involved in the conversion (Imanaka and Aiba, 1976). Molar enthalpies can, in part, be taken from thermodynamic tables or, in part, calculated by, for example, using the obser-

TABLE 5.3. Metabolic process enthalpies for fermentations.

Process	$-\Delta H_R^{(X)}$ [kcal/gX]	$-\Delta H_R^{(S)}$ [kcal/gS]
Hexoses $\xrightarrow{O_2} X$	2.6	1.4
Hexoses $\rightarrow P + X$	1.5–2.7	0.117–0.162
Hydrocarbons $\xrightarrow{O_2} X$	6.8	6.8

vation of Kharash (1929) that each mole of oxygen consumed in combustion results in a release of 0.444 MJ. The following equation can be written in the case of aerobic batch cultures of *Sacch. cerevisiae* for the rate of heat produced:

$$r_{H_v} = \left(\frac{-\Delta H_S}{M_S}\right)\frac{\mu \cdot x}{Y_{X|S}} + \left(\frac{-\Delta H_N}{M_X}\right)\mu x - \left(\frac{-\Delta H_x}{M_X}\right)\mu x - \left(\frac{-\Delta H_P}{M_P}\right)q_P \cdot x \tag{5.143}$$

where r_{H_v} is the rate of heat of fermentation $[J \cdot h^{-1} \cdot l^{-1}]$ and $-\Delta H_S$, $-\Delta H_N$, $-\Delta H_X$, $-\Delta H_P$ are the heats of combustion of substrate, ammonia, cells, and product $[J\ mol^{-1}]$. Rearranging Equ. 5.143 results in an expression for the heat of fermentation per unit mass of cell produced $\Delta H_R^{(X)}$ $[J \cdot g^{-1}$ of cell$]$ (Volesky et al., 1982)

$$-\Delta H_R^{(X)} = K_1 + \frac{K_2}{Y_{X|S}} - K_3\frac{q_P}{\mu} \tag{5.144}$$

where

$$K_1 = \frac{-\Delta H_N + \Delta H_X}{M_X} \tag{5.145a}$$

$$K_2 = \frac{-\Delta H_S}{M_S} \tag{5.145b}$$

$$K_3 = \frac{-\Delta H_P}{M_P} \tag{5.145c}$$

Similarly, heat of fermentation per unit mass of substrate consumed $\Delta H_R^{(S)}$ $[J \cdot g^{-1}$ of substrate$]$ can be expressed as

$$-\Delta H_R^{(S)} = \left(K_1 + \frac{K_2}{Y_{X|S}} - K_3\frac{q_P}{\mu}\right)Y_{X|S} \tag{5.146}$$

Similar results were obtained by recent work with microcalorimetry (Brettel et al., 1981a,b; Lovrien et al., 1980; Oura, 1973).

Imanaka and Aiba (1976), using the concept of heats of combustion, came to the same result as that expressed in Equ. 5.140, by direct measurements of the evolution of volumetric heat H_v $[J \cdot l^{-1}]$.

5.5 Multisubstrate Kinetics

In real situations, there are complex cases such as in biological waste water treatment and fermentation technology (complex media with multiple carbon sources, e.g., molasses, worts, metabolic intermediates, vitamins, etc.) that cannot be treated with the simple model equations $\mu = \mu(s)$ given in Sect. 5.3. In the course of a growth process in a complex medium, the valuable, easily utilized components are exhausted after a short time. For use of the remaining

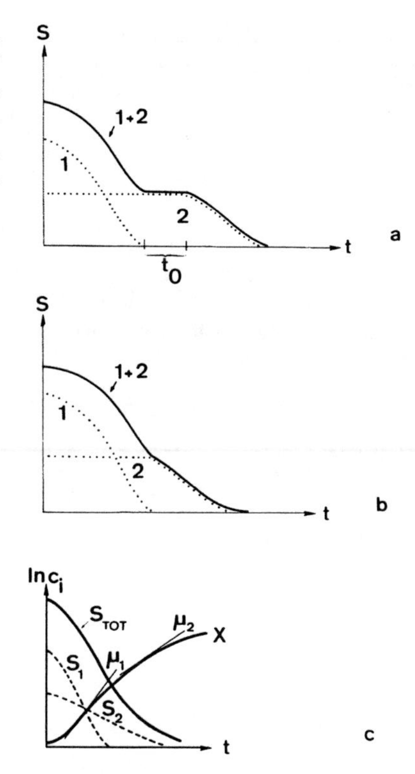

FIGURE 5.48. Multisubstrate kinetics for a two-substrate reaction: (a) Concentration/time diagram for strictly sequential substrate utilization (diauxic growth). (b) Partly overlapping and partly sequential substrate use. (c) Simultaneous substrate utilization. Sum-type kinetics are often applied in (b) and (c).

components, the cellular enzymes responsible for their breakdown must first be synthesized. The cells pass through many transitions; a series of growth phases will result, each with a successively decreasing growth rate.

Figure 5.48 summarizes practical situations in bioprocessing (A. Moser, 1981). Classification of situations can be achieved by distinction between sequential and simultaneous utilization, with a transition case of overlapping utilization. While sequential or consecutive consumption of substrates (Monod, 1942, 1949) can often be analyzed in two separate growth phases, the simultaneous utilization encountered in biological waste water treatment is more difficult for mathematical modeling.

5.5.1 Sequential Substrate-Utilization Kinetics

A general equation for sequential substrate utilization appears to be given by the relation

$$\mu = \mu(s_1) + \mu(s_2) \cdot f_{\mathrm{repr}}(s_1, t) \tag{5.147}$$

where the factor f_{repr} is a formal expression for catabolic repression operative on the use of s_2 as long as s_1 is present in the medium (see Equ. 5.106). A similar equation for the diauxic growth according to Equ. 5.147 is given by

Imanaka et al. (1972) resp. A. Moser (1978b) where the simple relationship

$$f_{repr}(s_1) = \frac{s_2}{s_1 + s_2} \tag{5.148a}$$

resp.

$$f_{repr}(s_1) = \frac{1}{1 + s/K_R} \tag{5.148b}$$

is used to represent regulation (cf. Equ. 5.106). Bergter and Knorre (1972) used another function for this purpose

$$\dot{f}_{repr}(s_1, t) = \frac{1}{t_{L2}}\left(\frac{K_S}{K_S + s} - f_{repr}\right) \tag{5.149}$$

that incorporates a diauxic lag time t_{L2}. This is a formal analogy to the formulation of biological inertia of Romanovsky et al. (1974), in which growth rate is related to ribosomal concentration c_R, where $\bar{c}_R$ is a stationary value

$$\frac{dc_R}{dt} = \frac{1}{t_L}(\bar{c}_R - c_R) \tag{5.150}$$

The opposing inhibitory effects of the substrate in catabolism are taken into consideration by an interaction term in the kinetic equations (Aris and Humphrey, 1977; Knorre, 1977; Yoon, Klinzing, and Blanch, 1977)

$$\mu = \mu_1(s_1, s_2) + \mu_2(s_2, s_1) \tag{5.151}$$

This case will be outlined in Sect. 5.5.3.

Strictly sequential S utilization is encountered, for example, in beer brewing, where glucose, maltose, and maltotriose are used one after the other (Budd, 1977). A simple model describing the kinetics of wort sugar uptake during batch fermentation of brewers' wort by yeast can be found in the literature (Fidgett and Smith, 1975); it operates with the concept of a critical concentration used as a switch from one limiting substrate to another, and it uses a simple "on-off" concept for individual rates. Basically, sequential utilization operates with only one limiting substrate at a time, expressed as an equation with additive terms (see Equ. 5.147). Sometimes, however, terms appear multiplicative in the final equation. This discrepancy will be discussed in Sect. 5.5.3.

Another concept for a formal kinetic description of two-substrate utilization with partial overlapping uses the idea of a critical substrate concentration (A. Moser, 1978b, 1981) that is directly measurable (Wöhrer et al., 1982). The value $s_{1,crit}$ is the concentration at which there is, for a short time, simultaneous utilization of s_1 and s_2 so that the factor

$$f_1 = \frac{1}{1 + s_1/s_{1,crit}} \tag{5.152}$$

can be substituted in Equ. 5.147 instead of f_{repr}.

5.5.2 Simultaneous Substrate-Utilization Kinetics

The case of simultaneous uptake is common in biological waster water treatment and elsewhere, and the following equation represents a general approach (Wuhrmann, von Beust, and Ghose, 1958)

$$r_{tot} = \sum_i r_i \tag{5.153}$$

or, using enzyme kinetics (Atkinson et al., 1969; Wilderer, 1976),

$$q_{S,tot} = \sum_i \left(q_{S,max,i} \frac{s_i}{K_{S,i} + s_i} \right) x \tag{5.154}$$

in which $K_{S,i}$ is a constant for the ith reaction and $q_{S,max,i}$ is the contribution of the ith reaction to the maximal rate such that (Shehata and Marr, 1971)

$$q_{S,max} = \sum_i q_{S,max,i} \tag{5.155}$$

The appearance of an overall reaction order one as the sum of several uptake rates was explained by such approaches (Wuhrmann et al., 1958), and it was later quantified (Wolfbauer et al., 1978). This fact is indicated in Fig. 5.48c. However, it seems to be possible to use the simple Monod equation for modeling, and an increased value of apparent K_S will be the consequence. A case of parameter estimation with double substrate limitation is shown in Fig. 5.49 using the Walker plot (see Fig. 4.21b) for the integrated form of the rate equation (Wilderer, 1976). Difficulties arise due to overlapping and to the phenomenon of biosorption (cf. Sect. 5.3.9), leading to a zero-order behavior even if the overall order of S utilization is unity (see Equ. 5.36).

Other groups have also worked successfully with similar first-order summary kinetics to represent the components in biological waste water treatment, S_i (for example, biological oxygen demand, BOD, or chemical oxygen demand, COD) (Eckenfelder and Ford, 1970; Grau, Dohanyos, and Chudoba, 1975; Joschek et al., 1975; Krötzsch et al., 1976; Oleszkiewicz, 1977; Tucek et al. 1971). In the relationship due to Joschek et al. (1975)

$$-r_S = k_S \cdot S \cdot X \cdot f(T) \tag{5.156}$$

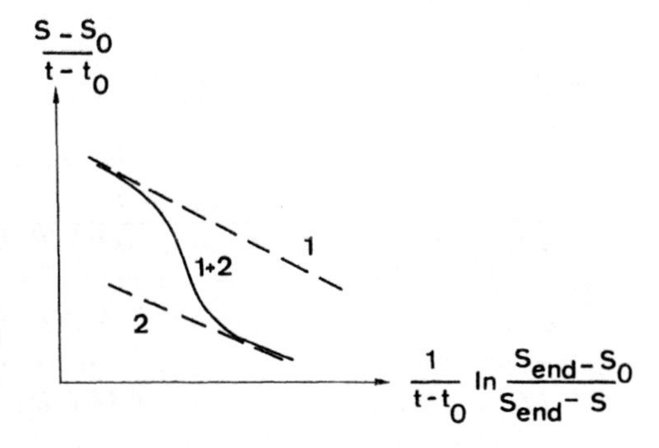

Figure 5.49. Plot of two-substrate kinetics in a Walker plot (cf. Fig. 4.21b). (Wilderer, 1976)

the rate constants show summary kinetics with the following relationship to the kinetic parameters (Moser and Lafferty, 1977)

$$k_S = \frac{\mu_{max}}{Y_{X|S} \cdot K'_S} \, [1/g \cdot h] \tag{5.157}$$

where the endogenous metabolism is indirectly reflected in an elevated K'_S value (Moser and Steiner, 1975). The equation due to Grau et al. (1975)

$$-r_S = k \cdot X \left(\frac{s}{s_0} \right)^n \tag{5.158}$$

with the integrated form for the case $n_S = 1$

$$s = s_0 \cdot \exp\left(-\frac{k \cdot x \cdot t}{s_0} \right) \tag{5.159}$$

uses a correction with variables a and b for metabolizable and nonmetabolizable substrate in the form

$$s = \frac{\text{BOD}}{a} - b \tag{5.160}$$

and is in agreement with the empirical relationship of Tucek, Chudoba, and Madera (1971) for the case $n_S = 2$.

To a first approximation, Equ. 5.159 can be written

$$\frac{s}{s_0} = 1 - \frac{k \cdot x \cdot t}{s_0} \tag{5.161}$$

Eckenfelder and Ford (1970) used the equation

$$\frac{S_0}{S} = 1 + k \cdot X \cdot t \tag{5.162}$$

whereas Krötzsch et al. (1976) expressed the relationship between substrate and cell mass with

$$-r_S = k_n \left(\frac{s}{x} \right)^n \tag{5.163}$$

and used a Contois kinetic approach.

5.5.3 Generalizations in Multisubstrate Kinetics

Generally, in multicomponent media, substrates will be utilized sequentially. Physiologically, two different responses can be distinguished, known as catabolite repression and inhibition, both resulting in a diauxic growth pattern. Remarkable differences in growth behavior can be observed comparing batch and continuous operation in a CSTR (Harder and Dijkhuizen, 1975), due to the existence of much lower steady-state concentrations of s_i in CSTRs so that the effects of both repression and inhibition may be absent or reduced.

Yoon et al. (1977), Aris and Humphrey (1977), and Knorre (1976) derived generalized Monod equations for multisubstrate systems, based on the following sequence of reactions:

$$X + a_1 S_1 \underset{k_{-1}}{\overset{k_1}{\rightleftharpoons}} X' \tag{5.164a}$$

$$X + a_2 S_2 \underset{k_{-2}}{\overset{k_2}{\rightleftharpoons}} X'' \tag{5.164b}$$

$$X' \xrightarrow{k_3} 2X \tag{5.164c}$$

$$X'' \xrightarrow{k_4} 2X \tag{5.164d}$$

where X' and X'' are different intermediary states of the cells and a_i is the stoichiometric coefficients. Applying the steady-state approximation, the following equation results

$$\mu(s_1, s_2) = \mu_{\max,1} \frac{s_1}{K_{S1} + s_1 + a_2 s_2} + \mu_{\max,2} \frac{s_2}{K_{S2} + s_2 + a_1 s_1} \tag{5.165}$$

with a complex meaning of parameters K_{S1} and K_{S2} according to the mechanistic interpretation of K_m (cf. App. III).

Further $\mu_{\max,1} = k_3$ and $\mu_{\max,2} = k_4$. The stoichiometric coefficients are given by

$$a_2 = \frac{k_2(k_{-1} + k_3)}{k_1(k_{-2} + k_4)} \tag{5.166a}$$

and

$$a_1 = \frac{1}{a_2} \tag{5.166b}$$

and indicate that each substrate exhibits a competitive inhibition effect toward the utilization of other substrates. Noncompetitive inhibition was considered by Lee (1973). For n multiple substrates, Yoon et al. (1977) derived Equ. 5.167 from these considerations

$$\mu = \sum_i \left[\mu_{\max,i} \frac{s_i}{\left(K_i + \sum_j a_{ij} s_i\right)} \right] \tag{5.167}$$

and applied such kinetics to batch and continuous two-substrate situations.

Another generalization of the Monod-type kinetics was suggested by Tsao and Hanson (1975) and by Tsao and Yang (1976). They assumed the existence of growth-enhancing substrates S_I and of essential substrates S_E, resulting in an equation

$$\mu(s_I, s_E) = \left(\mu_{\max,0} + \sum_i \mu_{\max,i} \frac{s_{I,i}}{K_{S_{I,i}} + s_{I,i}} \right) \prod_i \frac{s_{E,i}}{K_{S_{E,i}} + s_{E,i}} \tag{5.168}$$

The effect of growth-enhancing substrates is given as the sum of Monod-type expressions, while essential substrates form a product. The value of μ_{max} will be the sum of $\mu_{max,0}$ and all $\mu_{max,i}$. When all S_I are missing, a value of $\mu_{max,0}$ is still possible. By this approach a type of equation can be explained by assuming glucose and O_2 as essential substrates.

With an aerobic bioprocess it is evident that, in addition to substrate, O_2 can be rate limiting. The curve for biomass concentration increasing with time in a batch run is significantly influenced by the oxygen transfer rate quantified by the volumetric transfer coefficient $k_{L1}a$. This situation of an external transport limitation is depicted in Fig. 5.50 and in App. II (cf. also Sect. 4.5.1). Kinetic modeling in this case is successful with the aid of a double-substrate-limitation function (Megee et al., 1972; Ryder and Sinclair, 1972) as shown by Reuss and Wagner (1973)

$$\mu(s, o) = \mu_{max} \frac{s}{K_S + s} \cdot \frac{o}{K_O + o} \tag{5.169}$$

This equation was incorporated into the mass balance equation for O_2 where oxygen transfer was also considered

$$r_O = k_L \cdot a(o_L^* - o_L) - \frac{1}{Y_{X|O}} \cdot \mu(s, o) \cdot x \tag{5.170}$$

Equation 5.169 can be used as a formal kinetic approach without assuming a mechanism.

The question that generally arises is whether more than one substrate can exert control in a given system. Bader et al. (1975) and Bader (1978,1982) showed that for the growth rate to be able to respond to a controlling substrate or to both substrates simultaneously, the stoichiometric line must intersect the transition line $\alpha = \beta$ $(s_1/K_{S1} = s_2/K_{S2})$. The stoichiometric line relates the steady-state values of the substrates, which line, rearranged in terms of dimensionless substrate concentrations, become in dimensionless form

$$\beta_0 - \bar{\beta} = \frac{Y_{X|S_1} \cdot K_{S1}}{Y_{X|S_2} \cdot K_{S2}} (\alpha_0 - \bar{\alpha}) \tag{5.171}$$

This is shown in Fig. 5.51. For this intersection to occur, the following inequality must be satisfied

$$Y_{X|S_1} \frac{K_{S1}}{Y_{X|S_2}} \cdot K_{S2} > \frac{\beta_0}{\alpha_0} > 1 \tag{5.172}$$

This is in agreement with the requirements proposed by Sykes (1973) and defines the shaded region in Fig. 5.51, indicating that both double-*S* limitations and a switch between limiting substrates are quite common. However, Equ. 5.172 holds only for kinetic models such as the Monod and exponential models approaching saturation asymptotically. To handle this problem, it is desirable to overlay curves of constant μ on Fig. 5.51 for some of the models discussed earlier. Two different philosophies can be distinguished: interactive and noninteractive models.

An *interactive model* is based on the assumption that if two substrates are

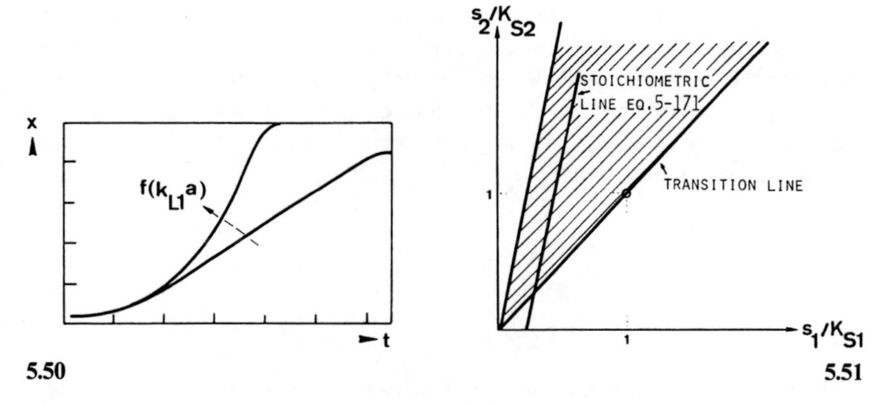

FIGURE 5.50. Linear growth as a consequence of transport limitation in case of O_2 quantified with $k_{L1}a$ (after Reuss and Wagner, 1973) in a concentration/time graph.

FIGURE 5.51. Dimensionless plot showing the feed conditions required for double-substrate limitations to be possible (shaded area). The stoichiometric line must cross the transition line with $s_1/K_{S1} = s_2/K_{S2}$ (Reprinted with permission from *In Microbial Population Dynamics*, Bader, 1982. Copyright CRC Press, Inc., Boca Raton, FL.).

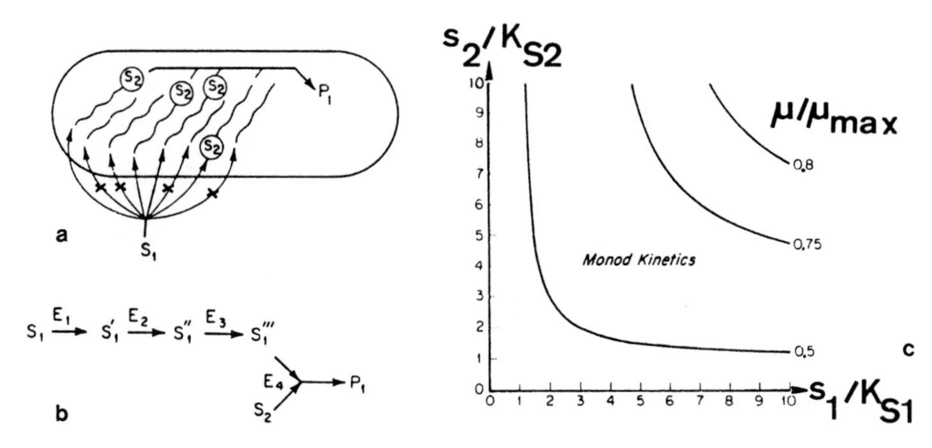

FIGURE 5.52. Conceptual representations of the interactive model. (a) S_1 is converted to P_1 by an enzyme that requires S_2 as a cofactor. (b) Substrates S_1 and S_2 from two parallel pathways are combined by enzyme E_r to produce a product P_1 that is required for growth. (c) Plots of lines of constant dimensionless specific growth rate μ/μ_{max} as a function of two dimensionless substrate concentrations for interactive models of the Megee type (cf. Equ. 5.169) with Monod kinetics (Reprinted with permission from *In Microbial Population Dynamics*, Bader, 1982. Copyright CRC Press, Inc., Boca Raton, FL.)

present in less than limiting concentrations, both substrates will affect the overall rate. The simplest model is constructed by simply multiplying two single-S-limited models together: Figure 5.52a shows the conceptual representation of such an approach.

For strictly sequential substrate utilization, Bader (1978) has proposed a "noninteractive" model that will seldom be encountered in its purest form. A *noninteractive model* basically implies that μ is limited by only one substrate at a time. Therefore, the growth rate will be equal to the lowest rate that would be predicted from the separate single-S models. For Monod-type kinetics, this would be written as follows:

$$\mu = \mu_{\max,1} \frac{s_1}{K_{S1} + s_1} \quad \text{for} \quad \frac{s_1}{K_{S1}} < \frac{s_2}{K_{S2}} \tag{5.172}$$

$$\mu = \mu_{\max,2} \frac{s_2}{K_{S2} + s_2} \quad \text{for} \quad \frac{s_2}{K_{S2}} < \frac{s_1}{K_{S1}} \tag{5.173}$$

Figure 5.53a represents the concept of this approach, examples of which may be found in the literature (Ryder and Sinclair, 1972; Sykes, 1973).

Comparisons of both approaches for Monod kinetics are shown in Figs. 5.52c and 5.53b. Noninteractive models are, by their nature, discontinuous functions at the transition line from one substrate limitation to another, predicting

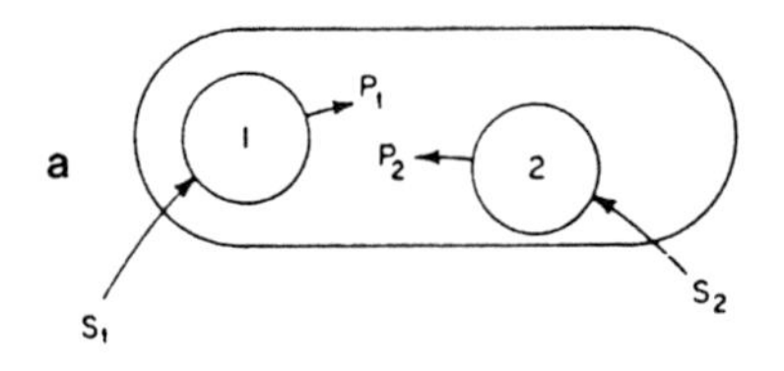

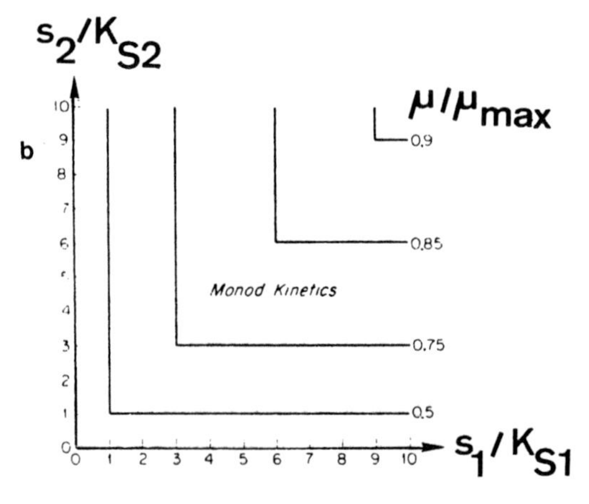

FIGURE 5.53. (a) Conceptual representation of the noninteractive model. Systems 1 and 2 operate independently of one another. (b) Plots of lines of constant dimensionless specific growth rate $\mu/\mu_{\max}$ as a function of two dimensionless substrate concentrations for noninteractive models of the Megee type with Monod kinetics (Reprinted with permission from *In Microbial Population Dynamics*, Bader, 1982. Copyright CRC Press, Inc., Boca Raton, FL.)

higher values of μ in the region where s_1/K_{S1} and s_2/K_{S2} are small. Interactive models are continuous functions but may err on the side of predicting lower values of μ when α and β both are small. It seems unlikely that any two cellular subsystems would be totally independent of each other, even though the degree of interaction may be rather small. Both types exist for certain types of substrates. Finally, it has been shown by Bader (1982) that the region of simultaneous limitation by two substrates is extremely small, so that double-S limitation can be regarded as a rare event, difficult to realize. Nevertheless, a closed analytical solution of multi-S kinetics is needed for engineering calculations.

5.6 Mixed Population Kinetics

Mixed populations of microorganisms (Aris and Humphrey, 1977; Jannasch and Mateles, 1974; Yoon and Blanch, 1977) occur, for example, in waste water treatment, in which the variation of organism composition with time plays a central role. The connection between various environmental circumstances (such as substrate choice and organism species composition) is shown in Figs. 5.54 and 5.55.

Even in natural self-purifying systems like rivers, a succession of populations occur: chemoorganotrophic bacteria, ciliates, nitrifiers, algae, and fish (cf. Fig. 5.54). In biological waste water treatment reactor systems, a plurality of consecutive reactions is known as a result of activities of the biocoenosis, shown in Fig. 5.55 (Wilderer and Hartmann, 1978). Mixed populations are found in aerobic and anaerobic sludge digestion; beer, corn, cheese, and yogurt production; the human skin; the alimentary canal; aquatic environments; and the soil. A similar but simpler situation is found in the case of nitrification or denitrification of biological waste water: Both have been described as a two-step microbial process (Eggers and Terlouw, 1979; Tanaka et al., 1981). Another case of mixed population interactions is the rumen fermentation (Czerkawski, 1973), for which mathematical models have also been developed.

The initial supply of organic substances in waste water plants is utilized especially for carbon-containing components by bacteria (population X_1) as shown in Fig. 5.55. The nitrogen-containing components in a conventional operation are utilized later by nitrifying (population X_2) and denitrifying (population X_3) cells. The phosphorus-containing compounds are stored in particular populations due to multiple interactions between aerobic, anoxic, and anaerobic zones. Bacteria also are themselves food for bacteria-consuming organisms such as protozoa and ciliates. These are all coupled in a food chain, population X_i, collectively referred to as the "biocoenosis" or ecosystem.

In analogy to natural systems, some technical processes involving mixed populations have recently been introduced. In a situation where the optimal strategy for the production of ethanol from sugars involves batch fermentation of highly concentrated sugar solutions, the use of dual organisms that possess

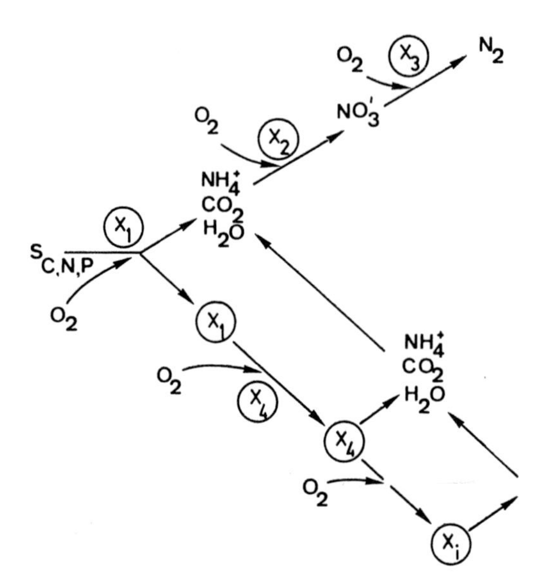

FIGURE 5.54. Relative predominance of microbial organisms utilizing organic substances in waste water: sarcoidia (x_1), holophytic flagellates (x_2), holozoic flagellates (x_3), bacteria (x_4), ciliates (x_5), suctoria (x_6), stalked ciliates (x_7), rotifers (x_8).

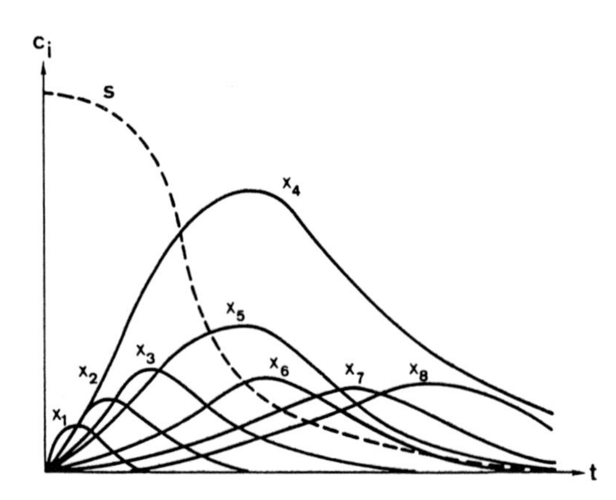

FIGURE 5.55. Reaction with a mixed population of microorganisms in a low-loaded waste water purification operation as a function of the time course of environmental changes (mixed substrate, $S_{C,N,P}$). X_1, saprophytic bacteria; X_2, nitrification bacteria; X_3, denitrifying bacteria; X_4 to X_i, organisms in the food chain. (Adapted from Wilderer and Hartmann, 1978.)

different substrate and product inhibition characteristics can bring about improved ethanol productivity. The extent of any improvement depends crucially on the functional form of the inhibition relationships and on the values of the kinetic parameters (e.g., Jones and Greenfield, 1981).

5.6.1 Classification of the Types of Microbial Interactions

Various terms are used to denote various types of interactions; however, there is so much overlap in the meaning of these terms that they do not fit all categories of the interaction pattern (Bungay and Bungay, 1968; Meers, 1973; Noack, 1968). Table 5.4 shows a classification scheme of all combinations of interactions known, that is, competition, commensalism, amensalism,

TABLE 5.4. Classification of pairwise interactions between microbial populations based on the signs of the entries a_{ji} and a_{ij} from the community matrix A ($i = j$) (May, 1973).

		Effect of species j *on species* i (*sign of* a_{ij})		
		−	0	+
Effect of species i *on species* j (*sign of* a_{ji})	−	− −: competition	−0: amensalism	− +: predation
	0	0 −: amensalism	00: neutralism	0 +: commensalism
	+	+ −: predation	+0: commensalism	+ +: mutualism

mutualism, and predation. The influence of species i on species j is determined by defining an element a_{ij} in the so-called "community matrix" (Bailey and Ollis, 1977). If a_{ij} is positive, species j has a positive effect on growth of species i; an inhibitory effect is described by a negative value of a_{ij}.

Competition occurs when a community of two or more species are mutually limiting because of their joint dependence on a common factor external to them. *Commensalism* is the case where the growth of one species is promoted by the presence of a second species in a population, the growth of the second species being unaffected by the presence of the first. *Mutualism* is similar to commensalism, but both organisms grow faster in the presence of the other than they do separately. It could be caused by the production of growth factors or products that serve as nutrients. *Symbiosis* is similar and occurs if the mutualistic partnership is necessary for survival of one species. *Synergism* is a third type of mutualism in which the formation of specific products is greater in mixed than in pure cultures. *Amensalism* is the situation where the growth of one species is repressed because of the presence of a toxic substance produced by another; it represents the opposite of commensalism. Neutralism, which is relatively rare, means that the two species have no observable effect on one another. *Predation* occurs when an organism totally engulfs and digests another organism, which thereby loses the ability to reproduce itself. *Parasitism*, which for microbial interrelationship is difficult to distinguish from predation, occurs when one organism feeds or reproduces at the expense of tissues or body fluids of another.

Supplementary to this classification scheme, a distinction is often made between open and closed environments, that is, continuous and discontinuous reactors. An open environment leads to population stability (homeostasis), although oscillations may be observed due either to periodic fluctuations in the environment or to interactions between microbial species. The factors influencing the survival of given species are different in closed and open systems (Meers, 1973).

5.6.2 Kinetic Analysis of Microbial Interactions

The behavior and mathematical model of a mixed population are basically similar to those of a pure culture. The Monod model can often be used as a global approach, even though the kinetic parameters μ_{max}, K_S, and Y are not strictly constant. Examples are biological waste water treatment (Gaudy and Gaudy, 1972) and biogas fermentation (Andrews, 1971; Chen and Hasimoto, 1978). However, several distinct steps can sometimes be distinguished, for example, in the biogas process the fermentative, acetogenic and methanogenic reaction. Thus the following scheme is the starting point for the setup of a mathematical model:

$$S_i \xrightarrow{x_1} \text{fatty acids} + H_2 + CO_2 \xrightarrow{x_2} \text{acetic acid} + H_2 + CO_2 \xrightarrow{x_3} CH_4 + CO_2 \tag{5.174}$$

5.6.2.1 Competition (Between Two Organisms Growing on a Single Substrate)

With the simplest model including two species (x_1, x_2) competing for the same limiting substrate (s), the principle of selection can be demonstrated ("survival of the fittest"—organism with the greater specific growth rate). The scheme can be written

$$S \begin{matrix} \nearrow X_1 \\ \searrow X_2 \end{matrix} \tag{5.175}$$

The kinetic model, using the Monod equation for $\mu_i = f(s)$, will be for a CSTR

$$r_{X_1} = \mu_1(s)x_1 - Dx_1 \tag{5.176}$$

$$r_{X_2} = \mu_2(s)x_2 - Dx_2 \tag{5.177}$$

$$r_S = -\frac{1}{Y_{X_1|S}}\mu_1(s)x_1 - \frac{1}{Y_{X_2|S}}\mu_2(s)x_2 + D(s_0 - s) \tag{5.178}$$

Figure 5.56 shows the steady-state values of this model as a function of D. Evidently no coexistence is possible. However, for lower dilution rates $(D < D_s)$ the first species survives, whereas for a middle range of D $(D_s < D < D_c)$ the second species survives. Above the critical dilution rate D_c both organisms wash out. The plot of the kinetics $D = \mu_1(^2s)$ and $D = \mu_2(^3s)$ indicates that the switch from the first surviving species to the second is due to the intersection of both μ characteristics (Gutke, 1980, 1982). This model is the basis for studies of ecosystems and of the problems of the evaluation or selection of mutants in a CSTR for genetic optimization studies.

The principle factors governing the kinetic behavior are the differences in μ and K_S of both species, for example, mutants at different dilution rates. The dependencies $\mu(s)$ can be divided into two main categories. The first type (cases I, II, IV, V in Fig. 5.57) is characterized by the fact that the $\mu(s)$ curves for both mutants have no common point apart from the origin. In the second type (cases III, VI), the two curves intersect (see Fig. 5.57). It is clear that the curves will cross each other if $\mu_{max,2} < \mu_{max,1}$ and $K_{S2} < K_{S1}$. All of these cases are illustrated in Fig. 5.58, where correspondingly denoted areas reflect the magnitude of the ratios $\mu_{max,1}/\mu_{max,2}$ and K_{S1}/K_{S2} (Sikyta et al., 1979). Thus,

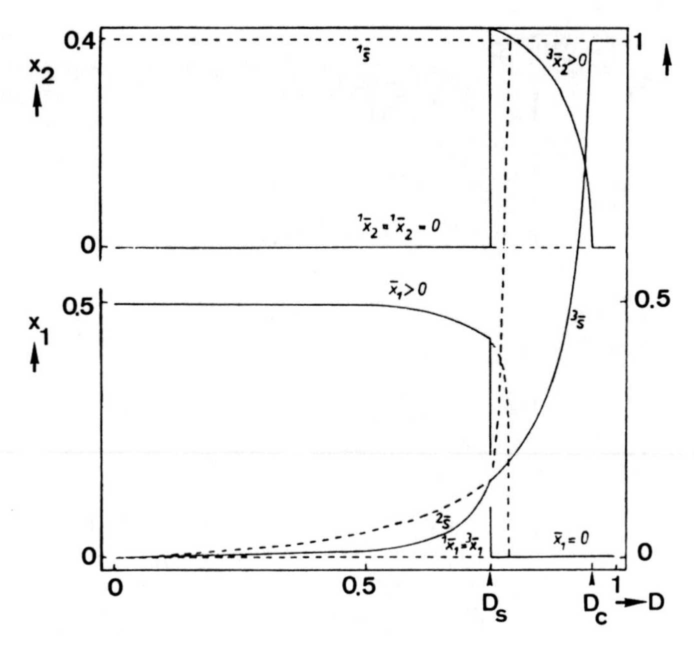

FIGURE 5.56. Steady-state concentrations of two competing microbial species x_1 and x_2 and of the limiting substrate s in dependence of dilution rate D in a CSTR according to Equs. 5.176 through 5.178. Parameter values for simulation: $\mu_{max,1} = 0.8\ h^{-1}$, $K_{S1} = 0.01\ g \cdot l^{-1}$; $\mu_{max,2} = 1 \cdot h^{-1}$; $K_{S2} = 0.05\ g \cdot l^{-1}$; $Y_{X2|S} = 0.5$. For explanation see text.

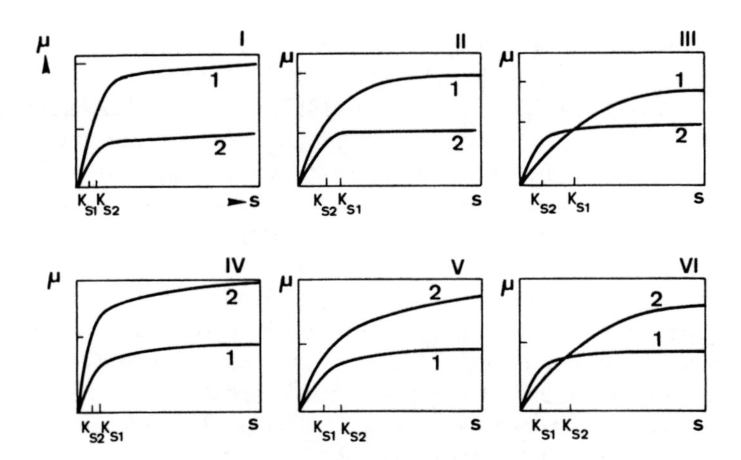

FIGURE 5.57. Theoretical dependencies of specific growth rate μ on the limiting substrate concentration s for a population including two microorganisms (e.g., mutants). A first type (cases I, II, IV, V) exhibits no common point apart from the origin, while a second type (cases III, VI) shows an intersection between both curves. (From Sikyta et al., 1977.)

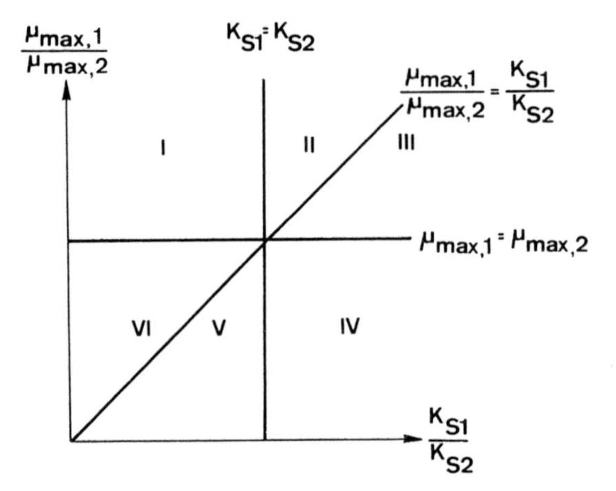

FIGURE 5.58. Generalized plot of functions from Fig. 5.57 with indications of ranges of cases I through VI. (Sikyta et al., 1979).

with known values μ_{max} and K_S, Fig. 5.58 can serve for the identification of a given trend and for the prediction of population changes. The dilution rate can be used as a controlling factor. The point of intersection in Fig. 5.58 can be found by putting μ_1 equal to μ_2 and solving for s:

$$s_{intersect} = \frac{\mu_{max,2} \cdot K_{S1} - \mu_{max,1} \cdot K_{S2}}{\mu_{max,1} - \mu_{max,2}} \tag{5.179}$$

A chemostat will operate stably at $D = \mu_{intersect}$ such that both species will coexist. With $D < \mu_{intersect}$ the situation is the same as before, with species 1 washing out. With $D > \mu_{intersect}$, species 1 will grow faster and species 2 will wash out.

Growth of two competing species in a closed environment was analyzed by Volterra (1931), who established many of the fundamentals of mathematical ecology. Powell (1958) and H. Moser (1957) have provided mathematical analyses of the fate of contaminants or of mutants in CSTRs. For the survival of competing cells, the surface/volume ratio is also important, as it affects the value of μ_{max} but not of K_S (Veldkamp, 1975). Several papers concern mixed-culture steady-state studies in the chemostat (e.g., Jannasch and Mateles, 1974; Veldkamp and Jannasch, 1972) and also the dynamics (e.g., Aris and Humphrey, 1977; Lee et al., 1976). Stability analyses show that possible steady states depend on the relative disposition of the two growth curves and on the position of the point whose coordinates are the nutrient feed concentration and dilution rate. Qualitative phase portraits are drawn for each of the 31 distinct types of situations. The operation of a periodically forced CSTR with two competing populations was carried out to give criteria for the stability of the resulting cycles and to show conditions under which stable periodic trajectories of coexistence can be achieved (Stephanopoulos et al., 1979). Similarly, a CSTR with cell recycle of heterogeneous populations was examined with the result that cell recycle can change the outcome of species competition. Selective recycling of one species can reverse this outcome or stabilize coexistence by its selective effect on cell residence time (Weissmann and Benemann, 1979). Other trends to be reported are the multivariable feedback control of a competing mixed-culture system (Wilder et al., 1980), the computer-aided analysis of mixed cultures (Ohtaguchi et al., 1979), the mathematical description of competition between two and three species under dual S limitation in a CSTR (Gottschal and Thingstad, 1982), and the anslysis of a two-stage CSTR system as an attractive alternative (Stephanopoulos and Frederickson, 1979). The advantageous use of multistage systems was recommended earlier by Veldkamp and Jannasch (1972). The complexity of microbial interactions often observed, however, needs a much more elaborate kinetic modeling, including categories in addition to competition. By varying the concentration of the medium components, various interacting systems can be created (e.g., Tseng and Phillipps, 1981).

5.6.2.2 Commensalism and Amensalism

Many microbes produce substances that promote the growth of other species; this phenomenon is probably the most widespread form of commensalism. It follows the scheme

$$\begin{matrix} S_{\mathrm{I}} \rightarrow X_1 \\ S_{\mathrm{II}} \rightarrow X_2 \end{matrix} \qquad P \tag{5.180}$$

The combinations of possible commensal relationships, however, are legion, and the interactions may well be competitive as well as commensalistic, but are rarely studied in full detail. When during the growth of one species the environment is altered in such a way that another species is inhibited, due either to the removal of essential nutrients or to the formation of toxic substances, published data do not necessarily specify the type of antagonism or amensalism that has occurred. The toxic products likely to cause amensalism may be divided into inorganic and organic compounds. Often, the negative growth rate of species i is caused by the pH change from the growth of species j. Kinetic model equations for pure commensalism without inhibition employ Monod-type relations for $\mu_{\mathrm{i}} = f(s_{\mathrm{i}})$ and the assumption that the nutrient for the dependent species is produced with no loss of yield from the nutrient for the independent species. The following balance equations for a CSTR were derived by Reilly (1974):

$$r_{\mathrm{X}_1} = (\mu_1 - D)x_1 \tag{5.181}$$

$$r_{\mathrm{X}_2} = (\mu_2 - D)x_2 \tag{5.182}$$

$$r_{\mathrm{S}_1} = D(s_{10} - s_1) - \frac{\mu_{\mathrm{i}} x_1}{Y_{\mathrm{X1}}} \tag{5.183}$$

$$r_{\mathrm{S}_2} = Ds_2 + \frac{\mu_1 x_1}{Y_{\mathrm{X1}}} - \frac{\mu_2 x_2}{Y_{\mathrm{X2}}} \tag{5.184}$$

For this simple system, three different steady states are possible.

The problem of a commensalistic model with feedback inhibition and activation by produced substances was also treated by Reilly (1974). Inhibition was assumed to be competitive

$$\mu_i = \mu_{max,i} \frac{s_i}{K_{S,i} + s_i + K_{S,i} \cdot I_i / K_{I,i}} \tag{5.185}$$

and activation to be additive

$$\mu_j = \mu_{max,j} \frac{s_j}{K_{S,j} + s_j} + \mu_{max,A_j} \frac{a_j}{K_{A,j} + a_j} \tag{5.186}$$

with a being the concentration of activator and $i = j$.

This complicated commensalistic system is shown to be less stable, with limit-cycle response occurring after dilution rate changes, as would be expected from a model with more parameters. Though virtually any feedback from the dependent to the independent species will cause some overshoot after a step change in D, the most pronounced oscillatory behavior is caused by feedback inhibition and feedforward activation. Limited agreement with experimental data was obtained even though the analysis was somewhat limited due to the complexity of the system and the large number of differential equations. Similar results were obtained by Sheintuch (1980), who examined the dynamics of commensalistic systems with self- and cross-inhibition. Multiplicity of steady states was observed as well as oscillatory states with these complex kinetics and in the case of a reactor with biomass recirculation. Stability and dynamics are summarized in a qualitative phase plane by this author.

Population dynamics and stability analyses in the case of commensalism and other interactions are widespread in the literature (e.g., Lee et al., 1976; Miura et al., 1980). A modified approach to commensalistic modeling in the case of yeast growth and nicotinic acid production employed Monod-type functions for species 1 and 3 but the following equation for μ_2 (Tseng and Phillips, 1981):

$$\mu_2 = \mu_{max,2} \frac{s_2}{K_{S2} + s_2} \cdot \frac{p}{K_P + p} \tag{5.187}$$

with p being the concentration of nicotinic acid. This paper concluded that mixed cultures are very complex, and the interacting mechanisms may not be as simple as commensalism or competition or both. Equation 5.185 was also used in modeling pure commensalism as single interaction of the type given in the scheme of Equ. 5.180, with the result that the five eigen-values are all real and negative, and therefore the steady-state solution is a stable model (Miura et al., 1980).

5.6.2.3 Mutualism

Mutualism, sometimes called "protocooperation" if it is not obligatory for survival, is the case if both species grow faster in the presence of the other than they do in pure culture. It can be caused by the production of growth factors or products P_i that serve as nutrients:

$$S_I \rightarrow X_1 \begin{smallmatrix} \nearrow P_2 \nwarrow \\ \searrow P_1 \nearrow \end{smallmatrix} X_2 \leftarrow S_{II} \tag{5.188}$$

Following this scheme, the specific growth rates (μ_1 and μ_2) can be expressed by

$$\mu_1 = \mu_{max,1} \frac{s_1}{K_{S1} + s_1} \cdot \frac{p_2}{K_{P2} + p_2} \tag{5.189}$$

and

$$\mu_2 = \mu_{max,2} \frac{s_2}{K_{S2} + s_2} \cdot \frac{p_1}{K_{P1} + p_1} \tag{5.190}$$

The balance equations in the case of mutualism in a CSTR are

$$r_{Xi} = \mu_i x_i - D x_i \tag{5.191}$$

$$r_{Si} = -\frac{1}{Y_{X_i}} \mu_i x_i + D(s_0 - s) \tag{5.192}$$

$$r_{P1} = q_{p1} \cdot x_1 - \frac{1}{Y_{2/P1}} \mu_2 x_2 - D \cdot p_1 \tag{5.193}$$

$$r_{P2} = q_{p2} \cdot x_2 - \frac{1}{Y_{1/P2}} \mu_1 x_1 - D \cdot p_2 \tag{5.194}$$

As in the case of commensalism, the steady-state solution is a stable node (Miura et al., 1980). When competition is accompanied by mutualism (and/or commensalism), two kinds of steady-state solutions are obtained owing to the values of system parameters, one being a stable node and the other a stable focus or an unstable saddle point. Damped oscillations tend to occur when either x_1 or x_2 has a lower growth ability than the other.

In a somewhat more complex case, x_1 grows on S, which is toxic at high concentrations to x_2, and x_2 produces a nutrient for x_1. It may be modeled as follows by using the same biomass balance as before but different S balances

$$r_S = D(s_0 - s) - \frac{\mu_1 x_1}{Y_{1S}} - \frac{\mu_2 x_2}{Y_{2S}} \tag{5.195}$$

$$r_p = D \cdot p + q_p x_2 - \frac{\mu_1 x_1}{Y_{X|P}} \tag{5.196}$$

with kinetic equations for S inhibition and co-utilization of P (cf. Equ. 5.190) (Meyer et al., 1975). Taguchi et al. (1978) simulated this model and received good agreement with experiments. The model behavior, determined with parameters taken from the pure cultures, demonstrates the coexistence of both species at lower dilution rates, above which the singular point of coexistence could not be obtained (the sole survivor being the strain that overcomes the competitor). It seems generally valid that a mutualistic interaction in a CSTR will be established only over a limited range of initial conditions of x_i and s (Meyer et al., 1975; Miura et al., 1980).

5.6.2.4 Predation (Predator–Prey Interactions)

A food chain of mixed populations indicated by the scheme (cf. Fig. 5.55)

$$S \dashrightarrow X_1 \dashrightarrow X_2 \dashrightarrow X_i \tag{5.197}$$

contains bacteria as the first members in an aquatic and terrestrial environment, while protozoa and ciliates represent other members of a biocoenosis. Thereby a prey (x_1) is consumed by the predator (x_2).

Gause (1934) was the first to systematically study the interactions between ciliates and their prey in a closed laboratory environment. The topic was also studied extensively by Volterra (1931) who, together with Lotka (1925), developed early mathematical models. The original Lotka–Volterra analysis considered μ_1 and μ_2 constants, but normally they would depend on their respective substrates. Thus

$$\mu_1 = f(s) = \mu_{\max,1} \frac{s_1}{K_1 + s_1} \tag{5.198}$$

and

$$\mu_2 = f(x_1) = \mu_{\max,2} \frac{x_1}{K_2 + x_1} \tag{5.199}$$

The model equations for simple prey–predator interactions, initially according to Lotka and Volterra and later refined by Bungay and Bungay (1968), are as follows:

$$r_{X1} = \mu_1 x_1 - k_1 x_1 \cdot x_2 \tag{5.200}$$

and

$$r_{X2} = k_2 x_1 \cdot x_2 - k_3 x_2 \tag{5.201}$$

where

k_1 = killing efficiency constant based on encounters

k_2 = predator growth constant, proportional to yield coefficient $Y_{X1/X2}$

k_3 = specific dead rate

For a CSTR the model equations become (Curds, 1971)

$$r_{X1} = \mu_1 x_1 - D x_1 - \frac{\mu_2}{Y_{X1/X2}} x_2 \tag{5.202}$$

$$r_{X2} = \mu_2 x_2 - D x_2 \tag{5.203}$$

and

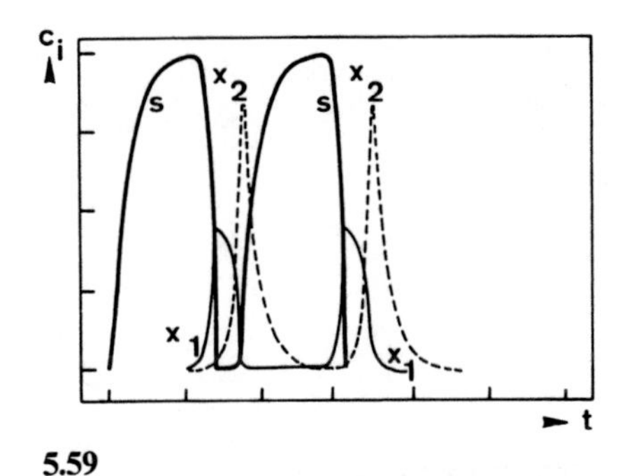

5.59

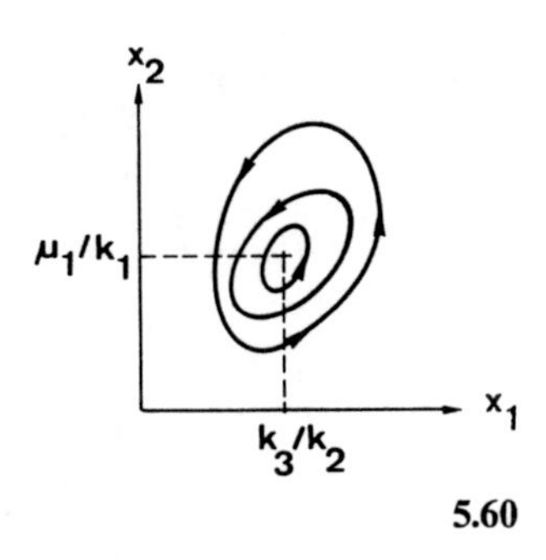

5.60

FIGURE 5.59. Computer simulation of the behavior of mixed populations (bacteria, x_1, and ciliates, x_2) exhibiting predation on the basis of Equs. 5.202 to 5.204 ($s_0 = 200$ mg$\cdot$l^{-1}; $D = 0.1$ h^{-1}; $\mu_{max,1} = 0.6$ h^{-1}; $\mu_{max,2} = 0.43$ h^{-1}; $K_{S1} = 4$ mg$\cdot$l^{-1}; $K_{S2} = 12$ mg$\cdot$l^{-1}; $Y_{X1|S} = 0.45$; $Y_{X2|S} = 0.54$). Limit cycle oscillations are the result. (Reprinted with permission from *Water Res.*, Vol. 5, Curds, Copyright 1971, Pergamon Journals Ltd.)

FIGURE 5.60. Closed cycle relationship between predator x_2 and prey x_1 from classical theory following a phase plane analysis. For more information see Lotka, 1925, 1956. (See Equ. 5.205.)

$$r_S = -\frac{\mu_1}{Y_{X|S_1}} x_1 + D(s_0 - s) \tag{5.204}$$

with Equ. 5.200 for the kinetics. A computer simulation of this set of predator model equations is illustrated in Fig. 5.59 (Curds, 1971), which can also be used for stability analysis. Three possible states may exist in this system, depending on growth parameters: stable oscillations (limit cycles), damped oscillations about the steady-state value, or an asymptotic approach to the steady-state value.

If Equ. 5.200 is divided by Equ. 5.201 and integrated to Equ. 5.205

$$-k_3 \log x_1 + k_2 x_1 - \mu_1 \log x_2 + k_1 x_2 = \text{constant} \tag{5.205}$$

then each value of the integration constant gives a closed cycle for the relationship between x_1 and x_2 on an $x_1 - x_2$ plane. This phase plane analysis, shown in Fig. 5.60, demonstrates that the populations of x_1 and x_2 oscillate around a vortex point, the equilibrium point on the x_1/x_2 axis given by the ratio μ_1/k_1 or k_3/k_2 (Canale, 1969). Biologists and also mathematicians have shown considerable interest in the Lotka–Volterra model. Experimental data suggest the existence of two singular points as derived by Sudo et al. (1975). Consequently, the model shown by Equs. 5.200 and 5.201 with constant values of k_i cannot be fully adequate. Canale (1970) developed a modified predator–prey model, indicated earlier with Equs. 5.198 and 5.199. Although Canale cited the emergence of a limit cycle on the phase plane as responsible for the oscillations, the experimental data cannot be attributed totally to the limit

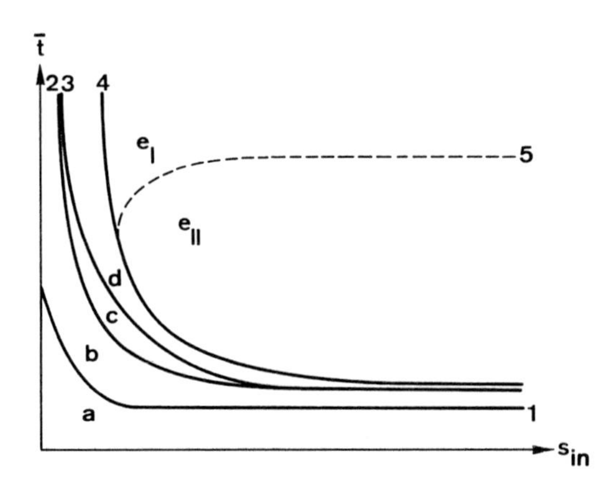

FIGURE 5.61. Predicted behavior of a predator–prey system in single-stage CSTR culture with respect to mean residence time $\bar{t}$ and concentration of limiting substrate s_{in} for bacterial prey in the influent. A modification is represented by incorporating a multiple saturation predator rate of Equ. 5.206 (after Jost et al., 1973, and Tsuchiya et al., 1972). The curves 1–5 separate the following regions of the operating diagram: (a) total washout, (b) x_2 washed out, (c) no oscillations, (d) damped oscillations, (e_I) periodic oscillations, and (e_{II}) sustained oscillations.

cycle. In this situation Sudo et al. (1975) proposed a more sophisticated model by distinguishing between two parts of bacterial concentration: bacterial food unavailable to the protozoa due to flocs and bacteria available because of dispersion. Better agreement was found using a model incorporating microbial floc formation and disintegration.

In the models, oscillations are inherent to the system. However, Tsuchiya et al. (1972) and Jost et al. (1973) have demonstrated that under certain conditions (combinations of dilution rate and S-concentration data) it is also possible that predator–prey systems in a CSTR do not exhibit oscillatory behavior. Results of stability analysis are shown in Fig. 5.61 (curves 1–4 with regions a–e_I) representing an operating diagram for the general dynamic features for this case (after Jost et. al., 1973). The diagram is based on Monod's saturation model for growth of both x_1 and x_2 (cf. Equ. 5.198) and shows regions of different characteristics for given values of kinetic constants when plotting $1/D$ versus $s_0 = s_{in}$. However, deviations from this model behavior are known in experiments, leading to an important model incorporating the multiple saturation predation rate of Equ. 5.206

$$\mu_2 = \mu_{max,2} \frac{x_1^2}{(K_2 + x_1)(K_3 + x_1)} \tag{5.206}$$

This equation reduces to the Monod-type Equ. 5.199 for x_1 much larger than K_2 and K_3 and has the meaning that μ_2 varies as x_1^2 when x_1 is small. The revised CSTR model comprising Equ. 5.206 can be analyzed according to the operating diagram $1/D$ versus s_0. Figure 5.61 with curve 5 and regions

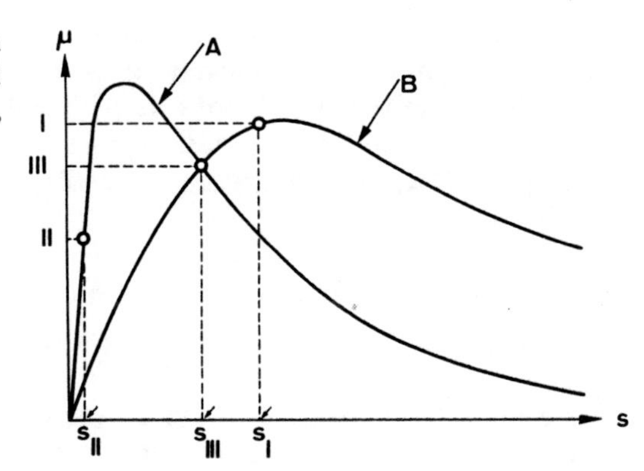

FIGURE 5.62. Mixed population bioreactor problems illustrated in a μ/s plot. (From Humphrey, 1977b.)

a through d and e_{II} shows the behavior that is in accordance with the experimental findings: that sustained oscillations disappear when $\bar{t}$ increases.

It is commonly accepted by ecologists that environmental heterogeneity exerts a stabilizing influence on predator–prey interactions. This fact led to investigations of combined wall growth in a CSTR with predation (Ratnam et al., 1982). Data obtained showed that bacteria but not protozoa were attached strongly to the walls, and that wall growth had a significant effect on the dynamics. A reasonable fit of all the experimental data was achieved by combining a wall growth model (Topiwala and Hamer, 1971) with the multiple saturation model for predator growth (cf. Equ. 5.206).

Beyond this field of fascinating possibilities for application and research activities, some bioreactor operation problems should be referred to as typical for mixed populations. An infinite number of stable steady states can appear, for example, in the case of phenol degradation (Humphrey and Yang, 1975). As shown in Fig. 5.62, the fermentation system contains two separate and distinct species capable of metabolizing phenol. Species A is inhibited by rather low levels of phenol, while species B is inhibited only at rather high concentrations. The system behaves in a random way with respect to effluent phenol level and microbial flora. At operating point I in Fig. 5.62, the system should settle on species B with s_I in the effluent. In the case of a low throughput rate (point II), species A should dominate with low concentrations in the effluent (s_{II}). Virtually any ratio of A to B is possible at the steady state if operation is maintained at point III. Further, if fluctuations occur in the influent, the effluent concentration and cell mass relationship will wander all over, depending in part on where they were last.

This phenomenon is thought to be much more general in biology than expected (Humphrey, 1977a). It occurs with a single species growing on two substrates when one metabolic pathway is repressed by the substrate of the other. Another case is when a single species grows on a single substrate but has two enzyme systems for handling the primary metabolite, one constitutive and the other inducible.

These infinitely variable problems in bioreactor operation are waiting to be solved. There have been relatively few systematic tests of the principles of interaction of mixed populations. Experimental work on the basis of the understanding from kinetic model theories should be encouraged by studies utilizing a CSTR, a CSTR cascade, or a CPFR.

5.7 Dynamic Models for Transient Operation Techniques (Nonstationary Kinetics)

Transient reactor operation plays an increasingly important role in bioprocessing and has to some extent already been considered (classification, see Fig. 3.31; fed-batch culture, see Fig. 3.37; situation, see Fig. 4.4; guidelines to solution, see Sect. 4.2 and Fig. 4.5; structured cell model concept, see Fig. 4.7; application, see Chap. 6). Both "balanced" and "frozen" conditions have also been considered in Fig. 3.34. A biosystem is in balanced condition when the mechanism is fully adapted, as in a quasi-steady-state (if $\tau_e \gg \tau_r$). All different equations can be reduced to algebraic equations. A biosystem is in frozen condition of the initial state (if $\tau_e \ll \tau_r$) and the mechanism may be neglected due to the fact that the slowest step is rate determining ("rds concept"). By this procedure, equations are reduced to parameters so that the number of equations is reduced (e.g., the case of dropwise addition of substrate). This is the case of steady state CSTR.

Intermediate periodic operation lies between both extremes and covers systems in which the response time is of the same order as the imposed function cycle time so that "resonance" is possible. Here the state of the mechanism changes dynamically and is not directly related to the environmental conditions at the moment considered. Intermediate periodic operations encompass: semibatch periodic operations, selectivity in periodically forced CSTRs, cycled reactors with heterogeneous catalysts, and non-steady-state biofilm reactors (e.g., Rittmann and McCarty, 1981). Lag, overshoot phenomena, and oscillations can typically occur. Biological reactor operation will basically fall into this latter class of operation, and balanced growth must be distinguished from steady-state growth (Barford et al., 1982).

5.7.1 Definitions of Balanced Growth and Steady-State Growth

5.7.1.1 Balanced Growth

The concept of balanced growth was introduced by Campbell (1957) to describe a metabolic state of a culture in terms of the distributed concentration (x_i) or total mass (X_i) of a metabolic variable. According to Campbell's definition, growth is balanced when the specific rate of change of all such variables is constant. That is,

$$\frac{1}{x_i} \cdot \frac{dX_i}{dt} = \frac{1}{X_i} \cdot \frac{dX_i}{dt} = \text{constant} \tag{5.207a}$$

This definition applies to a culture of microorganisms. In the case of an individual microorganism, a broader definition is required (Campbell, 1957).

$$\frac{1}{x_{i,\text{ave}}} \cdot \frac{\Delta x_i}{\Delta t} = \frac{1}{X_{i,\text{ave}}} \cdot \frac{\Delta X_i}{\Delta t} = \text{constant} \tag{5.207b}$$

where

$$t = n \cdot t_g$$

$$n = \text{any integer}$$

$$t_g = \text{generation time [h]}$$

$$t_g = \log 2 | \mu$$

In addition to the growth of a single microorganism, many other microbial growth phenomena require a broader definition of balanced growth. An example of this is periodic behavior or sustained oscillations (e.g., Heinzle et al., 1982). If such sustained oscillations are to be regarded as balanced growth, the definition of balanced growth may be broadened, again using Equ. 5.207b but with

$$t = n \cdot t_p \tag{5.208}$$

where n is any integer and t_p is the period of oscillation [h]. These considerations are a simple example of the important concept of regime analysis (cf. Fig. 4.5).

Consideration also needs to be given to the question of how closely balanced growth can be approached under batch growth conditions. While the ideal experimental batch curve is often referred to (Herbert, 1961), the number of reported experimental examples of balanced growth in batch culture is extremely small (Barford and Hall, 1979b). In most cases, the experimental data are not sufficient to allow any conclusive judgment as to whether the growth is balanced. Therefore all metabolic rates are to be considered ($\mu, q_S, q_O, q_C, q_P, q_{Hv}, \ldots$). While batch cultures that approach balanced growth have been reported (Barford and Hall, 1979b), it is possible that perfectly balanced growth can never be achieved in a batch experiment. This is not because of the limited time scale, but because the composition of the extracellular medium in a batch culture is constantly changing and could be expected to induce corresponding changes in cell composition.

Several important concepts are illustrated by these observations. First, it is essential that a range of metabolic variables be considered before any well-based decision can be made as to whether balanced growth has been achieved. In addition, and most important, some metabolic processes achieve a balanced growth condition more rapidly than others (e.g., μ and RQ before q_O and q_C). Consequently, the variable that is used as the criterion for the attainment of balanced growth must be carefully chosen and explicitly stated.

5.7.1.2 Steady-State Growth

It is possible to consider steady-state growth as an extension of balanced growth. That is, in addition to the definitions outlined previously, a further requirement is that the overall time rate of change of the distributed concentration (x_i) or total mass (X_i) of any metabolic variable be zero. That is,

$$\frac{dx_i}{dt} = \frac{dX_i}{dt} = 0 \tag{5.209}$$

Consequently, while balanced growth is a necessary condition for steady-state growth, it is not a sufficient condition. Likewise, steady state is a sufficient condition for balanced growth but is not a necessary one.

A final consideration from the definition of steady state is that no reference to time scale is included. How long must a culture exhibit no time rate of change of metabolic variables to be considered at steady state? Clearly this should be related to the number of generation times without significant variation in the magnitude of major metabolic variables, but how many generation times and which metabolic variables? While tacitly it has been generally accepted that 15 to 20 generation times would be excessive for the achievement of steady state, many examples of much longer periods of time necessary for steady states are available (Heineken and O'Connor, 1972). Furthermore, very few experimental studies precisely state what criteria were used in the determination of steady state (see Barford et al., 1982).

5.7.2 Mathematical Modeling of Dynamic Process Kinetics

As previously outlined, mathematical modeling uses both the methodological concepts of unstructured and structured models.

5.7.2.1 Unstructured Models

In general, unstructured models can be considered a good approximation in two distinct cases (relaxed steady state and quasi-steady periodic operation, as shown in Fig. 3.34). However, unstructured models can successfully be applied to batch processing with lags (e.g., Knorre, 1976) and to the dynamics of stirred tanks (Agrawal et al., 1982) by introducing a formal kinetic approach for the lag time t_L (see Equ. 5.72) and the variation in yield coefficient (cf. Equ. 2.41). The number of model parameters is increased by t_L, but the number of process variables is only $3(S, X, t)$. Unstructured but modified modeling of process kinetics has been successfully applied in the case of biological waste water treatment, where nonstationary behavior occurs as a consequence of fluctuations in feed stream or shock loading. Mona et al. (1979) modified Monod-type kinetics and adapted them to real situations by creating load-dependent kinetic parameters ($q_{S,max}$ and K_S):

$$q_{S,\max} = k_1 \cdot B_x \tag{5.210}$$

and

$$K_S = k_2 \cdot B_x \tag{5.211}$$

where B_x is the mass loading rate per unit biomass (sludge)

$$B_x = \frac{F \cdot s_0}{V \cdot x} [M_i M_x^{-1} T^{-1}] \tag{5.212}$$

Further, these authors introduced time lags τ_{L1} and τ_{L2}, responsible for the time dependence

$$\hat{q}_{S,\max} + \tau_{L1} \frac{d\hat{q}_{S,\max}}{dt} = k_1 \cdot B_x \tag{5.213}$$

and

$$\hat{K}_S + \tau_{L2} \frac{d\hat{K}_S}{dt} = k_2 \cdot B_x \tag{5.214}$$

with $\hat{q}_{S,\max}$ and $\hat{K}_S$ being the deviation variables defined as

$$\hat{q}_{S,\max} = q_{S,\max}(t) - \bar{q}_{S,\max} \tag{5.215}$$

and

$$\hat{K}_S = K_S(t) - \bar{K}_S \tag{5.216}$$

The basis of this formal kinetic modeling is shown in Fig. 5.63 together with the definitions of τ_{L1} and $\bar{q}_{S,\max}$.

A formal kinetic approach for product formation in transient conditions

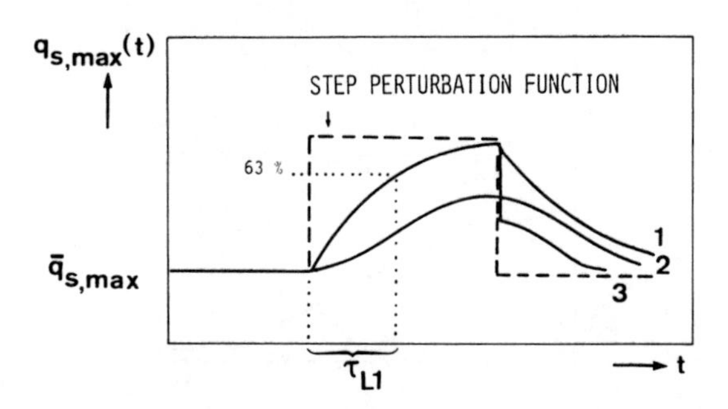

FIGURE 5.63. The time-dependent change in the maximum rate of substrate use, $q_{S,\max}$, as an illustration of nonstationary behavior (dynamic kinetics) of a bioprocess such as this example of waste water purification: (1) easily metabolizable, synthetic medium, (2) difficult to metabolize medium such as starch, (3) mixture of easily metabolized substrate and suspended particles in waste water. τ_{L1} = delay time for $q_{S,\max}$. (Adapted from Mona et al., 1978.)

was presented by Gutke et al. (1980)

$$k_P = k_P(\mu, \dot{\mu}) \tag{5.217}$$

As the physiological state can be characterized in the most simple way by the specific growth rate μ, the nonstationary state is quantified by the time derivation of specific growth rate ($\dot{\mu}$), being constant (c_μ) a given time period during a shift-down

$$\dot{\mu} = -c_\mu \tag{5.218}$$

with

$$\mu(t) = \mu_0 - c_\mu \cdot t \tag{5.219}$$

The behavior of this model is illustrated in the computer simulation in Fig. 5.64.

Another attempt at generalized rate equations was realized by Harder and Roels (1981), applying the characteristic-time concept (cf. Sect. 4.2). A model describing the dynamics of product formation was derived using Equ. 5.117 together with Monod-type kinetics for r_s, incorporating endogenous metabolism with the aid of Equ. 5.80, resulting in an expression for steady state

$$q_P = \frac{Q}{Y_{X|S}} \mu + Q \cdot m_S \tag{5.220}$$

with Q being the product formation activity function, which is related in steady state ($\bar{Q}$) to μ

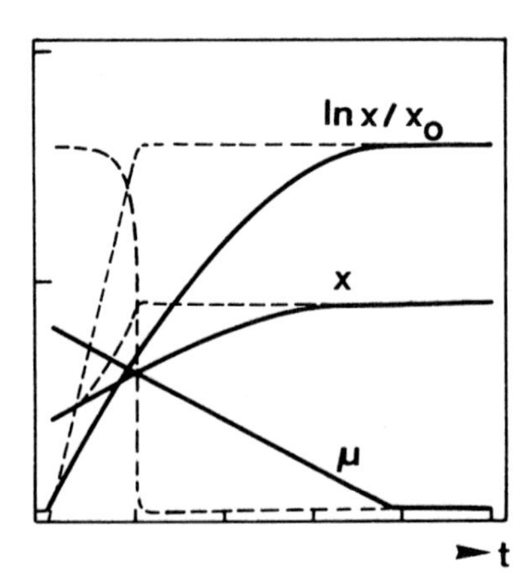

FIGURE 5.64. Kinetics of biomass concentration, logarithmic biomass concentration, and specific growth rate according to Equs. 5.218 and 5.219 (solid lines; $x_0 = 0.2$ g/l, $\mu_0 = 0.4\ h^{-1}$, $c = 9.1\ h^{-2}$) and for comparison according to the well-known Monod model (dashed lines; $x_0 = 0.2$ g/l, $\mu_{max} = 0.8\ h^{-1}$, $K_S = 0.01$ g/l, $s_0 = 0.49$ g/l, $Y_{X|S} = 0.5$). (From Gutke et al., 1980.)

$$\bar{Q} = f(\mu) \tag{5.221a}$$

or

$$\bar{q}_Q = \mu \cdot \bar{Q} \tag{5.221b}$$

From Equ. 5.220 and 5.221a it follows that

$$\bar{Q} = \frac{Y_{X|S} \cdot \bar{q}_{P,max}}{\mu + m_S \cdot Y_{X|S}} \tag{5.222}$$

The extension to dynamic situations is achieved by the formulation of a rate of adaptation of Q to environmental conditions with the aid of the intrinsic balance equation

$$\dot{Q} = \frac{1}{x}(r_Q - r_X \cdot Q) \tag{5.223}$$

with r_Q the rate of synthesis of Q. The difference between actual rate of Q synthesis and the steady-state rate is assumed to follow

$$q_Q = \bar{q}_Q + f(Q - \bar{Q}) \tag{5.224a}$$

or

$$q_Q = \bar{q}_Q - K_{adapt}(Q - \bar{Q}) \tag{5.224b}$$

where K_{adapt} is a positive constant (constant of adaptation) given by

$$K_{adapt} = -\frac{\partial f}{\partial Q_{Q=\bar{Q}}} \tag{5.225}$$

This constant can be estimated from shift-down or shift-up experiments.

Finally, combining Equs. 5.221b, 5.223, and 5.224b for the rate of change of Q, one has the result

$$\dot{Q} = -(K_{adapt} + \mu)(Q - \bar{Q}) \tag{5.226}$$

The relaxation time $t_{charact}$ for the adaptation to a new steady state during a shift in continuous culture is given by

$$t_{charact} = \frac{1}{(\mu + K_{adapt})} = \frac{1}{\mu(1 + K_{adapt}/\mu)} \tag{5.227}$$

If K_{adapt} is large, a new steady state will be reached almost instantaneously with no lag of the organism. If K_{adapt} is small, the time constant for adaptation will be equal to $1/\mu$, that is, dilution through growth will control the adaptational process.

Figure 5.65 represents the apparent relationship between Q and μ according to Equ. 5.226 in dynamic situations. As can be seen, when K_{adapt} is large the steady-state relationship (see Equ. 5.116) is obtained.

Esener et al. (1981) applied unstructured modeling to fed-batch processing

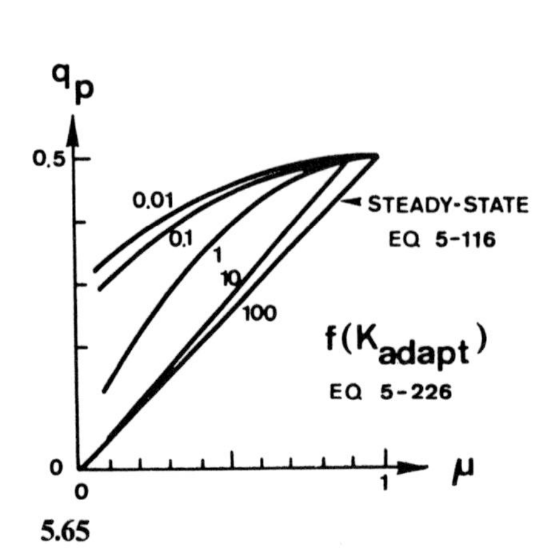

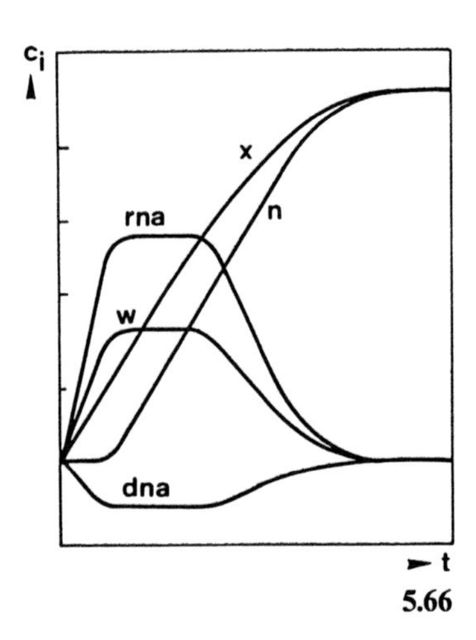

FIGURE 5.65. Dynamic relationship between specific rate of product formation q_P and specific growth rate μ for various rates of the exponential decrease of the specific growth rate expressed with the aid of varied values of the constant K_{adapt} (see Equ. 5.226). The steady-state relationship is obtained at large K values. (From Harder and Roels, 1982.)

FIGURE 5.66. Typical plot of growth kinetics and macromolecular composition of microorganisms during the course of a batch fermentation according to Herbert (1961). Concentrations of dry biomass (x), individual cell weight (w), number of cells per volume (n), and RNA and DNA per unit weight.

on the basis of Equ. 3.92 for the concentration of X and S. If such an unstructured model is combined with the elemental and energy-balance principles (see Sect. 2.4), the time dependence of O_2 consumption and CO_2 and H_v production can be calculated with great success for batch, continuous, and fed-batch cultivation. The results of this investigation showed that in those cases where detailed kinetics play a role (i.e., in the shift from a process controlled by the maximum specific growth rate of the organism to one where growth is limited by the addition rate of the substrate), the model provides only a poor fit to the phenomena actually observed. Structured models, therefore, should be more adequate for the description of a system during highly transient periods.

5.7.2.2 Structured Models

In structured model building one must select the parameters that are the most relevant for the description of the physiological state of the organism. Information from molecular biology, chemistry, microbiology, and biochemistry is required for this purpose. Logical first choices would be the DNA, RNA, carbohydrate, and protein contents of the cells for describing their physiological state. All of these variables can be experimentally determined, and their dependence on steady-state growth rate is well established, as shown in Fig. 5.66 (Herbert, 1961). It must be noted, however, that even the consideration of these four relevant components is not sufficient to describe the activities and qualities of the organisms fully; for example, no information can be obtained about the geometrical structure of the cells or the dependence of diffusional processes on such structures (Kossen, 1979). Thus, a structured model should be formed in such a way that it provides information only about the most relevant processes and variables (see Sect. 2.3). For a comprehensive description, too many parameters would need to be incorporated into the model. Such complicated models (e.g., Fig. 2.22) are mathematically very complex to manipulate, and most of the parameters often lose their biological significance. For such models, numerous critical experiments have to be performed to determine the parameters.

Simpler models can be obtained by considering a few variables as an extension of the unstructured models, which consider one biotic variable (Esener et al., 1982b). These models, in which the activity of biomass is specified by more than one and up to three or four variables, are called "compartmental models." They have moderate mathematical complexity and are easier to verify experimentally. The first formal compartmental model was due to Williams (1967, 1975), who showed that even a simple model with two compartments could describe most of the experimentally observed phenomena, at least qualitatively. Some inconsistencies of this model were later corrected by Roels and Kossen (1978) and Roels (1983). Ramkrishna et al. (1967) and Fredrickson et al. (1970) have contributed greatly to the theoretical aspects of structured modeling.

The building of chemically structured models consists of (a) setting up of mass balances for the relevant components of the biomaterial over an appropriately chosen boundary, (b) postulation of the relevant kinetic expressions for reactions taking place within the biotic phase and its boundaries, and (c) evaluation of the constraints imposed by elemental balances and the application of thermodynamic principles. With reference to Fig. 5.66, it can be concluded that RNA and carbohydrate fractions change most in response to changes in steady-state growth rate. Thus, it might be desirable to consider these components explicitly in a structured model. Protein content changes only slightly over the same range. It is difficult to deal with carbohydrate as a separate compartment: The absolute intrinsic concentration of carbohydrate is very low (about 5% dry weight) and hence variations in it are difficult to detect analytically. Therefore, in the model to be presented, RNA, carbohydrate, and other small cellular molecules are lumped into one compartment, which will be referred to as the "K compartment." Since RNA is the main constitutent, the amount in the K compartment may be expected to be proportional to the RNA content. The other compartment is called the G compartment and contains the genetic material and all the rest of the cell constituents, that is, proteins, DNA, structural material, and so on. This approach more or less assumes that the RNA concentration (and thus the

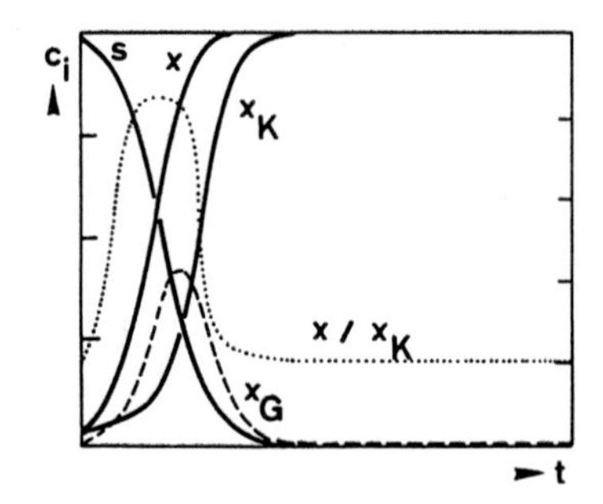

FIGURE 5.67. Concentration/time plot of structured biomass model according to Williams (1967): Biomass G and K compartment (x_G, x_K) and total biomass (x) and substrate (s) simulated on the basis of Equs. 5.228d–f with $k_1 = 0.0125$ and $k_2 = 0.025$.

RNA synthesis) is the bottleneck. Since RNA plays a central role in the synthesis of proteins, this approach to structured modeling seems reasonable. A diagram of the two-compartment model was shown in Fig. 4.8.

The following reactions can be formulated, based essentially on the work of Williams (1967) and on its modifications and extensions by Roels and Kossen (1978) and Roels (1978):

$$S \xrightarrow{G} Y_{SK} K \qquad \text{rate } r_{SK} \tag{5.228a}$$

$$K \xrightarrow{K} Y_{KG} G \qquad \text{rate } r_{KG} \tag{5.228b}$$

$$G \longrightarrow K \qquad \text{rate } r_{GK} \tag{5.228c}$$

The G compartment is assumed to be synthesized from the K compartment under the catalytic action of the K compartment. The latter compartment is thought to be synthesized from substrate under the catalytic action of the G compartment. This is a great simplification of the complexity of the cellular processes, but is significantly closer to reality than the unstructured approach. A computer simulation of the Williams model is shown in Fig. 5.67, illustrating the basic behavior of the two-compartment model (Williams, 1967).

The next step is the postulation of rate expressions for r_{SK}, r_{KG}, and r_{GK}. For a description of the state of the culture as a function of time, the approach advocated by Harder and Roels (1982) can be used without problems when the kinetics and the stoichiometry of the processes involved are defined:

1. Conversion of the substrate to the K compartment. (rate r_{SK} and stoichiometry Y_{SK}). From Equ. 228a the following relationship is valid:

$$r_{SK} = f_1(s) \cdot f_2(w_G) \cdot x \tag{5.229}$$

   The rate of substrate consumption is assumed to be of saturation type, that is,

$$r_{SK} = \frac{k_{SK} \cdot s}{K_S + s} \cdot \frac{w_G}{K_G + w_G} \cdot x \tag{5.230}$$

   Here k_{SK} represents the maximum value r_{SK} can achieve and w_G is the mass fraction of component G.

2. Transformation of the K compartment into the G compartment (rate r_{KG} and stoichiometry Y_{KG}). The following relationship is proposed:

$$r_{KG} = f_3(w_G) \cdot x \tag{5.231}$$

The value of r_{KG} can be described by

$$r_{KG} = k_{KG} \cdot w_K \cdot w_G \cdot x = k_{KG} \cdot w_K(1 - w_K)x \tag{5.232}$$

since $w_K + w_G = 1$, by definition. Here, the rate of K to G is assumed to be influenced by the intrinsic concentrations of K and G as well as by the total biomass concentration. An interesting observation here is that when $G = 0$, Equ. 5.232 predicts no G formation. This has a biological explanation: When $G = 0$, enzymes and DNA necessary for G synthesis are absent.

3. Turnover of the compartments of the biomass is modeled by

$$r_{GK} = m_G \cdot w_G \cdot x = m_G(1 - w_K)x \tag{5.233}$$

where m_G is the specific turnover rate of compartment G (maintenance rate). It is assumed to be a depolymerization process and is first order in the total amount of G compartment ($w_G x$). The specific rate of depolymerization is m_G. The yield constant for the formation of the K compartment from the G compartment is assumed to be unity, that is, no mass is lost during the depolymerization of the G compartment to precursors.

The balance equations for the rate of change of the substrate concentrations, the biomass concentration, and the fraction of the G compartment are now obtained by the application of the formalism treated by Harder and Roels (1982). The resulting equations are

$$\frac{ds}{dt} = -q_{S,\max}\frac{s}{K_S + s} \cdot \frac{w_G}{K_G + w_G} x + F_S \tag{5.234}$$

with F_S the flow of substrate to the system $[\mathrm{kg \cdot m^{-3} \cdot h^{-1}}]$,

$$\frac{dx}{dt} = Y_{SK} \cdot q_{S,\max}\frac{s}{K_S + s} \cdot \frac{w_G}{K_G + w_G} x + (Y_{KG} - 1)f_3(w_G)x + F_X \tag{5.235}$$

with F_x the net flow of unstructured biomass dry weight to the system $[\mathrm{kg \cdot m^{-3} \cdot h^{-1}}]$, and

$$\frac{dw_G}{dt} = -Y_{SK} \cdot q_{S,\max}\frac{s}{K_S + s} \cdot \frac{w_G}{K_G + w_G} \cdot w_G + f_3(w_G) \times \{w_G + Y_{KG}(1 - w_G)\} - m_G w_G \tag{5.236}$$

These equations contain the transport contributions F_S and F_X, which depend on the mode of operation and are summarized in Table 5.5. A most important feature of Equs. 5.234 through 5.236 is the fact that the intrinsic balance equation is independent of the mode of operation.

The necessity for a model of such simplicity, while sufficient knowledge is available for the construction of a model of much greater realism, may not be

TABLE 5.5. State equations of the Williams two-compartment model according to Harder and Roels (1982).

	F_S	F_X
For batch operations	0	0
For CSTR with feedback of fraction $(1 - w_D)$ of biomass leaving system	$D(s_{in} - s_{ex})$	$-w_D \cdot D \cdot x$
For fed-batch culture	$F_S(t)$	0

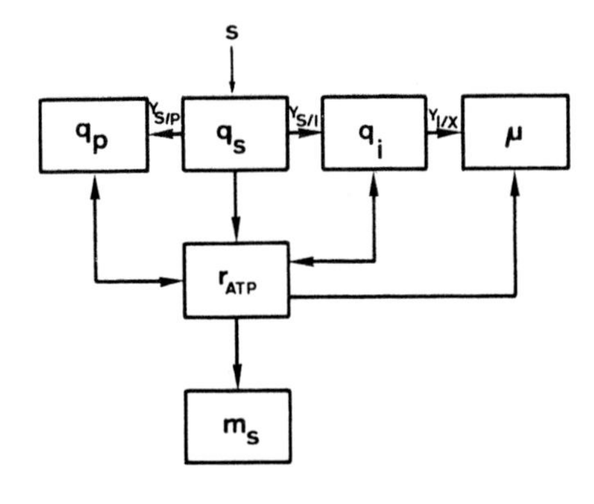

FIGURE 5.68. Block diagram showing a simple model of metabolism; I, intermediary product or precursor. (From Roels and Kossen, 1978.)

obvious. Two factors should be considered:

1. A number of regulatory mechanisms at the level of energy generation and consumption operate with such small relaxation times that a pseudo-steady-state hypothesis with respect to these mechanisms is justified. Hence, the introduction of these details seems to be unnecessary.
2. A minimum of complexity is desirable because the complex model often proves very difficult to verify and may fit experimental results without having any relationship with the behavior of the organism. Roels and Kossen (1978) gave a block diagram of a simple approach to metabolism, shown in Fig. 5.68. Only after obtaining experimental evidence that the simple model should be rejected because of insufficient fit of the data or unrealistic parameter values should additional complexity be introduced.

If the relations obtained for r_S are compared, the structured model can be seen to have a maintenance term dependent on the growth rate in the form of the following equation:

$$m_S = \frac{k_{KG}}{Y_{SK}} \cdot \frac{D + m_G}{(k_{KG} \cdot Y_{KG})^2} [k_{KG} Y_{KG} - (D + m_G)](1 - Y_{KG}) \qquad (5.237)$$

The two-compartment model exhibits many features observed in batch and continuous culture experiments. The method is certainly promising as an adequate approach to transients.

The particular model presented, however, must be considered as a preliminary proposal because many of the kinetic assumptions do not rest on solid biochemical facts about the internal regulation of the cell. Furthermore, there are difficulties in identifying the compositional nature of the K and G compartments in terms of structural components of the cell. It is clear that a more thorough study of known regulation phenomena and an empirical study of transient situations, for example in continuous culture, is needed, especially in the CSTR.

5.8 Kinetic Models of Heterogeneous Bioprocesses

5.8.1 Biofilm Kinetics

While pseudohomogeneous rates of biokinetics follow the equation for r_i $[\mathrm{kg \cdot m^{-3} \cdot h^{-1}}]$

$$r_i = f(k_1, k_2, s_i) \qquad (5.238)$$

with k_1 and k_2 defined in Equ. 4.25, heterogeneous rates are expressed as rates r_i' $[\mathrm{kg \cdot m^{-2} \cdot h^{-1}}]$ related to the area of active biomass a_s $[\mathrm{m^2 \cdot m^{-3}}]$

$$r_s' = f(k_1, k_2, k_3, d_P, s_i) \qquad (5.239)$$

The parameters k_3 and d_P are responsible for heterogeneity, d_P being the particle diameter and k_3 a process kinetic constant, defined by Atkinson (1974) as

$$k_3 = \sqrt{\frac{k_1'}{D_{S,\mathrm{eff}}}} \qquad (5.240)$$

with $k_1' = a_S \cdot k_1$. The parameters k_1, k_2, and k_3 are referred to as coefficients of the "biological rate equation" (BRE) (Atkinson, 1974). Parameter k_3 represents the interactions between reaction (k_1) and physical internal transports ($D_{S,\mathrm{eff}}$, related to d_P by the two-film theory given in Equ. 3.30a).

The BRE (cf. Equ. 5.239) is valid for heterogeneous microbiological and biochemical systems, including fermentations and enzyme engineering systems. The model uses the same type of equation as Equ. 4.60 with limiting boundary conditions Equ. 4.61a and b, with the exception that the substrate utilization rate is referred to the surface area of the biological mass, r_S', which is a quantity more easily measured. The connection with the specific rate, q_S $[\mathrm{hr^{-1}}]$, is through the relationship

$$q_S = r_S' \frac{A_S}{V \cdot \rho} = r_S' \frac{a_S}{\rho} \qquad (5.241)$$

where a_S is the specific surface area of the biological mass, which has a density ρ. A graphical representation of the BRE is depicted in Fig. 5.69, plotting r_S' versus s in analogy to the Monod-type diagram (cf. Fig. 5.13). Figure 5.69 can be applied directly to biofilm kinetics (r_S').

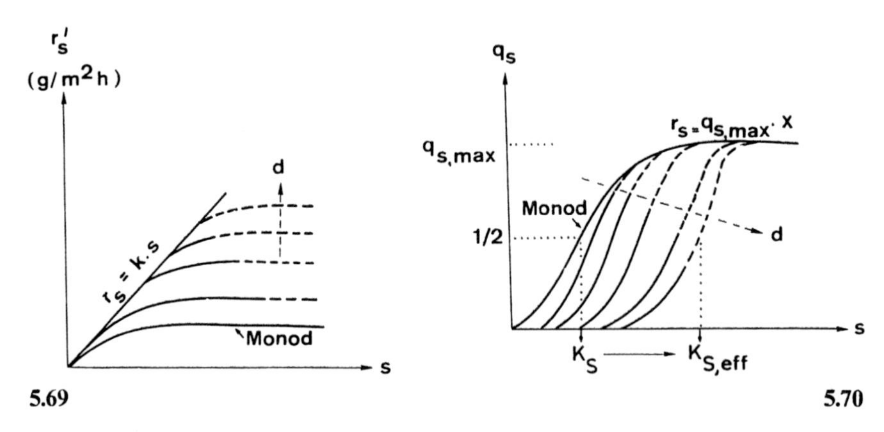

FIGURE 5.69. Biological film kinetics: Plot of the rate of substrate use, r'_S, referred to the effective biological film surface area, as a function of the substrate concentration, s, at various thicknesses of a biological film, d. A realistic limiting case (——) is found at low s where formal first-order kinetics in s are found. (Adapted from Atkinson, 1974; with permission of Pion, London.)

FIGURE 5.70. Kinetics of heterogeneous bioprocess: Diagram of the specific rate of substrate use, q_S, as a function of the substrate concentration, s, varying the thickness of the biocatalytically active mass, d, for floc and film geometries. Monod kinetics appear as the boundary case at low d. Unrealistic cases are indicated with dashed lines. (Adapted from Atkinson, 1974; with permission of Pion, London.)

The complete computer simulation of the BRE is given by dotted lines for various film thicknesses d, and the area of experimental verification is indicated with solid lines. Obviously, high S concentration cannot be attained with thick films. As can be seen from Fig. 5.69, the appearance of a simple first-order reaction rate in cases of biofilm processing can often be understood as an observed case of pseudokinetics.

With biological flocs, reaction rates should be plotted as illustrated in Fig. 5.70 (Atkinson, 1974). With increasing film thickness, and with larger particles, the curve moves to higher S values. Again, dotted lines represent the computer simulation, while solid lines correspond to experimental situations.

The BRE shows that only one kinetic equation can exist for various geometries of the biological mass. In the past, attempts were often made to formally utilize zero-order equations with microbial particles (flocs) and to formally use first-order equations with microbial films, and this discrepancy was the starting point of Atkinson's work.

For both types, one may recognize Monod-type relationships that omit D_S. For biological films, one actually has a realistic zero-order equation (Equ. 5.244). Higher substrate concentrations are attainable with flocs, so that here zero order also dominates.

In Fig. 5.71 various limiting cases for microbial particles (flocs) are presented

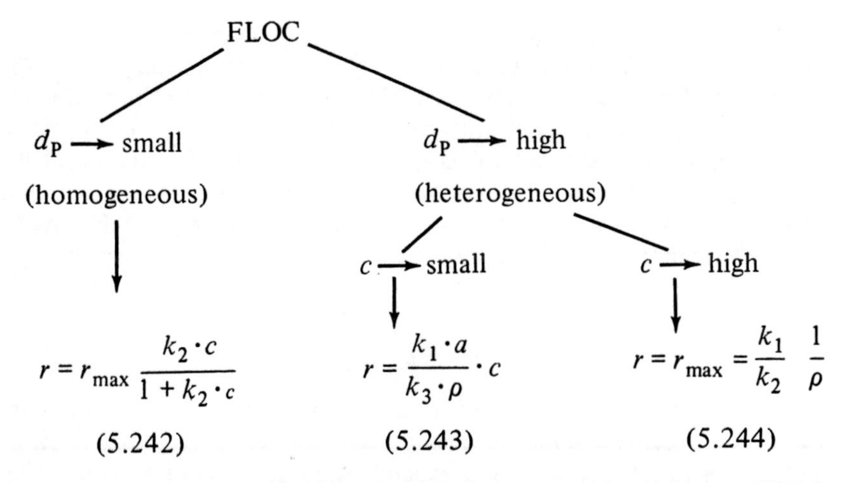

FIGURE 5.71. Equations derived from the biological rate equation (Atkinson, 1974) that can be used for the quantification of different cases of "biofloc processing."

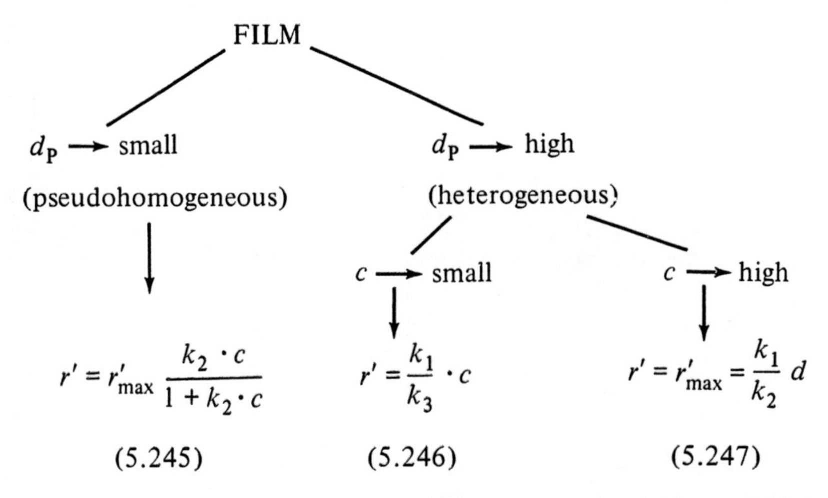

FIGURE 5.72. Equations derived from the biological rate equation (Atkinson, 1974) that can be used for the quantification of different cases of "biofilm processing."

along with Equs. 5.242 through 5.244 derived from the BRE. The same are shown in Fig. 5.72 for the thin-film geometry (Equs. 5.245–5.247).

The BRE in the form of Equ. 5.239 can be rewritten to give

$$r'_i = f(k_2 \cdot s, k_3 \cdot d_p) \tag{5.248}$$

equivalent to

$$r'_i = f\left(\frac{s}{K_S}, \Phi\right) \tag{5.249}$$

using the Thiele modulus ϕ (cf. Equs. 4.74 and 4.80) instead of Equ. 5.240 (see also Sect. 4.5.2). The consequence of this type of modeling is that the complexity of interactions between biological reactions (k_r) and physical transports (k_{TR}) can be significantly reduced with the aid of the concept of efficiency η_r interpreted in Equ. 4.53 as a function of ϕ or Da_{II} (cf. Fig. 4.30). With the η concept, the truly heterogeneous nature of bioprocessing is simplified to the pseudohomogeneous case, where the algebraic factor η_r is only a multiplicative term in the pseudohomogeneous rate. Thus, for Monod-type kinetics we have

$$r_{i,\mathrm{eff}} = r_{i,\max}\frac{s}{K_S + s}\cdot x\cdot\eta_r \tag{5.250}$$

In the literature are many articles on porous diffusion, especially in connection with carrier-bound enzymes or cells (for example, Pitcher, 1978). These are directly connected to the principles expressed in Sect. 4.5 concerning the influence of internal and external mass transport. The results are presented in the same graphical form as Fig. 4.36 in which the effectiveness factor of the reaction η_r is presented as a function of the Thiele modulus. For formulating an appropriate moduls one needs knowledge of the difficult to measure D_{eff} value. The following equation has shown itself useful in that the volume-based reaction rate r_S is obtainable directly from the experimental measurements (Pitcher, 1978) (cf. Equ. 4.74)

$$\Phi = \frac{d^2}{D_{\mathrm{eff}}}\cdot\frac{1}{s}\cdot r_{S,\mathrm{eff}} \equiv \phi^2\cdot\eta_r \tag{5.251}$$

with $r_{S,\mathrm{eff}}$ the observed effective rate per unit volume of porous carrier $[\mathrm{kg\cdot m^{-3}\cdot h^{-1}}]$ (Pitcher, 1975). Since the effectiveness factors for intermediate K_S/s values lie between the curves for the limiting cases of $n = 0$ and $n = 1$,

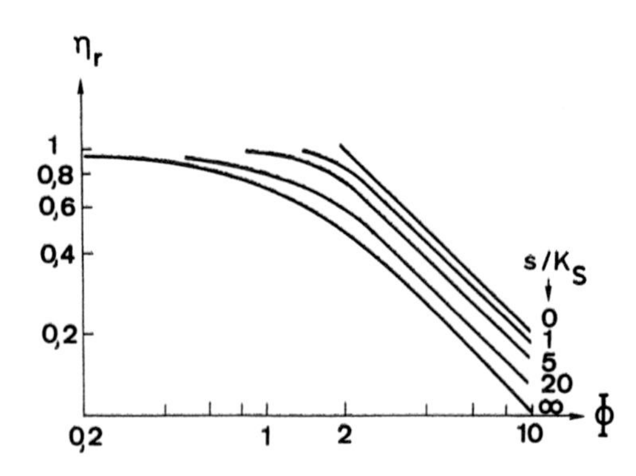

FIGURE 5.73. Plot of the effectiveness factor of reaction, η_r, as a function of the modified Thiele modulus, Φ (Equ. 5.251) for enzymatic reactions. Variation of the K_S value are shown in the case of a flat-plate geometry of the solid phase (Pitcher, 1978). This plot is analogous to Fig. 4.36.

we see from Fig. 5.73 that η_r is indeed relatively insensitive to the remaining intrinsic parameters K_S/s and k_2. Figure 5.73 is an equivalent form to demonstrate the BRE, as shown in Figs. 5.69 and 5.70. Variations of the plot in Fig. 5.73 are described in the literature with modified definitions of a modulus (e.g., Shieh, 1981) or change in the functional dependence, as shown in Fig. 5.70. Numerical solutions are recommended (Moo-Young and Kobayashi, 1972) for cases involving complex kinetic equations such as substrate or product inhibition.

In addition to the process kinetic formulations of Atkinson, which are derived from the theory of mass transport with simultaneous biological reaction, other kinetic equations, macrokinetic in origin, are also known.

With reliance on the external, apparent nature of the influence of mass transport, as shown in Fig. 5.70, where with an increasingly thick biological mass there is a higher K_S value, an equation based on Monod kinetics has been proposed (Powell, 1967), already shown in Equ. 5.43, where

$$K_{eff} = K_S + K_D \tag{5.252}$$

K_{eff} is the sum of the Monod constant K_S and a formal kinetic coefficient for the volumetric mass transfer resistance, K_D, in the solid phase. The behavior of this approach at varied ratio K_D/K_S is shown in Fig. 5.74. The application of this equation has been encouraged by Humphrey (1978).

A second alternative for a kinetic equation has been proposed by Harremoës (1977, 1978): It is a kinetic description of the plot in Fig. 5.70, and is an equation formally one-half order in S (cf. Equ. 4.87)

$$r_S = k_{1/2} \cdot s^{1/2} \cdot X \tag{5.253}$$

The integrated form of such an equation has already been discussed (Equ. 5.35). To illustrate the range of applicability of each of these approaches ($n = 0, 1$) and the transitions between the cases, it is convenient to use dimensionless quantities A, B, E, which are simply the ratios between the three different bulk reaction rates, interrelated as $A^2 = E \cdot B$.

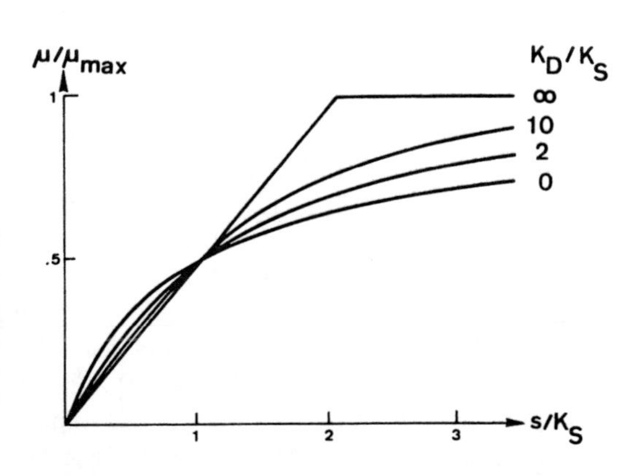

FIGURE 5.74. Normalized kinetic plot of the Powell approach to internal transport limitation (cf. Equ. 5.43) with variation of the ratio K_D/K_S. Monod-type kinetics appear when this ratio is zero (after Humphrey, 1978).

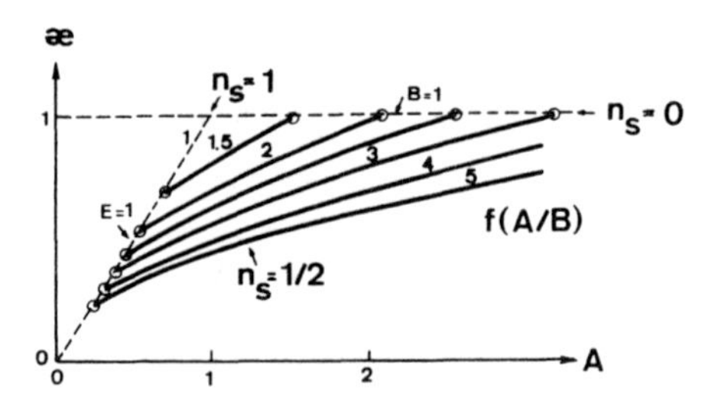

FIGURE 5.75. Dimensionless plot of reaction rate $\kappa = r'/k_0'$ versus substrate concentration in the bulk liquid A (cf. Equ. 5.254a). Zero-, first-, and half-order reactions ($n_S = 0, 1, 1/2$) are demonstrated in their range of applicability. Note how $n_S = 1/2$ may erroneously be interpreted as the saturation effect of Monod kinetics. B and E are explained in Equs. 5.254b and c (Harremoës, 1977).

$$A = \frac{k_1'}{k_0'} \cdot s \tag{5.254a}$$

$$B = \frac{k_{1/2}^2}{k_0^2} \cdot s \tag{5.254b}$$

$$E = \frac{k_1^2}{k_{1/2}^2} \cdot s \tag{5.254c}$$

Figure 5.75 shows the bulk reaction rate of the biofilm κ in dimensionless form as a function of the bulk concentration A with different biofilm characteristic values of A/B (Harremoës, 1978). Even though the depth of penetration may be small compared with the entire film thickness, the effect of internal diffusion resistances is to mask the true zero-order kinetics (see Equ. 5.111), yielding an apparent reaction order of one-half.

Formulations involving simultaneous internal and external limitations are to be found in the literature (Frouws, Vellanqa, and Dewilt, 1976; Hamilton, Gardner, and Colton, 1974) and are referenced in Sect. 4.5.3.

5.8.2 UNSTRUCTURED MODELS OF PELLET GROWTH

Fungal growth occurs in two morphologically distinct forms: mycels (filamentous hyphae dispersed in fluid) and pellets (stable, spherical agglomerations). In submerged cultures, growth in filamentous form follows the exponential law (Metz and Kossen, 1977; van Suijdam et al., 1982). Individual hyphae grow only at the tips (see Equ. 5.109) and have a linear extension rate. Exponential growth is maintained by continuous branching of the hyphae. This can be expressed by the "hyphal growth unit," the mean length of hyphae per growing tip. The combination of a linear extension rate with a constant hyphal growth unit concept results in exponential growth, where the unit is independent of μ (van Suijdam and Metz, 1981).

The growth of pellets is not always exponential because of mass transfer limitations (Metz and Kossen, 1977; Pirt, 1975; Whitaker and Long, 1973). Increasing the size of pellets results in a significant decrease in reaction rate, and the apparent K_S value increases (e.g., Kobayashi et al., 1973). To describe the growth of pellets, one can use either the exponential law or the Monod relationship. However, the logistic equation (Kendall, 1949) in the form of Equ. 5.74 or the Gompertz equation (Chiu and Zajic, 1976)

$$r_X = k_i \cdot X \cdot \exp(-k_j \cdot t) \tag{5.255}$$

is often preferred. In Equ. 5.255, k_i and k_j are constants with the dimension $[h^{-1}]$.

Most often, the model used to describe the growth of pellets is the "cube root equation" (cf. Metz and Kossen, 1977; Pirt, 1975; Trinci, 1970). Considering the cell mass to be related to the pellet radius, R, and the number of pellets, N, according to Equ. 5.256

$$X = \frac{N}{V} \rho \frac{4\pi}{3} R^3 \tag{5.256}$$

the cube root law is often expressed in two different ways:

$$X_t^{1/3} = \alpha_1 X_0^{1/3} \cdot t \tag{5.257a}$$

$$X_t^{1/3} = X_0^{1/3} + k \cdot t \tag{5.257b}$$

with α_1 a proportionality constant and with

$$k = \mu \cdot \tilde{d}_P \left(\frac{4\pi}{3} \rho N\right)^{1/3} \tag{5.258}$$

where $\tilde{d}_P$ is the thickness of a peripheral zone around the pellet; the zone alone contributes to growth (Pirt, 1975).

The appropriate rate equation for this law is

$$r_X = 3 \cdot \alpha_2 \cdot \tilde{d}_P \cdot \mu \cdot X^{2/3} \tag{5.259}$$

where α_2 $[kg^{1/3} \cdot m^{-2}]$ is a constant.

The cube root law results from a combination of exponential growth with mass transport limitations in the solid phase, which is expressed in the concept of $\tilde{d}_P$ as shown by Pirt (1975).

The predictions of different kinetic model equations for pellet growth are compared in Fig. 5.76 (van Suijdam et al., 1982). The main objection against the models presented is that they are autonomous, that is, the biomass rate equation is not related to the concentration of the limiting substrate, and that mass transfer limitations are not explicitly considered in these macrokinetics. To introduce these phenomena, a combination of mass transfer limitation effects and biokinetics must be used. Such an integrated model for pellet growth was developed (Metz, 1975) and extended (van Suijdam et al., 1982), based on balance equations for X, S, and O_2 in liquid and solid phase.

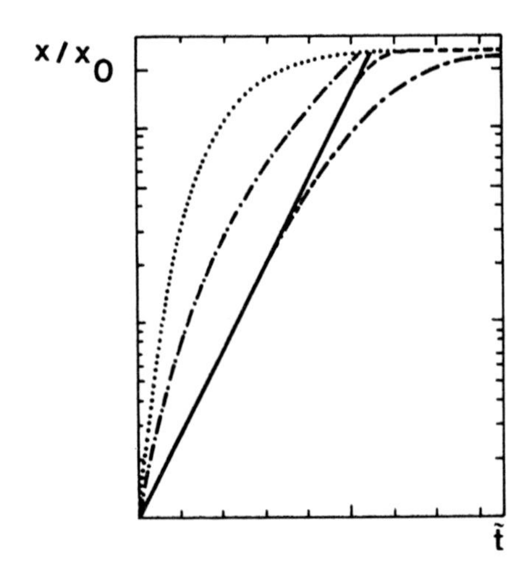

FIGURE 5.76. Comparison of various rate equations in integrated form in a normalized plot of biomass concentration x/x_0 versus time $\tilde{t} = \mu_{max} t$ in case of pellet growth: exponential law $r_X = \mu_{max} x$ (——); Monod equation, see Equ. 5.38 (– – –); logistic law, see Equ. 5.75 (-·-·-·); Gompertz' law, Equ. 5.255 (·····); cube-root law, Equ. 5.257 and 5.259. Parameter values $x_{max}/x_0 = 250$, $k_i = 5.5$, $\alpha_2 \tilde{d}_P = 1.5\ \mathrm{kg}^{1/3} \cdot \mathrm{m}^{-1}$ (van Suijdam et al., 1982).

5.8.3 LINEAR GROWTH

The characteristics of a linear growth pattern as the consequence of an enzyme system with constant activity have been thoroughly investigated and modeled by Knorre et al. (1978a,b). It is thought that linearity is caused by constant activity of enzymes in systems where substrate concentration is limited. Linearities always indicate the presence of some limitations; even their exact nature cannot be readily determined. Beyond a lack in nutrients, linear growth phases can also be the result of transport limitations. Thus, from a systematic point of view, these data repesent pseudokinetics.

5.9 Pseudokinetics

Pseudokinetic parameters result from a kinetic study whenever the model (either knowingly or unknowingly) is simpler than the real situation. Generally, falsification can be the result of the undetected influence of other reactions or of transport phenomena (macrokinetics).

Pseudokinetic phenomena become evident only when process kinetic analysis is carried out with mathematical models. Most bioprocesses are basically heterogeneous systems. Generally, pseudohomogeneous rates measured in L phase analyses are used, because they are thought to reflect directly the intrinsic reaction rate of metabolism in the solid phase (biomass). Even under steady-state conditions, however, this assumption is not necessarily valid, because different concentration levels may create different reaction rates. Barford and Hall (1979a) demonstrated that external overall flux measured in the L phase does not even approximately reflect the true internal fluxes of metabolism. This fact suggests caution in accepting any model as a "true" reflection of the in vivo state.

Many factors are known to influence the shape of Monod-type biokinetics (A. Moser, 1981; Fiechter, 1982); most of them alter K_S. Apparent K_S values can be the result of:

- Multi-S limitation
- Insufficient mixing in the liquid phase
- External transport limitation
- Internal transport limitation
- Ionic strength (Hornby et al., 1968)
- High cell concentration (Contois kinetics)
- Endogenous metabolism
- Non-stationary processing
- Product inhibition
- Biosorption

One might also think that the K_S value depends on the size of the cells (transport limitation by D_S), and that only in some ideal cases is K_S the same as that appropriate to the enzyme or mitochondria (Hartmeier, Bronn, and Dellweg, 1971; Kessick, 1974).

In case of 2-S limitation, the apparent K_S value is given by

$$K_{S,1,app} = K_{S1}\left(1 + \frac{K_{S2}}{s_2} + K_{S2}\cdot\frac{s_1}{s_2}\cdot K_{S1}\right) \tag{5.260}$$

Analogous cases of kinetic equations for heterogeneous situations show similar behavior, explaining apparent K_S values by external mass transfer limitation:

Hornby et al. (1968)

$$K_{S,app} = K_S + \frac{r'_{max}}{k_{L2}} \tag{5.261}$$

where r'_{max} is the maximum reaction rate per unit external surface area

Schuler et al. (1972)

$$K_{S,app} = K_S + \left[\frac{r'_{max}\cdot K_S}{k_{L2}(s_0 + K_S)}\right] \tag{5.262}$$

Kobayashi and Moo-Young (1971)

$$K_{S,app} = K_S + \frac{3\cdot r'_{max}}{4\cdot k_{L2}} \tag{5.263}$$

Finally, a case of pseudokinetics appears in case of global measurement techniques, as in biological waste water treatment.

The BOD, COD, and TOC are widely used. These values, however, are not correlated strictly (e.g., Aziz and Telbut, 1980). As a specific example the BOD_5 value should be mentioned. This measurement (in a time interval of 5 days) reflects not only the O_2 used for metabolizing the primary substrate, as illustrated in Fig. 5.77, but also other O_2 demands (for endogenous metabolism, for other strains in the food chain of biocoenosis, cf. Fig. 5.55, and for metabolism of N-containing substances). As indicated in Fig. 5.77, only the BOD_{PL} (after 1 day) is a true value for S consumption and thus is the reliable value for design purposes (Wilderer et al., 1977):

$$BOD_1 = Y_{O|S}(s - s_o) \tag{5.264a}$$

or

$$q_O = Y_{O|S} \cdot q_S \tag{5.264b}$$

where the yield coefficient $Y_{O|S}$ is called "specific O_2 utilization."

5.10 Kinetics of Sterilization

Techniques such as sterilization, pasteurization, or disinfection operate on the basis of the effects of heat, radiation, or chemical agents on viability of cells (bacteria, viruses), while filtration and membrane techniques act mainly mechanically. Even though the mechanisms of desactivation are quite different, similar basic concepts can be applied in most of these cases.

5.10.1 Basic Kinetic Approaches in Sterilization Kinetics

The death rate for a population of microbial cells is commonly found to obey a logarithmic law, and this is expressed by formal first-order kinetics

$$-\frac{dN}{dt} = k_d \cdot N \tag{5.265}$$

where N is the concentration of viable cells [number/ml] and k_d is the specific death rate constant [h^{-1}].

Figure 5.78 is a simple plot of cell number versus time at a given temperature according to the logarithmic death rate. To achieve a certain level of sterility (N_{st}) it is clear that a time t_{st} is needed. A normal value for N_{st} is about 10^{-4} cells/ml.

A consequence of the first-order law is that $t \to \infty$ for $N_{st} = 0$. However, the first-order concept cannot predict the behavior of a single cell; it is only a statistical approach to the behavior of a large population. Integrating Equ. 5.265 with initial conditions $N = N_0$ and $t = t_0$ gives

$$\ln \frac{N}{N_0} = -k_d \cdot t \tag{5.266}$$

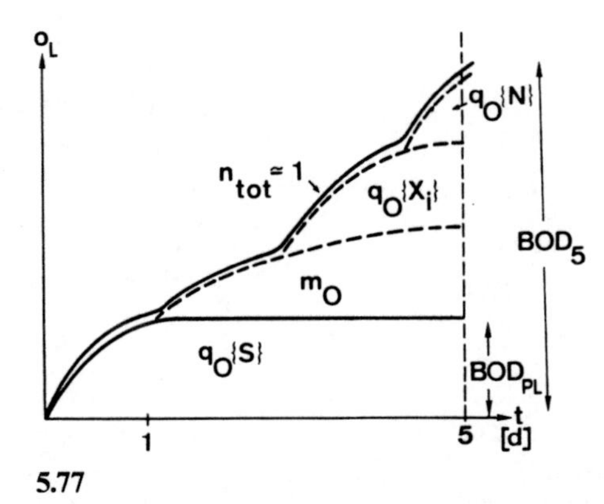

5.77

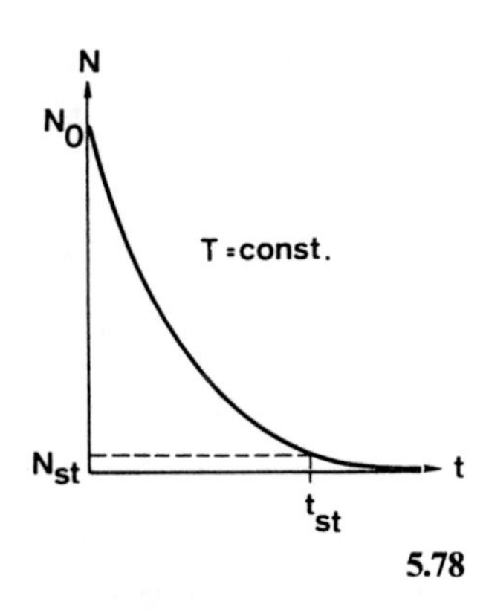

5.78

FIGURE 5.77. Schematic diagram of the time course of biological oxygen demand represented in a graph of oxygen concentration o_L versus time, t, in days, showing a succession of different steps: O_2 consumption due to waste degradation ($q_O\{s\}$); maintenance requirement (m_O); O_2 consumption by the "food-chain" organisms x_i ($q_O\{x_i\}$); O_2 utilization due to nitrogen removal at high residence time ($q_O\{N\}$). Interpretation of BOD_5 values thus shows that these values are of low significance for engineering design purposes. As a substitute, a "plateau BOD" (BOD_1) was proposed to occur after one day (Wilderer et al., 1977).

FIGURE 5.78. Plot of the decrease of cell number N due to sterilization effects in time t at a given and constant temperature. The "sterilization niveau," N_{st} (normally 10^{-4} cells/l) is reached at t_{st}, the time needed for sterilization.

which can be plotted to give the linear form shown in Fig. 5.79. Death rate constant k_d is a function only of temperature. Deviations from the simple logarithmic death rate are observed, especially with bacterial spores, as illustrated in Fig. 5.80. Another case is included in the figure, where some activation by temperature occurs before the number of spores decreases. For the inactivation of spores, a death model was proposed by Prokop and Humphrey (1970) in which a resistant type of spore N_r proceeds to death (N_d) via a sensitive intermediate state N_S. Thus

$$N_r \xrightarrow[n=1]{k_{d,1}} N_S \xrightarrow[n=1]{k_{d,2}} N_d \tag{5.267}$$

The corresponding differential equations are

$$\frac{dN_r}{dt} = -k_{d,1} \cdot N_r \tag{5.268}$$

and

$$\frac{dN_S}{dt} = k_{d,1} \cdot N_r - k_{d,2} \cdot N_S \tag{5.269}$$

The solution takes the form

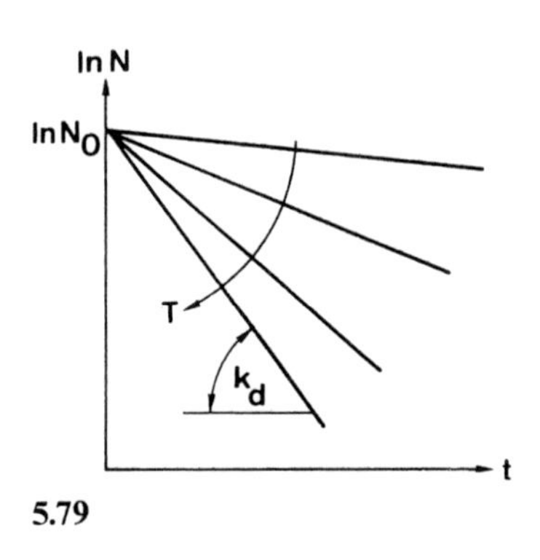

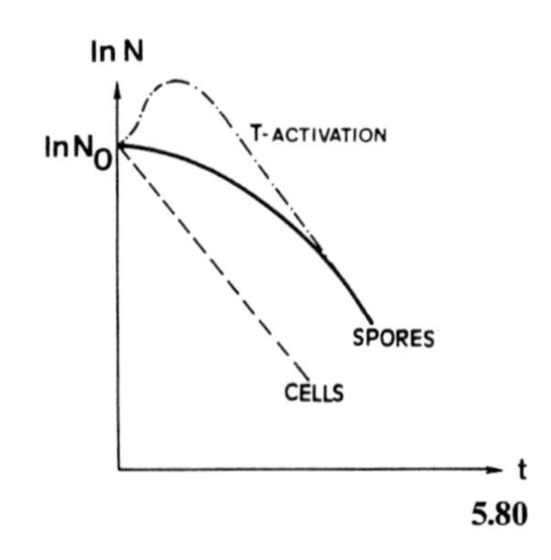

FIGURE 5.79. Typical plot of sterilization kinetics with viable cells, according to Equ. 5.265 with varying temperature. Death rate constants k_d can be estimated from the slope of the lines.

FIGURE 5.80. Typical plot of sterilization kinetics with bacterial spores compared with vegetative cells. A deviation is known due to the activation of spores by temperature.

$$\frac{N}{N_0} = \frac{k_{d,1}}{k_{d,1} - k_{d,2}} \left[\exp(k_{d,2} \cdot t) - \frac{k_{d,2}}{k_{d,1}} \exp(-k_{d,1} \cdot t) \right] \tag{5.270}$$

Alternative expressions for the kinetic constants k_d are sometimes used in different technologies (see Equs. 5.8 and 5.9). Supplementary to the "*D* value" and "*Z* value," an "*F* value" is known, representing the value of t_{st} at 121°C.

Another case of logarithmic behavior appears when air is sterilized (filtered) by fibrous media, and the "log-penetration law" is valid. This law relates filter effectiveness (N/N_0) to the filter thickness L and a factor K_F

$$\ln \frac{N}{N_0} = -K_F \cdot L \tag{5.271}$$

The filter constant K_F depends on fiber diameter, volume fraction of fibers in the filter, and a fiber effectiveness factor (Chen, 1955).

5.10.2 Multicomponent Systems in Food Technology

Almost all reactions taking place in foodstuffs, such as quality loss due to the loss of vitamins, enzymes, or color (chlorophyll); to the death of living cells or microbial spores; or to other chemical reactions, follow formal first-order kinetics. Reactions associated with browning are often described by zero-order kinetic reactions (Saguy and Karel, 1980).

Thus, the technical handling problems can schematically be described with the aid of Fig. 5.81, where a series of components exhibit different k values. As indication for sterility in food technology, the heat-resistant enzyme per-

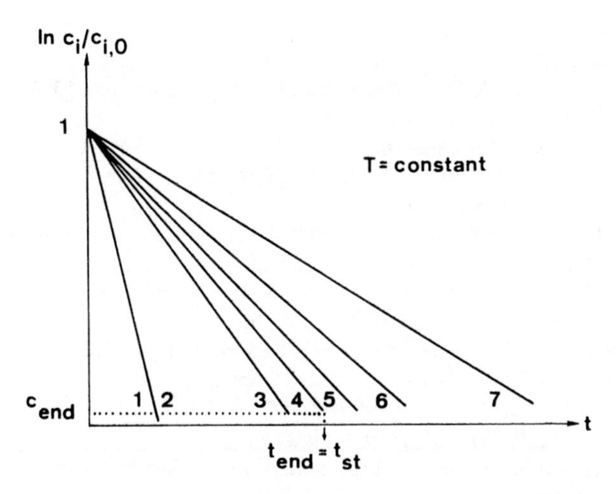

FIGURE 5.81. Schematic representation of a bioprocess in a multicomponent food-processing system in a normalized concentration/time diagram. The kinetic rate constant, k, of reactions 1–7 is (A. Moser, et al. 1980b):

	k [s^{-1}]
1. Microbial cells	$10^1 - 10^{13}$
2. Enzymes	$10^1 - 10^{13}$
3. Spores	0.5–20
4. Thermoresistant enzymes (peroxidase)	0.37
5. Chlorophyll	0.2
6. Vitamins	0.02
7. Nonenzymatic browning (Meillard) reactions	0.008

oxidase is often used with $k = 0.37\ s^{-1}$ at $T = 121°C$ or the spores of *Bacillus stearothermophilus* ($k = 2.9\ min^{-1}$ at $T = 121°C$) are used for fermentations.

To obtain sterile foodstuffs that can be stored for long periods of time, cells, spores, and enzymes must be destroyed while at the same time vitamine content, color, and aroma are not substantially affected. This is possible using kinetic data and a reactor with an average processing time $\bar{t} = t_{St}$. Since a first-order reaction is involved, the optimal reactor configuration is a continuous tube reactor (Moser, Kosaric, and Margaritis, 1980b).

BIBLIOGRAPHY

Adair, G.S. (1925). *J. Biol. Chem.*, 63, 529.

Agrawal, P., Lee, C., Lim, H.C., and Ramkrishna, D. (1982). *Chem. Eng. Sci.*, 37, 453.

Agrawal, P., Lim, H.C. (1984). *Adv. Biochem. Eng. Biotechnol.*, 30, 61.

Aiba, S. (1982). *Adv. Biochem. Eng.*, 23, 85.

Aiba, S., and Hara, H. (1965). *J. Gen. Microbiol.*, 11, 41.

Aiba, S., Humphrey, A.E., and Millis, N.F. (1973). *Biochemical Engineering*. New York: Academic Press.

Aiba, S., and Shoda, M. (1969). *J. Ferment. Technol.*, 47, 790.
Aiba, S., et al. (1968). *Biotechnol. Bioeng.*, 10, 845.
Aiba, S., et al. (1969a). *AIChEJ*, Amer. Inst. Chem. Eng. Journal, 15, 624.
Aiba, S., et al. (1969b). *Biotechnol. Bioeng.*, 11, 1285.
Andrews, J.F. (1968). *Biotechnol. Bioeng.*, 10, 707.
Andrews, J.F. (1969). *J. Sanit. Eng. Div., ASCE*, 95, 95.
Andrews, J.F. (1971). *Biotechnol. Bioeng. Symp.*, 2, 5.
Andreyeva, L.N., and Biryukov, V.V. (1973). *Biotechnol. Bioeng. Symp.*, 4, 61.
Aris, R., and Humphrey, A.E. (1977). *Biotechnol. Bioeng.*, 19, 1375.
Asai, T., and Kono, T. (1983). *Adv. Ferment.*, 83 (Suppl. Proc. Biochem.), 212.
Asai, T., et al. (1978). *J. Ferment. Technol.*, 56, 369.
Atkinson, B. (1974). *Biochemical Reactors.* London: Pion.
Atkinson, B., et al. (1969). *Trans. Inst. Chem. Engrs.*, 45, T257.
Aziz, J.A., and Tebbut, T.H.Y. (1980). *Water Res.*, 14, 319.
Bader, F.G. (1978). *Biotechnol. Bioeng.*, 20, 183.
Bader, F.G. (1982). In Bazin, M.J. (ed.). *Microbial Population Dynamics.* Boca Raton, Fla.: CRC Press.
Bader, F.G., et al. (1975). *Biotechnol. Bioeng.*, 17, 279.
Bailey, J.E., and Ollis, D.F. (1977). *Biochemical Engineering Fundamentals.* New York: McGraw-Hill.
Bajpaj, R.K. and Reuss M. (1980). *J. Chem. Tech. Biotechnol.*, 30, 332.
Bajpaj, R.K., and Ghose, T.K. (1978). *Biotechnol. Bioeng.*, 20. 927.
Bajpaj, R.K., and Reuss, M. (1980). Paper presented at 6th Internat. Fermentation Symp., London, Ontario.
Bajpàj, R.K., and Reuss M. (1982). *Biotechnol. Bioeng.*, 23, 717.
Barford, J.P., and Hall, R.J. (1979a). In Proceedings 7th Australian Conference on Chemical Engineering, p. 27.
Barford, J.P., and Hall, R.J. (1979b). *J. Gen. Microbiol.*, 114, 267.
Barford, J.P., et al. (1982). In Bazin, M.J. (ed.). *Microbial Population Dynamics.* Boca Raton, Fla.: CRC Press.
Barnes, R., Vogel, H., and Gordon, I. (1969). *Proc. Nat. Acad. Sci.* (*Wash.*), 62, 263.
Bayer, K., and Meister, G. (1982). *Eur. J. Appl. Microbiol. Biotechnol.*, 15, 36.
Bazin, M.J. (1982). In Bushell, M.E., and Slater, J.H. (eds.). *Mixed Culture Fermentations.* London–New York: Academic Press, Chap. 2.
Bazua, C.D., and Wilke, C.R. (1977). *Biotechnol. Bioeng. Symp.*, 5, 105.
Bergter, F. (1972). *Wachstum von Mikroorganismen.* Jena, East Germany; G. Fischer Verlag.
Bergter, F. (1978). *Z. Allgem. Mikrob.*, 18(2), 143.
Bergter, F., and Knorre, W. (1972). *Z. Allgem. Mikrob.*, 12(8), 613.
Bertalanffy, L., von (1942). *Theoretische Biologie.* Berlin: Borntraeger.
Blackman, F.F. (1905). *Ann. Bot.*, 19, 281.
Bleecken, St. (1979). *Populationsdynamik einzelliger Mikroorganismen, Modellbildung und -anwendung.* (Fortschritte der Experimentellen und Theoretischen Biophysik), vol. 23. (Beier W., ed.) Leipzig: Georg Thieme.
Bol, G., et al. (1980). In *Proceedings of the 6th International Fermentation Symposium,* Adv. in Biotechn., vol. I, p. 339.
Borzani, W., and Baralle, S.B. (1983). *Biotechnol. Bioeng.*, 25, 3201.
Borzani, W., and Hiss, H. (1979). *Biotechnol. Bioeng.*, 21, 2149.

Botts, J., and Morales, M. (1953). *Trans. Faraday Soc.*, 49, 696.
Boyle, W.C., and Berthouex, P.M. (1974). *Biotechnol. Bioeng.*, 16, 1139.
Brettel, R., et al. (1981a). *Eur. J. Appl. Microbiol. Biotechnol.*, 11, 205.
Brettel, R., et al. (1981b). *Eur. J. Appl. Microbiol. Biotechnol.*, 11, 212.
Bronn, W.K. (1971). *Chem. Ing. Techn.*, 43, 70.
Brown, D.E., and Halsted, D.J. (1975). *Biotechnol. Bioeng.*, 17, 1199.
Brown, D.E., and Vass, R.C. (1973). *Biotechnol. Bioeng.*, 15, 321.
Budd, J.A. (1977). *Eur. J. Appl. Microbiol.*, 3, 267.
Bu'Lock, J.D. (1975). In Smith, J.E., and Berry, D.R. (eds.). *The Filamentous Fungi*, Vol. 1. London: Edward Arnold Ltd., Chap. 3.
Bungay, H.R., III, and Bungay, M.L. (1968). *Adv. Appl. Microbiol.*, 10, 269.
Busby, J.B., and Andrews, J.F. (1975). *J. Water Pollut. Contr. Fed.*, 47, 1055.
Calam, C.T., and Russell, D.W. (1973). *J. Appl. Chem. Biotechnol.*, 23, 225.
Calam, C.T., et al. (1971). *J. Appl. Chem. Biotechnol.*, 21, 181.
Caldwell, I.Y., and Trinci, A.P.J. (1973). *Arch. Mikrobiol.*, 88, 1.
Campbell, A. (1957). *Bact. Rev.*, 21, 263.
Canale, R.P. (1969). *Biotechnol. Bioeng.*, 11, 887.
Canale, R.P. (1970). Biotechnol. Bioeng., 12, 353.
Chen, C.V. (1955). *Chem. Rev.*, 55, 595.
Chen, Y.R., and Hasimoto, A.G. (1978). *Biotechnol. Bioeng. Symp.*, 8, 269.
Chiu, Y.S., and Zajic, J.E. (1976). *Biotechnol. Bioeng.*, 18, 1167.
Chiu, S.Y., et al. (1972). *Biotechnol. Bioeng.*, 14, 179.
Constantinides, A., et al. (1970a). *Biotechnol. Bioeng.*, 12, 803.
Constantinides, A., et al. (1970b). *Biotechnol. Bioeng.*, 12, 1081.
Contois, D.E. (1959). *J. Gen. Microbiol.*, 21, 40.
Cooney, Ch.L., et al. (1969). *Biotechnol. Bioeng.*, 11, 269.
Crooke, Ph.S., and Tanner, R.D. (1980). *Appl. Math. Modelling*, 4, 376.
Curds, C.R. (1971). *Water Res.*, 5, 793.
Cysewski, G.R., and Wilke, Ch.R. (1978). *Biotechnol. Bioeng.*, 20, 1421.
Czerkawski, J.W. (1973). *Process Biochem.*, October, 25.
Dabes, J.N., Finn, R.K., and Wilke, C.R. (1973). *Biotechnol. Bioeng.*, 15, 1159.
Dagley, S., and Hinshelwood, C.N. (1938). *J. Chem. Soc.*, 1930.
Dean, A.C.R., and Hinshelwood, C.N. (1966). *Growth, Function and Regulation in Bacterial Cells.* Oxford: Clarendon Press.
Demain, A.L., et al. (1979). In 29th *Symposium of the Society of General Microbiology*, Cambridge: Cambridge University Press. (Bull A., ed.).
Dialer, K., and Löwe, A. (1975). In Winnacker, K., and Küchler, L. (eds.). *Chemische Technologie*, Vol. 7, 3rd ed., Munich: C. Hanser.
Eckenfelder, W.W., and Ford, D.L. (1970). *Water Pollution Control.* Austin, Tex.: Pemberton Press.
Edwards, V.H. (1969). *Biotechnol. Bioeng.*, 11, 99.
Edwards, V.H. (1970). *Biotechnol. Bioeng.*, 12, 679.
Edwards, H.V., and Wilke, Ch.R. (1968). *Biotechnol. Bioeng.*, 10, 205.
Eggers, E. and Terlouw, T. (1979). *Water Res.*, 13, 1077.
Esener, A.A., et al. (1981). *Biotechnol. Bioeng.*, 23, 1851.
Esener, A.A., et al. (1982a). In Proceedings of the 2nd IWTU, *Intern. Symp. Waste Treatment Utilization.* Moo-Young, M., et al. (eds.).: Pergamon Press, Oxford, New York Vol. 2.

Esener, A.A., et al. (1982b). *Biotechnol. Bioeng.*, 24, 1749.
Esener, A.A., et al. (1983). *Biotechnol. Bioeng.*, 25, 2803.
Falch, E. (1968). *Biotechnol. Bioeng.*, 10, 233.
Fennema, Q.F. (1975). *Principles of Food Science, Part II*. New York and Basel: Marcel Dekker, p. 54ff.
Fidgett, M., and Smith, E.L. (1975). *J. Appl. Chem. Biotechnol.*, 25, 355.
Fiechter, A. (1982). *In Biotechnology—A Comprehensive Treatise* (Rehm, H.J., and Reed, G., eds.) Verlag Chemie, Weinheim, 1, Chap. 7.
Fishman, V.M., and Biryukov, V.V. (1973). *Biotechnol. Bioeng. Symp.*, 4, 647.
Follman, H., Märkl, H., and Vortmeyer, D. (1977). Dechema Annual Conference, June 23–24, Frankfurt, Lecture No. 9.15.
Forrest, W.W. (1972). In Norris, J.R., and Ribbons, D.W. (eds.). *Methods in Microbiology*, Vol. 6.B. London–New York: Academic Press, p. 285.
Frederickson, A.G., et al. (1970). *Adv. Appl. Microbiol.*, 13, 419.
Frouws, M.J., Vellenga, K., and Dewilt, H.G.J. (1976). *Biotechnol. Bioeng.*, 18, 53.
Fujimoto, Y. (1963). *J. Theoret. Biol.*, 5, 171.
Fukuda, H., et al. (1978). *J. Ferment. Technol.*, 56, 351.
Gaden, E.L., Jr. (1955). *Chem. Ind. (Lond.)*, 154, 192.
Gaden, E.L., Jr. (1959). *J. Biochem. Microbiol. Technol. Eng.*, 1, 413.
Garfinkel, D. (1969). *In Concepts and Models in Biomathematics* (Heinmets F., ed.) New York: Marcel Dekker p. 1.
Gaudy, A.F., and Gaudy, E.T. (1972). *Adv. Biochem. Eng.*, 2, 97.
Gause, G.F. (1934). *The Struggle for Existence*. Baltimore: Williams & Wilkins.
Ghose, T.K., and Tyagi, R.D. (1979). *Biotechnol. Bioeng.*, 21, 1401.
Giona, A.R., et al. (1976). *Biotechnol. Bioeng.*, 18, 473.
Glende, M., and Reich, J.G. (1972). *Acta Biol. Med. Germ.*, 29, 595.
Gottschal, J.C., and Thingstad, T.F. (1982). *Biotechnol. Bioeng.*, 24, 1403.
Grady, C.P.L., et al. (1972). *Biotechnol. Bioeng.*, 14, 391.
Grant, C.A., et al. (1978). *Biotechnol. Bioeng.*, 19, 1817.
Grau, P., Dohanyos, M., and Chudoba, J. (1975). *Water Res.*, 9, 637.
Grm, B., et al. (1980). *Biotechnol. Bioeng.*, 22, 255.
Gutke, R. (1980). In UNEP/UNESCO/ICRO training course, *Theoretical Basis of Kinetics of Growth, Metabolism and Product Formation of Microorganisms*. Jena: Science Academy of East Germany, Central Institute for Microbiology and Experimental Therapy (ZIMET), Vol. 1, pps. 30, and 112.
Gutke, R. (1982). In Biotechnological Fundamentals of Continuous Biomass Production, Manual for International Training Course. Jena (ZIMET) and Leipzig (Inst. Techn. Chem.) East Germany: Academy of Science, p. 39 and p. 58.
Gutke, R., et al. (1980). *Biotechnol. Lett.*, 2, 315.
Hamilton, B.K., Gardner, C.R., and Colton, C.K. (1974). *AIChEJ*, 20, 503.
Harder, W., and Dijkhuizen, L. (1975). In Dean, A.C.R., et al. (eds.). *Continuous Culture*, Vol. 6. London: SCI; Chichester: E. Horwood, Chap. 23, p. 297.
Harder, A., and Roels, J.A. (1982). *Adv. Biochem. Eng.*, 21, p. 56.
Harremoës, P. (1977). *Vatten*, 2, 122.
Harremoës, P. (1978). In Mitchell, R. (ed.) *Water Pollution Microbiology*, Vol. 2. J. Wiley, N.Y. p. 71.
Harris, R.S., and Karmas, E. (1975). *Nutritional Evaluation of Food Processing*, 2nd ed. Westport, Conn.: AVI Publishing, p. 211ff.

Han, K. and Levenspiel, O. (1987). *Biotechnol. Bioeng.* 30, in press.
Hartmeier, W., Bronn, W.K., and Dellweg, H. (1971). *Chem. Ing. Techn.*, 43, 76.
Heckershoff, H., and Wiesmann, U. (1981). *Chem. Ing. Techn.*, 53, 268.
Hegewald, E.M., et al. (1978). In Sikyta, B. et al., eds. *Proceedings of the 7th Symposium on Continuous Culture of Microorganisms.* Czech. Acad. Sciences Prague: p. 717 (1980).
Hegewald, E.M., et al. (1981). *Biotechnol. Bioeng.*, 23, 1563.
Heidel, D., and Schultz, J.S. (1978). *Biotechnol. Bioeng.*, 20, 301.
Heijnen, J.J., et al. (1979). *Biotechnol. Bioeng.*, 21, 2175.
Heineken, F.G., and O'Connor, R.J. (1972). *J. Gen. Microbiol.*, 73, 35.
Heinzle, E., et al. (1982). In Proc. IFAC Symposium "*Modelling & Control of Biotechnical processes*" (Halme, A., ed.) p. 57 Pergamon Press, Oxford.
Herbert, D. (1958). In Malek, I., ed., *Proceedings of the 1st Symposium on Continuous Culture of Microorganisms.* Publ. House Czech. Acad. Sciences Prague: p. 45.
Herbert, D. (1961). In Elsworth, R. ed. *Proceedings of the 2nd Symposium on "Continuous Culture of Microorganisms."* Soc. Chem. Ind. London: SCI Monograph, 12, p. 21.
Hess, B. (1973). In *Symposium of the Society of Experimental Biology.* Cambridge: Cambridge University Press, p. 105.
Hess, B., and Boiteux, A. (1973). In Change, B., Pye, E.K., Ghosh, A.K., and Hess, B. (eds.). *Biological and Biochemical Oscillations.* New York: Academic Press.
Hill, A.V. (1910). *J. Physiol. (Lond.)*, 40, 4.
Higgins, J. (1967). *Ind. Eng. Chem.*, 59, 19.
Hinger, K.J., and Blenke, H. (1975). *Chem. Ing. Techn.*, 47, 976.
Hinshelwood, C.N. (1946). *The Chemical Kinetics of the Bacterial Cell.* Oxford: Clarendon Press.
Hiss, H., and Borzani, W. (1983). *Biotechnol. Bioeng.*, 25, 3079.
Holzberg, I., et al. (1967). *Biotechnol. Bioeng.*, 9, 413.
Hoppe, G.K., and Hansford, G.S. (1982). *Biotechnol. Lett.*, 4, 39.
Hornby, W.E., Lilly, M.D., and Crook, E.M. (1968). *Biochem. J.*, 107, 669.
Howell, J.A., and Atkinson, B. (1976). *Biotechnol. Bioeng.*, 18, 15.
Humphrey, A.E. (1972). *Chem. Reaction Eng., Adv. Chem. Ser.*, 109, 603.
Humphrey, A.E. (1977a). *Encycl. Chem. Proc. Des.*, 4, 359.
Humphrey, A.E. (1977b). *Chem. Eng. Prog.*, 85.
Humphrey, A.E. (1978). *Am. Chem. Soc. Symp. Ser.*, 72.
Humphrey, A.E., and Yang, R.D. (1975). *Biotechnol. Bioeng.*, 17, 1211.
Imai, H., et al. (1979a). *J. Ferment. Technol.*, 57, 333.
Imai, H., et al. (1979b). *J. Ferment. Technol.*, 57, 453.
Imanaka, T., and Aiba, S. (1976). *J. Appl. Chem. Biotechnol.*, 26, 559.
Imanaka, T., and Aiba, S. (1977). *Biotechnol. Bioeng.*, 19, 757.
Imanaka, T., et al. (1972). *J. Ferment. Technol.*, 50, 633.
Imanaka, T., Kaieda, T., and Taguchi, H. (1973) *J. Ferment. Technol.*, 51, 423.
Jannasch, H.W., and Mateles, R.I. (1974). *Adv. Microbiol. Physiol.*, 11, 165.
Jerusalimsky, N.D. (1967). In Powell, E.O., et al. (eds.). *Microbial Physiology and Continuous Culture.* London: Her Majesty's Stationery Office, p. 23.
Jerusalimsky, N.D., and Neronova, N.M. (1965). *Dokl. AN USSR*, 161, 1437 (Russ.).
Johnson, F.H., Eyring, H., and Polissar, M.J. (1954). *The Kinetic Basis of Molecular Biology.* New York: John Wiley.

Jones, R.P., and Greenfield, P.F. (1981). *Biotechnol. Lett.*, 3, 225.
Joschek, H.I., Dehler, J., Koch, W., Engelhardt, H., and Geiger, W. (1975). *Chem. Ing. Techn.*, 47, 422.
Jost, J.L., et al. (1973). *J. Theoret. Biol.*, 41, 461.
Kafarow, W.W. (1971). *Kybernetische Methoden in der Chemie und Chemischen Technologie*. Berlin: Akademie Verlag.
Kafarow, W.W., Winarow, A.J., and Gardejew, L.S. (1979). *Modelling of Biochemical Reactors*. Moscow: Lesnaja Promyshlenost (Russ.).
Kargi, F. (1977). *J. Appl. Chem. Biotechnol.*, 27, 704.
Kargi, F., and Shuler, M.L. (1979). *Biotechnol. Bioeng.*, 21, 1871.
Kawashima, E. (1973). *J. Ferment. Technol.*, 51, 41.
Kendall, D.G. (1949). *J. Roy. Statist. Soc.*, B11, 230.
Kessick, M.A. (1974). *Biotechnol. Bioeng.*, 16, 1545.
Kharash, M.S. (1929). *Bur. Stand. J. Res.*, 2, 359.
Knorre, W.A. (1968). *In* Malek, I., Beran, K., Fencl, Z., Munk, V., Ricica, J., and Smrckova, H. (eds.) *Continuous Cultivation of Microorganisms*, Proceedings of 4th Symposium, Prague, New York: Academic Press, p. 225.
Knorre, W.A. (1976). In *Mathematische Modellbildung in Naturwissenschaft und Technik*. Berlin: Akademie Verlag, p. 221.
Knorre, W.A. (1980). In Beier, W., and Rosen, R. (eds.). *Biophysikalische Grundlagen der Medizin*. Stuttgart, New York: G. Fischer Verlag, p. 132.
Knorre, W.A., et al. (1978a). *Zeit. Allgem. Mikrob.*, 18(4), 255.
Knorre, W.A., et al. (1978b). *Zeit. Allgem. Mikrob.*, 18(8), 609.
Knowles, G., et al. (1965). *J. Gen. Microbiol.*, 38, 263.
Kobayashi, T., and Moo-Young, M. (1971). *J. Gen. Microbiol.*, 13, 893.
Kobayashi, T., et al. (1973). *Biotechnol. Bioeng.*, 15, 27.
Koga, S., et al. (1967). *Appl. Microbiol.*, 15, 683.
Konak, A.R. (1974). *J. Appl. Chem. Biotechnol.*, 24, 453.
Kono, T. (1968). *Biotechnol. Bioeng.*, 10, 105.
Kono, T., and Asai, T. (1968a). *J. Ferment. Technol.*, 46, 391.
Kono, T., and Asai, T. (1968b). *J. Ferment. Technol.*, 46, 398.
Kono, T., and Asai, T. (1968c). *J. Ferment. Technol.*, 46, 406.
Kono, T., and Asai, T. (1969a). *Biotechnol. Bioeng.*, 11, 19.
Kono, T., and Asai, T. (1969b). *Biotechnol. Bioeng.*, 11, 293.
Kono, T., and Asai, T. (1969c). *J. Ferment. Technol.*, 47, 651.
Kono, T., and Asai, T. (1971a). *J. Ferment. Technol.*, 49, 128.
Kono, T., and Asai, T. (1971b). *J. Ferment. Technol.*, 49, 133.
Kosaric, N., et al. (1984). *Acta Biotechnolog.*, 4, 153.
Kossen, N.W.F. (1979). *Symp. Soc. Gen. Microbiol.*, 20, p. 327.
Kossen, N.W.F., and Roels, J.A. (1978). In Bull, M.J., ed., *Progress of Industrial Microbiology*, Vol. 14. Elserfer 95.
Kremen, A. (1971). *J. Theoret. Biol.*, 31, 363.
Krötzsch, P., Kürten, H., Daucher, H., and Popp, K.H. (1976). *Chem. Ing. Techn.*, MS, 351.
Labuza, Th.P. (1980). *Food Technol.*, February, 67.
La Motta, E.J. (1976). *Biotechnol. Bioeng.*, 18, 1029, 1359.
Lee, I.H. (1973). Ph.D. Thesis, University of Minnesota, Minneapolis.
Lee, I.H., et al. (1976). *Biotechnol. Bioeng.*, 18, 513.

Leffler, J.E. (1966). *J. Org. Chem.*, 31, 533.
Levenspiel, O. (1972) Chemical Reaction Engineering, 2nd ed., John Wiley & Sons, New York.
Levenspiel, O. (1980). *Biotechnol. Bioeng.*, 22, 1671.
Lodge, R.M., and Hinshelwood, C.N. (1943). *J. Chem. Soc.*, 213.
Lotka, A.J. (1925). *Elements of Mathematical Biology.* New York: Dover. (Reprinted 1956.)
Lovrien, R., et al. (1980). *Biotechnol. Bioeng.*, 22, 1249.
Luedeking, A., and Piret, E.L. (1959). *Biotechnol. Bioeng.*, 1, 393.
Lumry, R., and Eyring, H. (1954). *J. Phys. Chem.*, 58, 110.
Luong, J.H.T. (1982). *Eur. J. Appl. Microb. Biotechnol.*, 16, 28.
Luong, J.H.T., and Volesky, B. (1980). *Canad. J. Chem. Eng.*, 60, 163.
Maaloee, O., and Kjeldgaard, N.O. (1965). *The control of Macromolecular Biosynthesis.* New York: Benjamin.
Mahler, H.R., and Cordes, E.H. (1966). *Biological Chemistry.* New York: Harper International.
Marr, A.G. et al. (1963). *Ann. N.Y. Acad. Sci.*, 102, 536.
Mason, T.J., and Millis, N.F. (1976). *Biotechnol. Bioeng.*, 18, 1337.
May, R.M. (1973). *Stability and Complexity in Model Ecosystems.* Princeton, N.J.: Princeton University Press, p. 25.
Meers, J.L. (1973). *CRC Crit. Rev. Microbiol.*, 2, 139.
Megee, R.D., Drake, J.F., Fredrickson, A.G., and Tsuchiya, H.M. (1972). *Canad. J. Microbiol.*, 18, 1733.
Megee, R.D., III, et al. (1970). *Biotechnol. Bioeng.*, 12, 771.
Metz, B. (1975). Ph.D. Thesis, Technical University, Delft, Netherlands.
Metz, B., and Kossen, N.W.F. (1977). *Biotechnol. Bioeng.*, 19, 781.
Meyer, J.S., et al. (1975). *Biotechnol. Bioeng.*, 17, 1065.
Meyrath, J. (1973). *Mitteil. Versuchsanstalt für Gärungsgewerbe Wien*, 5/6, 95.
Meyrath, J., and Bayer, K. (1973). In Dellweg, H. (ed.). *Proceedings of the 3rd Symposium Technical Microbiology.* Berlin: Inst. f. Gärungsgewerbe und Biotechnologie, p. 117.
Minkevich, I.G., and Eroshin, V.K. (1973). *Folia Microbiol.*, 18, 376.
Miura, Y., et al. (1980). *Biotechnol. Bioeng.*, 22, 929.
Mochan, E., and Pye, E.K. (1973). *Nature New Biol.*, 242, 177.
Mona, R. (1978). Ph.D. Thesis, Technical University, Zurich, No. 6088.
Mona, R., et al. (1979). *Biotechnol. Bioeng.*, 21, 1561.
Monk, P.R. (1978). *Proc. Biochem.*, December, 4.
Monod, J. (1942). *Recherches sur la Croissance des Cultures Bacteriennes.* Paris: Hermann & Cie.
Monod, J. (1949). *Ann. Rev. Microb.*, 3, 371.
Monod, J., Changeux, J.P., and Jacob, F. (1963). *J. Molec. Biol.*, 6, 306.
Moreira, A.R., et al. (1979). *Biotechnol. Bioeng. Symp.*, 9, 179.
Morgan, M.S., and Edwards, V.H. (1971). *Chem. Eng. Progr. Symp., Ser.*, 114, 67, 51.
Moser, A. (1974). Gas Wasserfach/Wasser-Abwasser, 115, 411.
Moser, A. (1977). Thesis of habilitation Technical University, Graz, Austria.
Moser, A., Lafferty, R.M. (1977). *Zbl. Bakt. I. Abt. Ref.*, 252, 60.
Moser, A. (1978a). In *Proceedings of the 1st European Congress on Biotechnology*, Interlaken, Switzerland, Part 1, p. 88.
Moser, A. (1978b). *Gas Wasserfach/Wasser-Abwasser*, 119, 242.

Moser, A. (1978c). In P. Galzy et al., Chair of Genetics & Microbiology (ed.) *Proceedings of the 6th International Special Symposium on Yeasts*, July 2–8 Montpellier, France, p. 16. Administrative Secretariat Midi Contacts.

Moser, A. (1980a). UNEP/UNESCO/ICRO training course, *Theoretical Basis of Kinetics of Growth, Metabolism and Product Formation of Microorganisms*. Jena: Science Academy of East Germany, Central Institute for Microbiology and Experimental Therapy (ZIMET), Part II, p. 27.

Moser, A. (1980b) In Moo-Young, M., et al. (eds.). *Proceedings of the 2nd International Symposium on Waste Treatment and Utilization*. Oxford, New York: Pergamon Press, Vol. 2, 177.

Moser, A. (1981). *Bioprozesstechnik*. Vienna–New York: Springer-Verlag.

Moser, A. (1983a). *Acta Biotechnolog.*, 3, 195.

Moser, A. (1983b). In *Proceedings Biotechnology '83, International Conference on Commercial Applications and Implications of Biotechnology*. London: Online, Northwood p. 961.

Moser, A. (1984a). In Rehm, H.J., and Reed, G. (eds.). *Biotechnology—A Comprehensive Treatise*, Vol. 2. Deerfield Beach, Fla., and Basel, Verlag Chemie Weinheim, Chap. 16.5.

Moser, A., and Schneider, H. (1988) Bioprocess Engineering *4*, in press.

Moser, A. (1984b). *Acta Biotechnolog.* 4, 3.

Moser, A., Kosaric, N., and Margaritis, A. (1980b). Paper at 30th Canad. Chem. Eng. Conf., Edmonton, Alberta, 19–22 October.

Moser, A., and Steiner, W. (1975). *Eur. J. Appl. Microbiol.*, 1, 281.

Moser, A. (1980). Paper at 6th Internat. Ferm. Symp., London, Ontario. July 20–25.

Moser, F. (1977). *Verfahrenstechnik*, 11, 670.

Moser, H. (1958). *The Dynamics of Bacterial Populations Maintained in the Chemostat*. Carnegie Institution, Washington D.C. Publ. no. 614.

Moo-Young, M., and Kobayashi, T. (1972). *Canad. J. Chem. Eng.*, 50, 162.

Mou, D.G., and Cooney, Ch.L. (1976). *Biotechnol. Bioeng.*, 18, 1371.

Netter, H. (1969). *Theoretical Biochemistry*. Edinburgh: Oliver & Boyd.

Nino, G.J., and Ross, L.W. (1969). *Biotechnol. Bioeng.*, 11, 719.

Noack, D. (1968). *Biophysikalische Prinzipien der Populationsdynamik in der Mikrobiologie* (Fortschritte der experimentellen und theoretischen Biophysik Vol. 8, Beier, W., ed.) Leipzig: Georg Thieme.

Nyholm, N. (1976). *Biotechnol. Bioeng.*, 18, 1043.

Ohtaguchi, K., et al. (1979). *Internat. Chem. Eng.*, 19, 313, 591.

Oleskiewicz, J.A. (1977). *Prog. Water Technol.*, 9, 777.

Oura, E. (1973). In Patomäki, L., and Kiuru, A. (eds.). *Proceedings of the 1st National Meeting. Biophysics and Biotechnology*. Finland: ALKO, Helsinki p. 142.

Pamment, N.B., et al. (1978). *Biotechnol. Bioeng.*, 20, 349.

Pardee, A.B. (1961). *Symp. Soc. Gen. Microb.*, 11, 19.

Peringer, P., et al. (1974). *Biotechnol. Bioeng.*, 16, 431.

Perret, C.J. (1960). *J. Gen. Microbiol.*, 22, 589.

Peters, H. (1976). In Hartmann, L. (ed.). *Karlsruher Berichte zur Ingenieurbiologie*. Karlsruhe: Inst. f. Ingenieurbiologie und Biotechnologie, p. 9. Univ. Karlsruhe.

Petrova, T.A., et al. (1977). *Z. Allgem. Mikrob.*, 17(7), 531.

Pickett, A.M. (1982). In Bazin, M.J. (ed.). *Microbial Population Dynamics*. Boca Raton, Fla.: CRC Press, Chap. 4.

Pirt, S.J. (1965). *Proc. Roy. Soc. B.*, 163, 224.
Pirt, S.J. (1975). *Principles of Microbe and Cell Cultivation.* Oxford: Blackwell Scientific Publishing.
Pirt, S.J., and Righelato, R.C. (1967). *Appl. Microbiol.*, 15, 1284.
Pirt, S.J., et al. (1983). *J. Chem. Tech. Biotechnol.*, 33B, 35.
Pitcher, W.H. (1978). *Adv. Biochem. Eng.*, 10, 1.
Powell, E.O. (1958). *J. Gen. Microbiol.*, 18, 259.
Powell, E.O. (1967). In Powell, E.O., et al. (eds.). *Microbial Physiology* and *Continuous Culture.* London: Her Majesty's Stationery Office, p. 34.
Prokop, A., and Humphrey, A.E. (1970). In Bernardo, M.A. (ed.). *Disinfection.* Marcel Dekker, New York: p. 61.
Prosser, J.I. (1982). In Bazin, M.J. (ed.). *Microbial Population Dynamics.* Boca Raton, Fla.; CRC Press, Chap. 5.
Ramkrishna, D., et al. (1966). *J. Ferment. Technol.*, 44, 210.
Ramkrishna, D., et al. (1967). *Biotechnol. Bioeng.*, 9, 129.
Ratnam, D.A., et al. (1982). *Biotechnol. Bioeng.*, 24, 2675.
Reich, J.G., et al. (1972). *Eur. J. Biochem.*, 26, 368.
Reilly, P.J. (1974). *Eur. J. Biochem.*, 16, 1373.
Reuss, M., and Wagner, F. (1973). *Proceedings of the 3rd Symposium on Technology and Microbiology.* Berlin Dellweg, H., ed., p. 89, Inst. f. Gärungsgewerbe und Biotechnologie.
Reuss, M. (1977). *Fort. Verfahrenstechnik*, 15F, 549.
Reuss, M., and Bajpaj, R.K. (1982) *J. Chem. Technol. Biotechnol. 32*, 81.
Ricica, J. (1969). In Perlman, D. (ed.). *Fermentation Advances.* New York: Academic Press, p. 427.
Rittmann, B.E., and McCarty, P.L. (1981). *J. Environ. Eng. Div., ASCE*, 107(EE4), 831.
Roels, J.A. (1978). *Proceedings of the 1st European Congress on Biotechnology*, Interlaken, Switzerland, Dechema Monograph, 82, 221.
Roels, J.A. (1980). *Biotechnol. Bioeng.*, 22, 2457.
Roels, J.A. (1982). *J. Chem. Tech. Biotechnol.*, 32, 59.
Roels, J.A. (1983). *Energetics and Kinetics in Biotechnology.* Amsterdam, New York, Oxford: Elsevier Biomedical Press.
Roels, J.A., and Kossen, N.W.F. (1978). In Bull, M.J. (ed.). *Progress in Industrial Microbiology*, Vol. 14. Amsterdam: Elsevier, p. 95.
Romanovsky, J.M., et al. (1974). *Kinetische Modelle in der Biophysik.* Jena: G. Fischer. (German translation by W.A. Knorre and A. Knorre).
Rowley, B.I., and Pirt, S.J. (1972). *J. Gen. Microbiol.*, 72, 553.
Ryder, D.N., and Sinclair, C.G. (1972). *Biotechnol. Bioeng.*, 14, 787.
Ryu, D.Y., and Humphrey, A.E. (1972). *J. Ferment. Technol.*, 50, 424.
Ryu, D.Y., and Mateles, R.I. (1968). *Biotechnol. Bioeng.*, 10, 385.
Saguy, I., and Karel, M. (1980). *Food Technol.*, February, 78.
Schaezler, D.J., McHarg, W.H., and Busch, A.W. (1971). *Biotechnol. Bioeng. Symp.*, 2, 107.
Schmid, R., and Sapunov, V.N. (1982). *Non-Formal Kinetics. Monographs in Modern Chemistry*, Vol. 14. Deerfield Beach, Fla., and Basel: Verlag Chemie Weinheim.
Schneider, H., Moser, A. (1986). Bioprocess Engng. 2, 129.
Schügerl, K. (1983). In Proceedings, NATO ASI, *Mass Transfer with Chemical Reactions*, Izmir Turkey, 1981, Vol. 1(72), p. 415.

Shuler, M.L., et al. (1972). *J. Theoret. Biol.*, 35, 67.
Schultz, J.S., et al. (1974). AIChEJ, 20, 417.
Schultz, K.L., et al. (1964). *Arch. Mikrobiol.*, 48, 1.
Segel, I.H. (1975). *Enzyme Kinetics.* New York: Wiley-Interscience.
Setzermann, U., et al. (1982). *Acta Biotechnolog.*, 2, 325.
Shehata, T.E., and Marr, A.G. (1971). *J. Bacteriol.*, 107, 210.
Sheintuch, M. (1980). *Biotechnol. Bioeng.*, 22, 2557.
Shieh, W.K. (1980a). *Biotechnol. Bioeng.*, 22, 667.
Shieh, W.K. (1980b). *Water Res.*, 14, 695.
Shu, P. (1961). *Biotechnol. Bioeng.*, 3, 95.
Shuler, M.L., and Domach, M.M. (1982). In Blanch, H.W., et al. (eds.). Amer. Chem. Soc. Winter Symp. *Kinetics and Thermodynamics in Biological Systems.* Boulder, Colo.: Amer. Chem. Soc.
Shuler, M.L., et al. (1979). *Ann. N.Y. Acad. Sci.*, 326, 35.
Siimer, E. (1978). *Biotechnol. Bioeng.*, 20, 1853.
Sikyta, B., et al. (1977). In Meyrath, J., and Bu'Lock, J.D. (eds.). *FEMS Symposium 4* New York: Academic Press, 119.
Sinclair, C.G., and Topiwala, H.H. (1970). *Biotechnol. Bioeng.* 12, 1069.
Sonnleitner, B., and Fiechter, A. (1983). *Trends in Biotechnol.* 1, 74.
Stephanopoulos, G., and Fredrickson, A.G. (1979). *Biotechnol. Bioeng.*, 21, 1491.
Stephanopoulos, G., et al. (1979). *AIChEJ*, 25, 863.
Sudo, R., et al. (1975). *Biotechnol. Bioeng.*, 17, 167.
Suga, K., et al. (1975). *Biotechnol. Bioeng.*, 17, 185.
Sykes, R.M. (1973). *J. Water Pollut. Contr. Fed.*, 45, 888.
Taguchi, H., et al. (1978). *J. Ferm. Technol.*, 56, 158.
Takamatsu, S., et al. (1974). *J. Ferment. Technol.*, 52, 190.
Takamatsu, T., et al. (1981). *J. Ferment. Technol.*, 59, 131.
Talsky, G. (1971). *Angew. Chem.*, 83, 553.
Tanaka, H., Uzman, S., and Dunn., I.J. (1981). *Biotechnol. Bioeng.*, 23, 1683.
Tanner, R.D. (1970). *Biotechnol. Bioeng.*, 12, 831.
Teissier, G. (1936). *Ann. Physiol. Physiochim. Biol.*, 12, 527.
Tempest, D.W. (1976). *In Continuous Culture 6: Applications and New Fields* (Dean, A.C.R. et al., eds.) Society of Chemical Industry, London: Chichester: E. Harwood Ltd., p. 349.
ten Hoopen, H.J.G., et al. (1980). *Adv. Biotechnol.*, 1, 315.
Terui, G. (1972). In Sterbacek, Z. (ed.). *Microbial Engineering.* London: Butterworth, p. 377.
Theophilou, J., Wolfbauer, O., and Moser, F. (1978). Gas-WasserFach/Wasser-Abwasser 119, 135.
Thijssen, H.A.C. (1979). *Lebensm.-Wiss. -Technol.*, 12, 308.
Toda, K. (1976). *Biotechnol. Bioeng.*, 18, 1117.
Toda, K. (1980). *Biotechnol. Bioeng.*, 22, 1805.
Toda, K. (1981). *J. Chem. Tech. Biotechnol.*, 31, 775.
Topiwala, H.H., and Hamer, G. (1971). *Biotechnol. Bioeng.*, 13, 919.
Trilli, A. (1977). *J. Appl. Chem. Biotechnol.*, 27, 251.
Trinci, A.P.J. (1970) *Arch. Microbiol.*, 73, 353.
Tsao, G.T., and Hanson, Th.P. (1975). *Biotechnol. Bioeng.*, 17, 1591.
Tsao, G.T., and Yang, C.M. (1976a). *Biotechnol. Bioeng.*, 18, 1827.

Tsao, G.T., and Yang, C.M. (1976b). In Dellweg, H., ed. Proc. 5th Internat. Ferm. Symp., Berlin, Abstract No. 5.03. Inst. f. Gärungsgewerbe und Biotechnologie, June 28–July 3.
Tseng, M., and Phillipps, C.R. (1981). *Biotechnol. Bioeng.*, 23, 1639.
Tseng, M.C., and Wayman, M. (1975). *Canad. J. Microbiol.*, 21, 994.
Tsuchiya, H.M., et al. (1966). *Adv. Chem. Eng.*, 6, 125.
Tsuchiya, H.M., et al. (1972). *J. Bacteriol.*, 110, 1147.
Tucek, F., Chudoba, J., and Madera, V. (1971). *Water Res.*, 5, 647.
van Dedem, G. (1975). *Biotechnol. Bioeng.*, 17, 1301.
van Dedem, G., and Moo-Young, M. (1973). *Biotechnol. Bioeng.*, 15, 419.
van Suijdam, J.C., and Metz, B. (1981). *Biotechnol. Bioeng.*, 23, 111.
van Suijdam, J.C., et al. (1982). *Biotechnol. Bioeng.*, 24, 177.
van Uden, N. (1971). *Z. Allgem. Mikrob.*, 11(6), 541.
van Uden, N., Abranches, P., and Cabeca-Silva, C. (1968). *Arch. Microbiol.*, 61, 381.
van Uden, N., and Vidal-Leiria, M.M. (1976). *Arch. Microbiol.*, 108, 293.
Vavilin, V.A. (1982). *Biotechnol. Bioeng.*, 24, 1721, 2609.
Volesky, B., et al. (1982). *J. Chem. Tech. Biotechnol.*, 32, 650.
Veldkamp, H. (1975). In Dean, A.C.R. (ed.). *Continuous Culture*, Vol. 6. London: SCI, Chap. 24.
Veldkamp, H., and Jannasch, H.W. (1972). *J. Appl. Chem. Biotechnol.*, 22, 105.
Verhoff, F.H., et al. (1972). *Biotechnol. Bioeng.*, 14, 411.
Verhulst P.F. (1845). Recherches Mathématiques sur la Loi d'Accroissement de la Population. Nouv. Mém. de l'Acad. Roy. des Sciences et Belles-Lettres de Bruxelles, 18, p. 1.
Volterra, V. (1931). *Lecons sur la Theorie Mathematique de la Lutte pour la Vie.* Paris: Gauthier–Vallars.
von Bertalanffy, L. (1932, 1942). *Theoretische Biologie*, West Berlin: Borntraeger.
von Bertalanffy, L., et al. (1977). *Biophysik des Fliessgleichgewichtes.* East Berlin: Akademie Verlag.
Wayman, M., and Tseng, M.C. (1976). *Biotechnol. Bioeng.*, 18, 383.
Webb, J.L. (1963). *Enzyme and Metabolic Inhibitors*, Vol. 1. New York: Academic Press.
Webster, I.A. (1983). *Biotechnol. Bioeng.*, 25, 2981.
Weissmann, J.C., and Benemann, J.R. (1979). *Biotechnol. Bioeng.*, 21, 627.
Whitaker, A., and Long, P.A. (1973). *Proc. Biochem.*, November, 27.
Wilder, C.T., et al. (1980). *Proc. Biochem.*, 22, 89.
Wilderer, P. (1976). In *Karls. Ber. Ingenieurbiol.*, Vol. 8, Hartmann, L., ed. Inst. f. Ingenieurbiologie und Biotechnologie, Univ. Karlsruhe 1–145.
Wilderer, P., Engelmann, G., and Schmenger, H. (1977). *Gas-Wasser-Fach/Wasser-Abwasser*, 118, 357.
Wilderer, P., and Hartmann, L. (1978). In *Moderne Abwasserreinigungsverfahren* (Münch. Beitr. Abw., Fisch. Flussbiol.) Oldenburg Verlag München Vol. 29, p. 9.
Williams, F.M. (1967). *J. Theoret. Biol.*, 15, 90.
Williams, F.M. (1975). In Patten, B.V. (ed.). *System Analysis and Simulation in Ecology*, Vol. 7. New York: Academic Press, Chap. 3, p. 197.
Wolfbauer, O., Klettner, H., and Moser, F. (1978). *Chem. Eng. Sci.*, 33, 953.
Wöhrer W., and Röhr, M. (1981). *Biotechnol. Bioeng.*, 23, 567.
Wuhrmann, K., Beust, F. von, and Ghose, T.K. (1958). *Schweiz. Hydrol.*, 20, 284.

Yagil, G., and Yagil, E. (1971). *Biophys. J.*, 11, 11.
Yano, T., et al. (1966). *Agr. Biol. Chem. (Jap.)*, 30, 42.
Yoon, H., and Blanch, H.W. (1977). *J. Appl. Chem. Biotechnol. Bioeng.*, 19, 1193.
Yoon, H., Klinzing, G., and Blanch, H.W. (1977). *Biotechnol. Bioeng.*, 19, 1193.
Young, T.B., Bruley, D.F., and Bungay, A.R. (1970). *Biotechnol. Bioeng.*, 12, 747.
Young, T.B., and Bungay, A.R. (1973). *Biotechnol. Bioeng.*, 15, 377.

CHAPTER 6

Bioreactor Performance: Process Design Methods

Bioprocess design, as suggested in Fig. 2.14 (including especially calculation of the conversion for prediction of the mode of reactor operation that will lead to optimal production), represents the stage where kinetic and bioreactor data are integrated. In practical situations, the dominant problems are often maintenance of sterility, improvement in the strain of microorganism, and isolation of the product. All of these greatly affect economic considerations. Once in operation, a plant using a stirred vessel can usually be modified only with respect to the operating conditions: Changing to a different type of reactor is not an option. In the planning of a new operation, selection of both the optimal mode of operation and the optimal reactor type is a foremost consideration. Selection will be even more important in the future, when large volume/low priced processes become economically competitive (cf. Fig. 1.2).

The problems posed by the interaction of the reactor type and the physiology of the organism have only slowly been recognized (Finn and Fiechter, 1979; Melling, 1977; A. Moser, 1983a and b), cf. Sect. 3.3.11 and Leegwater et al. (1982).

The present chapter will deal with general methods that are suitable for making predictions in developing a process on the basis of kinetic data usually obtained from measurements made in a discontinuously operated stirred vessel. The design equation may be written in a qualitative way as

$$\zeta_i = f(r_i, \mathrm{H_vTR}, \mathrm{OTR}, \mathrm{RTD}, J \text{ resp. } t_C \ldots) \tag{6.1}$$

Equation 6.1 is a generalized form of the conservation of mass equation (cf. Equ. 2.3).

The influence of the OTR on kinetics (and therefore on productivity) was indicated in Equs. 5.169 and 5.170. Therefore, we will discuss here primarily those process engineering factors involved in various different types of reactors and operations. Last but not least, some unconventional reactors such as membrane (dialysis) reactors and synchronous culture techniques will be discussed.

6.1 The Ideal Single-Stage, Constant-Volume Continuous Stirred Tank Reactor, CSTR (Pseudohomogeneous L-Phase Reactor Model)

The continuous culture technique using a homogeneous single-stage stirred tank reactor was introduced long ago in bioprocessing. The mathematical foundation of continuous culture theory has been given in the basic papers of Monod (1942, 1950) and of Novick and Szilard (1950). Many reviews and several books have appeared on the theoretical and practical aspects of this technique (Dean et al., 1972; Fiechter, 1982; Herbert et al., 1956; Malek and Fencl, 1966; Malek and Ricica, 1969; A. Moser, 1985a; Powell et al., 1967; Ricica, 1973; Tempest, 1970).

6.1.1 Performance of the CSTR with Simple Kinetics

In a completely mixed continuous flow reactor, the composition is thought to be uniform throughout the reactor, and is the same as in the exist stream (cf. Fig. 3.30). Applying the law of conservation of mass (cf. Equ. 2.3) to a CSTR yields Equ. 3.90 for the steady state. The dilution rate D is reciprocal to the mean residence time of a fluid ($\bar{t}$).

The material balance in general form becomes

$$\text{Accumulation} = \text{input} - \text{output} \pm \text{reaction} \tag{6.2}$$

and for biomass X and substrate S

$$\frac{dx}{dt} = D \cdot x_{\text{in}} - D \cdot x_{\text{ex}} + r_{\text{X}} \tag{6.3a}$$

$$\frac{ds}{dt} = D \cdot s_{\text{in}} - D \cdot s_{\text{ex}} - r_{\text{S}} \tag{6.3b}$$

At steady state with $\bar{x} = x_{\text{ex}} = x_{\text{R}}$ and $\bar{s} = s_{\text{ex}} = s_{\text{R}}$,

$$r_{\text{X}} = D \cdot \bar{x} \tag{6.4}$$

$$r_{\text{S}} = D(s_{\text{in}} - \bar{s}) \tag{6.5}$$

As a consequence of Equ. 6.4, the definition of growth rate μ in a CSTR at steady state is given by $\mu = D$ (Equ. 3.91). This fact is most important for practical application of the CSTR in microbial culture techniques. It means that this reactor configuration functions as a differential reactor (cf. Sect. 4.4.1), and it enables the direct measurement of a biological reaction rate (e.g., growth rate) without mathematical manipulation of the measurements.

The practical consequence from this statement is that μ can be controlled by the amount of nutrient fed to a constant-volume CSTR. This system is called "realstat" or "chemostat," a name that refers to the constant chemical environment characteristic of the steady state. The nutrient medium is designed so that one essential nutrient is the limiting substrate.

However, the simplest type of continuous culture system is a "turbidistat." In this device, the cell concentration in the vessel is maintained at a constant value by monitoring the optical density of the culture. The disadvantage of this simple technique is the difficulty in adequately monitoring the cell concentration. The basic behavior of both types of control apparatus for CSTR operation becomes clear by analyzing CSTR reactor performance using the balance equation previously discussed. Introducing the simple Monod equation (cf. Equ. 5.38) and using a yield coefficient $Y_{X|S}$ (cf. Equ. 2.14a) result in the following set of equations, which represent the basic behavior of CSTRs:

$$\bar{s} = K_S\left(\frac{D}{\mu_{max} - D}\right) \tag{6.6}$$

$$\bar{x} = Y_{X|S}(s_0 - \bar{s}) \tag{6.7}$$

A plot of these equations is given in Fig. 6.1, the dilution rate D being used as the variable in Fig. 6.1a and the mean residence time $\bar{t}$ in Fig. 6.1b. Figure 6.1b is preferred by process engineers, while Fig. 6.1a is used predominantly by microbiologists.

A modified turbidostat scheme has recently been proposed, as have other control schemes such as pH auxostat, CTR, OTR, and OUR control (Agrawal and Lim, 1984).

The critical washout point (D_{crit} or D_c) can be estimated from

$$D_{crit} = \mu_{max}\frac{s_0}{K_S + s_0} \tag{6.8}$$

and is somewhat smaller than μ_{max}.

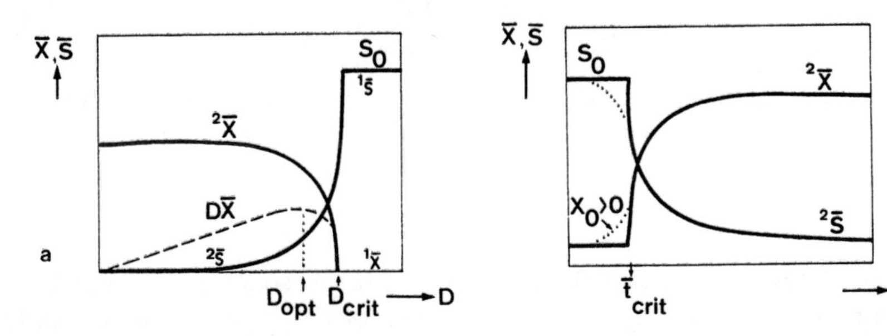

FIGURE 6.1. Theoretical behavior of a continuous culture of microorganisms with simple Monod kinetics in a stirred vessel (chemostat) with $x_0 = 0$: Stationary concentration of the cells ($\bar{x}$) or substrate ($\bar{s}$) as a function of the rate of dilution, $D = F/V$ (a) or mean residence time $\bar{t} = 1/D$ (b). In (b), the case for $x_0 > 0$ is also shown by dotted lines.

The productivity r_{CSTR} for continuous processing in a single CSTR is

$$r_{i,CSTR} = D \cdot \bar{c}_i \tag{6.9}$$

where $\bar{c}_i$ is the steady-state concentration of component i ($\bar{x}$, $\bar{s}$, or $\bar{p}$) in a bioreactor at a given dilution rate D. Substituting Equ. 6.6 into Equ. 6.9 for biomass production gives

$$r_X = D \cdot Y_{X|S}\left[s_0 - K_s\left(\frac{D}{D_c - D}\right)\right] \tag{6.10}$$

The productivity has been plotted as a function of D or $\bar{t}$ in Fig. 6.1. The dilution rate corresponding to optimum productivity D_{opt} can be calculated for a single CSTR by setting the first derivative of Equ. 6.10 equal to zero. Thus

$$D_{opt} = D_c\left[1 - \left(\frac{K_S}{K_S + s_0}\right)^{1/2}\right] \tag{6.11}$$

As expected, this maximum productivity does not coincide with the point of maximum conversion of substrate. This fact was previously pointed out in Fig. 2.12 with batch processes.

Finally, the ratio of the productivity in a CSTR to that in a DCSTR becomes, with some simplifications ($D_c \approx \mu_{max}$, $K_S/s_0 \approx 0$, $x_{max}/x_0 \gg 1$),

$$\frac{r_{X,CSTR}}{r_{X,DCSTR}} = \frac{x_{max}}{1/\mu_{max}\ln(x_{max}/x_0) + t_0} = \ln\frac{x_{max}}{x_0} + t_0 \cdot \mu_{max} \tag{6.12}$$

Using this equation, it is clear that continuous operation in a CSTR is superior to batch processing, especially when organisms with maximum growth rates of more than about 0.2 are used. Inserting the Michaelis–Menten equation (Equ. 2.54) for a poison-free enzymatic reaction into the performance equation of a single CSTR (Equ. 3.90) gives

$$\bar{t}_{CSTR} = \frac{(s_0 - \bar{s})(K_m + \bar{s})}{r_{max} \cdot \bar{s}} \tag{6.13}$$

An equation that does allow a direct evaluation of the kinetic parameters r_{max} and K_m can be obtained by rearranging Equ. 6.13 into a linear form:

$$\bar{s} = -K_m + r_{max}\frac{\bar{s}}{s_0 - \bar{s}} \cdot \bar{t} \tag{6.14}$$

A plot of this linear form is shown in Fig. 6.2 (Levenspiel, 1979). To evaluate the kinetic parameters μ_{max} and K_S from a set of CSTR data, Equ. 6.6 can be rearranged using Equ. 6.2 to give a linear form

$$\frac{1}{s} = \frac{\mu_{max}}{K_S} \cdot \bar{t} - \frac{1}{K_S} \tag{6.15}$$

A graphical representation is in Fig. 6.3.

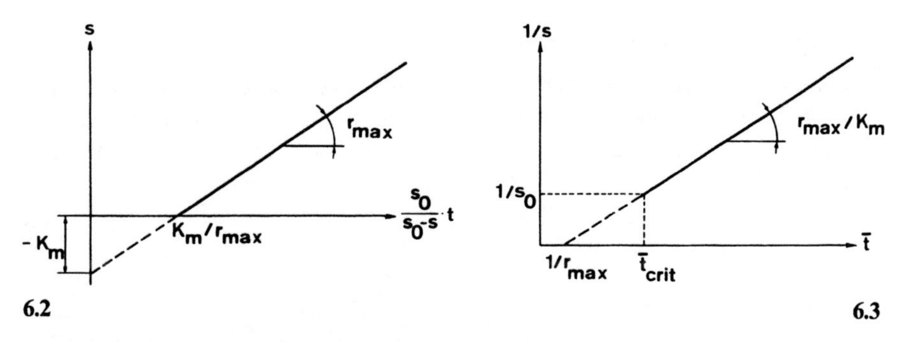

FIGURE 6.2. Linear plot for enzyme kinetics according to Equ. 6.14, which allows estimation of kinetic parameters r_{max} resp. $r_{S,max}$ and K_m from CSTR operation (Levenspiel, 1979).

FIGURE 6.3. Linear plot for Monod kinetics according to Equ. 6.15 in the form of a double reciprocal plot (cf. Fig. 4.24c) applied to a CSTR with $D = 1/\bar{t} = \mu$, which is suitable for parameter estimation (μ_{max} and K_S).

A special method for determining μ_{max} is used of the washout conditions in a CSTR ($D \gg D_c$). The balance equation of biomass in this case with $s \gg K_S$ during washout experiments is

$$x = x_0 \cdot \exp(\mu_{max} - D)t \tag{6.16a}$$

or when transformed

$$\ln x = \ln x_0 + (\mu_{max} - D)t \tag{6.16b}$$

Thus, μ_{max} can easily be calculated by performing nonlinear or linear regression on the $\ln x$ versus t data (e.g., Esener et al., 1981b). The performance equation shows that everything—washout, optimum cell concentration, maximum production rate—depends on K_S and s_0.

In the case of a CSTR with $x_0 \neq 0$, the balance equation is (Levenspiel, 1979)

$$\mu_{max} \cdot \bar{t} = \frac{(x - x_0)[Y(s_0 + K_S) - (x - x_0)]}{Y \cdot s_0 \cdot x - x(x - x_0)} \tag{6.17}$$

Since $x_0 \neq 0$, no washout occurs and there is no restriction on $\mu_{max} \cdot \bar{t}$.

6.1.2 Performance of the CSTR with Complex Kinetics

Modifications of biokinetic models can be included in the CSTR performance equation.

6.1.2.1 Lysis and Maintenance

Modifications have been introduced to deal with the discrepancy between model behavior and chemostat experiments. Including the specific lysis rate k_d and the maintenance coefficient m_S (cf. Equs. 5.80 and 5.76), the enlarged model is

$$\frac{dx}{dt} = \mu(s)\cdot x - k_d x - Dx \tag{6.18}$$

$$\frac{ds}{dt} = -\frac{1}{Y_{X|S}}\cdot \mu(s)x - m_S\cdot x - D(s_0 - s) \tag{6.19}$$

which has the following nonwashout steady state:

$$({}^2\bar{x}, {}^2\bar{s}) = \left(Y_{X|S}\frac{D}{D + k_d + m_S\cdot Y_{X|S}}\left\{s_0 - K_S\frac{D + k_d}{\mu_{max} - D - k_d}\right\},\right.$$
$$\left. \times\, K_S\frac{D + k_d}{\mu_{max} - D - k_d}\right) \tag{6.20}$$

Hence, lysis as well as maintenance may cause the curve of biomass concentration against the dilution rate D to approach zero as the dilution rate tends to zero (see Figs. 6.4 and 6.5). Furthermore, the lysis causes a decrease in the washout point

$$D_c = \mu_{max}\frac{s_0}{s_0 + K_S} - k_d \tag{6.21}$$

instead of Equ. 6.8 for Monod's model.

The specific lysis rate can be determined by the analysis of batch experiments, but the maintenance coefficient cannot be so determined. The estimation of this coefficient is possible by analysis of chemostat experiments. The theoretical basis for this is given with Equ. 6.20. According to this relation, the plot of the quotient $(s_0 - \bar{s})/\bar{x}$ versus the reciprocal dilution rate $1/D$ gives a straight line:

$$\frac{1}{Y} = \frac{s_0 - {}^2\bar{s}}{{}^2\bar{x}} = \frac{1}{Y_{X|S}} + \frac{1}{D}\left(\frac{k_d}{Y_{X|S}} + m_S\right) \tag{6.22}$$

We call the quotient $\bar{x}/(s_0 - \bar{s})$ the phenomenological yield coefficient Y; it equals the "true" yield coefficient if the lysis and maintenance rates fall. If for different dilution rates D the steady-state values $\bar{x}$ and $\bar{s}$ are measured with known s_0 and k_d, one can determine "true" yield coefficient $Y_{X|S}$ as well as the maintenance coefficient m_S by linear regression on the basis of Equ. 6.22.

A more structured model using a viability concept has been presented in Fig. 5.24. The complexity of maintenance has been stressed in the literature by introducing the definition of "coefficient of apparently non-finalized substrate consumption" (Goma et al., 1979) and by reporting the difficulties involved in experimental verification of maintenance in the case of incorrect

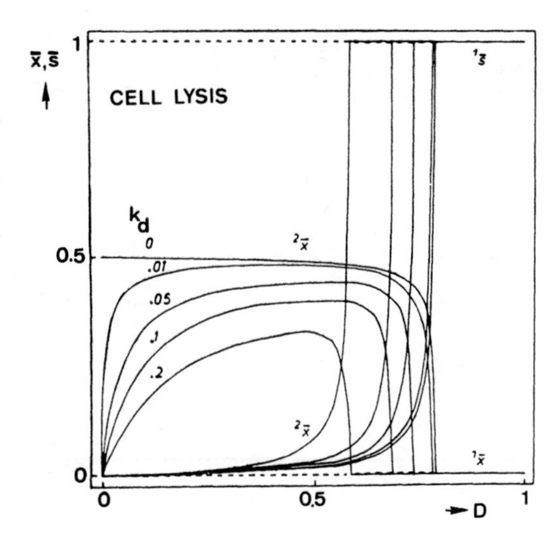

FIGURE 6.4. Computer simulation of CSTR operation showing the steady-state concentrations ($\bar{x}$ and $\bar{s}$) in dependence of dilution rate D with variation of specific rate of cell lysis k_d according to Equs. 6.18 and 6.19, with $m_S = 0$. Parameter values: $\mu_{max} = 0.8\ h^{-1}$; $Y_{X|S} = 0.5$; $K_S = 0.01\ g \cdot l^{-1}$; $s_0 = 1\ g \cdot l^{-1}$.

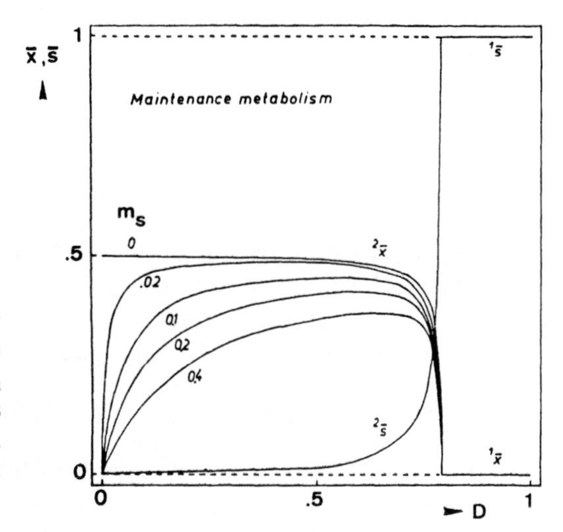

FIGURE 6.5. Effect of maintenance metabolism on behavior of CSTR as seen by the influence of maintenance coefficient m_S on steady-state concentrations ($\bar{x}, \bar{s}$) as a function of dilution rate D, with $k_d = 0$. (See Equs. 6.18 and 6.19.)

modeling by neglect of product formation during anaerobic growth (Esener et al., 1981b). Other interesting papers concerning maintenance can be found in literature (Kuhn et al., 1980; Solomon and Erickson, 1981).

6.1.2.2 Incomplete Mixing and Wall Growth

Another deviation often observed is the so-called wall growth, which results in plots similar to Fig. 3.12. Topiwala and Hamer (1971) analyzed this effect of the adherence of microorganisms to glass or metal surfaces. When the part of adhered biomass that cannot wash out is denoted by x_W, the following modification of Monod's model results:

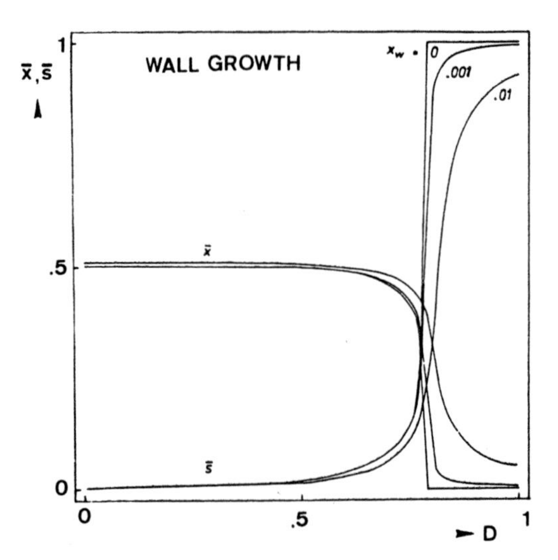

FIGURE 6.6. Effect of wall growth on CSTR behavior in a plot of $\bar{x}$ and $\bar{s}$ versus D with varied values of x_w. (See Equ. 6.23.)

$$\frac{dx}{dt} = \mu(s)x - D(x - x_W) \tag{6.23a}$$

and

$$\frac{ds}{dt} = \frac{1}{Y_{X|S}}\mu(s)x + D(s_0 - s) \tag{6.23b}$$

The steady-state values of this model as a function of D are shown in Fig. 6.6. Note that the biomass concentration does not vanish above the washout point (see also Sinclair and Brown, 1970; Toda and Dunn, 1982). Imperfectly mixed bioreactor systems have been summarized recently by A. Moser (1985b).

6.1.2.3 Product Formation

The microbial product-formation rate in a single CSTR is (Pirt, 1975)

$$\frac{dp}{dt} = q_P \cdot x - D \cdot x \tag{6.24}$$

The change of product concentration in the culture over time depends on the term of synthesis ($q_P \cdot x$) and on the term of outflow of product ($-D \cdot x$).

In the stationary state with $dp/dt = 0$

$$\bar{p} = \frac{q_P \cdot \bar{x}}{D} \tag{6.25}$$

where $\bar{p}$ and $\bar{x}$ are the steady-state concentrations of product and biomass, respectively. If the product is strictly "growth linked" (cf. Equ. 5.116), the product concentration in the stationary state $\bar{p}$ and the output rate $D \cdot \bar{p}$ will

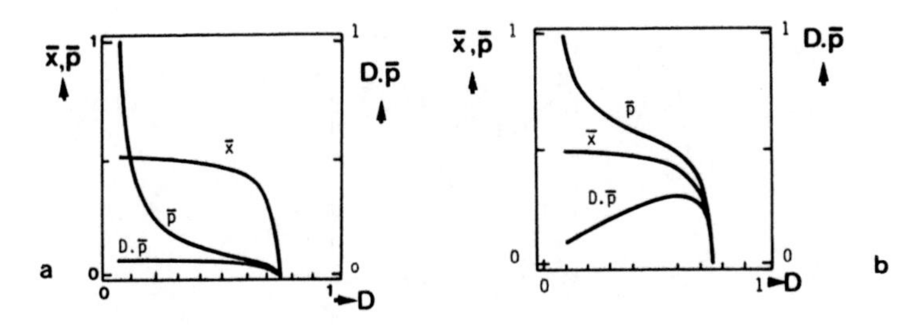

FIGURE 6.7. Growth-associated (a) and non-growth-associated (b) product formation, growth, and productivity $D \cdot \bar{p}$ in CSTR operation without P inhibition. $Y_{P|X} = 0.5$ and $k_P = 0.1$.

vary with D in the same manner as does biomass concentration (cf. Fig. 6.1). If the q_P is independent of the growth rate (cf. Equ. 5.120), the product concentration varies inversely with the dilution rate D. Then, over a wide range of dilution rates, the output rate is constant, as shown in Fig. 6.7a. However, as D tends to zero, eventually the assumption that k_P = constant becomes invalid because of the decay of enzyme activity, because some required substrate will be exhausted, or because of regulatory processes in the cell. If the product formation is partly "growth linked" and partly independent of the growth rate (cf. Equ. 5.122), then the product concentration in the steady state will vary with dilution rate D, as shown in Fig. 6.7b.

The product output rate, if D tends toward zero, is the "non-growth-linked" contribution k_P, multiplied by biomass concentration $\bar{x}$. From Equ. 6.25 we get

$$D \cdot \bar{p} = q_P \cdot x \tag{6.26}$$

Substituting the logistic form of product formation kinetics (see Equ. 5.122) for q_P gives

$$D \cdot \bar{p} = (Y_{P|X} \cdot \mu + k_P)\bar{x} \tag{6.27}$$

Rearranging with $\bar{x} = Y_{X|S} \cdot s_0$ yields

$$D \cdot \bar{p} = (Y_{P|X} \cdot \mu + k_P) Y_{X|S} \cdot s_0 \tag{6.28}$$

and the product output rate, in the case when D tends toward zero, becomes

$$D \cdot \bar{p} = Y_{X|S} \cdot s_0 \cdot k_P \tag{6.29}$$

The kinetic aspects of product formation in continuous culture are usually dealt with using the simple, formal kinetic approach of the Luedeking–Piret equation (Equ. 5.122). A modified kinetic approach to microbial product formation uses the concept that r_P not only is proportional to the actual biomass concentration but also depends on a second carbon source S_2 (Hegewald et al., 1978):

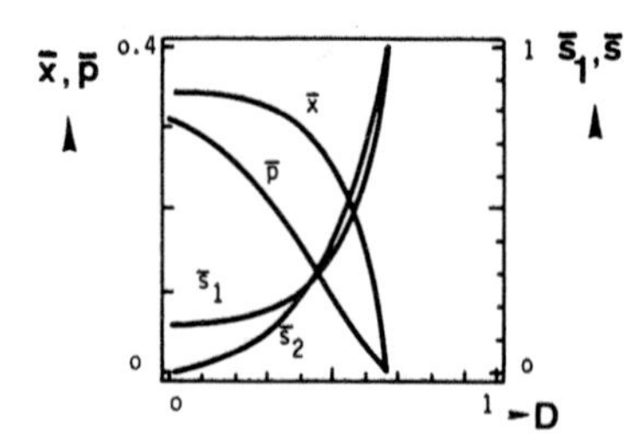

FIGURE 6.8. Prediction of steady-state values of CSTR operation from batch data using two substrates for growth and turimycin production with *Streptomyces hygroscopicus* on the basis of modeling the kinetics with the aid of Equ. 6.30 (Hegewald et al., 1978).

$$r_P = q_{P,max} \frac{s_2}{s_2 + K_{S2}} \cdot x \cdot f_{repr} - D \cdot p \tag{6.30}$$

This equation was later extended to supplementary O_2 limitation (Bajpaj and Reuss, 1980). The factor f_{repr} in Equ. 6.30 represents the effect of easily metabolized sugars in reducing antibiotic synthesis via repression or inhibition (cf. Equ. 5.106). Figure 6.8 shows the calculated prediction of CSTR behavior based on Equ. 6.30.

6.1.2.4 Product Inhibition

The difference between batch and continuous culture techniques with respect to the effect of product inhibition is that under the conditions of continuous cultures product is diluted, while in batch runs product is accumulated. Therefore, in batch cultures reaction rates eventually slow down. In chemostat cultures, however, oscillations of x and p appear due to periodic effect of, for example, pH control and/or permanent inflow and outflow of fresh medium. For the mathematical modeling and computer simulation of this problem, it is possible to formulate the following differential equations:

$$\frac{dx}{dt} = \mu(s, p) \cdot x - D \cdot x \tag{6.31a}$$

$$\frac{ds}{dt} = D(s_0 - s) - \mu(s, p)x \cdot \frac{1}{Y_{X|S}} \tag{6.31b}$$

$$\frac{dp}{dt} = q_P \cdot x - D \cdot p \tag{6.31c}$$

Extensive studies of oscillations induced by products are given by Knorre (1980) together with mathematical modeling and the problem of multiple steady states (see next section).

Inserting a generalized kinetic equation for pure product inhibition from the previous chapter (Equ. 5.63 with Equ. 5.105) into the basic performance equation of a single CSTR (cf. Equ. 3.90) and converting all concentrations

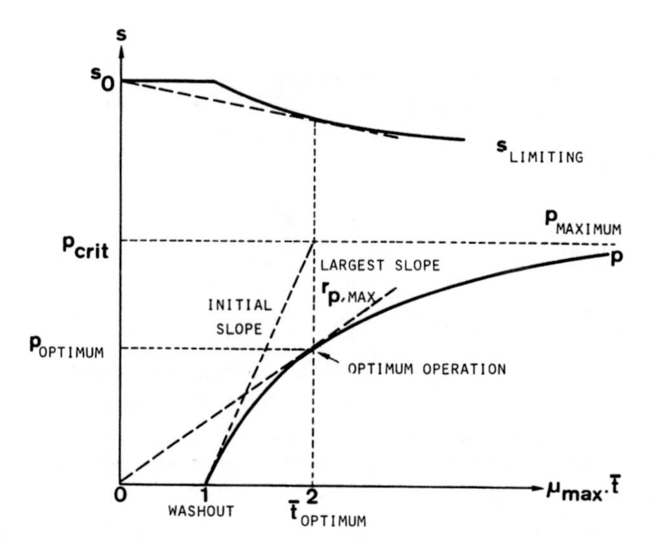

FIGURE 6.9. Product inhibition kinetics in a CSTR with $x_0 = 0$, demonstrated in a plot of concentration versus time (mean residence time $\bar{t}$) cf. Fig. 6.1b, (Levenspiel, 1980).

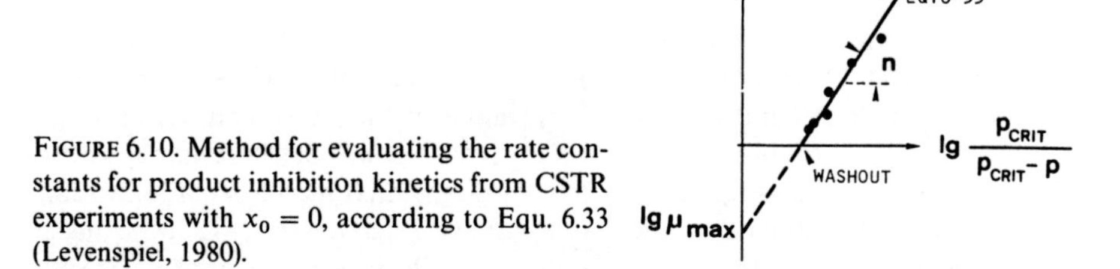

FIGURE 6.10. Method for evaluating the rate constants for product inhibition kinetics from CSTR experiments with $x_0 = 0$, according to Equ. 6.33 (Levenspiel, 1980).

into p (Levenspiel, 1980 and Han and Levenspiel, 1987), gives for the case of $x_0 = 0$

$$\mu_{max} \cdot \bar{t} = \left(1 - \frac{p}{p_{crit}}\right)^{-n} \quad \text{for } \mu_{max} \cdot t > 1 \tag{6.32}$$

The properties of Equ. 6.32 are displayed in Fig. 6.9 (cf. Fig. 6.1b), which shows that washout occurs at $\mu_{max} \cdot \bar{t} = 1$ and that the maximum production rate is dependent in a simple manner on p_{crit} and n.

To find the kinetic constants from CSTR experiments, first evaluate p_{crit} in a batch run using an excess of substrate and letting $t \to \infty$. Then rearrange Equ. 6.32 to give

$$\log \bar{t} = -\log \mu_{max} + n \log \frac{p_{crit}}{p_{crit} - p} \tag{6.33}$$

and plot as in Fig. 6.10. The slope and intercept of the best line through the data will then give the kinetic parameters μ_{max} and n in the case of pure product inhibition without substrate dependence. In situations where both substrate availability and product inhibition affect the rate, the complete rate equation (cf. Equ. 5.105) must be used, as previously outlined in Figs. 5.10b and 5.33. This is due to Equ. 6.34

$$\bar{t} = \frac{1}{k_{obs}} + \frac{K_S}{k_{obs}} \frac{1}{s} \tag{6.34}$$

6.1.2.5 Substrate Inhibition

The performance of CSTR is drastically changed by substrate inhibition kinetics. All systems previously presented have the property of equifinality, with only one unique steady state, and the attraction domain contains all states. But in nonlinear open systems, additional steady states and additional attraction domains are possible. A CSTR with S inhibition is an example where there are two stable steady states (^1x, ^1s and ^2x, ^2s) and where the phase plane is dissected in two nonoverlapping attraction domains.

The added factor $1/(1 + s/K_{I,S})$ in Equ. 5.88 represents the toxicity of the substrate at higher concentrations. Let us recall that the condition for calculation of the stationary state with nonvanishing biomass concentration is the relation $\mu(s) = D$. This equation has only one solution if $\mu(s)$ is a monotonic function. But with characteristics as in Equ. 5.88, there are two solutions. Together with the washout state (^1x, ^1s) we have three stationary states. Two of them are stable (^1x, ^1s and ^2x, ^2s), one of them is unstable (^3x, ^3s). Thus, we have a bistable system. The stationary values of the stable and the unstable stationary state are shown as a function of D in Fig. 6.11. Hysteresis may occur in shift experiments. Figure 6.12 shows how the final biomass concentration depends on the initial concentration. Figure 6.13 demonstrates that the phase plane is divided into two attraction domains. Both domains are touched by a separatrix in which the unstable stationary state lies. Note that, after an external disturbance, the system can cross over the separatrix and shift from one steady state to the other. This bistable behavior is a serious problem in, for example, waste treatment: It takes place if substrates such as alcohols, phenols, or hydrocarbons occur in such high concentrations that the utilization of these substrates is inhibited.

6.1.3 Stability Analysis and Transient Behavior of the CSTR

6.1.3.1 Stability Analysis

The steady-state behavior of a single CSTR can be calculated using Equ. 6.3a, b, representing the long-term behavior of the model system. However, nonlinear algebraic equation systems may have more than one solution. To

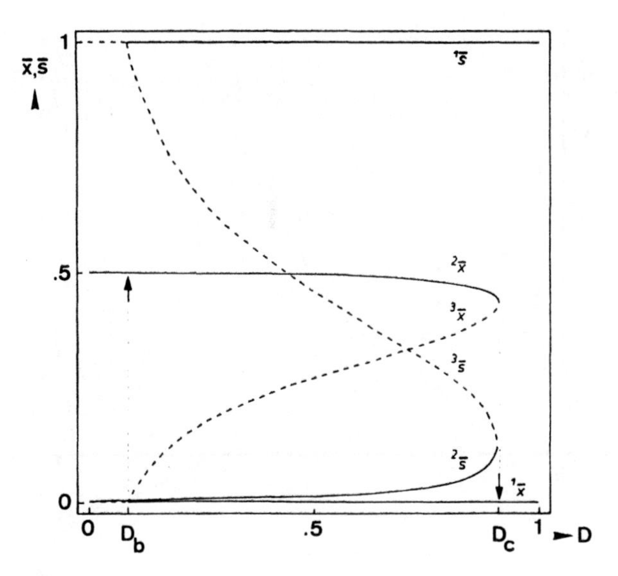

FIGURE 6.11. Effect of substrate inhibition on the behavior of CSTR: For $D_b < D < D_c$, two steady states ($^1\bar{x}$, $^1\bar{s}$) and ($^2\bar{x}$, $^2\bar{s}$) and an unstable steady state ($^2\bar{x}$, $^3\bar{s}$) coexist. The dotted line sketches the hysteretic behavior for a shift-up/shift-down cycle where D_b and D_c have been passed.

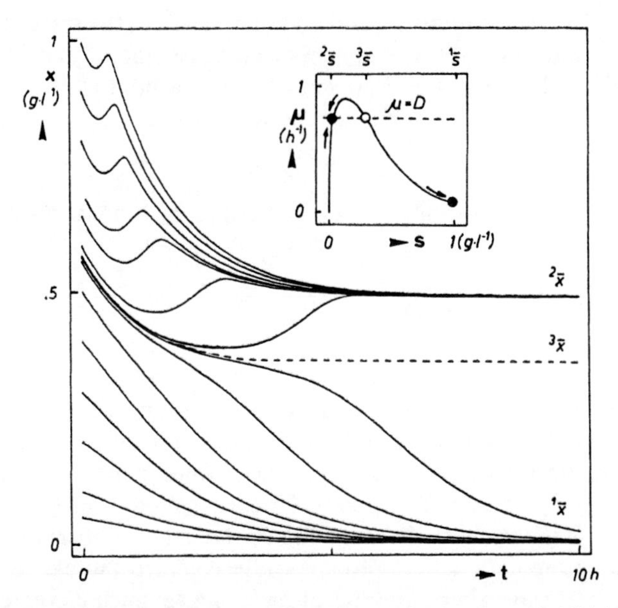

FIGURE 6.12. Bistable growth kinetics for growth limitation by an inhibiting substrate in a c/t plot. The inserted graph shows the corresponding kinetics. (Cf. Fig. 5.33.)

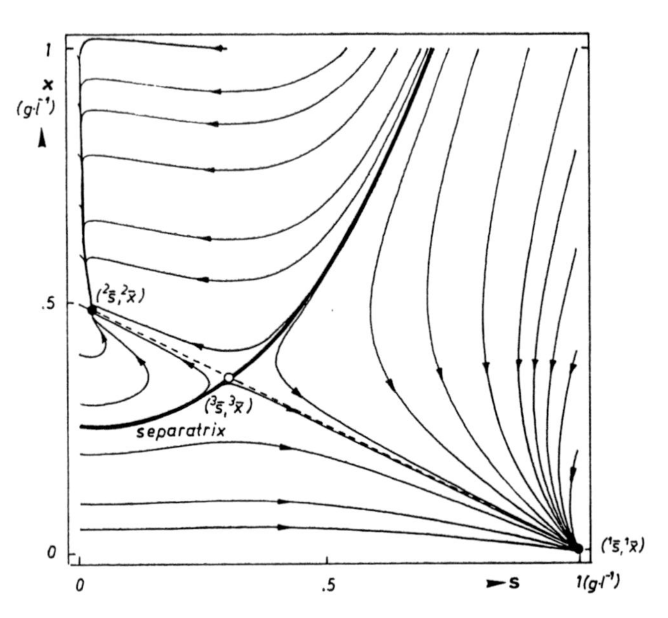

FIGURE 6.13. Bistability for growth limitation of an inhibiting substrate shown in a phase-plane diagram. The attraction domains of the two steady states are touched by a separatrix.

denote the various solutions we add an upper index to the symbols of steady-state concentrations ($\bar{x}, \bar{s}$). The system of Equs. 6.6 and 6.7 has two solutions (Gutke, 1980). The first solution we call the "washout state," because it is characterized by a vanishing biomass concentration:

$$({}^1\bar{x}, {}^1\bar{s}) = (0, s_0) \tag{6.35}$$

The second stationary solution is characterized by a nonvanishing biomass concentration:

$$({}^2\bar{x}, {}^2\bar{s}) = \left(Y_{X|S}\left\{s_0 - Y_{X|S}\frac{K_S \cdot D}{\mu_{max} - D}\right\}, \frac{K_S \cdot D}{\mu_{max} - D}\right) \tag{6.36}$$

Both solutions are included in Figs. 6.1a, 6.4, 6.5, and 6.11.

Starting the second step of analyzing the long-term behavior we may ask these questions: Are there two stationary states or is only one of them realizable in a practical case? What kind of changes occur in the system if the system in one of these states is disturbed? These questions can be answered by application of the stability theory, using analytical or numerical calculations or computer simulation. A stationary state is called "stable" if the system returns to this state after any kind of disturbance. Such a stable stationary state is called a "steady state." For our model one can demonstrate that the washout state (${}^1\bar{x}, {}^1\bar{s}$) is stable for dilution rates greater than a critical dilution rate D_c. The critical dilution rate is called "washout point": For dilution rates below this washout point, the washout state is unstable, and only a stationary state with nonvanishing biomass concentration is stable.

In contrast to closed discontinuous (batch) cultivation systems, in which the final concentrations depend on the initial concentrations, in an open system in continuous culture the final concentration is largely independently of the initial states, which phenomenon is called "equifinality." An appropriate plot for demonstration of equifinality is the plot of so-called trajectories in the phase plane, as shown in the next section and in Fig. 6.13.

General considerations concerning stability analysis of biochemical reaction systems have been presented by Stucki (1978). The problem of monostability, bistability, and multiplicity (multiple steady state) is further discussed in review articles and books (Aris and Humphrey, 1977; Bergter, 1972; Chi et al., 1974; Fiechter, 1982; Gutke and Knorre, 1980; Knorre, 1980; Koga and Humphrey, 1967; Prokop, 1978; Romanovsky et al., 1974; Russell and Tanner, 1978; Yang and Humphrey, 1975; Yano and Koga, 1973).

6.1.3.2 Transient Behavior of the CSTR

Apparently simple models are not always suitable for the description of the dynamics of bioreactor operations. As shown in Fig. 6.14, there are six classes of transient behavior of the CSTR:

- I x monotonically decreasing, s monotonically increasing
- II x monotonically decreasing, s undershooting
- III x overshooting, s undershooting
- IV x monotonically increasing, s monotonically decreasing
- V x monotonically increasing, s overshooting
- VI x undershooting, s overshooting

Figure 6.14 shows for all of these six classes one representative example in the plot of biomass (a) and substrate concentration versus time (b).

As previously outlined, a plot of trajectories in a phase plane x versus s is advantageous: In such a graph the time factor is eliminated. The sequence of states $[x(t), s(t)]$, which has passed from the initial state $[x(0), s(0)]$ to the steady state $(\bar{x}, \bar{s})$, is called the "trajectory." Figure 6.15 shows trajectories for various initial states. All trajectories tend to the same steady state $({}^2\bar{x}, {}^2\bar{s})$. But their transient behavior is different.

In considering the transient behavior of a CSTR after a shift in the dilution rate, shift-up and shift-down experiments must be distinguished (see Fiechter, 1982). Assuming that the culture before the shift is in a steady state (mathematically defined by Equ. 6.35 with $D = D_1$), consider the shift from D_1 to D_2: According to Equ. 6.35 (Fig. 6.1a), after a shift-up the biomass concentration must tend to a smaller value, but after a shift-down it must converge to a higher value. From Equs. 6.4 and 6.5, it follows that the balance relation between the biomass and the substrate concentration must be maintained

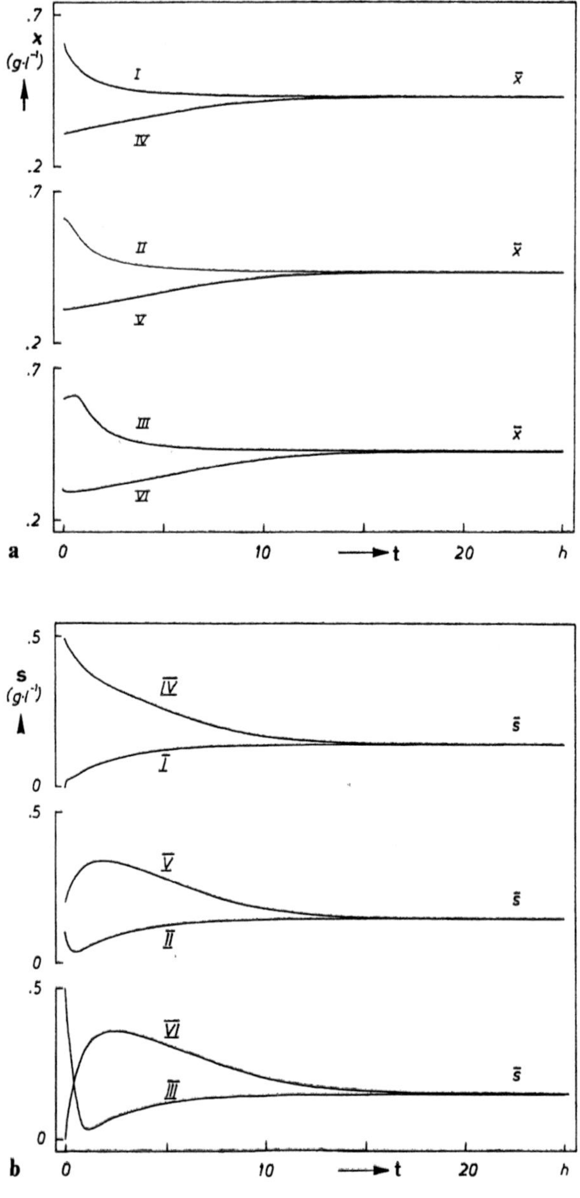

FIGURE 6.14. Transient behavior of biomass (a) and substrate concentration (b) from different initial states to the steady state using Monod kinetics with six classes of transients (I–VI).

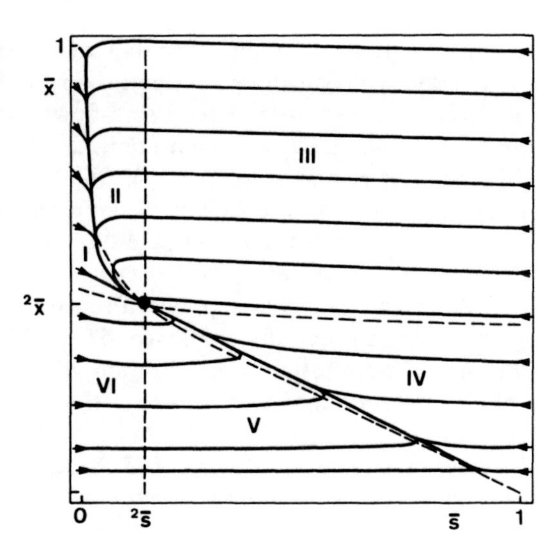

FIGURE 6.15. Trajectories in the phase plane for Monod kinetics with the six classes of transient behavior from Fig. 6.14.

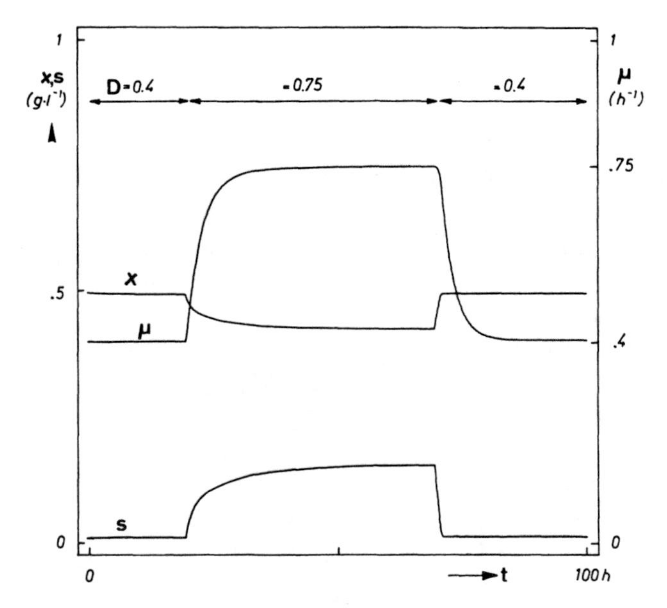

FIGURE 6.16. Monotonic transient behavior after shift-up and shift-down according to Monod kinetics, using parameter values as in Fig. 6.4.

after the shift. Thus, both steady states must lie on the straight line $\bar{x} = Y_{X|S}(s_0 - \bar{s})$. This line lies within the domains I and IV of the phase plane shown in Fig. 6.15. Thus, the response after the shift in the dilution rate is expected to be a smooth transient to the new steady state. This prediction on the basis of the Monod model is demonstrated by computer simulation in Fig. 6.16.

But this behavior is in contrast to that found experimentally: In experiments, an overshoot of substrate concentration and of specific growth rate are observed after a shift-up. After a shift-down the transient to the new steady state takes place faster and monotonically. The biological interpretation for this behavior is that in the case of shift-up the RNA concentration must increase to increase the capacity for protein synthesis appropriate to the higher supply of substrate. Thus, the specific growth rate follows the increase of substrate concentration with a time lag (t_L). Conversely, in the case of shift-down the substrate supply is decreased and the capacity for protein synthesis cannot be fully employed. Thus, the specific growth rate follows the decreasing substrate concentration without any lag. For discontinuous culture, we introduced a lag factor in a simple way in Equ. 5.72, and in a more generalized way by using Equ. 5.149. For $t_L = 0$ we return to the direct, unretarded Monod equation (Equ. 5.38). To formulate this we have a lag for shift-up but not for shift-down, and we write the following equations:

$$\mu(s,t) = \mu_{\max}\frac{s(t)}{K_S + s(t)} - t_L \cdot \dot{\mu} \qquad \text{for } \dot{\mu} > 0 \tag{6.37a}$$

$$\mu(s,t) = \mu_{\max}\frac{s(t)}{K_S + s(t)} \qquad \text{for } \dot{\mu} \leq 0 \tag{6.37b}$$

Note that this more general formulation of time lag is also suitable for modeling a lag in a discontinuous culture. Figure 6.17 shows the transient

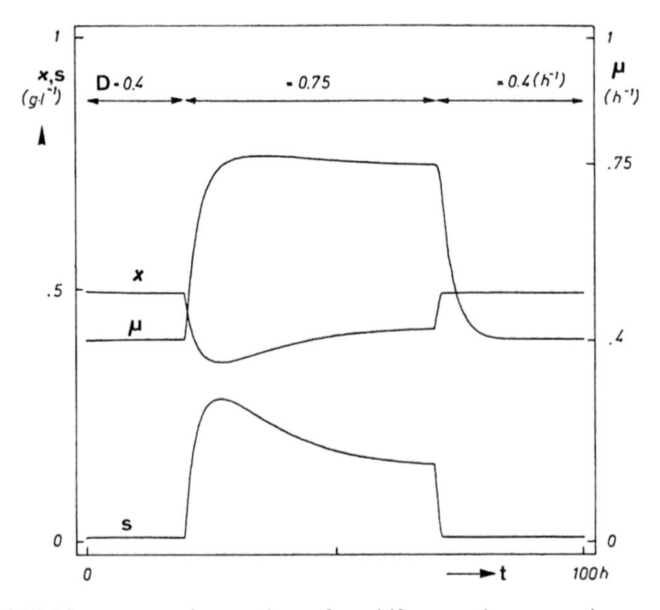

FIGURE 6.17. Nonmonotonic transient after shift-up and monotonic transient after shift-down according to Monod kinetics, modified by a lag time t_L.

behavior of the modified Monod model where Equ. 5.38 is replaced by Equ. 6.37a,b. The simulation shows an overshoot of s and μ after the shift-up, whereas after the shift-down we have the same monotonic response as shown in Fig. 6.16. Thus, the modification of Monod's model with Equ. 6.37a,b is suitable for simulating shift experiments.

6.2 Variable Volume CSTR Operation (Fed-Batch and Transient Reactor Operation)

Although continuous culture methods are commonly associated with physiological studies of microbial populations in the steady state, these methods have considerable advantage in studying behavior under transient conditions. Examination of transient behavior in a continuous culture provides valuable insight into the mechanisms of regulation in biological systems (Harrison and Topiwala, 1974). The effects of a continuously changing environment are significant in both natural and industrial microbiological systems. Without resorting to strain selection, medium optimization, or new reactor designs, a significant increase in product yield can sometimes be obtained by a very simple operational mode: transient reactor operation (Pickett et al., 1979a).

Process improvement obtained with periodic operation has been shown to depend on the reaction rate constants. In the case of consecutive competing chemical reactions (e.g., k_1, k_2, k_3), no yield or selectively improvement occurs if $k_1 \ll k_2$ or $k_1 \gg k_2$. With parallel reactions, a 20% increase over the steady-state operation has been observed (Dorawala and Douglas, 1971).

Although biological reactors generally operate in a batch or sometimes in continuous mode, some fermentations, termed "fed-batch cultures" (Yoshida et al., 1973), utilize a continuous periodic operation technique. This transient condition takes several forms. In "repeated" fed-batch culture (Pirt, 1974), complete medium is fed continuously to a batch culture; the resultant biomass is reduced by being pumped out at preset intervals, and the process is then repeated. In fed-batch culture, the dilution rate is therefore continually changing and represents the transient of the system. Some processes utilize a feed of the (limiting) carbon source only; this type of fed-batch culture has been termed "semibatch" culture (Yamane and Hirano, 1977). A further modification of fed-batch culture, called "extended" culture, has also been reported (Edwards et al., 1970). This technique is more controlled than other types of the fed-batch culture, since environmental sensors directly measure specific parameters of culture growth and regulate the addition of the limiting carbon source to the culture. Both semibatch and extended-batch types of culture usually only cover one fed-batch "cycle" and so differ from repeated fed-batch culture in the duration of the periodicity applied to the culture (see Figs. 3.31–3.33).

For quantification of variable-volume CSTR operations, see Equ. 3.86a,b and the performance equation derived for a variable-volume CSTR operation, Equ. 3.92.

Mathematically, in terms of volumetric flow rates, liquid densities ρ, and volume

$$\frac{d(V\cdot\rho_1)}{dt} = F_{in}\cdot\rho_{in} - F_{ex}\cdot\rho_{ex} \tag{6.38}$$

Normally, the density changes are small, giving

$$\frac{dV}{dt} = F_{in} - F_{ex} \tag{6.39a}$$

The total mass balance gives the rate of change of volume with inlet and outlet flow rates for a well-mixed constant-density system. A fed batch is a special case of a variable-volume CSTR operation: It has been defined as a bioreactor with inflowing substrate but without outflow. For this system, the equation becomes

$$\frac{dV}{dt} = F_{in} \tag{6.39b}$$

where F_{in} can be a function of time $F(t)$, according to Equ. 3.86. Equation 6.39b together with Equ. 3.92 provides the mathematical model for the fed-batch reactor, and the behavior calculated by computer simulation is shown in Fig. 3.37.

The variable-volume system considered here has three dependent variables: V, x_{ex}, and s_{ex}. The equations previously mentioned provide the required number of independent equations: Their simultaneous solution will yield the variation of V, x_{ex}, and s_{ex} with time. The assumptions and restrictions are as follows: (a) well-mixed reactor, (b) single limiting substrate, (c) balanced growth, (d) constant yield coefficient, and (e) Monod kinetics. The restriction that balanced cell growth must exist for the Monod relation to be valid puts a strong limitation on the possible rates of change that can be considered. For example, large differences in dilution rates would lead to rapid changes in substrate concentration, and the specific growth rate could not possibly adapt according to the Monod equation.

It is instructive to compare the fed-batch model with the model for a constant-volume chemostat (see Equs. 3.90 and 3.91). Equation 6.3a,b is formally identical with Equ. 3.92a,b, but it is important to note that the physical meaning of the terms is not identical. Comparing the term $-F_{ex}\cdot x_{ex}/V$ in Equs. 3.92 and 6.3 it can be seen that the origin of this term for the fed batch is the expression $x_{ex}\cdot dV/dt$; it thus represents a decrease in cell concentration due to the volume change that arises from inlet flow rate F_{in}. On the other hand, $-F_{in}\cdot x_{ex}$ in Equ. 6.3, the chemostat, is a washout term expressing the mass flow rate of cells that leave with the outgoing stream. A fed batch can be compared with a constant-volume chemostat whose feed rate is decreasing slowly.

Because the mathematical form of the fed-batch model Equ. 3.92a,b is identical to that of the chemostat Equ. 6.3a,b, it can be concluded that the fed batch will behave analogously. A dynamic steady state will be achieved for sufficiently low flow rates such that the specific growth rate is maintained exactly equal to the dilution rate F_{in}/V. This phenomenon has been identified previously, and the dynamic steady state has been termed a "quasi-steady state." It is characterized by a constant value of x_{ex}, which must exist because $\mu = F_{in}/V$. Since the volume is increasing steadily, μ must be maintained by a decrease in s_{ex}; therefore, ds_{ex}/dt is not zero. Computer simulations obtained by solving Equ. 3.92a,b together with Equ. 6.39b numerically show that the phenomena do indeed occur as described. The computer simulation shown in Fig. 3.37 indicates that substrate concentration steadily decreases during the quasi-steady state $\mu = D$, in order to maintain the quasi-steady state as the volume increases (Dunn and Mor, 1975). When the quasi-steady state is achieved, the equality of μ and F_{in}/V leads to the following relationship for s_{ex}:

$$s_{ex} = \frac{F_{in}}{V} \cdot \frac{K_S}{\mu_{max} - (F_{in}/V)} \tag{6.40}$$

For the special case of constant feed rate, F_{in}, the fed batch, has further properties that are worth noting:

1. From Equ. 3.92a, under the quasi-steady-state conditions, $\mu = F_{in}/V$, and it follows that

$$\frac{d(V \cdot x_{ex})}{dt} = F_{in} \cdot x_{ex} = \text{constant} \tag{6.41}$$

The rate of change of total biomass is constant.

2. From Equ. 6.39b, the volume must increase linearly with time:

$$V = V_0 + F_{in} \cdot t \tag{6.42}$$

3. The rate of change of μ during a quasi-steady-state period is

$$d\frac{F_{in}}{V} = -\frac{F_{in}^2}{V^2} = -\frac{F_{in}^2}{(V_0 + F_{in} \cdot t)^2} \tag{6.43}$$

At low t when $V = V_0$ this becomes

$$\frac{d\mu}{dt} = -\frac{F_{in}^2}{V_0^2} \tag{6.44}$$

at higher values of t or conditions when $V \gg V_0$

$$\frac{d\mu}{dt} = -\frac{1}{t^2} \tag{6.45}$$

Thus, it is seen that μ decreases most rapidly during the first period of the fed batch, the rate of decrease becoming slower with time. For the purpose of obtaining a particular desired rate of change of μ, the flow rate F_{in} could be changed progressively during the course of fermentation (Dunn and Mor, 1975).

Fed-batch operations are of great theoretical and practical importance. For basic research in process kinetics, a fed-batch process is advantageous—the extended culture offers the possibility of maintaining low S concentrations over long periods of time, and this facilitates measurement of kinetic parameters (Esener et al., 1981c; Lee and Yan, 1981). Several computer simulations have been carried out to demonstrate fed-batch productivities compared with the CSTR at constant volume (Keller and Dunn, 1978a) and to show the influence of such factors as fluctuations in volume or S feed (Keller and Dunn, 1978b). Computational results also provide quantitative sensitivity analysis that is very useful in determining the degree of precision necessary in applying process equations (Kishimoto et al., 1976). Furthermore, the fed-batch process provides the same information as the CSTR without the requirements for an outflow, without volume control, and without the necessity of shifting steady states. Bioprocessing on a larger scale profits from the fact that fed-batch cultures show the unique feature of transient conditions, with the growth rate under control between fixed values. There is evidence that the maximum rates of some processes can be achieved only transiently: Examples are productions of some bacterial antigens (Pirt et al., 1961), synthesis of beta-galactosidase (Knorre, 1968), biosynthesis of secondary metabolites such as penicillin (Bajpaj and Reuss 1981; Bajpaj and Reuss, 1980; Court and Pirt, 1976; Heijnen et al., 1979; Pirt, 1974; Wright and Calam, 1968), turimycin (Gutke and Knorre, 1981), streptomycin (Hegewald et al., 1978), and cephalosporin (Matsumura et al., 1981; Trilli et al., 1977). A typical and famous example of a drastic increase in productivity in case of secondary metabolite production is shown in Fig. 6.18 by choosing the adequate level of S dosing and OTR capacity (Bajpaj and Reuss, 1980, 1981). In addition to these case involving catabolite repression and/or inhibition (Demain et al., 1979; Gutke and Knorre, 1982), fedbatch processes are advantageous in the cases with complex kinetics in yeast technology (Aiba et al., 1976; Dairaku et al., 1981) and in waste water treatment in which substrate inhibition is also commonly encountered. The increase in cell mass, derived from Equ. 3.43, for the quasi-steady state is given by

$$V \cdot \frac{dx}{dt} = F \cdot Y \cdot s_0 \tag{6.46}$$

and the cell mass to be harvested at this condition is calculated by

$$x \cdot V = x_0 \cdot V + F \cdot Y \cdot s_0 \cdot t \tag{6.47}$$

Transient reactor operations have been summarized in review articles (e.g., Barford et al., 1982; Douglas, 1972; Pickett et al., 1979a; Pickett, 1982) and appear increasingly in the literature (e.g., Borzani et al., 1976; Chi and Howell, 1976; Cooney and Wang, 1976; Daigger and Grady, 1982; Klei et al., 1975; Parulekar and Lim, 1985; Pickett et al., 1980; Regan et al., 1971; Sherrard and Lawrence, 1975; Sundstrom et al., 1976; Yamane and Shimizu, 1984).

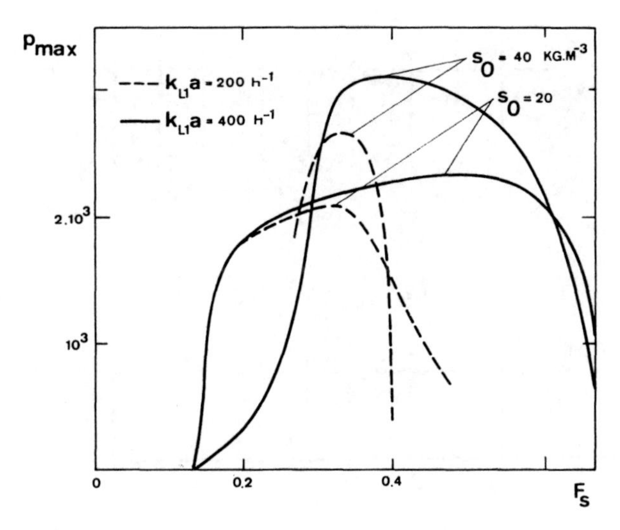

FIGURE 6.18. Computer simulation of product concentration p_{max} (units penicillin/ml) as a function of substrate feeding rate F_S [$g \cdot l^{-1}$ initial volume $\cdot h^{-1}$] with varied $k_{L1} \cdot a$ value and initial substrate concentration s_0. At high values of F_S, productivity decreases due to catabolite repression and O_2 limitation. (From Reuss et al., 1980.)

6.3 Multistage Single and Multistream Continuous Reactor Operation

The complexity of multistage systems increases with the number of stages. In practice, however, it is rarely necessary to use more than three stages, and the essential principles of multistage operation may be illustrated by a discussion of only two stages.

6.3.1 CLASSIFICATION

Herbert (1964) distinguished three main types of multistage systems, as shown in Fig. 6.19:

- Single-stream multistage (Fig. 6.19a)
- Multistream multistage (Fig. 6.19b)
- Multistage systems with recycle (Fig. 6.19c)

Single-stream systems are a cascade with a single medium inflow that is constant through all other stages. The characteristic features are

- Later stages cannot affect earlier stages
- The first stage behaves as a CSTR
- Dilution rate D cannot be changed in one stage without changing it in all others, because the flow is constant
- The dilution rate in individual stages depends only on the reactor volume

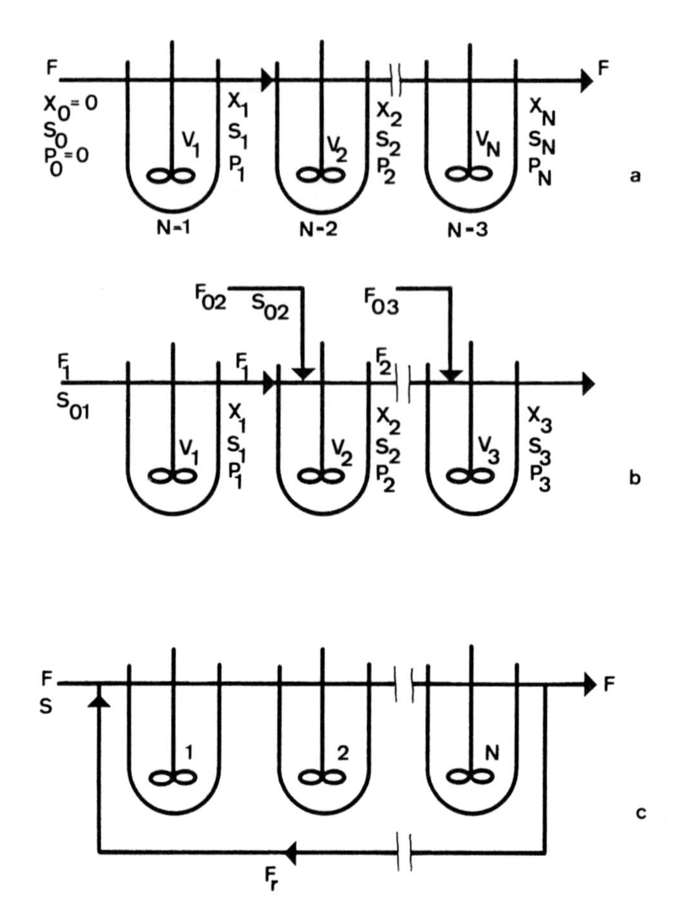

FIGURE 6.19. Different types of multistage CSTR operation: (a) single-stream multistage system, (b) multistream multistage system, (c) multistage with recycle.

$$F = D_1 \cdot V_1 = D_2 \cdot V_2 = D_3 \cdot V_3 \tag{6.48}$$

Multistream systems have multiple inputs of feed and therefore are characterized as follows:

- Earlier stages are again independent of later stages, and the first stage behaves as a CSTR
- The different medium feeds may be varied independently and individual dilution rates are independent process variables

6.3.2 Potentialities of Multistage Systems

Multistage systems have seldom been used in industry or research. From a theoretical viewpoint, the potential advantages are quite promising (Ricica, 1969a). In general, multistage systems provide different environments in each stage and thereby approximate the behavior of tubular reactors (cf. Fig. 3.30). Multistage systems, therefore, potentially offer the advantages of CPFRs (cf. Sect. 6.4), which can be summarized as follows:

- Maximum conversion is achieved by complete utilization of the substrate in the later stages and maximum productivity in the first stage. This is essential in the case of expensive substrates, for example, steroid transformation, or in the case of environmental problems, for example, waste water treatment.
- All physiological states of microbial cultures are accessible, including the "mature cell" stage necessary for some types of product formation such as secondary metabolities.
- Long residence times can be maintained. This is advantageous in the case of bioprocesses in the stationary growth phase with complex kinetics or with complex media.
- A certain product quality may be desired (e.g., bakers' yeast), and multistage processing for this purpose is already found in industry.
- Optimal environmental operating conditions (for temperature, pH, and p_{O_2}) can be maintained in such gradient reactors.

6.3.3 Single-Stream Multistage Operation

Consider a two-stage reactor system in which the first stage behaves like a single CSTR (Fig. 6.19a; Herbert, 1964; Malek and Fencl, 1966; Ricica and Necinova, 1967). The second stage may be treated in much the same way as the first, but the equations will be more complicated, since x and s enter the second stage. The balance equations for x and s are

$$\frac{dx_2}{dt} = D_2 \cdot x_1 - D_2 \cdot x_2 + \mu_2 \cdot x_2 \tag{6.49}$$

and

$$\frac{ds_2}{dt} = D_2 \cdot s_1 - D_2 \cdot s_2 - \frac{1}{Y_{X|S}} \cdot \mu_2 \cdot x_2 \tag{6.50}$$

where μ_2 follows simple Monod kinetics. In the steady state with $dx_2/dt = 0$

$$\mu_2 = D_2 \left(\frac{x_2 - x_1}{x_2} \right) \tag{6.51}$$

and with $ds_2/dt = 0$

$$\mu_2 = Y \cdot D_2 \left(\frac{s_1 - s_2}{x_2} \right) \tag{6.52}$$

Eliminating μ_2 from Equs. 6.51 and 6.52 gives

$$x_2 = Y(s_0 - s_2) \tag{6.53}$$

from which x_2 can be calculated when s_2 is known. To find s_2 we have, from Equ. 6.52 and Monod kinetics by eliminating μ_2,

$$x_2 = \frac{Y \cdot D_2}{\mu_{max} \cdot s_2}(K_S + s_2)(s_1 - s_2) \tag{6.54}$$

and eliminating x_2 from Equs. 6.54 and 6.53 gives

$$Y(s_0 - s_2) = \frac{Y \cdot D_2}{\mu_{max} \cdot s_2}(K_S + s_2)(s_1 - s_2) \tag{6.55}$$

Solving this equation for s_2 after substituting Equ. 6.6 gives a quadratic equation

$$(\mu_{max} - D_2)s_2^2 - \left(\mu_{max} \cdot s_0 - \frac{K_S D_1 D_2}{\mu_{max} - D_1} + K_S D_2\right)s_2 + \frac{K_S^2 \cdot D_1 \cdot D_2}{\mu_{max} - D_1} = 0 \tag{6.56}$$

in which s_2 is expressed as a function of s_0, D_1, D_2, and μ_{max}, K_S. The quadratic equation has two positive roots and gives two values for s_2, one smaller and one greater than s_1; the first is the correct value, and the second corresponds to the biologically imaginary but mathematically real case of "reverse" growth (with conversion of cells to substrate). Thus, from Equs. 6.53 and 6.56 the steady-state concentration of cell and substrate in the second stage can be calculated, and the solution is shown graphically in Fig. 6.20. Extension of this mathematical analysis to three or more stages can be continued, the equations for each fermenter being derived from those of the preceding one. For the Nth reactor in a cascade, the material balance equations using symbols shown in Fig. 6.19a are, therefore, given by

$$V_N \cdot \frac{dx_N}{dt} = F \cdot x_{N-1} - F \cdot x_N + (\mu_N x_N) V_N \tag{6.57}$$

$$V_N \cdot \frac{ds_N}{dt} = F \cdot s_{N-1} - F \cdot s_N - \left(\frac{\mu_N}{Y_{X|S}} x_N\right) V_N \tag{6.58}$$

$$V_N \cdot \frac{dp_N}{dt} = F \cdot p_{N-1} - F \cdot p_N + \left(\frac{\mu_N}{Y_{P|X}} x_N\right) V_N \tag{6.59}$$

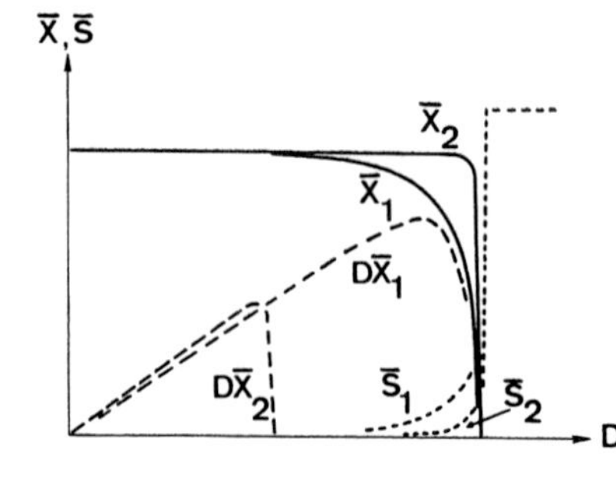

FIGURE 6.20. Theoretical relationships between stationary values of the cell mass concentrations $\bar{x}_1$ and $\bar{x}_2$ (or substrate concentrations $\bar{s}_1$ and $\bar{s}_2$) and the dilution rate in a cascade with $N = 2$. The productivities are designated $D\bar{x}$. (Adapted from Herbert, 1964.)

Thus, at steady-state conditions

$$x_N = \frac{D \cdot x_{N-1}}{D - \mu_N} \qquad (N \neq 1) \tag{6.60}$$

$$s_N = s_{N-1} - \frac{1}{D}\frac{Y_{P|X}}{Y_{X|S}} \cdot \mu_N \cdot x_N \tag{6.61}$$

$$p_N = \frac{1}{D}(D \cdot p_{N-1} + Y_{P|X} \cdot \mu_N \cdot x_N) \tag{6.62}$$

The number of stages N and the parameters of interest for a multistage system can be determined by a graphical method described by Deindoerfer and Humphrey (1959) or Luedeking and Piret (1959) in analogy to the chemical process design called the "periodic equilibrium curve." This method enables the calculation of initial estimates and forms the hypothetical model (cf. Fig. 2.18). In application, kinetic data from measurements in a DCSTR are plotted as r_X against x (Fig. 6.21). The productivity is always a combination of kinetic and mass conservation elements, and for the Nth reactor in a continuous process, from Equ. 3.89 (cf. Equ. 6.57)

$$\frac{dx_N}{dt} = \frac{F}{V}(x_N - x_{N-1}) \tag{6.63}$$

which is entered in Fig. 6.21.

For the first stage, with $x_{0,c}$ the straight line represented by Equ. 6.63 with the slope $F/V_1 = D_1 = \mu_1$ is drawn. The intercept of the line with the kinetic curve (labeled point 1) gives the concentration $\bar{x}$ for $N = 1$. An optimal reactor

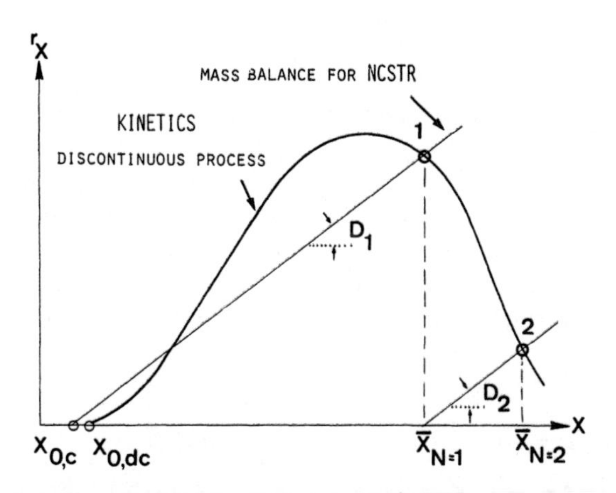

FIGURE 6.21. Graphical procedure for obtaining the dilution rate through the system, D, for a cascade of stirred vessels (NCSTR) in carrying out a continuous microbial growth process on the basis of discontinuous experimental data. The discontinuous process kinetics in the r_X/x plot are analogous to a plot from a chemical process.

system would operate with the maximum slope: $D_1 \to \mu_{max}$. However, due to the danger of washing cells out of the reactor, the actual optimum, D_{opt}, is selected according to Fig. 6.1a (cf. Equ. 6.11). The first intercept, the one at the lower region of the curve, represents an unstable operating point (the lag phase).

In the same way, additional curves can be drawn for $N = 2, 3 \ldots$ (Equ. 6.63), each with the starting concentration of the previous reactor in the series. The slope of the straight line represents the value of D for each case. When F is a constant for the whole system, the volume of each stage of the cascade may be optimized.

In developing a process with a continuous cascade of stirred vessels, it should be remembered that the productivity (in terms of product per unit volume per unit time) is not optimized. Rather, the conversion (in terms of substrate utilization) is maximized. To clarify this, Fig. 6.20 (similar to Fig. 6.1a) is the diagram of a two-stage process (Herbert, 1964). One sees here that the productivity of the two-stage CSTR ($\mathrm{Pr} = D \cdot \bar{x}_2$) is less than $D \cdot \bar{x}_1$, but the use of substrate ($\bar{s}_2$) is more complete than with a single stage CSTR ($\bar{s}_1$).

In considering the whole system it is useful to calculate its average rate, D_{ave}, defined as the total flow through the system divided by the total volume of culture in all fermenters. For a single-stream chain of N fermenters this is

$$D_{ave} = \frac{F}{N \cdot V} \tag{6.64}$$

The washout rate for the whole system, that is, the critical value of D_{ave}, is therefore

$$(D_{ave})_c = \frac{F_c}{N \cdot V} = \frac{D_c}{N} \tag{6.65}$$

in a third or subsequent fermenter. In fact, if calculations for a third and fourth stage are made, the resulting curves are scarcely distinguishable from curve x_2 in Fig. 6.20. This suggests that there will seldom be much practical advantage in using more than two stages, at least for the quantity production of cells. On the other hand, further stages might be important in obtaining cells of a desired quality. Also, using a two-stage process, a continuous culture may be maintained at a truely maximum growth rate; this is impossible in a single-stage process.

6.3.4 Multistream Multistage Operation

We may conclude from the previous section that the experimental possibilities of single-stream continuous systems are rather limited. There are many more possibilities with multistream systems, since these have multiple medium inflows that are independently variable and that may, if desired, contain a variety of different nutrients. This can produce very complicated experimental situations. We will consider only the simple case in which a single growth-limiting nutrient is used in a two-stage system. (Fig. 6.19b). Systems with different substrates fed to later systems are referred to (Herbert, 1964).

The dilution rates of the two fermenters are respectively $D_1 = F_1/V_1$ and $D_2 = F_2/V_2$. It is convenient mathematically to consider D_2 as the sum of two "partial dilution rates" D_{02} and D_{12}, due to the two different inflows to fermenter 2, and defined as

$$D_{02} = \frac{F_{02}}{V_2}, \qquad D_{12} = \frac{F_1}{V_2}, \quad D_{02} + D_{12} = D_2 \tag{6.66}$$

Again, the first stage is described with Equs. 6.6 and 6.7.

For the second stage, the mass balance equations for x and s are

$$\frac{dx_2}{dt} = D_{12} \cdot x_1 - D_2 \cdot x_2 + \mu_2 \cdot x_2 \tag{6.67}$$

and

$$\frac{ds_2}{dt} = D_{12} \cdot s_1 + D_{02} \cdot s_{02} - D_2 \cdot s_2 - \left(\frac{\mu_2}{Y_{X|S}}\right) x_2 \tag{6.68}$$

In the steady state, these equations become

$$\mu_2 = D_2 - D_{12} \frac{x_1}{x_2} \tag{6.69}$$

and

$$\mu_2 = \frac{x_2}{Y}(D_{12}s_1 + D_{02}s_{02} - D_2 s_2) \tag{6.70}$$

Eliminating μ_2, x_1 and rearranging gives

$$x_2 = Y\left(\frac{D_{12} \cdot s_{01}}{D_2} + \frac{D_{02} \cdot s_{02}}{D_2} - s_2\right) \tag{6.71}$$

from which x_2 can be calculated when s_2 is known. To find s_2 on the basis of Equ. 6.70 and the Monod relation for μ_2, which is eliminated, we get Equ. 6.72 after again eliminating x_2 and rearranging

$$(\mu_{max} - D_2)s_2^2 - \left(\frac{\mu_{max} \cdot D_{12} \cdot s_{01}}{D_2} + \frac{(\mu_{max} - D_2)D_{02} \cdot s_{02}}{D_2}\right.$$
$$\left. - D_{12} \cdot s_1 + K_S \cdot D_2\right)s_2 + K_S \cdot D_{02} \cdot s_{02} + K_S \cdot D_{12} \cdot s_1 = 0 \tag{6.72}$$

From this quadratic form, s_2 can be computed by inserting the value of s_1 from Equ. 6.6, and the steady-state behavior for any value of D_1 and D_2 can be derived; the only precondition concerns the values of kinetic parameters.

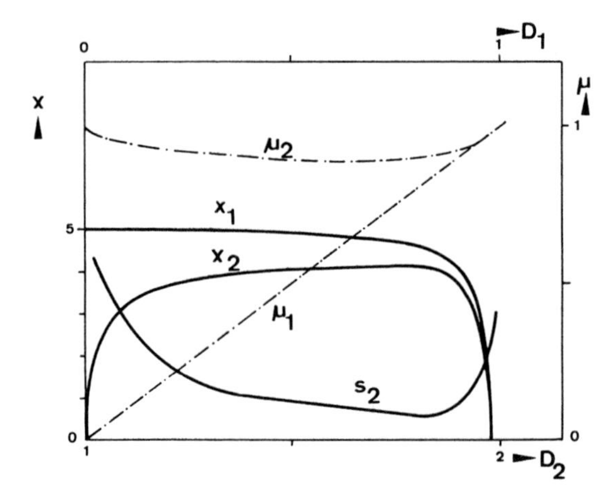

FIGURE 6.22. Two-stream two-stage CSTR operation in the steady state: Growth rates in first and second stage (μ_1, μ_2) and concentrations calculated with the help of Equs. 6.67 through 6.71 ($D_{02} = 1\ \text{h}^{-1}$). (From Herbert, 1964.)

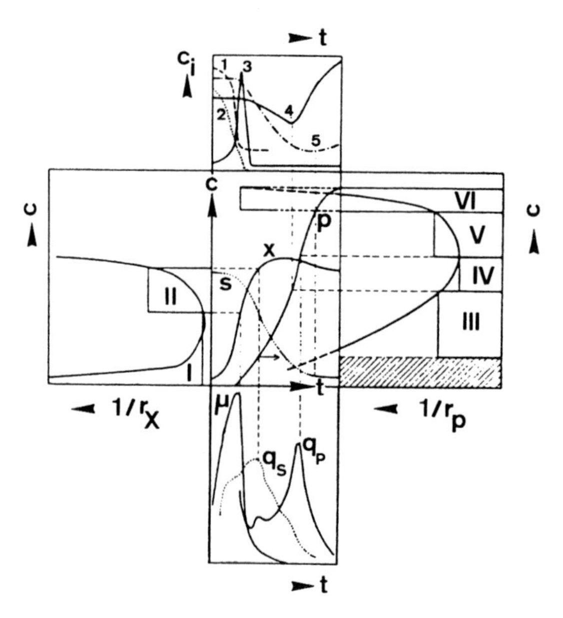

FIGURE 6.23. Multigraphical interpretation of streptomycin production according to Ricica (1969). The central graph represents the c/t plot of batch data, with c/t plots of individual components in the upper graph (1, α-amino nitrogen; 2, phosphorus; 3, pyruvic acid; 4, pH; 5, ammonia nitrogen). The quantification is illustrated in the lower part (μ, q_S, and q_p vs. t). The design of an NCSTR cascade with $N = 6$ is demonstrated in the graphs on the left side (I and II) and right side (III–VI) using plots of $1/r_i$ versus c_i. (cf. Fig. 6.32.)

The output for a single-stream chain of N vessels is

$$\text{Output} = \frac{F \cdot x_N}{N \cdot V} = D_{ave} \cdot x_N \tag{6.73}$$

The behavior of a two-stream, two-stage system is illustrated in Fig. 6.22. In this case, the medium flow to the second fermenter is assumed to be held at constant value while the inflow to the first fermenter is varied. The curves of Fig. 6.22 are calculated for a constant second-stage medium inflow of $D_{02} = 1.0$ and D_1 values from zero to D_c (0.98). These curves illustrate very clearly the stability of the system when operating at high second-stage growth rates, and the relative insensitivity of the second stage to changes in the first. It will be seen that alternations in the first-stage dilution rate over nearly the whole of its working range ($D_1 = 0.15$ to $0.9\ h^{-1}$) produce variations of less than 10% in the second-stage cell concentration and growth rate. The great flexibility of multistream systems, combined with their stability and the ease of operation at high flow rates, makes them superior to single-stream systems in nearly all respects.

Figure 6.23 illustrates the design procedure in case of antibiotic production in a six-stage cascade (Ricica, 1969a) using the design concept of Fig. 6.32.

6.4 Continuous Plug Flow Reactors (CPFR)

Reactors systems that have plug flow characteristics (that is, $N \geq 5$ or $Bo \geq 7$, see Sect. 3.3) differ from CSTR types in that in the former case there is a narrow residence time distribution. The whole mathematical theory of continuous cultivation discussed so far has concerned the well-mixed stirred tank reactor.

6.4.1 Performance Equations

An ideal case of a CPFR is represented by a narrow empty tube through which the unstirred liquid flows uniformly (see Fig. 3.38). The concentration profile was illustrated in Fig. 3.30. As washout is immediate, a constant inoculum or a permanent recycle stream is required for continuous operation of a CPFR. Since the composition of the fluid varies from position to position along the longitudinal axis z, the material balance must be made on a differential element of fluid $dV = A \cdot dz$, as referred to in Sect. 3.4 (cf. Equs. 3.94–3.96).

Equations 3.95a,b and 3.96a,b form a set of nonlinear differential equations that are difficult to solve analytically; they are more readily solved by computer techniques. In special cases, however, the equations can be integrated, for example, when the amount of biomass formed by the reaction is small relative to the entering amount and the value of x is nearly constant over the reaction length (x_{ave}). Thus, integration of Equ. 3.95a is possible, yielding the following expression with Monod kinetics

$$(s_i - s) + K_S \frac{\ln s_i}{s} = \frac{\mu_{max} \cdot x_{ave} \cdot A \cdot z}{Y \cdot F(1 + r)} \tag{6.74}$$

The small change of x along the reactor is approximated by $(x - x_i) = Y(s_i - s)$; substituting for $(s_i - s)$ from Equ. 6.74 gives Equ. 6.75:

$$(x - x_i) = \frac{\mu_{max} \cdot x_{ave} \cdot A \cdot z}{F(1 + r)} - K_S \cdot Y \cdot \ln \frac{s_i}{s} \tag{6.75}$$

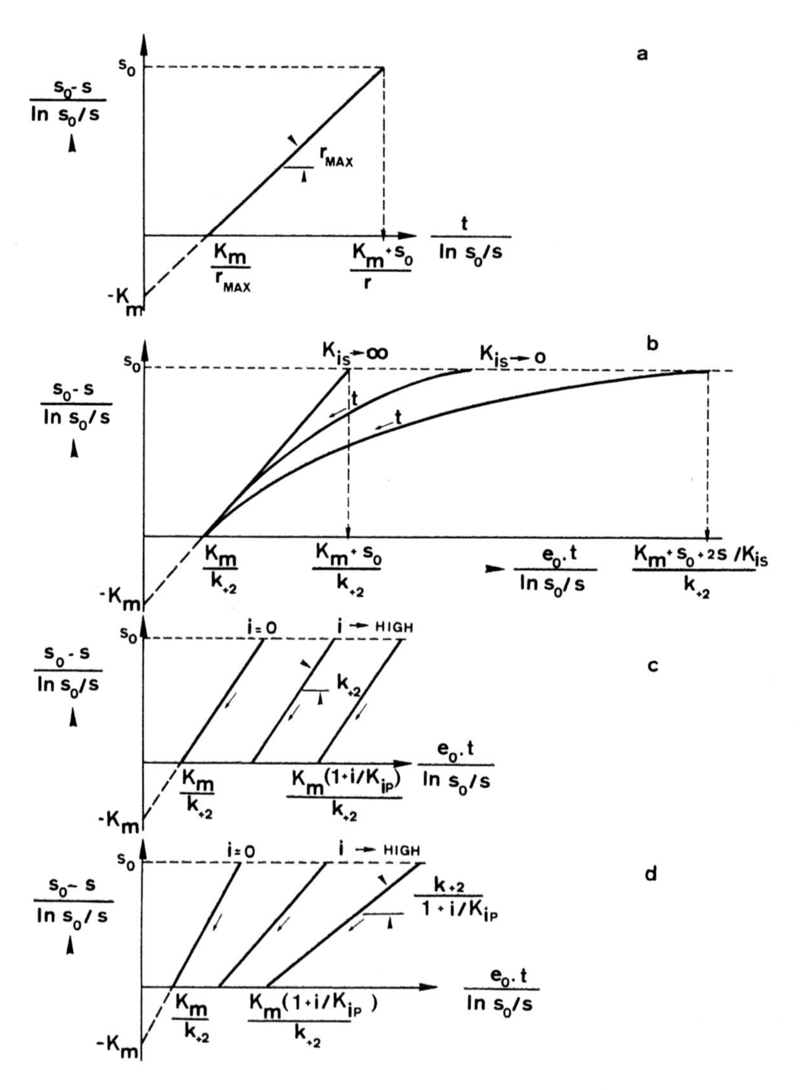

FIGURE 6.24. Parameter estimation from integral reactors (DCSTR and CPFR) in the case of simple enzyme kinetics (a), substrate inhibition (b), and competitive (c) and noncompetitive product inhibition (d), according to Levenspiel (1979).

For enzyme processing, Fig. 6.24 summarizes some graphical methods for parameter estimation from CPFR.

6.4.2 Potential Advantages of CPFR Operation

Tubular reactors offer some potential advantages over conventional stirred tank types of reactors (A. Moser, 1985b):

- High productivity and optimum conversion are realized simultaneously
- Mixing within tubular devices is more uniform eliminating dead spaces and resulting in more secure scale-up (Russell et al., 1974)
- Surface area-to-volume ratio is significantly higher resulting in facilatated transfer processes. As a consequence, mass transfer is achieved with comparable less power consumption in the case of horizontal devices (Moser, 1985b; see also Küng and Moser, 1986; Ziegler et al., 1977) and heat transfer is easy to be realized. This fact becomes crucial in extreme situations of bioprocessing e.g. with solid-substrates, photoreactions (maximum exposure to light), shear sensitive tissues etc.
- Bioprocessing occur with gradients of concentrations and/or temperature over the length of the tubes, so that adaption of technical manipulations to biological demands can be carried out (e.g. T-programming as a function of axial distance; substrate-dosing, CO_2-removal etc., in the case of inhibition and/or repression kinetics (cf. Equ. 5.106). The case of product inhibition is quantified in Fig. 6.25. This property is advantageous also for fundamental investigations of biological processes. Fig. 6.26 illustrates the interrelations between biokinetics and optimal bioreactor design in case of continuous operation (Moser, 1983b). Four cases of kinetics are shown together with the bioreactor thought to realize highest conversion, which in all cases is based on a CPFR-system.
- Horizontal versus vertical position is possible (tubular vs. tower reactors), each having advantages. Horizontal configurations exhibit the property,

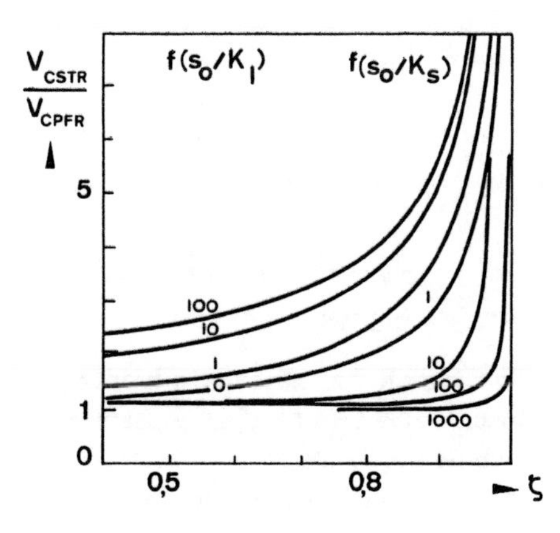

FIGURE 6.25. Demonstration of the advantage of CPFR over CSTR by calculating the volume ratio in dependence of conversion reached (ζ) as a function of biokinetics using the simple Monod equation and varying the ratio s_0/K_S and also product inhibition kinetics with variation of s_0/K_I. A. Moser, 1985c, reprinted with permission from Conservation and Recycling, vol. 8, No. 1/2, Pergamon Press, Oxford.)

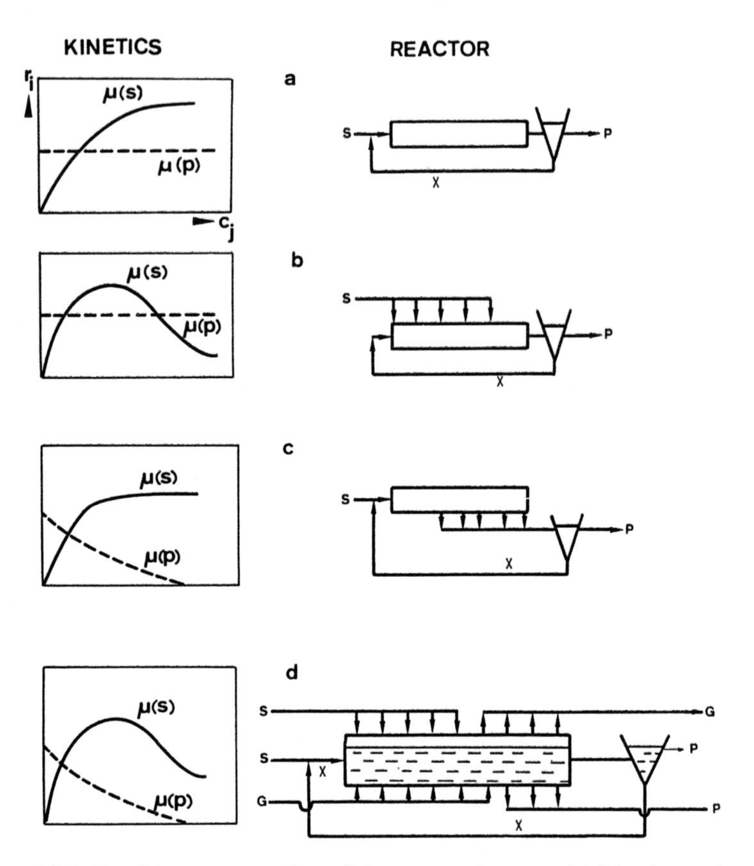

FIGURE 6.26. Graphic representation of the connex between biokinetics and corresponding optimal continuous bioreactor design in four cases (a–d) (Moser, 1983b).

that plug- flow is not disturbed by the CO_2 evolved and that hydrostatic pressure cannot become inhibiting or cannot create practical problems.

- Practical advantages exist due to the closed system of tubes (no aerosol formation in the case of waste treatment, easy operation under pressure and/or with pure O_2). The technical apparatus is easy to handle because the basic elements are commonplace (pipes, pumps, standard fittings). Also, wall growth seems to be more easily controlled due to the high liquid velocities.

6.4.3 Principal Properties and Design of CPFRs Compared with CSTRs

The CPFR has distinct advantages for those bioprocesses that call for a specific reaction time, as illustrated in Fig. 6.27. Examples of such processes include the formation of the products of secondary metabolism, sterilizations to destroy cells and spores, and the high temperature treatment of foodstuffs.

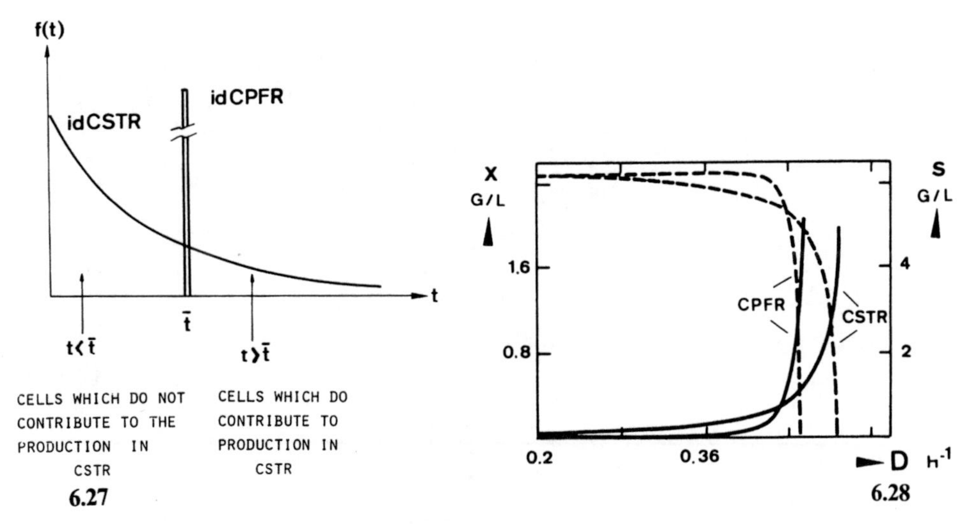

FIGURE 6.27. Schematic comparison between an ideal CSTR and an ideal CPFR in the case of a bioprocess maintaining a particular mean residence time $\bar{t}$ (production of secondary metabolites with maturation time $t_M = \bar{t}$); sterilization and heat treatment of foodstuffs with $\bar{t} = t_{St}$.

FIGURE 6.28. Comparison of CSTR and CPFR operation in a steady-state diagram of concentrations (x, ---; s, ——) versus dilution rate, both operating under the same recycle conditions ($r = 0.4$ and $\beta = 3$). (Adapted from Grieves et al., 1964.)

6.4.3.1 Processes Involving Enzyme Kinetics

Levenspiel (1979) and Powell and Lowe (1964) compared a chain of CSTRs with a tubular reactor by computing the biomass concentration obtained in the final stage of a cascade of five chemostats (with feedback). It was found that under certain circumstances a more complete utilization of growth-limiting substrate can be obtained in the series of CSTRs than is possible with the single CSTR. Similarly, Grieves et al. (1964) modeled a piston flow reactor with recycle, compared the performance with that of a CSTR with recycle (cf. Sect. 6.5), and concluded that if the objective of a process is cell mass production, the differences are slight, but if the objective is reduction in effluent substrate concentration, then a CPFR would be the optimum choice. With a large recycle factor and with efficient operation of the separator resulting in a large concentration factor, the differences in the critical residence time for CPFR and CSTR become much less pronounced, and the CPFR is able to provide a greater maximum production rate of microorganisms while yielding a considerably lower effluent substrate concentration than a CSTR. This behavior is illustrated in Fig. 6.28.

However, practical engineering experience with a tubular reactor, for example, for biological waste water treatment (F. Moser, 1977, 1980) shows that

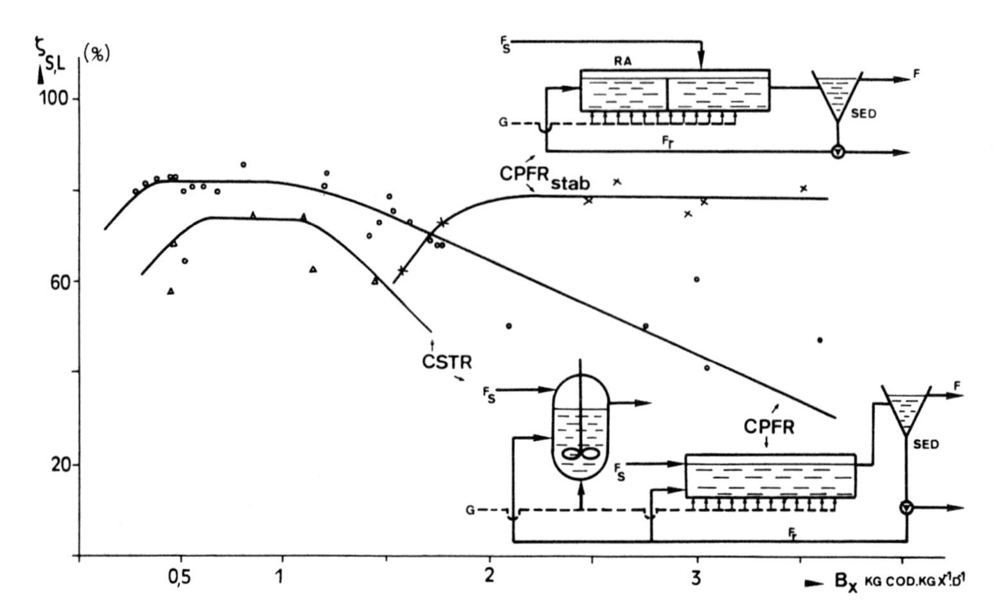

FIGURE 6.29. Conversion of substrate in the liquid phase ($\zeta_{S,L}$) as a measure of purification effect in waste water treatment versus mass loading rate per unit mass B_x in case of CSTR and CPFR operation: Compared are systems operating with one common sedimentation tank (SED), exhibiting only a small difference in conversion, and systems with separated SED, called "contact stabilization" or "sludge-reaeration," in which case the CPFR is superior to the CSTR. (Adapted from F. Moser, 1977.)

the CPFR is superior to a CSTR for *S* conversion only under certain conditions (see Fig. 6.29). Comparisons were carried out using a single sedimentation vessel (case A) and using separated sludge settling devices (case B). The theoretical advantage of a CPFR is experimentally realized only in case B, as this example is complicated by the phenomenon of biosorption (cf. Sect. 5.3.9). Biosorption does not play an essential role when fresh, unloaded sludge is used after sludge stabilization, as indicated on the inserts of Fig. 6.29 (case B).

For attaining a desired conversion, ζ_{ex}, one may calculate the necessary mean residence time from

$$t_{DCSTR} = \bar{t}_{CPFR} = -\int_{c_0}^{c_{ex}} \frac{1}{r}\, dc \equiv c_0 \int_{\zeta_0}^{\zeta_{ex}} \frac{1}{r}\, d\zeta \tag{6.77}$$

For the ideal CSTR, the concentration is constant ($c_R = c_{ex}$) and

$$\bar{t}_{CSTR} = \frac{c_{ex} - c_0}{r} \equiv -c_0 \frac{\zeta_{ex}}{r} \tag{6.78}$$

Equations 6.77 and 6.78 are generally applicable (for example, to sterilizations) and can be formulated for use in the case of enzyme kinetics involving substrate utilization as

$$\bar{t}_{\mathrm{CPFR}} = \int_{s_0}^{s_{ex}} \frac{Y(K_S + s)}{\mu_{max} \cdot s \cdot x} ds \tag{6.79}$$

or

$$t_{\mathrm{CSTR}} = \frac{(s_0 + s_{ex})\, Y(K_S + s)}{\mu_{max} \cdot s \cdot x} \tag{6.80}$$

Using $\bar{t} = V/F$, the volume necessary for a particular conversion in a CPFR or CSTR can be calculated. The result of this type of calculation is shown in Fig. 6.30 as the ratio of the volumes $V_{\mathrm{CSTR}}/V_{\mathrm{CPFR}}$ as a function of the K_S value (A. Moser et al., 1974). The various curves refer to differing values of s_0 and s_{ex}, that is, to differing conversions. From this graph, the differing influences of K_S may be seen. The K_S value determines whether a CPFR would be more advantageous than a CSTR for any particular continuous bioprocess. The range of numerical values for K_S is very low for enzyme reactions; even for microbial growth at low concentrations K_S is approximately 10 mg/l (Monod, 1942). In the literature there is a large variation in the K_S values given for bioprocesses from DCSTR and CSTR. The appearance of pseudokinetic parameters, especially K_S values, limits the reliability of predictions.

However, even at low K_S values, the CPFR continues to have a clear volume advantage (as may be seen in Fig. 6.30) when the attainment of either a high conversion or a low S concentration in the effluent ($s_{ex} \approx 20$ mg/l) is an important consideration (e.g., biological waste water treatment, (A. Moser, 1977; Wolfbauer, Klettner, and Moser, 1978). This consideration of low s_{ex} is not of such great significance in fermentation processes (except perhaps with expensive substrates), so the advantage of a CPFR is not nearly so pronounced (Finn and Fiechter, 1979).

When one looks at real processes, one can see that they actually show several different types of kinetic behavior. As shown in Fig. 6.31, reaction rates (or conversion) vary as a function of concentration: There are normal catalytic processes (curve a) and autocatalytic processes (curve b) that are dominant in microbial growth (cf. Equ. 2.7) (Levenspiel, 1972). Biotechnological processes are a combination of both (curve c). At the beginning they are autocatalytic; later, after the exponential growth phase, they shift to ordinary kinetics.

When one recalls the equation used to develop the CPFR and CSTR, one finds that it is possible to determine $\bar{t}$ graphically from a plot of $1/r$ versus c. From Equ. 6.78, $\bar{t}_{\mathrm{CSTR}}$ is represented by the area of the rectangle with sides equal to the final concentration, or the desired conversion. For $\bar{t}_{\mathrm{CPFR}}$, Equ. 6.77 applies, and the corresponding area is that under the curve. If one relies on Fig. 6.31 and chooses a suitable representation of the kinetics of one of the various processes (as shown in Fig. 6.32), $\bar{t}$ can be obtained. For a normal catalytic process such as is found in enzyme technology, the CPFR will always be advantageous (Lilly and Dunnill, 1971; Wandrey, Flaschel, and Schügerl, 1979). For most biotechnological processes, the optimum configuration thus

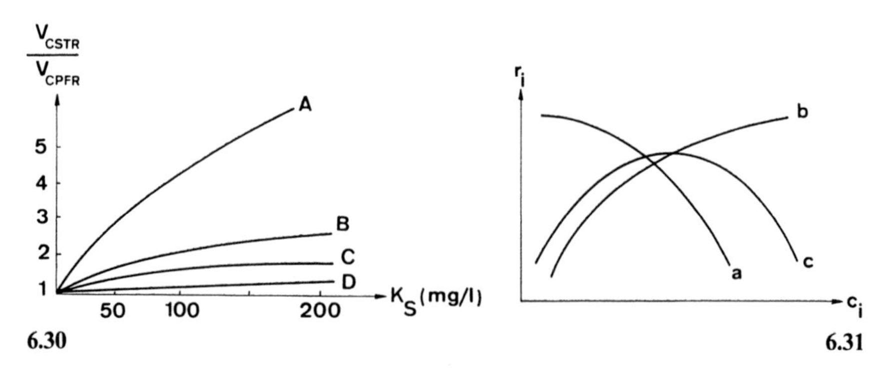

FIGURE 6.30. Calculated comparison of the reactor volume of a stirred vessel (V_{CSTR}) and a tube reactor (V_{CPFR}) (cf. Equs. 6.77 and 6.78) demonstrating the influence of the

case	s_0 [g/l]	s_{ex} [g/l]	ζ_S [%]
A	0.5	0.02	96
B	0.1	0.02	80
C	0.5	0.1	80
D	0.2	0.1	50

K_S value on the conversion of a continuous bioprocess with Monod kinetics (A. Moser et al., 1974).

FIGURE 6.31. Dependence of the reaction rate, the rate of formation or of consumption, r_i, on the concentration of component i, c_i, in (a) a normal catalytic process, for example, an enzymatic process, (b) an autocatalytic process such as pure biological growth, and (c) a biotechnological process such as fermentation or waste water treatment with combined growth and product formation.

would be combination of a CSTR followed by a CPFR in a second stage (Bischoff, 1966; Levenspiel, 1972; Topiwala, 1974; Yamane and Shimizu, 1982) or a cascade of CSTRs (Ricica, 1969a,b).

The graphical method of obtaining $\bar{t}$ shown in Fig. 6.32 can be used, but $\bar{t}$ may also be obtained with the aid of a mathematical representation of the kinetics (cf. Equs. 6.79 and 6.80). Changing variables from the concentration to the relative conversion (cf. Equ. 2.45) according to Topiwala (1974)

$$\zeta_{rel} = \frac{x}{Y \cdot s_0} \tag{6.81}$$

yields from Equ. 6.79

$$\begin{aligned} \bar{t}_{CPFR} &= \int_0^{\zeta_{ex}} \frac{1}{r_X} \cdot Y \cdot s_0 \cdot d\zeta \\ &= \int_0^{\zeta_{ex}} \frac{K_S/s_0 + (1-\zeta)}{\mu_{max}(1-\zeta)} d\zeta = \int_0^{\zeta_{ex}} f(\zeta) \cdot d\zeta \end{aligned} \tag{6.82}$$

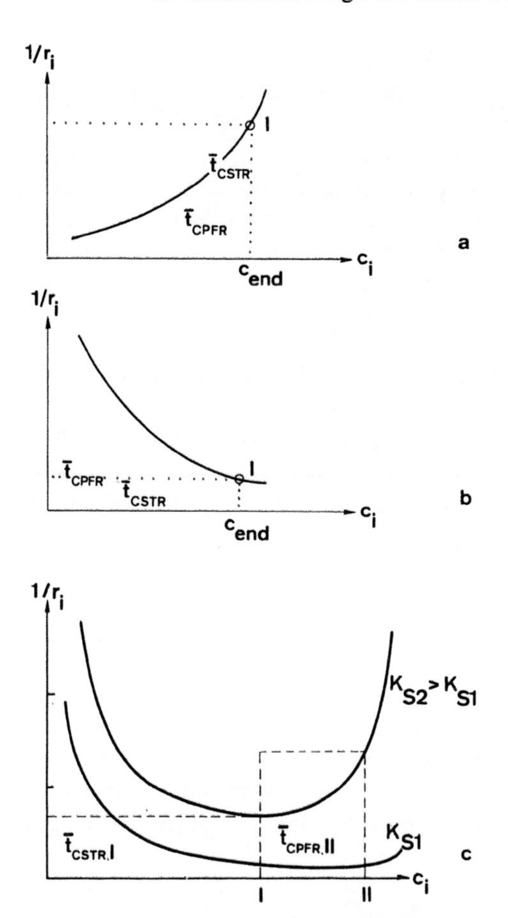

FIGURE 6.32. Plot of the reciprocal of the reaction rate, $1/r_i$, as a function of the concentration, c_i, as a measure of conversion. The processes shown in graphs a–c correspond to Fig. 6.31a–c, which may be used in evaluating the mean residence time for a continuous stirred vessel CSTR or a tubular reactor CPFR with Equs. 6.77 and 6.78. A combination of a CSTR with a CPFR is shown to be optimal in case c (stage I is a CSTR and stage II is a CPFR). (Figures 6.32a and 6.32b adapted from Levenspiel, 1972; Figure 6.32c adapted from Topiwala, 1974).

and from Equ. 6.80

$$t_{CSTR} = f(\zeta) \cdot (\zeta_{ex} - \zeta_0) \tag{6.83}$$

where

$$f(\zeta) = \frac{x_{max}}{r_X} \tag{6.84}$$

6.4.3.2 Fermentation Processes for Producing Secondary Metabolities

The kinetics of the production of secondary metabolites may be understood with the help of the "maturation time" concept, Equ. 5.136. For a continuous operation, the condition that $\bar{t} = t_M$ should be maintained, that is, the mean residence time spent by the suspended cells in the liquid phase, $\bar{t}$, must be equal to the ripening time, t_M.

As a consequence of this condition, in a CSTR all cells with $t < \bar{t}$ (cells with $t < t_M$) are not yet ripe enough for production. Only those cells with a lifetime $t \geq \bar{t}$ $(t \geq t_M)$ contribute to production. With the aid of the mathematical function that describes the residence time distribution in the liquid phase of a CSTR (Equ. 3.11), one can write

$$x(t) \equiv \bar{x} \equiv f(t) = D \cdot e^{-D \cdot t} \tag{6.85}$$

The product concentration can then be written from Equ. 5.136 as

$$\bar{p} = Y_{P|X} \cdot \int_{t_M}^{\infty} x(t) \cdot dt = Y_{P|X} \cdot \bar{x} \cdot e^{-D \cdot t_M} \tag{6.86}$$

In contrast, with a continuous tube reactor with $\bar{t} = t_M$, the maximum productivity can simply and directly be calculated from

$$\mathrm{Pr}_{max, CPFR} = D \cdot \bar{p} = D \cdot \bar{x} \cdot Y_{P|X} \tag{6.87}$$

A comparison of Equs. 6.86 and 6.87 easily shows that the productivity in a CPFR is greater than that in a CSTR (cf. Fig. 6.27).

An additional advantage of the CPFR in this case comes from the fact that some secondary metabolites (such as, for example, penicillin) are destroyed by an excessive amount of time in the reactor. This was indicated in Equ. 5.128. In a CSTR, all cells with $t > t_M$ and thus competent to produce the secondary metabolite would at the same time be subject to the negative effect of the destructive term, $-k_{P,d}$.

6.4.3.3 Sterilization Processes

The destruction of microbial cells follows formal first-order kinetics for the number of cells (cf. Equ. 5.267). The technical advantages of a CPFR process over a CSTR process can again be seen in situations where $t = t_{St}$ by considerations analogous to those of Fig. 6.27.

The T/t profile of a continuous reactor must be taken into consideration in any practical calculation for a sterilization, since k_d is temperature dependent (Equ. 5.3a). To determine the effective holding time at the sterilization temperature, the following equation is a basic consideration:

$$\ln \frac{N_1}{N_2} = k_{\infty} \cdot \int_{t_1}^{t_2} e^{-E_a/RT} dt \tag{6.88}$$

The integral for the heating and cooling phases may be graphically determined (Aiba, Humphrey, and Millis, 1973). Furthermore, in any real case, a deviation from ideal plug flow characteristics can be considered with the help of the dispersion model (Equ. 3.3); a reaction term representing the kinetics is added. Using a dimensionless number (Da_I, the Damkoehler number of first degree) for the kinetics

$$\mathrm{Da}_{\mathrm{I}} = \frac{k_{\mathrm{d}} \cdot L}{v} \tag{6.89}$$

and choosing the Danckwerts boundary conditions

$$\frac{dN}{dz} = 0 \qquad \text{at } z = 1 \tag{6.90a}$$

$$\frac{dN}{dz} + \mathrm{Bo}(1 - N) = 0 \qquad \text{at } z \to 0 \tag{6.90b}$$

one gets a solution of Equ. 3.3b with a reaction term for Bo → high expressed as

$$\frac{N(L)}{N_0} = \exp - \left(\mathrm{Da}_{\mathrm{I}} + \frac{\mathrm{Da}_{\mathrm{I}}^2}{\mathrm{Bo}} \right) \tag{6.91}$$

This equation can generate a nomogram to evaluate t_{St} as a function of Bo at various k_{d} values (Aiba et al., 1973).

A problem similar to sterilization occurs in the treatment of food with heat for purposes of preservation. However, the situation is more complex due to the multiple components present and the requirement for high quality standards (cf. Fig. 5.79). The technical advantages of a CPFR must be considered when new technology (such as, for example, the fluidized or spouted bed reactor) is introduced (Baxerres, Haewsungcharern, and Gibert, 1977).

6.4.4 Applications of CPFR

The CSTR cascade is often used as a substitute for CPFR behavior, and tubular reactors themselves are rarely found in bioprocessing. Exceptions are waste water treatment in oxidation ponds, river analysis (Metcalf and Eddy, 1972), sterilization technology, and fixed-bed reactors filled with immobilized enzymes for bioconversions (Lilly and Dunnill, 1972).

Tubular reactors as horizontal equipment have been summarized by Greenshields and Smith (1974) and by A. Moser (1983b, 1985b); applications vary from fermentation processing with flocs, including waste water treatment, to unconventional bioprocessing with immobilized cells, solid substrates, shear-sensitive tissue cell cultivation, and phototrophic organisms (Pirt et al., 1983). Figure 6.33 represents the advantages of a CPFR in case of waste water treatment by showing the drastic reduction in volume (F. Moser, 1977).

6.4.5 One-Phase (Liquid) Reactors with Arbitrary Residence Time Distribution and Micromixing

6.4.5.1 Graphical Methods

Reusser (1961) described a graphical procedure for predicting the yield of a nonautocatalytic process taking place in a continuous reactor in which the residence time distribution (RTD) may be freely chosen and the kinetics of the

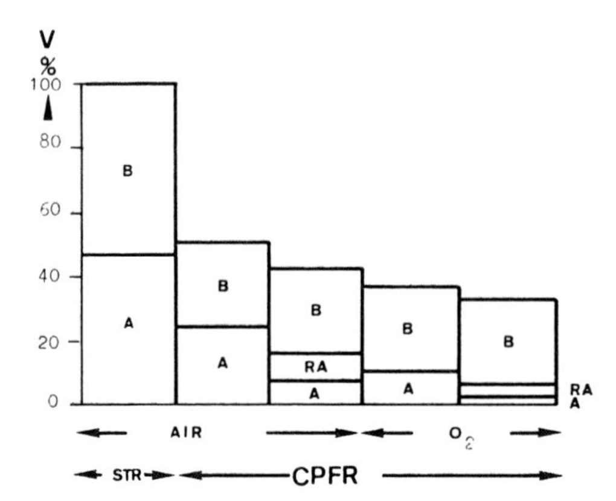

FIGURE 6.33. Demonstration of the advantage of CPFR used in biological waste water treatment by showing the volumes needed in aeration tank (A) and sedimentation tank (B). Two techniques are compared: the conventional tank (STR) with air and a tube reactor (CPFR) aerated with air or oxygen and using the concept of reaeration of the sludge (RA) (F. Moser, et al., 1979).

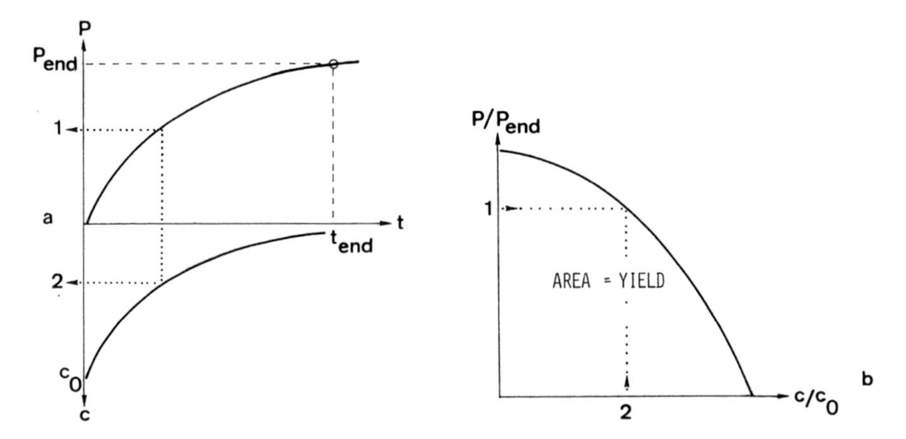

FIGURE 6.34. Calculation of the expected yield in a normal catalytic process: (a) Double plot of concentration/time for a discontinuous process $(p/p_{end})/(t/t_{end})$, and curve for a continuous reactor with variable residence time distribution. (b) Evaluation of yield. (Adapted from Reusser, 1961.)

equivalent discontinuous process are known. No mathematical formula is necessary in this method.

In Fig. 6.34a, two curves are plotted that show the time course of the discontinuous process kinetics and the residence time distribution in a continuous reactor of arbitrary RTD. Normalized values are used for ease of comparison (p_{end} = final concentration at time t_{end}; c_0 = total concentration of the pulse used to initiate the RTD measurement).

This plot can be used to eliminate that t axis: For each value of t the values from the kinetic and the RTD curve are read and are plotted against each other in a separate diagram (Fig. 6.34b). The area under the curve gives directly the yield to be expected in the steady-state condition of a continuous process.

This graphical method works only in cases where the reaction rate decreases

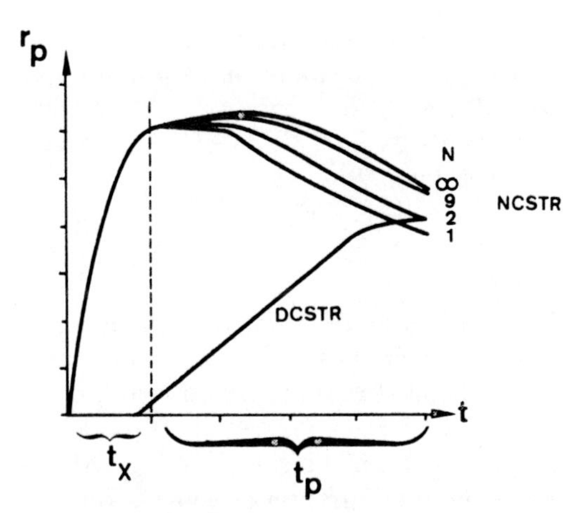

FIGURE 6.35. Predicted productivities of different novobiocin fermentation systems expressed in terms of means residence time of the overall process (cell growth in t_X and production in t_P) with variation of the number N in a cascade (NCSTR) compared with the productivity in a batch process (DCSTR). (From Reusser, 1961.)

with a decrease in the concentration (cf. Fig. 6.31a), and it is used in situations such as, for example, predicting the yield of an antibiotic-producing process. Predictions based on this concept are shown in Fig. 6.35 (Reusser, 1961).

For autocatalytic and biotechnological processes (Fig. 6.31b and c), a basically similar method can lead to the same goal; the solution, however, can only be obtained using mathematical formulas (cf. Equs. 6.92 and 6.94).

6.4.5.2 Calculation Methods

The simplest concept of a reactor is that each independent element of fluid volume behaves as a DCSTR. This is the case of micromixing with total segregation, $J = 1$ (cf. Fig. 3.1b). The fluid stream leaving the reactor is then an average of all the individual DCSTR elements, each of which is present for a different length of time in the entire system. In mathematical form, the output stream concentration is given by Equ. 6.92 (Danckwerts, 1958):

$$c_{ts} = \int_0^\infty c_{dc}(t) \cdot f(t)dt \tag{6.92}$$

where $f(t)$ is the residence time distribution measured with the pulse method (cf. Equs. 3.5 and 3.10) and c_{dc} is the concentration of the component in a DCSTR.

Using sterilization as an example (cf. Equ. 5.275), the concentration value for the DCSTR is expressed

$$N = N_0 \cdot e^{-k_d \cdot t_{St}} \tag{6.93a}$$

so that the sterilization process in, for example, a CSTR with a time distribution as in Equ. 3.11 may be calculated as

$$N_{St} = \int_0^\infty N_0 \cdot e^{-k_d \cdot t_{St}} \cdot \frac{1}{\bar{t}} \cdot e^{-t/\bar{t}} \cdot dt \tag{6.93b}$$

When the second extreme case of micromixing (from Fig. 3.1) is considered—maximal mixing (mm) with J = 0—the liquid volume elements in the reactor differ from one another in their "life expectancy" ($\lambda = 1 - \bar{t}$). From a differential conservation of mass equation (and simplification), one has (Zwietering, 1959)

$$\frac{dc}{dt} = r + \frac{f(t)}{1 - F(t)} \cdot (c - c_0) \tag{6.94}$$

For a first-order reaction, Equ. 6.94 is reduced to the form of Equ. 6.92.

These equations for total segregation and maximal mixing have been used for predicting microbial growth with Equs. 5.38 and 2.14a in a CSTR and in a CPFR (Fan et al., 1970, 1971; Tsai et al., 1969, 1971). The result is presented in Fig. 6.36; the plot is similar to Fig. 6.1b with $x_0 > 0$. In the region $0 < \mu_{max} \cdot \bar{t} < 2$, the concentration of x in a $CSTR_{mm}$ is higher than the corresponding concentration in a $CSTR_{ts}$ or in a CPFR. The influence of micromixing is clear here. Further, in agreement with Fig. 6.28, one sees that the utilization of substrate S beings earlier in a CPFR than in a CSTR and that somewhat higher X concentrations are present in a CPFR at a given value of K_S.

The question of micromixing is meaningless when $n_S = 0$; similarly, micromixing is meaningless when $n_S = 1$. Model studies of micromixing are therefore always undertaken using second-order chemical reactions ($n = 2$) (Danckwerts, 1958).

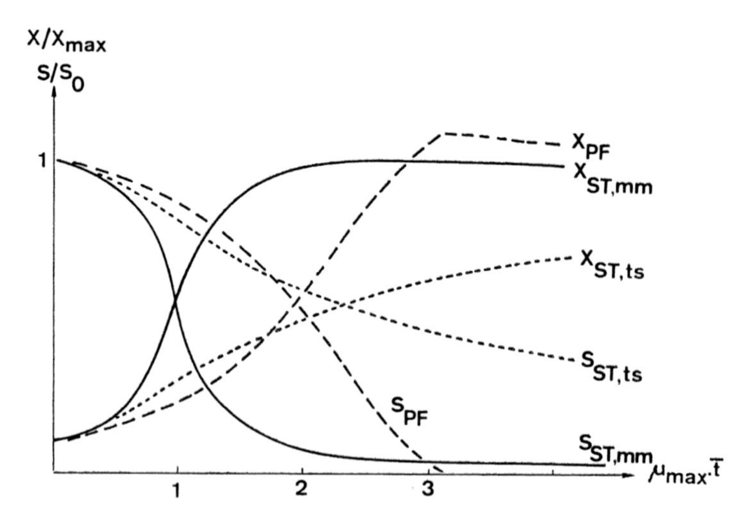

FIGURE 6.36. Plot of the dimensionless concentration of cell mass x and substrate s for a continuous culture as a function of the dimensionless mean residence time $\bar{t}$ as in Fig. 6.1b with $x_0 > 0$: Calculated comparison between a CSTR with maximum mixing (ST_{mm}) or one with total segregation (ST_{ts}) and a continuous plug flow reactor (PF), assuming Monod kinetics with a death rate k_d (Tsai et al., 1969).

6.5 Recycle Reactor Operation

In considering continuous operation, recycle loops are often included (Herbert,1964), with improved performance, that is, higher treatment rate, smaller vessel size, higher conversion. As a means for resource conservation and recovery as well as for environmental problems, recycle systems have become increasingly important.

6.5.1 Performance Equations of Recycle Reactors

6.5.1.1 The CSTR with Recycling of Cell Mass

Recycling cells from a CSTR effluent provides a means to continually inoculate the vessel and to add stability to the reactor, minimizing the effect of process perturbation. The productivity of a CSTR may be increased remarkably by recycling cells when the cells to be recycled are first concentrated by the factor $\beta = X_r/X$ in a sedimentation unit. With a recycling stream F_r, and with $r = F_r/F$, the conservation of mass equations are

$$\text{Change} = \text{inflow} + \text{recycle} - \text{outflow} + \text{growth}$$

and for x and s

$$\frac{dx}{dt} = D \cdot x_0 + r \cdot D \cdot x_r - D \cdot x_{ex}(1 + r) + \mu \cdot x \tag{6.95}$$

$$\frac{ds}{dt} = D \cdot s_0 + r \cdot D \cdot s - D \cdot s_{ex}(1 + r) - \frac{\mu \cdot x}{Y} \tag{6.96}$$

Hence in the steady state

$$\bar{s} = K_S \frac{\mu}{\mu_{max} - \mu} \tag{6.97}$$

$$\bar{x} = \frac{Y}{\mu}(s_0 - \bar{s}) \cdot D \tag{6.98}$$

and

$$x_{ex} = Y(s_0 - \bar{s}) \tag{6.99}$$

In contrast to the CSTR with no recycling, where $\mu = D$, the CSTR with cell recycling (rCSTR) realizes

$$\mu = D(1 + r - r \cdot \beta) \tag{6.100}$$

and in this case it is possible to operate with higher concentrations of cells

$$\bar{x}_{rCSTR} = \bar{x}_{CSTR}(1 + r - r \cdot \beta) \tag{6.101}$$

A schematic diagram of such a recycling system is shown in Fig. 6.37a. The equations are plotted in Fig. 6.37b together with the output of cells $D \cdot x_{ex}$; for

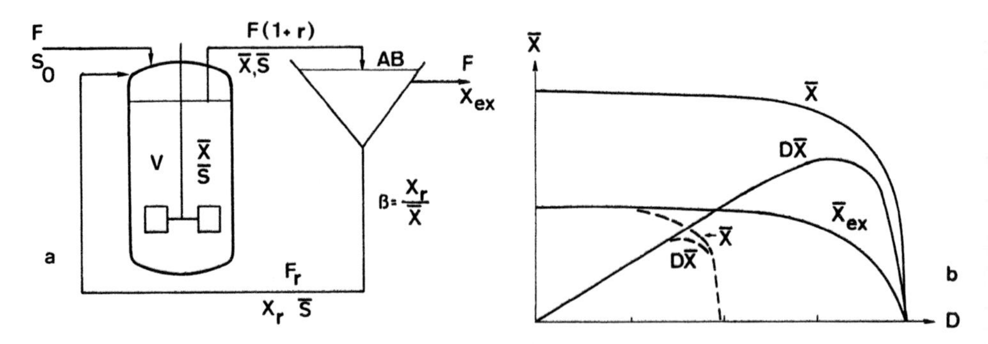

FIGURE 6.37(a). Flowchart for a continuous stirred vessel with recycling of cell mass (recycling stream F_r), with the cells concentrated in a settling vessel (concentration ratio, β). (b) Theoretical relation between cell mass concentrations, $\bar{x}$, and flow rate, D. For comparison, the value for a one-stage CSTR are shown (---).

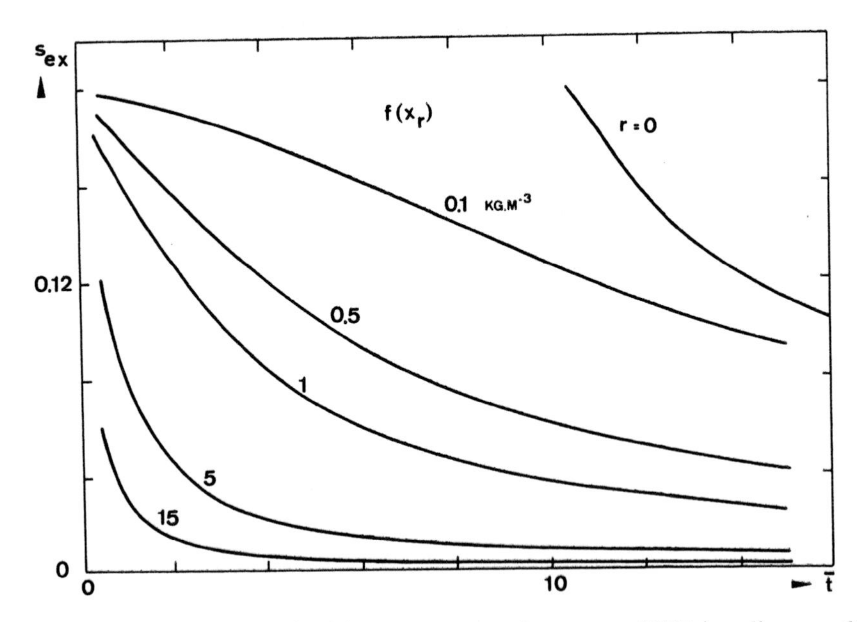

FIGURE 6.38. Effect of recycling biomass concentration x_r on a CSTR in a diagram of effluent substrate concentration s_{ex} versus mean residence time $\bar{t}$ (r = recycling ratio). (Adapted from Andrews, 1972.)

comparison, curves for cell mass concentration and output of cells in a single CSTR with no recycling are also plotted (dotted lines). As can be seen, a dilution rate greater than maximum growth rate may be employed, increasing overall productivity. The critical residence time is

$$\bar{t}_{crit} = \frac{1}{D_{crit}} = \frac{1 + r - r \cdot \beta}{\mu_{max}} \tag{6.102}$$

An estimate of the effect of cell mass concentration and recycling on process economics may be obtained by plotting effluent substrate concentration for several different values of x_r, as shown in Fig. 6.38. Increasing x_r has the effect of minimizing s_{ex} and also of making s_{ex} less sensitive or more stable with respect to variations in the flow rate of constant-volume reactors with residence times considerably below the washout value (Andrews, 1971).

6.5.1.2 Cascade of Reactors with Cell Recycling

For the case of cascade of reactors with cell recycling, Powell and Lowe (1964) derived a performance equation that is analytically intractable. Nevertheless, a simple formula of the critical dilution rate D_{crit} can be calculated in the case of a cascade of CSTRs as

$$D_{crit} = \left[\frac{1 - r}{(1 - r \cdot \beta^{1/N})N}\right] \mu_{max} \frac{s_0}{K_S + s_0} \tag{6.103}$$

illustrating how the critical dilution rate depends on the degree of feedback r, the number of stages N, and the concentration of limiting substrate.

Toda and Dunn (1982) emphasized that a combination of a backmix and a plug flow fermenter with recycle streams provides better performance than does a single CSTR for the continuous production of a substance that depends on the maturity of growing cells (see Equ. 5.136). A typical example is shown in Fig. 6.39, and it illustrates that the productivity of cell mass (Fig. 6.39a), of

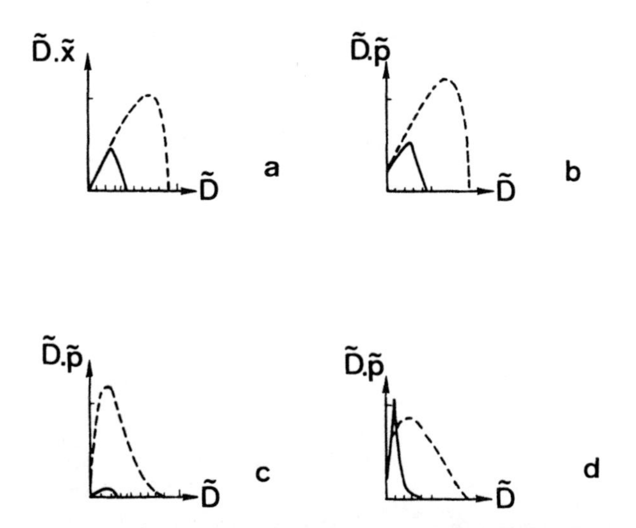

FIGURE 6.39. Productivity of fermentation product in combined systems of CSTR and CPFR (cf. Fig. 3.10d) in comparison with a single CSTR (----) in four different cases: (a) cell mass production following Monod kinetics, (b) Luedeking–Piret-type product formation, (c) productivity of repressible products, and (d) maturation time-dependent products. (Toda and Dunn, 1982.)

products formed according to Luedeking–Piret logistic kinetics (b) (Equ. 5.122) and of repressible substances produced (c) according to the kinetic model of van Dedem and Moo-Young (1973) is still greater in a CSTR. These facts represent additions to the concept shown in Fig. 6.32.

6.5.1.3 Plug Flow Reactors with Recycling

CPFRs containing microbial flocs can only be operated continuously on a recycling basis or with a CSTR as an inoculum reactor; otherwise washout will occur. For a CPFR with recycling of nonconcentrated cell mass, a performance expression was given by Levenspiel (1979). For $x_0 = 0$

$$\mu_{max} \cdot \bar{t} = (r + 1)\left(\frac{K_S}{s_0} \ln \frac{s_0 + r \cdot s}{r \cdot s} + \ln \frac{1 + r}{r}\right) \tag{6.104}$$

When material is to be processed to some fixed conversion ζ in a CPFR with recycling, a particular optimum recycle ratio r should exist. At this point reactor volume is minimized. The final solution is yielded by putting $\partial x/\partial r = 0$ in Equ. 6.104:

$$\frac{K_S}{s_0} \ln \frac{s_0 + r \cdot s}{r \cdot s} + \ln \frac{1 + r}{r} = \frac{1 + \mathrm{r}}{r} \cdot \frac{K_S}{s_0 + r \cdot s} + \frac{1}{r} \tag{6.105}$$

This equation can be solved by trial and error. The optimum r is found to be a function of K_S/s_0 and s/s_0. A graphical procedure for the determination of the optimum r is shown in Fig. 6.40 (Levenspiel, 1979). One has to try different values of $\zeta_{i,1}$ until the two dotted areas are equal. This means that the r^{-1} at the feed entering the reactor is equal to the average r^{-1} in the reactor, or mathematically

$$-\frac{1}{r_i}\bigg|_{\zeta_{i,1}} = \frac{\int_{\zeta_{i,1}}^{\zeta_{i,ex}} d\zeta_i/-r_i}{\zeta_{i,ex} - \zeta_{i,1}} \tag{6.106}$$

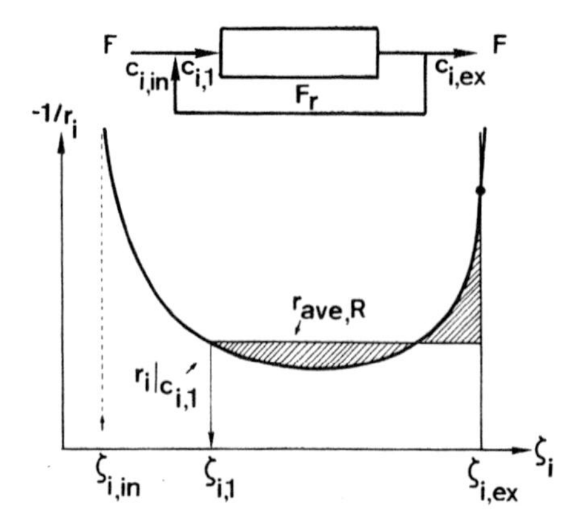

FIGURE 6.40. Optimization of conversion in a recycle reactor in a plot of reciprocal rate $1/r_i$ versus conversion ζ_i (cf. Fig. 6.32) by setting the shaded areas equal to each other according to Equ. 6.106 (Levenspiel, 1979).

The value of r corresponds to the concentration according to (cf. Fig. 6.40)

$$c_{i,1} = \frac{c_{i,0} - r \cdot c_{i,ex}}{1 + r} \tag{6.107}$$

At too high a recycle ratio, the average rate is higher than the rate in the feed entering the reactor, and vice versa. In Fig. 6.41 the optimum value r_{opt} is plotted as a function of K_S/s_0 and s/s_0. For a CPFR with recycling of concentrated cell mass, Grieves et al. (1964) derived a general equation that can be used for computer simulation of the basic behavior:

$$-\mu_{max} \cdot \bar{t} = (1 + r)\left\{\frac{(1 + r)K_S}{[r \cdot \beta(s_0 - s)/1 + r - r \cdot \beta] + s_0 + r \cdot s} \times \ln\frac{s \cdot r \cdot \beta}{s_0 + s \cdot r} - \ln\frac{1 + r}{r \cdot \beta}\right\} \tag{6.108}$$

As the critical residence time is approached, s increases much more sharply

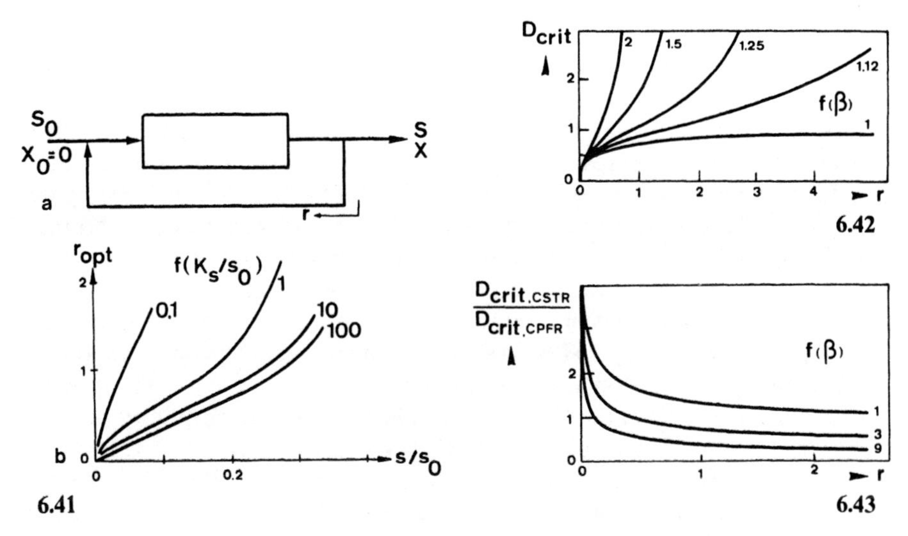

6.41

6.42

6.43

FIGURE 6.41. Graphical representation of the trial-and-error solution of Equ. 6.105 showing the dependence of optimum recycling ratio r_{opt} in a CPFR with recycling on substrate concentration s/s_0 as a function of varied K_S value. (Adapted from Levenspiel, 1979.)

FIGURE 6.42. Effect of biomass concentration on the washout of a CPFR shown in a graph of critical dilution rate D_{crit} in dependence on recycling ratio r at varied values of concentration factor β (Atkinson, 1974, with permission of Pion, London.)

FIGURE 6.43. Comparison of the washout flows of a CSTR and CPFR by showing the ratio of critical dilution rate D_{crit} depending on recycling ratio r with variation of the concentration factor β (Atkinson, 1974, with permission of Pion, London).

and x_{ex} decreases much more sharply than is the case for the CSTR (Fig. 6.26). Equation 6.108 indicates the general improvement in performance with increased recycling stream. The relationship between washout flow and recirculation can be deduced from Equ. 6.108 be setting s_0/s equal to unity (corresponding to zero conversion). According to Atkinson (1974)

$$D_c = \frac{1}{\bar{t}_{crit}} = \frac{\mu_{max}}{(1 + r)\ln[(1 + r)/\beta \cdot r]} \cdot \frac{s}{K_S + s} \tag{6.109}$$

The improvement in performance when β is increased is shown in Fig. 6.42, where both factors of the recycling ratio r, and β, the cell concentration factor, influence the performance of CPFR. A comparison of washout flow can be obtained from Equs. 6.102 and 6.109 (Atkinson, 1974):

$$\frac{D_{c,CSTR}}{D_{c,CPFR}} = \frac{1 + r}{1 + r - r \cdot \beta} \cdot \ln \frac{1 + r}{\beta \cdot r} \tag{6.110}$$

which is depicted in Fig. 6.43.

6.5.2 Applications of CRR

In fermentations and other bioprocesses, process stream recycling is intimately concerned with enhanced conversion. Such recycling schemes usually seek to reduce operating costs by increasing capital investment. The three major possibilities where recycling loops can be expected in the fermentation industry involve (a) processes employing gaseous substrates, including pure oxygen; (b) processes employing spent medium component and/or process water reutilization; and (c) processes designed to improve fermenter productivity by using elevated cell concentrations (Hamer, 1982).

In biological waste water treatment, the recycling of sludge has long been used (Metcalf and Eddy, 1979). As a consequence of recycling, reactor stability and substrate conversion are increased. Other important cases of the application of recycling operation are known from process design optimization for continuous ethanol production (e.g., Cysewski and Wilke, 1978). Generally, various systems for cell recycling are available: internal filtration, internal sedimentation, monostream, and external recirculation (Pirt, 1975). Recently a computer simulation study of ethanol fermentation with cell recycling showed several interesting facts (Lee et al., 1983). The cell growth equation with an inhibition term (cf. Sect. 5.3.5) predicts that the relative productivity of the recycle reactor will increase drastically (400–500%) as the bleed stream/feedstream ratio B/F decreases until the cell concentration in the reactor reaches the maximum value x_{max}.

Similar optimization studies were experimentally verified for ethanol production with *Zymomonas mobilis* (e.g., Charley et al., 1983). A two-stage CSTR system with cell recycling in the second stage resulted in high final alcohol concentrations (>100 g/l) at quite high overall volumetric productivity. Moreno and Goma (1979) used a cascade of eight CSTRs in a series and a yeast strain; with yeast the increase in cell concentration by recycling is not accompanied by a corresponding increase in ethanol productivity due to irreversible inhibitory effects on the recycled cells. Similar trends were observed with *Z. mobilis* by another group (Boks and Eybergen, 1981), where biomass production decreased while ethanol productivity did not change markedly.

Last, but not least, it should be mentioned that some other newly designed bioreactors are based on the concept of a CRR. A tubular loop reactor with a G-phase separator chamber or a cyclone was installed successfully in bench and pilot scale (Russell et al., 1974; Ziegler et al., 1977, 1980). Another loop reactor that also realizes plug flow in the L phase to some extent is the horizontal circular ring reactor (Herzog et al., 1983; Laederach, 1978). The cycle tube cyclone reactor is a vertical construction of a loop reactor that is a highly efficient aeration device with low power consumption (Liepe et al., 1978; Ringpfeil, 1980). These configurations are compared in a recent review, together with the fields of application (A. Moser, 1985b).

Finally, several other contributions to the literature suggest that the principle of recycle operation is being acknowledged in bioprocess research and development (Adler and Fiechter, 1983; Blenke, 1979; Bull and Young, 1981; Constantinides et al., 1981; Lippert et al., 1983; Seipenbusch and Blenke, 1980; Stieber and Gerhardt, 1981a,b).

6.6 Gas/Liquid (Two-Phase) Reactor Models in Bioprocessing

The formulation of a gas/liquid reactor model may be done by considering the analogous situation in chemical process engineering (cf. Sect. 4.5.4, especially Fig. 4.46). The importance of OTR enhancement in analogy to chemical processing (e.g., Nagel et al., 1977, 1978) can be stressed by showing Fig. 6.44. For a first-order reaction, the relative conversion ζ_{rel} is a function of both factors, the enhancement η_{TR} (Ha, cf. Equs. 4.115 and 4.112) and the ratio of hinterland Hl (cf. Equ. 3.59). In the case of a slow reaction ($\eta_{TR} = 1$), bioreactors with a high hinterland will give satisfactory conversion (e.g., bubble columns). Stirred tanks are more flexible ($10^2 < \text{Hl} < 10^3$) but do not have enough interfacial area to achieve high conversion in the case of fast reactions. Only when using bioreactors with low Hl values (high a values), for example, jet nozzle reactors and thin-layer reactors, can sufficient conversion be reached with "fast" reactions ($\eta_{TR} > 1$; Ha > 0.3). Thus, the choice of optimum bioreactor can only be made by a complete analysis and full understanding of this type of interaction between physics (transports in bioreactor) and biology (kinetics of metabolic reaction).

It was shown with the aid of modeling that theoretically only in aeration

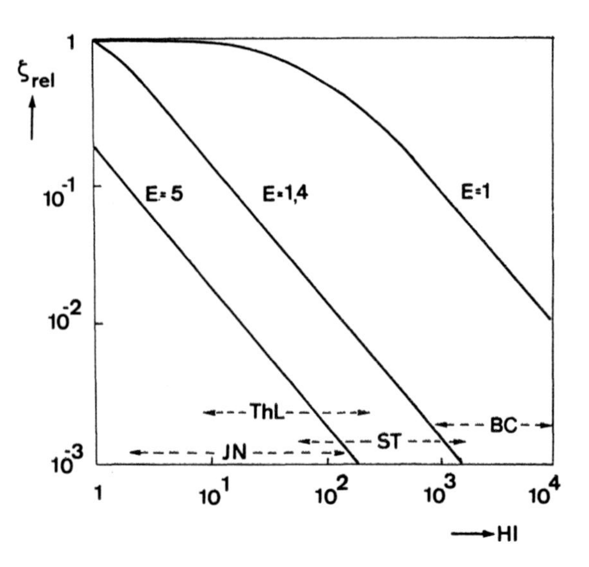

FIGURE 6.44. Dependence of the relative conversion ζ_{rel} on the degree of hinterland Hl in a G|L reactor (see Equ. 3.59) [bubble column, BC; stirred vessel, ST; thin-layer reactor, ThL; injector (jet nozzle) reactor, JN] for an aerobic bioprocess with various fast reactions, as represented by the oxygen transport enhancement factor, E (see Equ. 4.115). (From Nagel et al., 1972.)

systems with low k_{L1} values at high x and high $q_{O,max}$ in the case of no S limitation and low m_S in the region of $n_O = 1$ could bioenhancement contribute significantly to OTR (A. Moser, 1980b). The results of this modeling are in agreement with trends reported in the literature and are supported by recent investigations using the absorption system carbon-dioxide-phosphate buffer–carbonic anhydrase for the measurement of k_{L1} and a, which indicate that absorption enhancement is completely in line with ordinary theory (Alper et al., 1980b).

It must be stated here that it is hard to verify this effect in a reliable experimental method due to the difficulty of operating with microbial cells at fixed kinetics and of measuring k_{L1} and a separately (cf. Fig. 3.43) Küng and Moser (1988).

6.7 Biofilm Reactor Operation

Due to the adhesion capacity of microbes, biofilm formation must be taken into account in river analysis as well as in the chemostat or fixed beds (percolating or trickling filter). The significance of microbial films in fermenters has been extensively studied and reviewed (Atkinson, 1973, 1974; Atkinson and Fowler, 1974; Atkinson and Knights, 1975; Characklis, 1981; Harremoës, 1978).

6.7.1 POTENTIALITIES OF BIOFILM REACTORS

Process engineering possibilities of biofilms have been compared with those of floc processing in Table 3.1. As a consequence of these properties, process development using cell support systems seems to be promising (Atkinson and Kossen, 1978; Atkinson et al., 1980). The potential advantages are as follows (Moser, 1985a):

- As washout is not possible, high throughputs can be realized in reactors. Biomass holdup is independent of throughput.
- Due to the adhesion of cells on support surfaces, high biomass concentrations can be realized.
- Due to the adhesion to solid particles, the principle of fluidization can be introduced to bioreactor operation. This leads to the advantage of better transport processes, especially at L|S interface (turbulence) and in the S phase (abrasion or scouring).
- Sometimes the appearance of transport limitations or concentration gradients may be advantageous, as in these cases:
  a. With mixed populations—different species may occur at various depths within the particle, for example, nitrifiers near the surface and denitrifiers toward the center (Eggers and Terlouw, 1979).
  b. When using cells that must be "mature" before the product is produced (cf. Equ. 5.136)—retention of them within a particle may be beneficial, for example, citric acid and secondary metabolite production.
  c. When using S-inhibiting media, the most favorable S concentration may occur toward the center of the film.
- A special advantage of biofilm reactor operation in practice comes from the fact that the performance is independent of film thickness when thick films are used. However, as in thick biofilms generally S and O_2 limitations occur, which result in a loss of viability of the cells. Thereby, the adhesive bond to the support surface is weakened so that sloughing will occur, which will disturb the reactor performance.
- As a consequence of the foregoing problem, the control of biofilm thickness is of central importance. The thickness can be controlled by means of mechanical scraping or abrasion by friction as in the self-regulation effect at high hydrodynamic stress (cf. Sect. 3.6).
- A unique benefit of biofilm reactors for research purposes is the fact that due to the distinct separation between the L and the S phase (difference in density $\rho_S - \rho_L$), high relative velocities can be realized. As a result, external transport limitation can be excluded or easily studied simultaneously with internal transport limitations in the case of uniform and controlled biofilm thickness. In this respect, biofilm reactors are superior to the conventional STR.

Last, but not least, pellet processing should be mentioned here (Metz and Kossen, 1977; cf. Sect. 5.8.2). Usually investigators consider pellet formation as an undesirable feature, leading to inhomogeneous mycelium. Nevertheless, controlled pellet formation offers the advantage that pellet suspensions have a much lower viscosity than a filamentous broth (pulp). This leads to pellet processes that are energetically more efficient than a traditional pulp-like process. This was shown with a thermodynamic approach to calculation of the thermodynamic efficiency (see Equ. 2.29). The activity loss in the center of a pellet is very well compensated for on a macroscopic level by the reduced power requirement for such a process (Roels and van Suijdam, 1980).

6.7.2 Performance Equations of Biofilm Reactors

Complicated models for film bioreactors may be simplified when pseudohomogeneity can be assumed (see Sect. 4.2)—that is, when the S-concentration profile is flat throughout the cross section of the trickling filter.

6.7.2.1 Pseudohomogeneous Reactor Modeling of Fixed Bed Reactors

6.7.2.1.1 *Kinetics as the Rate-Determining Step*

There is a concentration gradient present in the fluid flowing over the length of the reactor at a rate F [m^3/h]. This requires dealing with the mass conservation for S using differential volume elements, as in Fig. 3.38 and Equ. 3.95 (with no recycling).

$$F(s + ds) = F \cdot s - r_S \cdot dV_X \tag{6.111}$$

or

$$F \cdot ds = -\frac{\mu}{Y} \cdot x \cdot dV_X \tag{6.112}$$

The important value for X is not the total mass of X; rather it is the active mass of X in the external surface of the film. The volume of the active biomass, V_X, may be calculated as the product of several factors: the differential thickness, dz; the cross-sectional area, A; the specific surface area of the falling body, a_S [m^2/m^3]; and the film thickness, d

$$dV_X = a_S \cdot d \cdot A \cdot dz \tag{6.113}$$

The final equation for S use as a function of depth may be written

$$-\frac{ds}{dz} = \frac{\mu}{Y} \cdot a_S \cdot X \cdot d \cdot A \cdot \frac{1}{F} \tag{6.114}$$

Using Monod kinetics, the mass conservation becomes

$$F \cdot ds + \frac{\mu_{max} \cdot x \cdot s \cdot a_S \cdot d \cdot A}{Y(K_S + s)} dz = 0 \tag{6.115}$$

With constant film thickness and constant K_S, the integration with bounds $z = 0$ at s_0 and $z = z$ at s_{ex} gives Equ. 6.116:

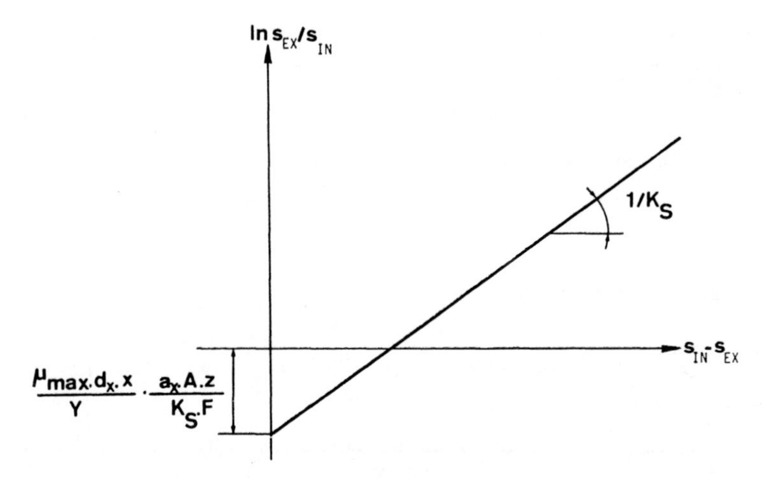

FIGURE 6.45. Graphical method of parameter estimation in case of Equ. 6.116 representing a pseudohomogeneous approach to biofilm processing, Kornegay and Andrews (1969) and Kornegay and Andrews (1968).

$$\ln\frac{s_{ex}}{s_{in}} = \frac{s_{in} - s_{ex}}{K_S} - \left(\frac{\mu_{max}\cdot d_x \cdot x}{Y}\right)\frac{a_x \cdot A \cdot z}{F \cdot K_S} \tag{6.116}$$

This solution describes the S consumption in the fixed bed with essentially plug flow behavior and shows that a plot of $\ln(s_{ex}/s_{in})$ versus $s_{in} - s_{ex}$ should give a straight line with a slope of $1/K_S$ and an intercept as indicated in Fig. 6.45 in agreement with Equ. 6.116. Thus, the parameters K_S and the term inside the parentheses of the second term on the right side of Equ. 6.116 can be evaluated from bench- or pilot-scale data in which s_{in} is varied at a fixed flow rate. Both values are pseudokinetic parameters, which depend upon liquid flow rate. Nevertheless, Equ. 6.116 can be used to successfully predict effluent conditions for various values of F, s_{in}, and total length of the fixed bed. Modifications of this solution are needed for the case where d_x is not constant, which will occur at substrate concentrations below about 300 mg/l. Here d_x will increase linearly with s (Harris and Hansford, 1976). Thus, instead of Equ. 6.116, the S balance must be modified, leading to a somewhat different solution given by

$$\ln\frac{s_{ex}}{s_{in}} = K_S\left(\frac{1}{s_{ex}} - \frac{1}{s_{in}}\right) - \frac{\mu_{max}\cdot d_x^0 \cdot x}{Y}\cdot\frac{a_x \cdot A \cdot z}{F} \tag{6.117}$$

where d_x^0 is a proportionality constant ($d_x = d_x^0 \cdot s$).

6.7.2.1.2 *Transport Factors as the Rate-Limiting Step*

A differential conservation of mass equation (Equ. 6.118) is found in the case where S utilization is determined by the rate of substrate transport through the liquid–solid interface (k_{L2} in Fig. 4.28). The latter is the more likely case with trickling filters

$$F \cdot ds = -k_{L2} \cdot a_S(s_L - s_S^*) \cdot A \cdot dz \tag{6.118}$$

With $s_S^* \rightarrow 0$, integration of Equ. 6.118 leads to

$$\ln \frac{s_{ex}}{s_0} = -\frac{k_{L2} \cdot a \cdot z}{F/A} \tag{6.119}$$

According to La Motta (1976)

$$k_{L2} = k_{L2}^0 \cdot v_{S,L}^{0.7} \tag{6.120}$$

with $v_{S,L}$ the surface flow velocity of the liquid phase, which is equal to (F/A), so that

$$\frac{s_{ex}}{s_0} = \exp\left[-\frac{(k_{L2}^0 \cdot a_S)z}{v_{S,L}^{0.3}}\right] \tag{6.121}$$

This result, which is based on a reactor model that assumes a mass transport limitation, is substantiated by empirically based model equations that are formally first order (cf. Equ. 5.159). These have the same form as Equ. 6.121, in which $(k_{L2}^0 \cdot a_S)$, the apparent value k_{app}, and the power of v_{SL} appear as experimentally determined coefficients.

It is interesting to note that many empirical models have been developed for fixed bed bioreactors, and that these models agree with this equation (e.g., Eckenfelder, 1966; Oleszkiewics, 1976, 1977). A practical design equation for trickling filter processes was developed by Kong and Yang (1979) by combining the approach of Kornegay and Andrews with the practical aspect of sludge age (sludge retention time) suggested by Kincannon and Sherrard (1974) and applied by Bentley and Kincannon (1976). Sludge age θ_x, which can easily be determined by any of the two independent process variables of the hydraulic loading rate B_v and S or by the organic loading rate B_x and S, can be employed to give

$$\frac{1}{\theta_x} = \frac{s_{in} - s_{ex}}{x} \cdot \frac{F}{V} Y - k_d \tag{6.122}$$

This equation can be plotted to give the values of Y and k_d from the slope and the intercept. All performance characteristics of trickling filters were shown to be a function of θ_x (e.g., removal, production, and volume index of sludge).

A mathematical model for percent removal of a pure, nonadsorbable, biodegradable substrate in a submerged biological filter was developed by Jennings et al. (1976), again using a Monod-type relation for biokinetics. Problems in the interpretation of removal kinetics are encountered in this case (Ottengraf, 1977).

Another pseudohomogeneous model of biofilm reactor operation was derived by Atkinson and Davies (1972) on similar assumptions to quantify the behavior of the CMMFF (cf. Fig. 3.42). For this case of a completely mixed reactor system, the balances on X and S, followed by algebraic combination and rearrangement, lead to the performance characteristics of fermenters containing immobilized cells:

$$A\left(\frac{s}{s_{in}}\right)^2 + B\left(\frac{s}{s_{in}}\right) - 1 = 0 \tag{6.123}$$

with A and B being coefficients defined, in the case where no thick film is present, as

$$A = \frac{s_{in}}{K_S}\left(1 - \frac{\mu_{max} \cdot V}{F}\right) \tag{6.124a}$$

and

$$B = 1 - \frac{s_{in}}{K_S} + \frac{\mu_{max}}{F}\left(\frac{s_{in}}{K_S} + \frac{x_{in}}{Y \cdot K_S}\right) \tag{6.124b}$$

The solution of Equ. 6.123 takes the form

$$\frac{s}{s_{in}} = f\left(\frac{F}{V \cdot \mu_{max}}, \frac{s_{in}}{K_S}, \frac{x_{in}}{Y \cdot K_S}\right) \tag{6.125}$$

and means that the conversion efficiency depends upon a dimensionless flow rate, inlet concentration, and biomass holdup. The algebra of Equ. 6.125 has been extended to include both internal and external transport limitation (Atkinson and Mavituna, 1983). The setup of true heterogeneous models for this situation of interacting biokinetics and external and internal transport was broadly discussed in Sects. 4.5 and 5.8.

The final equations for the solution of this interacting system are sometimes given in the form of the pseudohomogeneous rate Equ. 6.121, substituting for the factor $(k_{L2}^0 \cdot a_{L|S})$ the rate constant for biological reaction k_r multiplied with the effectiveness factor η_r (cf. Equ. 4.51)

$$\frac{s_{ex}}{s_{in}} = \exp - \frac{\eta_r \cdot k_r \cdot z}{v_{SL}} \tag{6.126}$$

The effectiveness factor η_r varies only with the biofilm thickness and the support particle size at constant voidage of the fluidized bed. This leads to the conclusion that an optimal biofilm thickness exists for fluidized bed biofilm reactors (Kargi and Park, 1982). The effect of biofilm thickness on the performance of a fluidized bed reactor was also examined by Shieh (1981), who showed the same results. Neither maintenance of thicknesses with 10 to 100 μm nor the highest biomass will be beneficial to optimum reactor operation.

Techniques for estimation of biofilm thickness were recently summarized by Charaklis et al. (1982). An ideal film thickness equal to the penetration depth of the limiting substrate or oxygen was stressed by Howell and Atkinson (1976). A capillary microelectrode technique has been developed by Bungay et al. (1969) for this purpose. The effect of growth rate on the biofilm buildup has been observed by Molin et al. (1982). All of these developments may help to elucidate problems of biofilm reactor operation.

The pseudohomogeneous biofilm model, given in Equs. 6.116 and 6.121, can be readily adapted to quantify the behavior of a rotating biological disk reactor. The setup of a similar balance equation for S using Monod-type kinetics yields for the case of rds kinetics

$$\frac{s_{ex}}{s_{in}} = \frac{1}{1 + (\mu_{max} \cdot d_x \cdot x \cdot a_x \cdot \bar{t})/[Y(K_S + s_{ex})]} \tag{6.127}$$

If rds = L|S transport, then, similarly to Equ. 6.121, the solution of the balance equation is

$$\frac{s_{ex}}{s_{in}} = \frac{1}{1 + k_{L2} \cdot a_{L|S} \cdot \bar{t}} \tag{6.128}$$

for each stage, where $\bar{t} = V/F$, the space time.

Wu et al. (1980) analyzed and modeled a biodisk system with six stages, and showed that the conversion efficiency is directly associated with $\bar{t}$ and is inversely affected by s_{in}, temperature T, and stage number N:

$$\frac{s_{ex}}{s_{in}} = 14.2 \cdot \frac{B_V^{0.55}}{\exp(0.32N)} \cdot s_0^{0.68} T^{0.24} \tag{6.129}$$

6.7.2.2 Unified Performance (Heterogeneous) Model of Fixed Bed Biofilm Reactor

Pseudohomogeneous models of biofilm reactors will fail in cases where deeper understanding and interpretation are needed. There are three basic requirements for a heterogeneous model for biofilm reactors (Rittmann, 1982):

1. Quantification of r_S, S utilization rate, in the case of interactions between reaction and internal or external transports. This problem of solving Equs. 4.76 and 4.101 simultaneously was discussed in Sect. 4.5. These solutions give steady-state S utilization but do not consider growth or decay of the biofilm. Adequate modeling of biofilm reactor operation must incorporate this fact.
2. Quantification of r_X, the growth. A specific decay coefficient k_d (cf. Equ. 5.76) has been shown to give experimentally valid results with fluidized beds (Andrews and Tien, 1982). Rittmann and McCarty (1980) have defined a steady-state biofilm as one having no net growth or decay for the entire biofilm depth. This dynamic constant thickness can be calculated by equating growth due to substrate flux n_S' with the maintenance decay of the entire biofilm. Thus

$$d_f = \frac{n_S' \cdot Y_{X|S}}{k_d \cdot x_f} \tag{6.130}$$

The significance of this prediction was questioned by Arcuri and Donaldson

(1981). Supplementary to this effect of maintenance decay, biomass is also lost from the biofilm through the action of shearing, in which the stress of water flowing past the biofilm pulls away part of the film. This effect can formally be quantified with an apparent decay coefficient k_τ [h^{-1}] (Rittmann, 1982) to be incorporated into Equ. 6.130 if shear losses become significant as, for example, in a fluidized bed biofilm reactor (FBBR). A key concept of the steady-state solution is the existence of a threshold concentration s_{min}, below which no significant steady-state biofilm activity occurs (see active depth δ_{crit}, Equ. 4.89). At bulk concentrations greater than s_{min}, n_S' and d_f are determined by simultaneously solving Equs. 4.77, 4.101, and 6.130 (Rittmann and McCarty, 1980). At high enough S, the steady-state flux becomes equal to the flux into a deep biofilm, where the reaction order $0.5 < n_S < 1$ (cf. Fig. 5.73).

3. Formulation of a proper reactor model and incorporation of kinetic expressions for r_S and r_X. Using the fundamental law of conservation of mass (cf. Equs. 4.76 and 4.101), the general partial differential equation of S concentration in a control volume is

$$dV \cdot \varepsilon_L \frac{\partial s}{\partial t} = -v_z \frac{\partial s}{\partial z} dV + D_{eff} \frac{\partial^2 s}{\partial z^2} dV - (\varepsilon_L \cdot r_S \cdot dV + a_f \cdot n_S' \cdot dV) \quad (6.131)$$

where ε_L is the bed voidage or porosity of the medium and the terms inside the parentheses represent removal reactions brought about by biofilm with $a_f \cdot n_S'$ and suspended cells r_S. Equation 6.131 can only be solved numerically, and numerous techniques (e.g., Remson et al., 1971) are available.

The development of such unified models was presented by Rittmann (1982) for different biofilm reactor types, that is, for a completely mixed reactor, once-through fixed bed reactor, and once-through FBBR without and with a recyclestage. Although Equ. 6.131 has been solved numerically for all reactor types, each reactor requires an appropriate value of s_{in}, a submodel to calculate n_S' and d_f, and an appropriate number of reactor segments (e.g., $N = 1$ for one CSTR and $N = 12$ for FBBR). These terms differentiate among the reactor models and are discussed in Rittmann's original paper.

An important consequence of this unified model approach is that it can be applied to all reactor configurations, since it is based on common principles for the dynamic interaction of kinetics and transports. These results demonstrate that simple loading criteria cannot predict the performance of a variety of biofilm operations because they ignore the interactions among r_S, r_X, and reactor balance.

Another advantage of quantifying the dynamic interactions is that transient conditions (e.g., conditions of active microbial growth) can be studied in various biofilm reactors, as was demonstrated by Andrews and Tien (1982). As this unsteady-state model requires the use of computer techniques, Andrews (1982) presented a method by introducing a variable transformation that eliminates the need for a computer solution. However, several limitations to the modeling results are known, so caution must be exercised in applying the unified model approach, that is, non-steady-state loading (fluctuations in s_{in} and v_z), dual-substrate limitation (e.g., Reuss and Buchholz, 1979), the effect of additional biomass on particle-settling velocity (which is different for, e.g., tower fermenter and fixed bed reactors), and some practical considerations (bed expansion, clogging, wall effects, and shearing). The use of non-monosized support particles is recommended as an alternative to a tapered tower fermenter, as normally a fixed bed bioreactor is superior due to almost constant cell concentrations throughout the bed (Andrews, 1982).

6.7.2.3 Fluidized Bed Biofilm Reactor (FBBR)

Recently, the application of fluidized beds to bioprocessing stimulated the modeling of FBBRs. In this complex case, however, models of the hydraulics must be combined with a model of biofilm kinetics, including mass transport-affected substrate conversion. This truly heterogeneous case of reactor operation will be presented here.

The unified model approach, according to Rittmann (1982), predicts that a once-through FBBR can achieve performance superior to mixed STR and fixed bed reactors because the biofilm is evenly distributed throughout the reactor while the liquid flow still exhibits plug flow behavior. This increases the overall effectiveness of the FBBR, in which effluent concentrations less than s_{min} will be obtained.

Because the principal advantage of the FBBR is the reduction in reactor size caused by the increased biomass concentration, an understanding of the factors affecting x is essential. A theoretical value for the reactor biomass may be determined by development of a hydraulic model on the basis of the principles of L|S fluidization. Mulcahy and La Motta (1978) developed a mathematical model for a FBBR. Prediction of S conversion requires again the simultaneous solution of the reactor flow equation coupled with biofilm effectiveness equations. Solution of these equations, however, is possible only after the parameters of bed porosity ε and biofilm thickness δ have been specified. An algorithm is proposed for this fluidization model that correlates these through a parameter referred to as the expansion index, that is related to the terminal settling velocity of the bioparticles.

The dispersed plug flow model was used to describe the flow behavior in the FBBR (cf. Equ. 2.3c, 3.3a, or 3.97). At steady state, the conservation equation, using dimensionless parameters as in Equ. 4.79, becomes for the L-phase expanded bed:

$$\frac{d\hat{s}}{d\hat{z}} - \mathrm{Bo}\frac{d^2\hat{s}}{d\hat{z}^2} + \eta_r \cdot r_S^+ = 0 \tag{6.132}$$

where η_r is the overall effectiveness factor, defined as in Equ. 4.51 and r_S^+ is the reaction rate on a per-unit expanded bed volume basis. r_S^+ is related to the observed rate of reaction on a per-unit biofilm volume basis $r_{S,eff}$ by the following simple expression

$$V_x \cdot r_{S,\text{eff}}|_z = H_x \cdot A \cdot r_S^+|_z \tag{6.133}$$

with V_x the biomass volume and $H_x \cdot A$ the expanded bed volume. Thus

$$r_S^+ = \eta_r \cdot \varepsilon_x \cdot r_{S,\text{ideal}} \tag{6.134}$$

where ε_x is the biomass holdup ($V_x/H_x \cdot A$), which can be conveniently expressed in terms of bed porosity ε and the operating parameter, the media volume V_m, as

$$\varepsilon_x = 1 - \varepsilon - \frac{V_m}{H_x \cdot A} \tag{6.135}$$

Substituting Equ. 6.134 for r_S^+ and Equ. 6.135 for ε_x into Equ. 6.132 results in an adequate reactor flow equation to be solved simultaneously with the biofilm effectiveness equation. The equation for η_r represents the solution of the previous problem of internal transport with simultaneous enzyme reaction, described in Equ. 4.81. The effect of external mass transport, or the rate of S conversion, is included through the boundary conditions at the L|S interface. Equation 4.81 and the accompanying boundary conditions expressed in dimensionless form are

$$\frac{1}{[\hat{x} + (d_m/2\delta)]^2} \cdot \frac{d}{d\hat{x}}\left[\left(\hat{x} + \frac{d_m}{2\delta}\right)^3 \frac{d(s/s_L)}{d\hat{x}}\right] - \phi^2 \frac{s/s_L}{(K_S/s_L) + s/s_L} = 0 \tag{6.136}$$

with

$$\frac{d(s/s_L)}{d\hat{x}} = \text{Bi}\left(1 - \frac{s}{s_L}\right) \quad \text{at } \hat{x} = 1 \tag{6.137a}$$

and

$$\frac{d(s/s_L)}{d\hat{x}} = 0 \quad \text{at } \hat{x} = 0 \tag{6.137b}$$

where $\hat{x} = r/\delta - d_m/2\delta$ = dimensionless bioparticle radial coordinate
d_m = support medium diameter
Bi = Biot number

(cf. Equ. 4.110). The overall effectiveness factor, comparing $r_{S,\text{eff}}$ to $r_{S,\text{ideal}}$, can then be obtained by using Equ. 6.136 with Equ. 6.137 after integration for the calculation of $r_{S,\text{eff}}$. The following expression, obtained in this way and written in dimensionless form, is the biofilm effectiveness equation

$$\eta_r = \frac{1 + K_S/s_L}{(d_m/2\delta)^2 + (d_m/2\delta) + 1/3} \int_0^1 \frac{s/s_L}{(K_S/s_L) + (s/s_L)} \left(\hat{x} + \frac{d_m}{2\delta}\right)^2 \cdot d\hat{x} \tag{6.138}$$

that is needed together with Equ. 6.132 for the quantification of the performance of FBBRs.

To predict S conversion within an FBBR using the mathematical model presented, a total of 13 system parameters and six empirical correlations must

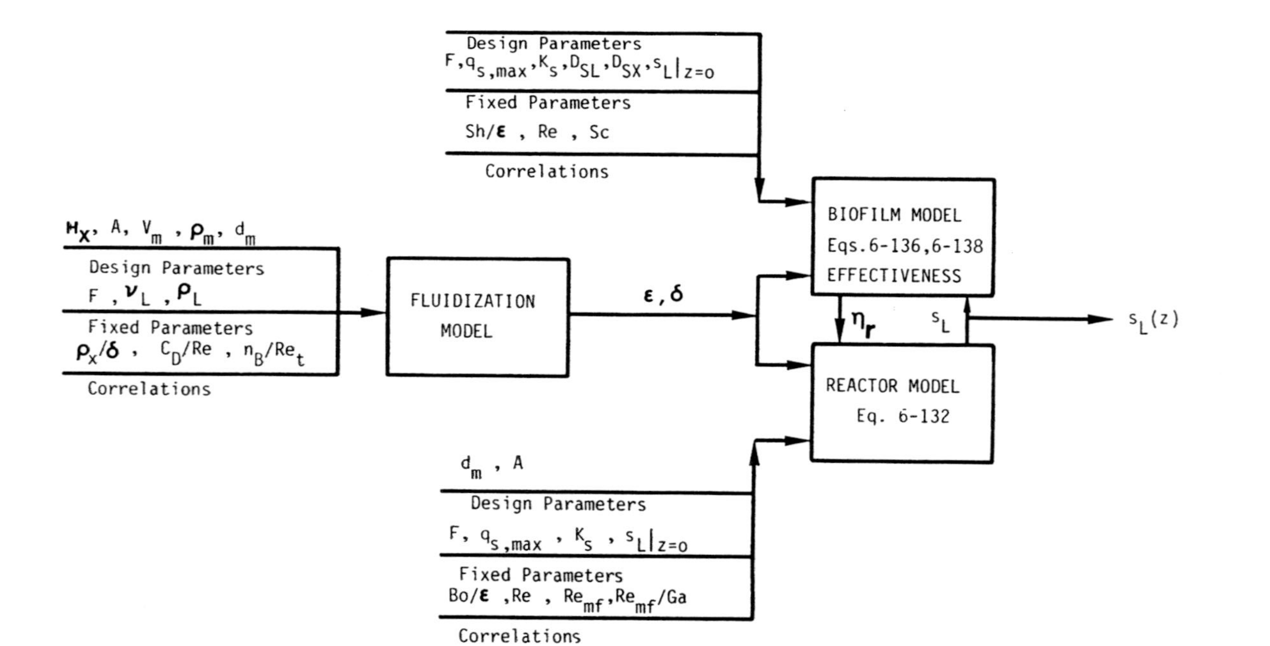

FIGURE 6.46. Block diagram of the fluidized bed biofilm reactor model according to Mulcahy and La Motta (1978), containing reactor flow equation, biofilm effectiveness equations, and fluidization model with 13 system (fixed and design) parameters and six empirical correlations.

Design parameters: reactor parameters—horizontal area, A; expanded bed height, H_X—and support media parameters—media density, ρ_m; media diameter, d_m; total volume of media, V_m. Fixed parameters: liquid phase parameters—diffusivity of substrate S in liquid, D_{SL}; liquid density, ρ_L; liquid viscosity, ν_L—biofilm parameters—diffusivity of substrate in biofilm, D_{SX}; maximum rate constant, $q_{S,max}$; Michaelis constant, K_S—and system dependent parameters—inflow concentration of substrate S, $s_{L|z} = 0$; inflow rate, F.

Empirical correlations:

1. Biofilm density—biofilm thickness correlation (ρ_x/δ).
2. Drag coefficient—Reynolds number correlation (C_D/Re).
3. Expansion index—terminal Reynolds number correlation, Equ. 6.139.
4. External mass transfer coefficient correlation, Equ. 4.1.
5. Axial dispersion coefficient correlation (Bo/ε resp. Re).
6. Minimum fluidization Reynolds number correlation (Re_{mf}/Ga).

be specified. The required parameters and correlations are presented in the legend of Fig. 6.46, where a block diagram of the FBBR model is presented (Mulcahy and La Motta, 1978).

The fluidization model indicated in Fig. 6.46 is not elaborated on here in detail. An algorithm is proposed by these authors, as already stated, correlating bed porosity ε determined through the expansion index n_B, which corresponds to the bioparticle terminal Re number Re_t as a measure of the terminal settling velocity v_t calculated from Newton's law in the following manner: With

$$n_B = 10.35 \cdot Re_t^{-0.18} \tag{6.139}$$

the resultant equilibrium bed porosity ε is

$$\varepsilon = \left(\frac{v}{v_t}\right)^{1/n_B} \tag{6.140}$$

The most common approach, however, in contrast to this strategy for the setup of a fluidization model, is first to define an empirical correlation for an isolated particle and then to extend it to cover multiparticle systems through inclusion of a correction factor dependent on ε, as demonstrated by Shieh et al. (1981). For an isolated spherical particle system with defined characteristics (d_p, ρ_p, ρ_L, and v), these workers gave the following expression

$$\varepsilon^{4.7} \cdot Ga = 18 \cdot Re + 2.7 \cdot Re^{1.687} \tag{6.141}$$

Developing an equation for the biomass concentration, again starting with an idealized bioparticle, Shieh et al. (1981) showed that

$$x = \rho_x(1 - \varepsilon)\left[1 - \left(\frac{d_m}{d_p}\right)^3\right] \tag{6.142}$$

where ρ_x is the dry density of biofilm and d_p is the diameter of bioparticle.

Examination of Equs. 6.141 and 6.142 reveals that the biomass concentration in the FBBR is mainly a function of the bioparticle diameter (or biofilm thickness) and of the hydraulic characteristics of the reactor that are under the control of the design engineer: that is, expanded bed height, d_m; reactor area perpendicular to flow, δ, which directly depends on v_{LS}, media volume; and expanded bed height. This proposed model was shown to be capable of predicting biomass concentration in the oxitron system FBBR. Support media size and biofilm thickness are two important variables affecting the biomass concentration. Shieh (1981) also developed a kinetic model for design purposes of the FBBR, assuming zero-order kinetics and complete S utilization in the outer shell of the biofilm, which may not be generally true. Integration of similar concepts as previously described (Equ. 6.43 with $r_S = k_0$ and a biofilm effectiveness equation) yields in this case

$$s_L^{0.55} = -k_{eff} \cdot \bar{t} + s_0^{0.55} \tag{6.143}$$

where $\bar{t}$ is the mean residence time and

$$k_{\mathrm{eff}} = 1.657(1-\varepsilon)(\rho \cdot k_0)^{0.55} \cdot D_{\mathrm{eff}}^{0.45} \Big/ \left(\frac{d_p}{2}\right)^{0.9} \tag{6.144}$$

Thus, the S concentration profile through the reactor can be described by a 0.55-order rate equation (cf. Equ. 5.254), and this result was successfully tested with data reported from Mulcahy and La Motta (1978) at different biofilm thicknesses. Maximum S conversion was reached with an optimal support particle size $d_{p,\mathrm{opt}}$. However, it is important to note that k_{eff}, a pseudo-homogeneous rate coefficient, is actually a parameter rather than a true rate constant, as it depends on the characteristics of both biofilm and substrate.

In contrast to the result from this work, Kargi and Park (1982) proposed a theoretical analysis that indicates no $d_{p,\mathrm{opt}}$ value but rather an optimal biofilm thickness for a given support particle size, that is, an optimal ratio of biofilm thickness/support particle size (δ/R_m) when ε_L is constant.

Using a Thiele modulus ϕ_δ according to Equ. 4.111 (cf. Fig. 4.45), the solution in this case has the form of the following equation (cf. Equ. 6.121) valid for FBBR design

$$\ln \frac{s_L}{s_{L,0}} = -\eta_r \cdot k_r \cdot \frac{H}{v_{SL}} \tag{6.145}$$

with H being the height of the bed and v_{SL} being the superficial velocity of the fluid [$\mathrm{cm \cdot s^{-1}}$].

6.7.3 Application of Biofilm Reactors

While types of biofilm model reactors have been summarized in Fig. 3.42, industrial-scale fermentations using biofilm operation are summarized here:

1. Biological waste water treatment in trickling filters (percolating filters or fixed bed reactors) and in rotating disk fermenters. A system somewhat related to trickling filters is the submerged filter, which, however, is considerably limited in application as only two-phase systems can be handled (L|S processing). Gaseous substances must be dissolved in the L phase before entering the filter.
2. The FBBR applied in biological waste water treatment (e.g., Atkinson, 1980), enzyme technology (e.g., Coughlin et al., 1975), and some fermentations (Baker et al., 1980). Higher values of mass transfer are attained together with increased interfacial areas ($a_{L|S} \sim 3 \cdot 10^3\ \mathrm{m^2/m^3}$) and biomass concentration ($x \sim 40\ \mathrm{kg \cdot m^{-3}}$), uniform distribution of solids, and so on (Baker et al., 1980).
3. The old "quick" vinegar process using packings of beechwood chips as support material for cells to increase the area of contact with the feed liquor trickling through.
4. Animal tissue culturing, in which cells are growing adhering to a surface in the presence of unstirred layers of medium (trays).

5. Bacterial leaching of ores for the recovery of metals by natural adhesion on solid materials.
6. In conventional STRs. Such STRs also imply biofilm formation (wall growth), a fact that is often ignored.
7. In natural aquatic systems, water distribution systems, waste water processing, heat exchangers, fuel consumption by ships, and even human disease. It should be noted that biofilms are emerging as a critical factor in these. The most common method of controlling biofilm accumulation in practice is chlorination, a process that is limited, due to its toxicity. Thus, research in biofilm processing has been stimulated (Charaklis, 1981). Often the term "fouling" is used to refer to the undesirable formation of deposits on surfaces that impede the flow of heat, increase the fluid frictional resistance, increase the rate of corrosion, and result in energy losses.

6.8 Dialysis and Synchronous Culture Operation

Conventional continuous or batch culture methods suffer from some restrictions that are disadvantageous for technical or economic reasons (e.g., accumulation of toxic and/or inhibition of metabolic end products, limited cell densities, etc.), or are disadvantageous for scientific research purposes (e.g., cultivations are related to the statistical mean of the cell population and not to the individual cell). Recent extensions of the continuous culture methods, that is, dialysis and synchronous culture techniques, are able to fill this gap.

6.8.1 Dialysis (Membrane) Reactor Operation

6.8.1.1 Potentialities

There are many different kinds of membrane processes, but all have certain features in common. In all of them, a fluid containing two or more components is in contact with one side of a membrane that is more permeable to one component (or a group of like components) than to other components—it is a selective membrane. The other side of the selective membrane is in contact with a fluid that receives the components transferred through the membrane. To cause the transfer of components, there must, of course, be a driving force of some kind. Such a force may be a transmembrane difference in concentration, as in *dialysis*; electrical potential, as in *electrodialysis*; or hydrostatic pressure, as in *reverse osmosis*, *ultrafiltration*, and *microfiltration*. The freedom to choose both the driving potential to be used and the diffusing species (solute or solvent) has led to a multitude of nomenclatures for membrane processing. Generally, processes involving the diffusion of solvent are termed osmosis or ultrafiltration, and those for the solute are termed dialysis.

Dialysis is among the oldest of the membrane processes, but it has found only a few industrial applications. However, greater industrial usage may develop for the newer processes of Donnan dialysis and ion-exchange dialysis (see Lacey, 1972).

In dialysis, that is, in the separation of solute molecules through their unequal diffusion through a semipermeable membrane because of a concentration gradient, the small-molecular-weight products are removed from the immediate environment of the bacterial cell (and ultimately from the intracellular enzyme site), which relieves feedback inhibition by a product that normally regulates its own production. As more product is withdrawn by dialysis, more substrate is consumed and more product is made: The fermentation thus becomes more efficient. As the cell population attains high density, the substrate is increasingly converted into product by maintenance rather than by growth metabolism, also thus improving fermentation efficiency. On the other hand, the application of dialysis increases costs, complicates operation, and eventually reduces efficiency because of membrane fouling. The potential advantages must compensate for the disadvantages if a dialysis process is to be useful.

Although membrane processes have been studied for more than a century, they have only recently become of interest for industrial separations (e.g., Applegate, 1984). This interest stems primarily from a basic advantage of membrane processes: They allow separation of dissolved materials from one another or from a solvent, with no phase change. Since the energy cost represents a sizeable portion of the total operating cost for most separations, the possibility of effecting worthwhile economics in energy cost by using membrane processes is attractive (e.g., Gerstenberg et al., 1980; Pye and Humphrey, 1979; Rautenbach and Albrecht, 1982). However, for total operating cost to be reduced, flux through the membranes has to be great enough to permit use of reasonably small areas of membrane, resulting in low equipment costs. Fermentation operation with dialysis culture systems was one of the first practices of use of cell retention. Laboratory equipment for cultivation of microorganisms with the removal of cell-free medium is a useful tool in the study of continuous culture.

6.8.1.2 Principles and Performance Equations

Four basic modes are generally possible in dialysis culture operation (Fig. 6.47): (a) continuous reservoir and continuous fermenter, where F_{res} and $F_{ferm} > 0$; (b) batch reservoir and batch fermenter, where F_{res} and $F_{ferm} = 0$; (c) batch reservoir and continuous fermenter, where $F_{res} = 0$ and $F_{ferm} > 0$; and (d) continuous reservoir and batch fermenter, where $F_{res} > 0$ and $F_{ferm} = 0$. The quantification of membrane performance is achieved with the aid of a formal approach based on Fick's law of diffusion using membrane permeability coefficient P_{mb}, which includes several unknown factors that must be measured experimentally:

$$P_{mb} = \frac{n_S}{A_{mb} \cdot \Delta s} \; [\mathrm{kg/kN \cdot d}] \tag{6.146a}$$

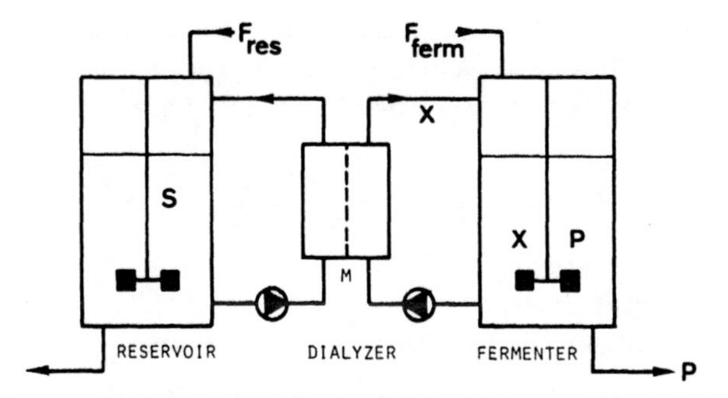

FIGURE 6.47. Typical experimental setup of dialysis culture using a reservoir for substrate feed S, a fermenter for bioconversion, and a chamber with a semipermeable membrane M to retain biomass X.

where n_S = rate of permeation [kg/h]
A_{mb} = membrane area [m^2]
Δs = substrate concentration gradient across the membrane

P_{mb} [kg/kN · h] can be estimated from a plot of L flux (n_S/A_{mb}) versus pressure on the feed side, according to

$$\frac{n_s}{A_{mb}} = J_1 = P_{mb}(\Delta p - \Delta \pi) \tag{6.146b}$$

Here $\Delta \pi$ is the osmotic pressure of feed (permeate). The flux rate of the solute (salt) obeys Fick's first law, that is,

$$J_2 = -D_2 \frac{(\Delta c_2)^m}{s_m} \tag{6.146c}$$

which normally is heavily influenced by concentration polarization. Thus, the solute flux is proportional to the solute difference inside the membrane $(\Delta c_2)^m$, which in case of concentration polarization is substituted by $c_2^m = K_2 \cdot c_2$, leading to an expression instead of Equ. 6.146c by using the bulk solute concentration difference

$$J_2 = D_2 \cdot K_2 \frac{\Delta c_2}{s_m} \tag{6.146d}$$

The performance equation of a membrane can be derived by introducing intrinsic rejection ($R_i = 1 - c_2^p/c_2^m$) and observed rejection ($R_{obs} = 1 - c_2^p/c_2$) with c_2^p being the bulk concentration at the permeate side:

$$\ln \frac{1 - R_{obs}}{R_{obs}} = 25 \frac{v}{u} \mathrm{Re}^{1/4} \cdot \mathrm{Sc}^{2/3} + \ln \frac{1 - R_i}{R_i} \tag{6.146e}$$

where v is the permeate velocity and u is the feed velocity. A plot of $(1 - R_{obs})/R_{obs}$ versus $J_1/u^{0.75}$ will give a straight line, with an intersect proportional to $R_i(\sim c_2^m)$.

The permeability coefficient reflects the overall resistance of the composite barrier, and therefore includes external transport limitations (cf. Sect. 4.5.1), which can be quantified using the two-film theory (cf. Equ. 3.30):

$$\frac{1}{P_{mb}} = \frac{1}{P^0_{mb}} + \sum_i \frac{\delta_i}{D} \tag{6.147}$$

where P^0_{mb} = true permeability
D = diffusion coefficient
δ_i = thickness of transport-limiting liquid films.

Thus, the rate of dialysis is strongly dependent on the liquid velocity at the membrane surface u. The thicknesses of the liquid films decrease as the bulk velocity near the membrane increases (see Equs. 4.1a and 6.120).

Mathematical analysis always consists of the writing of suitable balance equations by the insertion of kinetic terms. In the case of dialysis culture, balances must be made either on the reservoir chamber or the fermenter chamber. With use of Equ. 6.146a for membrane permeation, the basic balance equations for dialysis culture are as follows:

The balance in the reservoir in general form is

Accumulation = input − output + permeation

which gives for substrate and product (and not for cells)

$$V_{res}\frac{ds_{res}}{dt} = F_{res} \cdot s^0_{res} - F_{res} \cdot s_{res} + P_{mb,S} \cdot A_{mb}(s_{ferm} - s_{res}) \tag{6.148}$$

and

$$V_{res}\frac{dp_{res}}{dt} = F_{res} \cdot p^0_{res} - F_{res} \cdot p_{res} + P_{mb,P} \cdot A_{mb}(p_{ferm} - p_{res}) \tag{6.149}$$

A similar balance can be established in the fermenter for substrate, biomass, and product in general form:

Accumulation = input − output + permeation ± reaction

$$V_{ferm}\frac{ds_{ferm}}{dt} = F_{ferm}(s^0_{ferm} - s_{ferm}) - P_{mb,S} \cdot A_{mb}(s_{ferm} - s_{res}) - r_S \cdot V_{ferm} \tag{6.150}$$

and

$$V_{ferm}\frac{dx_{ferm}}{dt} = F_{ferm}x^0_{ferm} - F_{ferm}x_{ferm} + V_{ferm} \cdot r_X \tag{6.151}$$

and

$$V_{ferm}\frac{dp_{ferm}}{dt} = F_{ferm}(p^0_{ferm} - p_{ferm}) - P_{mb,P} \cdot A_{mb}(p_{ferm} - p_{res}) \cdot r_P \cdot V_{ferm} \tag{6.152}$$

These five mass balance equations plus model equations for kinetics (r_X, r_S,

FIGURE 6.48. Calculated comparison of dialysis and nondialysis continuous culture with respect to concentration of biomass x and substrate s (a), productivity $D \cdot \bar{x}$ (b), and process efficiency expressed as productivity ratio (c). Parameter values are: $V_{ferm} = 1$ l, $\mu_{max} = 1\ h^{-1}$, $K_S = 0.2\ gl^{-1}$, $Y_{X|S} = 0.5$, $s^0_{ferm} = s^0_{res} = 10\ gl^{-1}$, $F_{res} = 0.5\ lh^{-1}$, $p_{mb} \cdot A_{mb} = 420\ cm^3/h$. (Adapted from Schultz and Gerhardt, 1969.)

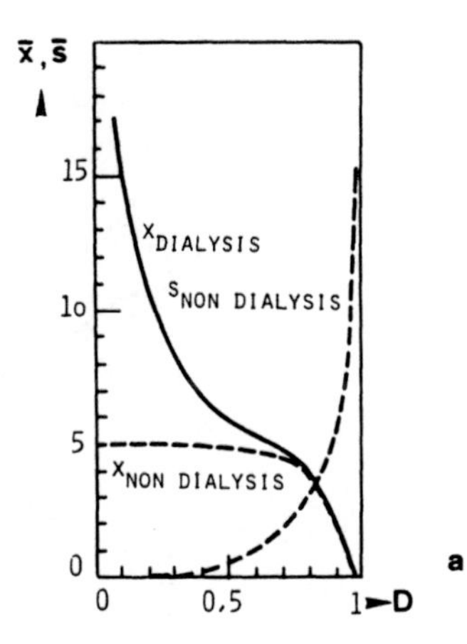

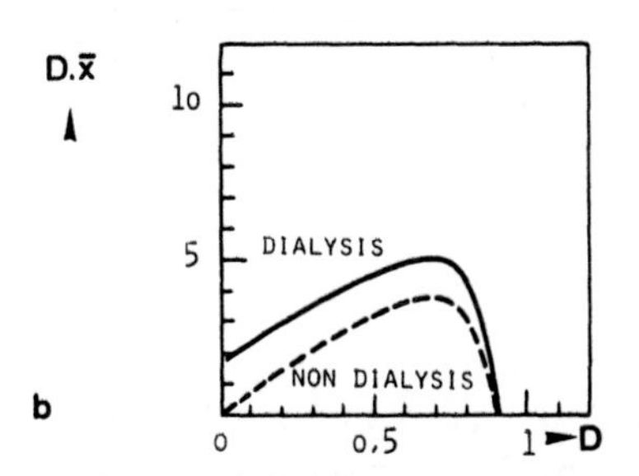

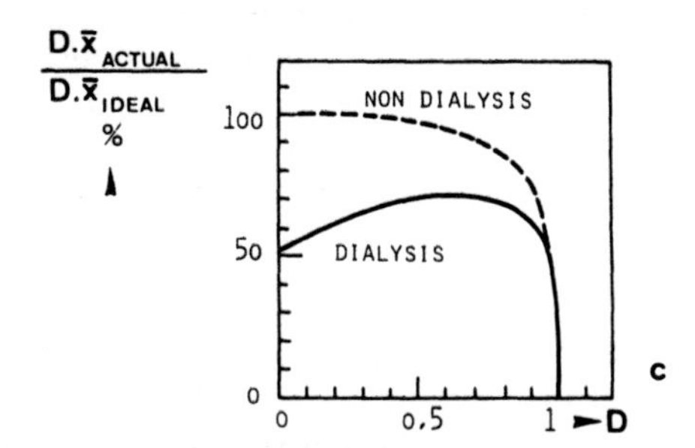

and r_P; see Chap. 5) are mathematically sufficient to determine dialysis culture operation (Coulman et al., 1977; Schultz and Gerhardt, 1969).

Case (a) of completely continuous dialysis operation is the only one that is essentially steady state in nature, so that Equs. 6.148 through 6.152 can be simplified as the time derivates are all equal to zero. Solutions are fully described in original papers, together with the behavior of x, s, and the critical dilution rate D_c as a function of dialysis. With this solution, a comparison between dialysis and nondialysis continuous culture can be realized, as shown in Fig. 6.48. The comparison is carried out with respect to cell concentration (a), productivity (b), and efficiency of cell production (c), which is defined as the actual production rate divided by the production rate equivalent to complete utilization of the substrate supplied.

The most striking and important difference between nondialysis and dialysis continuous culture is the much higher cell concentration attainable in dialysis culture, especially at low dilution rates. Further, it is seen that the maximum production rate in continuous dialysis culture is achieved at a lower dilution rate than in nondialysis continuous culture. This effect, however, is reached at the expense of lower efficiencies in converting substrate to cells.

Case (b) of completely batch operation of dialysis is the operational mode that is still predominantly used in process analysis. Mathematically, the values

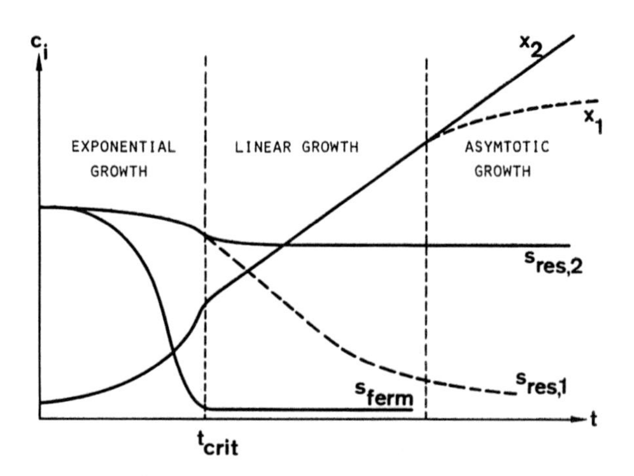

FIGURE 6.49. Expected changes in cell mass concentrations for dialysis culture systems operated with batch fermenter and batch reservoir (x_2 and $s_{res,2}$) or batch fermenter and continuous reservoir (x_1 and $s_{res,1}$). The critical time when substrate diffusion through the membrane becomes limiting is indicated as t_{crit}. Exponential, linear, and asymtotic growth can be observed. (Adapted from Schultz and Gerhardt, 1969.)

of F_{res} and F_{ferm} in Equs. 6.148 through 6.152 become zero. The expected growth pattern for the fully batch dialysis culture is best evaluated by computer and is shown in Fig. 6.49. The growth cycle exhibits two separate phases: exponential growth until the region of S limitation, and linear growth by S-diffusion rate limitation. With dialysis culture, the exponential growth phase is extended because of the additional nutrient via the membrane.

The case (c) of batch fermenter/continuous reservoir dialysis culture can be evaluated similarly. The behavior is also shown in Fig. 6.49, and the properties are analogous to those in case (b). Linear growth can be maintained, because s_{res} is constant due to membrane diffusion.

The limitations of dialysis fermentation are evident, since the inherently slow process of diffusion of both nutrients and metabolic products through the membrane is the rate-determining step. Inhibitory and/or toxic substances can accumulate and limit maximum cell concentration.

In the case of product formation in dialysis culture, the only additional information needed (see Equs. 6.148–6.152) is the permeability characteristics of the product through the membrane material and the kinetic model for product formation (see Sect. 5.4). The intricacies of fitting a mathematical model to the kinetics of product formation were illustrated for steroid conversion (Chen et al., 1965), in which phenomena such as substrate solubility, co-precipitation, feedback, and substrate inhibition have to be taken into account.

Analytical solutions are possible for particular situations, for example, in fully continuous operation. Here the final concentration of product that cannot diffuse across the membrane will be $p = P/V_{ferm}$. However, if the product is diffusible ($P_{mb,P} > 0$), then it will be distributed between the fermenter and the reservoir, $p = P/(V_{ferm} + V_{res})$. Thus

$$p_{res} = \frac{p_{ferm}}{F_{res}/(P_{mb,P} \cdot A_{mb}) + 1} \tag{6.153}$$

To obtain a large fraction of product in the reservoir effluent free from cells, both the permeability $P_{mb,P} \cdot A_{mb}$ and F_{res} must be large in comparison with F_{ferm}. This type of operation results in a low concentration of product, which creates difficulties in product recovery.

From the preceding analyses it is apparent that the behavior of reactor systems is more closely linked to the operation of the fermenter chamber than to operation of the reservoir. Therefore, the first step in designing a dialysis culture is to choose between the security and flexibility of batch cultures and the uniformity and economy of continuous techniques.

6.8.1.3 Membrane Reactors and Applications

Dialysis processes were originally developed as simple flasks with internal filtration (Gerhardt and Gallup, 1963). Later, modifications (e.g., Dostalek and Häggstrom, 1982) and modern designs (rotating microfilter, Sortland and Wilke, 1969; the rotorfermenter, Margaritis and Wilke, 1978a,b; hollow fiber reactors, e.g., Kan and Shuler, 1976, 1978) were developed. Some reviews should be mentioned (e.g., Applegate, 1984; Meiorella et al., 1981; A. Moser, 1985a) in which different reactors are compared and discussed in terms of their fields of application (fermentation and enzyme technology).

In several microbial processes, economic efficiency can be drastically increased when a technical solutions to the problems of discharging accumulated inhibitory or toxic products or metabolites, or of using substrates consisting of dispersed solids, can be found. Examples are the production of ethanol from carbohydrates (Charley et al., 1983; Kosaric et al., 1980; Rogers et al., 1980), utilization of whole or deproteinized whey for the production of food yeasts or nitrogenous feed supplement for ruminants (Gerhardt and Gallup, 1963, Lane, 1977), production of salicylic acid from naphthalene (Abbott and Gerhardt, 1970a), threonine biosynthesis (Abbott and Gerhardt, 1970b), and continuous aseptic production of phytoplankton (Marsot et al., 1981).

A number of alternative approaches are available as technical solutions for the above-mentioned problems (e.g., Hamer, 1982; Lilly, 1982; Meiorella et al., 1981): vacuum membrane retractive and extractive fermentation, and biphasic processing (Mattiasson, 1983; Reisinger et al., 1987) and ion-exchange resin culture techniques. Experimental tests of dialysis culture processes are described in the literature for ammonium–lactate fermentation of whey (Stieber et al., 1977), and an improved mathematical model was later developed (Stieber and Gerhardt, 1979) incorporating P inhibition. Recently, the mathematical model was modified to incorporate a second feedstream of cells, substrate, and product into the fermenter. The behavior of this dialysis culture system with a cell-feed flow from a prefermenter or from cell recycling was examined and the process was improved (Stieber and Gerhardt, 1981a,b). Results under steady-state conditions showed that these dialysate-feed systems are a new and useful way to immobilize living cells to produce a metabolite at a high rate for a prolonged time. The substrate consumed by the cells is converted to product via maintenance metabolism only and is sterilized by dialysis. Similar considerations in the case of the rotor fermenter (RF) led to an expression comparing cell productivity in this membrane reactor with that received in a CSTR. The cell productivity (cf. Equ. 6.9) for the rotor fermenter and CSTR at the same level of s and assuming the same Y value is expressed as a ratio

$$\frac{r_{X,RF}}{r_{X,CSTR}} = 1 + \frac{F_F}{F_B} \tag{6.154}$$

where F_B is the volumetric cell bleed rate [l/h] in the rotor fermenter and F_F is the filtrate flow rate [l/h].

As shown by Equ. 6.154, high values may be obtained in the RF by employing a high ratio of feed to bleed rate. Thus, by suitable choice of flow rates, the RF may fulfill the functions of both an ordinary fermenter and a centrifuge cell separator (Margaritis and Wilke, 1978b). With similar arguments, an expression can be derived giving the productivity ratio for ethanol as product:

$$\frac{r_{P,RF}}{r_{P,CSTR}} = 1 + \frac{F_F}{F_B} \cdot \frac{p_{RF}}{p_{CSTR}} \tag{6.155}$$

As seen from Equ. 6.155, if $p_{RF} \geq p_{CSTR}$ and $F_F \geq F_B$, then the productivity ratio $R_p > 1$. Thus, by suitable choice of operating conditions for $F_F > F_B$ so that F_F is maximized while $p_{RF} > p_{CSTR}$ is maintained, about 10 times greater ethanol productivity was experimentally found in the rotor fermenter from a CSTR, thus illustrating the advantage of membrane reactor operation.

6.8.2 Synchronous Culture Operation

Contemporary knowledge of cell metabolism is based on studies of asynchronous populations that are often of poorly defined origin and from in vitro rather than in vivo studies. As an alternative, it might be worthwhile to think of growth and metabolism in terms of the cell cycle and post-cycle patterns. It thus becomes possible to relate presently incompatible results from batch and continuous cultures. For instance, if we consider the culture in terms of the cell, the growth curve can be rearranged to bring out a possible connection between growth and secondary metabolism, which Bu'Lock (1965) and Bu'Lock et al. (1965) described as the "tropho- and idiophase" (see Dawson, 1972a).

For many research purposes it is of great importance to have a synchronous culture of microorganisms. Such situations appear with, for example, mutation experiments, and are also of general interest in respect to some inadequacy of continuous culture theory. Results from a CSTR at steady state relate to the cell in terms of dilution rate, which represents a reasonable, yet average, value of the homogeneous proliferation that exists in the single-stage CSTR, but which is not necessarily a true reflection of microbial behavior. In recent years the advent of cell synchrony as a refinement in continuous techniques has added new dimensions to the study of microbial growth (Cameron and Padilla, 1966; Zeuthen, 1964). The potential advantage of synchronous culture is directly related to the fact that in this technique the cells of the population are all in the same stage of cell development (physiological state), doing the same thing at the same time. The population serves as an amplification of the cell and thus it becomes possible to study cell behavior by observing the behavior of the population. Different methods are known, described as *synchronous/synchronized pulsed* and *phased culture* techniques (Dawson, 1972a). Table 6.1 summarizes the characteristics of the various fermentation techniques, which are illustrated in Fig. 6.50.

In "phased" culture, the cell population is synchronized, and growth rate, nutrient supply per cell, population numbers, and temporal activities become experimental parameters routinely involved in the operation of the technique. Phased cultures may be examined systematically like steady states in the chemostat, but instead of the single-point (average) determinations of the latter, patterns of cycle period ("cell cycle") activity that portray the replicative performance of the cell during the doubling time period are obtained. Phased culture permits us to rationalize, in a homogeneous population, the performance of the cell under experimentally controlled conditions that can, if necessary, be repeated indefinitely or varied at will (Dawson, 1980b). Dawson (1972a,b, 1980a,b) has reviewed some of the significant results obtained with phased cultures on the basis of cell cycle and post-cycle activities, an analysis that the experimental dimensions of the chemostat cannot approach.

TABLE 6.1. Main characteristics of techniques available for cultivating microbes.

Type	*Method*	*System*	*Growth rate*	*Technique*
Asynchronous	Batch	Closed	Changing	Traditional "Bactogen"
	Continuous	Open	Constant	Chemostat Turbidostat
Synchrony	Batch	Closed	Changing	Synchronous Synchronous
	Continuous	Open	Constant	Pulsed Pulsed

From Dawson, 1972a.

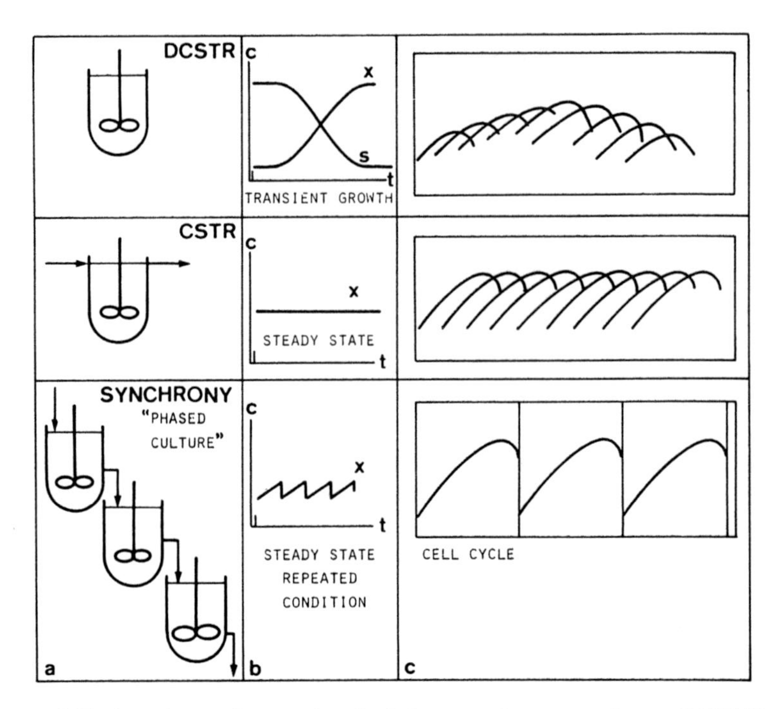

FIGURE 6.50. Experimental setup for obtaining asynchronous cultures (DCSTR and CSTR) and synchronous (phased) cultures (a), characterization of overall growth conditions (b), and illustration of growth of individual cells (c). A comparison of the different approaches shows clearly that in the DCSTR and CSTR, cells in the populations are randomized at all stages, while cells adjust their growth rate to the dosing interval and become synchronized to it in the "phased culture technique" (Dawson 1980a).

Two general classes of methods exist for attaining synchrony in batch and/or continuous culture:

1. Methods whereby cells at the same stage of division are selected or separated from a randomly dividing population (usually by a mechanical or physical procedure)
2. Methods in which the entire population is manipulated by applied constraints of the environment (which may be nutritional, physical, physiological, or growth inhibitory) to align the cells and produce the inoculum for a synchronized culture

A modified method was proposed by Kjaergaard and Jørgensen (1979) and is probably better than others. The principle of the apparatus is that of a normal chemostat, with the exception that the substrate fed with the dilution rate contains no carbon source and that the carbon source is fed automatically to the vessel at certain intervals in volumes that are negligible compared with the reactor volume. The background to this procedure is that cells grow only when a carbon source is present. As soon as the carbon source is added, the cells start to assimilate and cell concentration increases according to

$$x = x_0 \cdot \exp[(\mu - D)t] \tag{6.156}$$

The amount of carbon source is adjusted in such a manner that only one doubling in biomass can take place, which means that the sugar must be consumed after the time $\ln 2/\mu$.

The time period with no growth must be sufficient that x again is brought to x_0, which means that

$$x = x_0 \exp\left[\left(1 - \frac{D}{\mu}\right)\ln 2\right]\exp(-D \cdot t) \tag{6.157}$$

which gives

$$t = \left(\frac{1}{D} - \frac{1}{\mu}\right)\ln 2 \tag{6.158}$$

The total time between two additions of carbon source is therefore

$$\frac{\ln 2}{\mu} + \left(\frac{1}{D} - \frac{1}{\mu}\right)\ln 2 = \frac{\ln 2}{D} \tag{6.159}$$

This system will give a continuous synchronous culture in which the cell growth rate is independent of dilution rate ($\ln 2/D$) as long as $D < \mu$. A new generation of cells is produced every $\ln 2/D$.

For microbial biomass production phased cultures would appear to have a 50% production advantage by permitting the culture volume to be harvested at doubling time intervals, rather than at replacement or retention time intervals (which correspond to approximately 1.5 times the doubling time).

6.9 Integrating Strategy as General Scale-Up Concept in Bioprocessing

The integrating strategy has already been presented in Chap. 2 as a fundamental approach for simple but adequate bioprocess modeling (cf. Fig. 1.4). Especially in Fig. 2.13, the integrating point of view was represented by a block diagram of a bioreactor, indicating the problems involved. Agitation (micromixing), macromixing, and aeration will influence the internal and external transport phenomena that ultimately determine the extracellular environment. Simultaneously, this environment is modified by the feeding of substrate, which is used for production of biomass and products. An important aspect in this scheme is the feedback from biomass and product to the environment. It is well known that most secondary metabolites are produced with filamentous microorganisms, and the properties of the broth can drastically change. All of the transport phenomena of the bioreactor will depend ultimately on physical properties of the fermentation fluid, including viscosity, shear gradients, density, gas holdup, and so on. Thus, the integrating strategy deals with the joint aspects of balancing, stoichiometry, thermodynamics, and kinetics of reactions and transports.

6.9.1 Stoichiometry (Balancing Methods) Applied in Bioprocess Design

As a consequence of the increased worldwide competition in bioprocessing compared with chemical processing, considerable attention has been directed toward improvement in the efficiency of technical bioprocesses. In this situation, formulating balanced stoichiometric equations is a powerful technique.

Taking a general case of (cf. Equ. 2.8):

$$\nu_S(C_a H_b O_c) + \nu_N NH_3 + \nu_O \cdot O_2 \rightarrow \nu_X(C_\alpha H_\beta O_\gamma N_\delta) + \nu_P(C_{\alpha'} H_{\beta'} O_{\gamma'} N_{\delta'}) + \nu_W \cdot H_2O + \nu_c \cdot CO_2$$

the elemental balances are (cf. App. I):

$$C: \quad a \cdot \nu_S = \alpha \cdot \nu_X \rightarrow \alpha' \cdot \nu_P + \nu_C$$

$$H: \quad b \cdot \nu_S + 3 \cdot \nu_N = \beta \cdot \nu_X + \beta' \cdot \nu_P + 2 \cdot \nu_W$$

$$O: \quad c \cdot \nu_S + 2 \cdot \nu_O = \gamma \cdot \nu_X + \gamma' \cdot \nu_P + \nu_W + 2 \cdot \nu_C$$

$$N: \quad \nu_N = \delta \cdot \nu_X + \delta' \cdot \nu_P$$

Generally, there are two many unknowns (e.g., 15). In case the composition of S, X, and P are known $(a, b, c; \alpha, \beta, \gamma, \delta; \alpha', \beta', \gamma', \delta')$ there remain only six unknowns in four equations, which can then be solved using supplementary information, for example, on the basis of gas analysis (O_2, CO_2).

It was mentioned earlier that if the elementary analysis of the cells is known (i.e., the stoichiometric coefficients in Equs. 2.1 and 2.8), then the mass balance equation may be calculated without actually measuring the oxygen used or CO_2 formed. In this way, oxygen uptake and growth yield can be theoretically predicted (Roels, 1980a). A macroscopic analysis based on elemental and energy balances successfully describes relationships important for bioprocess design. The existence of limits to the oxygen and substrate yield factors is shown by Heijnen and Roels (1981) on this basis. For substrates of low degree of reduction, the energy content of the substrate poses a limit to $Y_{X|S}$. For substrates of a high degree of reduction, a limit is posed by the carbon available in the substrate. Simple models for the energetics of growth on substrates with differing degrees of reduction are derived (Roels, 1980b), and a quantitative description of growth on mixed substrates has been given (Geurts et al., 1980) showing a relationship between Y_{ATP} and the P/O ratio.

Generally there are two points of interest. The first is, what effect does the conversion yield have on the cost of production? The second is, how much room is there for improvement; how do actual yields compare with theoretical values?

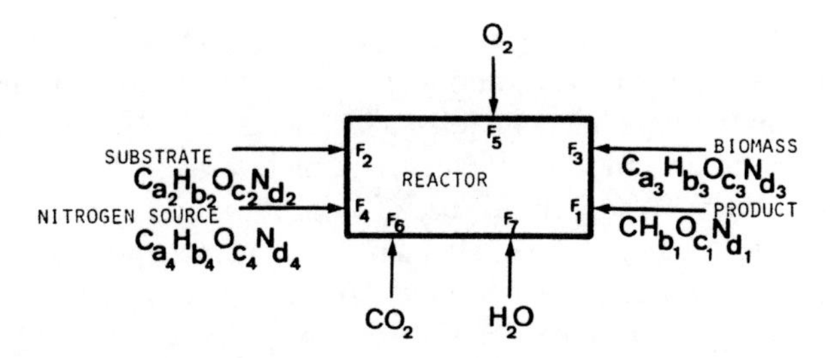

FIGURE 6.51. Diagrammatic representation of a steady-state bioprocess in balance area (reactor) following the macroscopic principle by analyzing elemental composition of significant process variables (substrate, nitrogen source, biomass, product, O_2, CO_2, H_2O). (Adapted from Roels, 1980a.)

On the basis of the preceding analysis of writing mass and energy balances (cf. Sects. 2.2.3 and 2.4.2, as outlined in Fig. 6.51 together with Appendix I), there exist a number of interesting papers in the literature. In the absence of information on the biosynthetic pathway, an alternative approach based on reaction stoichiometry was applied to penicillin synthesis (Cooney, 1979; Cooney and Acevedo, 1977). The theoretical maximum yield of penicillin from glucose was calculated; the actual conversion yield could be improved substantially by increasing and sustaining the specific rate of penicillin production and by minimizing maintenance metabolism. The actual yield of penicillin from glucose was shown to be an order of magnitude lower than the theoretical value. Increasing efficiency of glucose utilization for penicillin will markedly decrease the oxygen demand. This is important not only to reduce production costs, but also to increase the capacity of existing equipment, since it is often the OTR ability that is rate limiting in production.

Applying the strategy in formulating a simple unstructured model according to Fig. 2.19, Heijnen et al. (1979) developed elementary and enthalpy balances for the penicillin production process. This macroscopic analysis used carbon, hydrogen, nitrogen, oxygen, sulfur, phosphorus, and enthalpy balances for 11 relevant process variables (compounds), so that according to the element-species matrix (cf. Appendix I), at least five kinetic equations are needed to complete the model. On the basis of this fermentation model, the authors presented a series of simulation studies and showed some interesting results. They emphasized that a growth-coupled penicillin production provides an adequate description of most of the observed phenomena, which has often led to the assumption of non-growth-associated or age-dependent penicillin productivity. There appears to be no experimental evidence necessitating introduction of an age dependence or a time delay in modeling the kinetics of penicillin production.

Other important results concern the glucose balance and the effect of glucose feed rate schemes in fed-batch cultures. It was shown by this analysis that 20% of the glucose is used in the production of mycelial dry matter, 10% is used for penicillin synthesis, and as much as 70% is used in maintenance processes. In obtaining high penicillin yields, the glucose feed scheme is shown to be of crucial importance. A scheme using an increasing feed rate of glucose as a function of time is superior to use of a constant or a decreasing feed rate. However, in the investigations of Heijnen et al., only one point in each scheme of feed rate was simulated, so a complete picture is not given.

Recently the impact of sugar-feeding strategies was investigated, and results showed that maximum productivities are more or less independent of the feeding scheme (Bajpaj and Reuss, 1981). A strategy proposed by Mou (1979) was employed in which the biomass growth rate is controlled at one preset value during growth phase μ_{Gr} and at another preset value during production phase μ_{Pr} by supplying a rapidly metabolizing sugar. Thus, in principle, there are three external parameters that must be set to define an experiment: μ_{Gr}, μ_{Pr}, and X_{tr}, which is the transition concentration of biomass and which serves as an indicator for the onset of the production phase. During the fast growth period, empirical correlation using CO_2 production data accurately reflects the cell concentration to ± 1 g/l and the instantaneous value of μ to ± 0.01 h^{-1}:

$$X_t = Y_{X|C} \cdot \int_0^t r_C \cdot dt \quad [\text{g cells}] \tag{6.160}$$

Since the cells are growing at a low growth rate during the transition and production phases, maintenance activity and endogeneous metabolism are significant, and it is not possible to use Equ. 6.160. Instead, overall and instantaneous carbon-balancing equations allow successful on-line calculation of the cell concentration and instantaneous growth rate (Mou and Cooney, 1982a,b):

$$X_t = \frac{1}{C_X}(C_S + C_{PAA} - C_{CO_2} - C_P) \quad [\text{g cells}] \tag{6.161}$$

where C_S, C_{PAA}, C_{CO_2}, C_P, and C_X are carbon content in substrate consumed, phenylacetic acid as penicillin precursor fed, CO_2, penicillin, and cells. These equations, together with a feedback control method, enable the computer to control the production phase growth rate with minimum error (0.002 h^{-1}).

The examples presented have shown that the construction of simple unstructured models based on some notion of relevant kinetic equations in combination with the very important concept of elemental composition and enthalpy balance is of great help in understanding factors relevant in the optimization and control of even such complex fermentation processes as the production of antibiotics.

Another important field of successful application of the balancing method is biological waste water treatment. A unified approach toward mathematically describing activated sludge waste water treatment and to describing the biokinetic and stoichiometric relationships involved, has led to many significant insights into the principles of treatment process designs and operation

(Sherrard, 1977). Determination of the quantities of nutrients that must be added to nutrient-deficient waste water is important in the successful design and operation of treatment facilities. Nutrients must be added to provide balanced growth media and to avoid algal growth (in the case of excess nutrients) or incomplete conversion or takeover of filamentous bacteria (in the case of insufficient nutrients). Stoichiometric equations may be obtained at any given value of mean residence time so that the stoichiometric coefficients of all components are found and the requirements of oxygen, nitrogen, and phosphorus can be determined (Sherrard and Schroeder, 1976). Beyond these predictions of requirements for inorganic nutrients and O_2, the sludge production and effluent quality on further treatment processes can be rigorously analyzed as a function of the mode of process operation (Sherrard, 1980). A similar fundamental procedure of developing a reaction scheme, handling kinetic expressions, and reducing the system to important reactions and components was applied to the case of biological nitrogen removal (Irvine et al., 1980).

6.9.2 Interactions Between Biology and Physics via Viscosity of Fermentation Media

The rheological classification of liquids is normally given by a general expression, where the shear stress applied τ [force $\cdot$ length^{-2}] and the resulting shear rate $\dot{\gamma}$ [time^{-1}] are correlated as follows (e.g., Metz et al., 1979):

$$\tau = \tau_0 + K(\dot{\gamma})^m \qquad (6.162)$$

where K = power law constant or consistency index
τ_0 = yield stress
m = power law index or flow behavior index

This is known as the power law.

The following rheological classification of fluids is possible (Fig. 6.52):

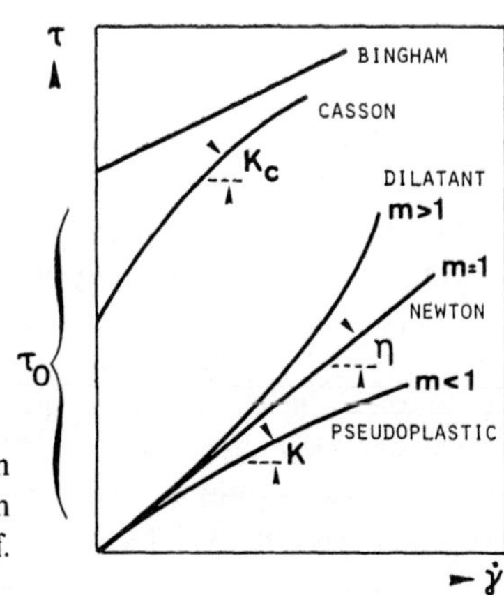

FIGURE 6.52. Rheogram of fermentation fluids in the form of shear stress τ versus shear rate $\dot{\gamma}$ in case of Newtonian and non-Newtonian fluids (cf. Equ. 6.162).

1. Newtonian fluids, with $m = 1$ and $\tau_0 = 0$. Thus, from Equ. 6.162

$$\tau = \eta \cdot \dot{\gamma} \tag{6.163}$$

where η is the viscosity (dynamic) [Ns m^{-2}].

2. Non-Newtonian fluids, of which there are several cases:

a. Power law fluids, where $\tau_0 = 0$

$$\tau = K(\dot{\gamma})^{m} \tag{6.164}$$

with $m < 1$ for pseudoplastic fluids and dilatant liquids with $m > 1$.

b. Bingham plastic fluids, where

$$\tau = \tau_0 + \eta \cdot \dot{\gamma} \tag{6.165}$$

The Casson equation is sometimes a very successful alternative approach:

$$\tau^{1/2} = \tau_0^{1/2} + K_c(\dot{\gamma})^{1/2} \tag{6.166}$$

where K_c is Casson viscosity [(N $sm^{-2})^{1/2}$]. This equation is often preferred to the power law equation in the case of mycelial suspensions (see Roels et al., 1974). Here, apparent viscosity is a function of shear rate:

$$\eta_{app} = \frac{\tau}{\dot{\gamma}} = K_c^2 + \frac{\tau_0}{\dot{\gamma}} + 2K_c\left(\frac{\tau_0}{\dot{\gamma}}\right)^{1/2} \tag{6.167}$$

It can be shown that many fermentation media behave like non-Newtonian fluids. Thereby (apparent) viscosity can be a function of either S and/or P and/or X concentration, exhibiting Newtonian and non-Newtonian behavior. The estimation of the rheological properties of biological fluids is summarized in the literature (e.g., Charles, 1978; Metz et al., 1979). Reuss et al. (1980) incorporated rheological considerations into process modeling. When viscosity is taken into account, clearly any correlation that does not contain the variation of viscosity during the fermentation must fail. An approach recently developed by Zlokarnik (1978), originally proposed for water and water–salt solutions, can be successfully used for this purpose (cf. Equ. 3.81):

$$k_{L1} \cdot a\frac{V_L}{F_G} = f\frac{P_G}{F_G \cdot \rho_L(v \cdot g)^{2/3}} \tag{6.168a}$$

The first group (N_{OTR}) combines $k_{L1} \cdot a$ with liquid volume and gas flow rate F_G; the second group includes power of the impeller P_G, liquid density ρ_L, F_G, kinematic viscosity ($v = \eta/g$), and acceleration due to gravity g. As can be seen from Fig. 6.53, this correlation is adequate for experimental observations of $k_{L1} \cdot a$ during fermentations of *Penicillium chrysogenum* and *Aspergillus niger*. The viscosity of fermentation broth in this case was shown to be primarily a function of biomass concentration. Recently, an alternative equation for this purpose (Stiebitz and Wolf, 1987) was derived, which is superior to Equ. 6.168a and which is valid in the range $2 < \eta < 16$ m Pa·s (up to 50% glucose concentration in H_2O):

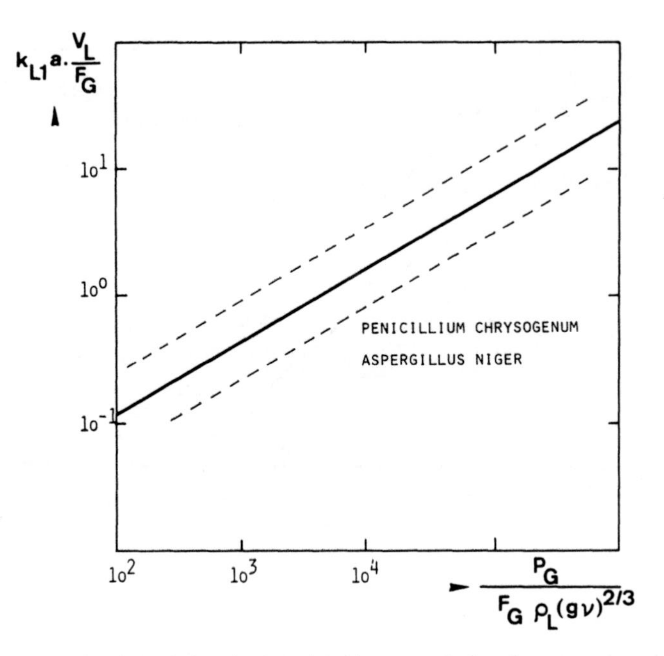

FIGURE 6.53. Application of chemical engineering correlation for oxygen mass transfer according to Equ. 6.168a (cf. Equ. 3.81) to technical bioprocessing in case of fermentations with *Penicillium chrysogenum* and *Aspergillus niger* under various conditions. The dotted lines indicate observed deviations in the bioprocess. (Adapted from Reuss et al., 1980)

$$\frac{k_{L1}a \cdot V_L}{F_G \cdot H_0}\left(\frac{\nu^2}{g}\right)^{1/3} = c_1\left[\frac{P_G}{F_G \cdot \rho_L \cdot H_0 \cdot g}\right]^{c_2} \tag{6.168b}$$

with $c_1 = 5.9 \cdot 10^{-5}$ and $c_2 = 0.52$.

Equation 6.164 was used where the shear rate $\dot{\gamma}$ in the bioreactor was estimated from a known correlation with agitation speed n in a STR (Calderbank and Moo-Young, 1959; Metzner, 1957):

$$\dot{\gamma}_{ave} = k_n \cdot n \tag{6.169}$$

with k_n a constant (being about 10).

Reuss et al. (1980) showed then that the consistency index K, or η_{app}, as well as the flow index m in Equ. 6.164 was correlated with the biomass concentration x following the empirical expressions

$$\eta_{app} = k_x \cdot x^p \tag{6.170}$$

and

$$m = \frac{1}{1 + k_m \cdot x^q} \tag{6.171}$$

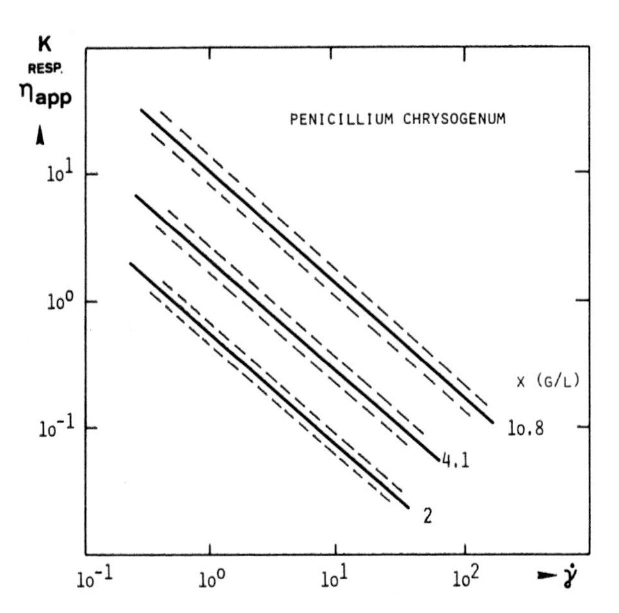

FIGURE 6.54. The correlation acc. to Fig. 6.53 can be verified by using a concept of the dependence of apparent viscosity η_{app} on shear rate $\dot{\gamma}$ as a function of biomass concentration according to Equ. 6.170. (Reuss et al. 1980).

Results of comparison of different measurement systems (turbine impeller and helical ribbon impeller) are shown in Fig. 6.54 at varying concentrations of *P. chrysogenum* biomass.

On the basis of this approach, the complete process model can be written in the following equations for a fed-batch culture (cf. Sect. 6.2 and Bajpaj and Reuss, 1980):

$$r_X = \mu(x, s, o) - \frac{x}{V} \cdot \frac{dV}{dt} \tag{6.172}$$

$$r_S = -\frac{1}{Y_{X|S}} \mu(x, s, o) - \frac{1}{Y_{P|S}} q_P(x, s, o) - m_S x + F_S - \frac{s}{V} \cdot \frac{dV}{dt} \tag{6.173}$$

$$r_P = q_P(x, s, o) - k_{P,d} \cdot p - \frac{p}{V} \cdot \frac{dV}{dt} \tag{6.174}$$

and

$$r_O = -\frac{1}{Y_{X|O}} \cdot \mu(x, s, o) - \frac{1}{Y_{P|O}} \cdot q_P(x, s, o) - m_O x - \frac{o}{V} \cdot \frac{dV}{dt} + k_{L1} \cdot a(o^* - o) \tag{6.175}$$

together with Equs. 6.168, 6.170, and 6.171 for the interaction between viscosity and OTR and with kinetics according to

$$\mu(x, s, o) = \mu_{\max} \frac{s}{K_{S,X} \cdot x + s} \cdot \frac{o}{K_{O,X} \cdot x + o} \tag{6.176}$$

and

$$q_P(x, s, o) = q_{P,\max} \frac{s}{K_P + s(1 + s/K_R)} \cdot \frac{o^r}{K_{O,P} x + o^r} \tag{6.177}$$

where K_R is the formal inhibition constant quantifying metabolite repression for penicillin production (cf. Equ. 5.106).

This process model can be used to predict optimum productivity for penicillin production, as depicted in Fig. 6.18. Productivity increases at low feeding rates of sugar (S limitation), but decreases due to catabolite regulation (K_R!) at higher feeding rates F_S. The influence of $k_{L1} \cdot a$ is demonstrated by the dotted lines in the figure. If $k_{L1} \cdot a$ is diminished as a consequence of increasing v due to increased x, productivity goes down. This model was later used to search for an optimum feeding strategy (cf. Sect. 6.2), and must be regarded as a bioprocess model based on the interactions of transport and kinetics, where the balance equations represent a general concept of scale-up instead of conventional concepts based on power input, mixing times, Re numbers, impeller tip velocity, or other factors of pure physical meaning. Recently, in situ viscosimeters were developed, which fact will enhance the application of the modeling approach described (Björkman, 1987).

6.9.3 Influence of Mycelium—The Morphology Factors ("Apparent Morphology")

Roels et al. (1974) have developed a model for the rheological behavior of filamentous suspensions using the Casson equation by applying a measuring technique consisting of observations of the torque exerted on a rotating turbine impeller. Some considerations analogous to the rheological description of polymer solutions—the so-called "excluded-volume-concept"—lead to the introduction of a morphology factor δ^*, which is a function of yield stress τ_0 and mycelial concentration x, defined as modified factor [Nm]:

$$\delta^* = \frac{\tau_0}{x^2} \text{ to } \frac{\tau_0}{x^{2.5}} \tag{6.178}$$

The results, using data on *P. chrysogenum*, are shown in Fig. 6.55. As can be seen, the power of x seems to be somewhat higher than 2, according to Equ. 6.178. Roels et al. (1974) found that the morphology index decreased during batch fermentation. As a consequence of this approach, the parameters K_c and τ_0 in the Casson equation can be expressed as dependent on δ^*:

$$\sqrt{\tau_0} = \sqrt{\delta^*} \cdot x \tag{6.179}$$

and

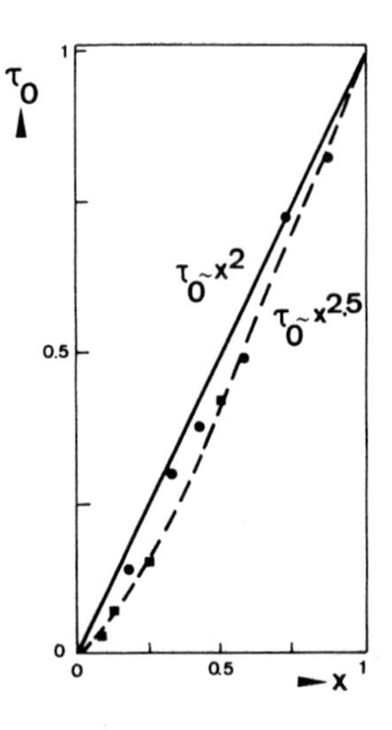

FIGURE 6.55. Relationship between yield stress τ_0 (cf. Equ. 6.162) and mycelial concentration x from different experimental measurements (●, ■) compared with Equ. 6.178). (Adapted from Roels et al., 1974.)

$$\frac{K_c}{\sqrt{\tau_0}} = \eta_0^{1/2}\left(\frac{1}{\sqrt{\delta^* \cdot x}} + c_1\right) \tag{6.180}$$

From Equ. 6.166, thus

$$\sqrt{\tau} = \sqrt{\delta^*} \cdot x\left(1 + \frac{K_c}{\sqrt{\tau_0}}\sqrt{\dot{\gamma}}\right) \tag{6.181}$$

where $K_c/\sqrt{\tau_0}$ is to be substituted from Equ. 6.180.

Even though there are objections to the theoretical model (hyphae treated as flexible chains forming spherical coils; network interaction between branched hyphae; randomizing effect of Brownian motion essential for validity of polymer rheology theory), this formal approach is quite useful (Metz et al., 1979).

A semiautomatic method for the quantitative representation of mold morphology was described (Metz, 1981) in which a variety of morphology indexes were shown to be useful, for example, the effective hyphal length, L_e; the total hyphal length, L_t; and the hyphal growth unit, L_u (cf. Equ. 5.109).

The influence of engineering variables such as agitation and mean energy dissipation $\bar{\varepsilon}$ on the morphology of filamentous molds was also elaborated on by the same group (van Suijdam and Metz, 1981), who concluded that an increase in ε gives a decreases in L_e, L_t, and L_u:

$$L_e = c \cdot \bar{\varepsilon}^{-0.25} \tag{6.182}$$

Very recently, the concept of "apparent" or "engineering" morphology has been presented (van Suijdam, 1986, van Suijdam and Dusseljee 1987), with account taken of the macroscopic nature of microbial suspensions. The aim is to have a parameter that

TABLE 6.2. Comparison of different "morphology factors" and approaches to bio-rheology.

Author	*Equation*	*Equ. no*	*Comments*
van Suijdam and Dusseljee (1987)	$\eta/\eta_0 = 1 + M_F \cdot x$	6.183a	M_F = global morphology factor "apparent morphology"
	$\eta/\eta_0 = 1 + (M_F \cdot x/1 - s \cdot x)$	6.183b	s = crowding factor
Einstein (1906; 1911)	$\eta/\eta_0 = 1 + 2.5 \cdot \phi_S$	6.184	valid if $\phi_S < 0.05$
Mooney (1951)	$\ln \eta/\eta_0 = 2.5 \cdot \phi_S (1 - s \cdot \phi_S)$	6.183d	ϕ_S = volume fraction of solids
Vand (1948)	$\eta/\eta_0 = 1 + 2.5 \cdot \phi_S + 7.25 \cdot \phi_S^2$	6.185	valid if $\phi_S < 0.15$
Eilers (1941)	$\eta/\eta_0 = 1 + \frac{2.5 \cdot \phi_S}{2(1 - \phi_S/\phi_{max})}$	6.186	
Shimmons et al. (1976)	$\eta = \eta_0 \cdot \frac{2}{1 - \phi_S} + 1.36 \cdot \phi_S$	6.187	
Reuss et al. (1979, 1982)	$\eta/\eta_0 = \frac{1}{1 - (h_S \cdot \phi_S)^{1/2}}$	6.188	$1/h_S \sim \phi_{max}$
Roels et al. (1974)	$\phi_c = \sqrt{\delta^* \cdot x}$	6.178	ϕ_c = excluded volume
Metz et al. (1979)	L_H	5.109	L_H = hyphae length
van Suijdam and Metz (1981)	L_e	6.182	
Nestaas and Wang (1981)	$\frac{t_{filt}}{V_{filt} \cdot V_C} = \left(\frac{K_{KC} \cdot \eta_{filt}}{32 \cdot A^2 \cdot \Delta p}\right) \frac{\rho_H \cdot \bar{V}_C}{(1 - \rho_H \cdot \bar{V}_C)^3 \cdot d_H^2}$	(6.189)	filtration characteristics: ρ_H = hyphae density $\equiv \frac{1}{\bar{V}}$; $\bar{V}$ = specific volume of mycelia; d_H = hyphae diameter; V_C = cake volume; K_{KC} = Kozeny-Carmen equation, factor; t_{filt} = filtration time; $\bar{V}_C$ = specific cake volume; η_F = viscosity filtrate; Δp = filtration pressure

- is directly measurable
- has a meaningful relation to process variables
- is based on theoretical principles
- is independent in its dimensions from broth properties, especially from biomass

In analogy to the Einstein equation, putting biomass x as a substitute for volume fraction of solids (cf. Table 6.2 with Equs. 6.183–6.189), a more general and flexible relation is achieved. The factor M_F, thus, is the "apparent" morphology, which is identical to the "intrinsic" viscosity or limiting viscosity number

$$M_F = [\eta] = \lim_{x \to 0} \frac{\eta/\eta_0^{-1}}{x} \tag{6.183c}$$

being the substitute for the factor 2.5 in Equ. 6.184, which was a shape factor. This approach, esp. Equ. 6.183b, is analogue to an equation (cf. Equ. 6.183d) derived by Mooney (1951). M_F, thus, can be estimated from the slope of the plot of η/η_0 versus x, respectively, η versus x. Deviations from the expected straight line appear due to high volume fraction of biomass.

These morphology factors incorporated in the interaction model presented (Equs. 6.172–6.177) and combined with kinetics and balancing methods will form an integrating strategy for successful bioprocess analysis and design, as shown in Fig. 2.15.

6.9.4 Structured Modeling of Bioreactors (OTR)

It has been mentioned previously (cf. Sect. 3.3.2.3) that micromixing in high-volume reactors can be handled adequately only by structured mixing models. These concepts are observed in case of measuring mixing time in technical scale units (cf. Equ. 3.19 in Table 3.1), but at the same time they exhibit influences on other engineering parameters.

Many correlations are known for mixing time (e.g., Brown, 1981; Hughmark, 1980; Joshi et al., 1982; Reuss et al., 1980), but these are valid, however, only in bench-scale units.

An exception are recent data reported concerning gas holdup ε_G (Stenberg, 1984):

$$\varepsilon_G \sim \left(\frac{P_G}{V_L}\right)^{0.21} \cdot 2 \frac{v_S}{1 + 41 \cdot v_S} \tag{6.190}$$

power consumption

$$P_G \sim n^{2.7} \cdot d_T^{4.65} \cdot v_S^{-0.3} \tag{6.191}$$

and O_2 mass transfer (with Equ. 6.190 for ε_G)

$$k_L a = \frac{\varepsilon_G}{d_B} \tag{6.192}$$

Thus, as a general conclusion it becomes clear that more sophisticated mixing models are needed for reactor quantification, for example, OTR, on a technical scale (cf. Fig. 3.10, 3.11, and especially Fig. 4.6 and 4.7).

Basically, two assumptions often used in structured (mixing) modeling are of general importance:

- The reactor vessel can be divided into a number of well-stirred or plug flow or even dead or scarcely agitated zones, called compartments, following the "unit cell" approach.
- A certain flow rate of liquid F_L or gas F_G is recirculated between various modules by the action of the impeller (pumping capacity of the stirrer F_P).

The general setup of structured models was graphically illustrated in Fig. 4.6 for a simple two-compartment model. According to the model presented, mass balances over the reactor must be formulated for significant process variables (e.g., concentrations s, x, o, p, h_v, and c).

The unsteady-state balance of the nth module is given by the following equation, written as a general expression:

$$\text{Rate of change} = \text{rate of input} - \text{rate of output by flow} \pm \text{rate of consumption or formation } r_i \quad (6.193a)$$

which takes the following form in case of mixed modules:

$$V_N \frac{dc_{i,N}}{dt} = F_{in} \cdot c_{i,N-1} - F_{ex} \cdot c_{i,N} \pm r_i \cdot V_N \quad (6.193b)$$

with V_N the volume of the nth compartment.

To determine the concentrations c_i leaving a set of perfectly mixed volume compartments in series including loops it is necessary to apply Equ. 6.193 to each compartment in sequence. This results in a set of coupled first-order differential equations that are to be solved including kinetics and stoichiometric coefficients, and assuming initial conditions resembling the experimental procedure in which an ionic tracer pulse is injected above the stirrer at $t = 0$ (thus $c = c_0$ in the compartment just above the stirrer and $c = 0$ (elsewhere). At $t = \infty$, the concentration is uniform ($c = c^*$).

Depending on the number of modules, the number of parameters increases. A typical set of equations in the case of a three-compartment model, in analogy to Fig. 4.6, will be of the following form including OTR:

$$F_p(c_2 - c_1) + k_L a \cdot V(c_1^* - c_1) - r_O \cdot V = 0 \quad (6.194a)$$

$$F_p(c_3 + c_1 - 2c_2) + k_L a \cdot V(c_2^* - c_2) - r_O \cdot V = 0 \quad (6.194b)$$

$$F_p(c_2 - c_3) + k_L a \cdot V(c_3^* - c_3) - r_O \cdot V = 0 \quad (6.194c)$$

Dividing each term by F_p we obtain a set of equations with a dimensionless parameter τ, which is the ratio of two characteristic times ($\tau = \bar{t}_p/t_{OTR}$). Here $\bar{t}_p$ is the mean internal residence time (V/F_p), representing the pumping capability of the stirrer, and $t_{OTR} = 1/k_L a$, representing the mass transfer capacity of the stirrer. Analogous concepts of characteristics times and characteristic rate constants are generally used in regime analysis (cf. Table 4.2, Sect. 4.2).

Similar structured approaches to mixing have been frequently used recently in modeling OTR in production-scale bioreactors. Clearly, the dissolved oxygen distribution in a reactor will depend on the interactions between mixing and mass transfer. Figure 4.7 represented a simple unstructured model for OTR in a stirred-tank reactor, taking into account the hydrodynamic phenomena associated with the stirrer (Warmoeskerken and Smith, 1982). Although this is a more mechanistic approach, the scale dependence is still a problem. The influence of geometry is not clear. Therefore, structured modeling is needed for scale-up purposes, including a model for gas holdup ε_G, liquid

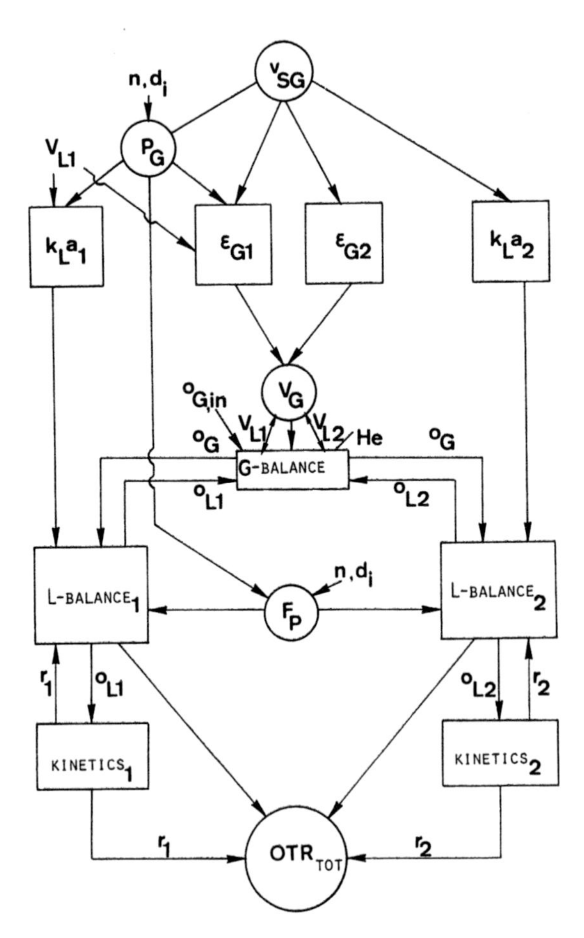

FIGURE 6.56. Block diagram of the two-compartment model for oxygen transfer in a production-scale bioreactor (cf. Fig. 4.6), including parallel work for mixed zone (1) and bubble zone (2). For symbols see nomenclature. (Adapted from Oosterhuis, 1984.)

circulation (t_c, CTD), the relation between gas bubble diameter d_B and k_L, or the coalescence and redispersion of bubbles (Oosterhuis, 1984).

Figure 6.56 gives an impression of the work related to this type of structured modeling, showing the balances and auxiliary correlations for $k_L a$, ε_G, P/V, F_p, and q_O, which are used to calculate the total oxygen transfer capacity of the reactor (OTR) as combined from mixed zone and bubble zone (cf. Fig. 4.6).

Finally, the effect of structured models due to imperfect mixing in bioreactors upon product formation, yield, and conversion will be mentioned here. It was shown by Oosterhuis (1984) that r_P can be considered to consist of two parts [t_1 = residence time in aerated compartment and $t_c = (V_1 + V_2)/F_p$)]:

$$r_P = 2 \cdot F_p \frac{c_1}{V_{tot}} + \frac{t_1}{t_c} r_{P,max} \tag{6.195}$$

In the case of a relatively low exchange flow F_p between compartments 1 and 2, the ratio t_1/t_c can be used as a scale-up or scale-down criterion.

6.10 Final Note

In place of general conclusions, Fig. 6.57 summarizes the overall strategy of bioprocess technology by graphical means. After the qualitative characteristics of the biocatalyst/substrate system are worked out in a microbiological laboratory, the first phase consists of quantitative measurements and mathematical model building of biokinetics in a "perfect bioreactor" (Chaps. 4 and 5). Quantitative measurements and model building for various types of reactors form the second phase (Chap. 3). Process development (Chap. 6) is concluded only with the formulation of a process model made up of transport, kinetics (including interactions), and scale-up elements based on additional pilot plant measurements.

All of the methods presented in this chapter agree with the strategy given in Figs. 1.4, 2.14, and 2.20. The necessary kinetic data come primarily from measurements made on discontinuous processes as a first working hypothesis. However, these data are not readily transferable to other modes of operation: The data reflect a strong coupling between the reactor operation and biological behavior. Maintaining a systematic approach, as suggested by the procedures described in this book, should lead step-by-step to a more reliable method for planning bioprocess operations.

Acknowledgment. Most of the computer simulations in Chaps. 5 and 6 were performed by the group of Prof. Dr. W. Knorre of ZIMET (Central Institute for Microbiology and Experimental Therapy, Biotechnology Division) at the Academy of Sciences of the German Democratic Republic in Jena (Prof. F.

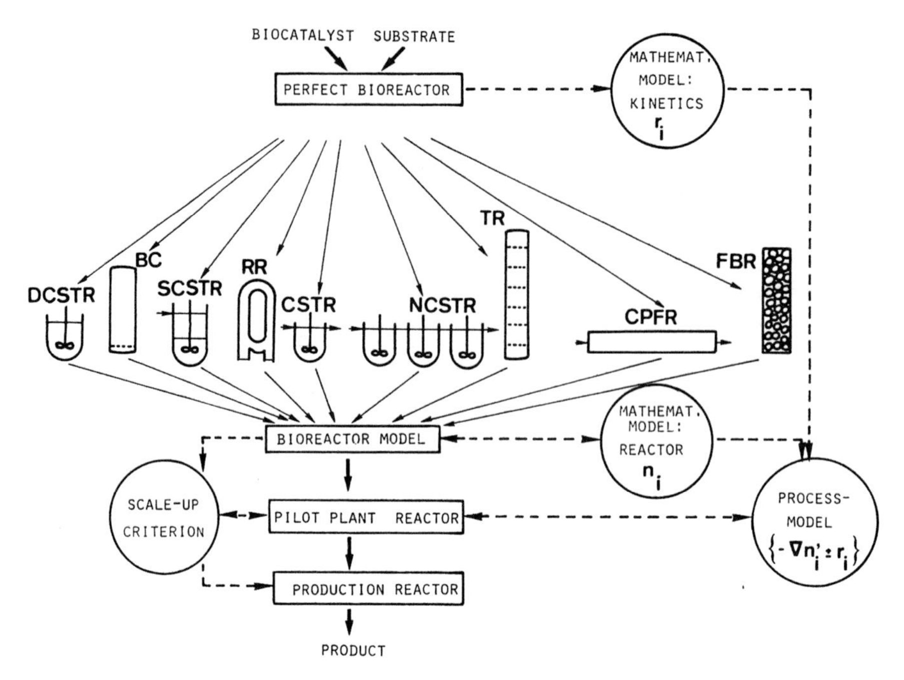

FIGURE 6.57. Summary diagram of work flow in the systematic development of a bioprocess of the presented integrating strategy. The diagram is based on the interaction between kinetics (Chap. 5) and transport (Chap. 3) processes, which are clarified during a kinetic analysis (Chap. 4). As a special situation, the design and utilization of new types of reactors are shown (discontinuous stirred vessel, DCSTR; bubble column, BC; semicontinuous stirred vessel, SCSTR; recycle reactor, RR; continuous stirred vessel, CSTR; continuous cascade, NCSTR; tower reactor, TR; continuous plug flow reactor, CPFR; fixed and fluidized bed reactor, FBR).

Bergter), based on a scientific collaboration with the author's group at the Technical University in Graz, Austria.

Bibliography

Abbot, B.J., and Gerhardt, P. (1970a). *Biotechnol. Bioeng.*, 12, 577.
Abbot, B.J., and Gerhardt, P. (1970b). *Biotechnol. Bioeng.*, 12, 603.
Adler, I., and Fiechter, A. (1983). *Chem. Ing. Techn.*, 55, 322.
Agrawal, P., and Lim, H.C. (1984). *Adv. Biochem. Eng.*, 30, 61.
Aiba, S., Humphrey, A.E., and Millis, N.F. (1973). *Biochemical Engineering.* New York and London: Academic Press.
Aiba, S., et al. (1976). *Biotechnol. Bioeng.*, 18, 1001.
Alper, E., et al. (1980a). *Chem. Eng. Sci.*, 35, 217.
Alper, E., et al. (1980b). *Chem. Eng. Sci.*, 35, 1264.
Andrews, J.F. (1971). *Biotechnol. Bioeng. Symp.*, 2, 5.
Andrews, J.F. (1982). *Biotechnol. Bioeng.*, 24, 2013.

Andrews, J.F., and Tien, C. (1982). Amer. Inst. Chem. Eng. Journal *AIChEJ*, 28, 182.
Applegate, L.E. (1984). *Chem. Eng.*, June, 64.
Arcuri, E.J., and Donaldson, T.L. (1981). *Biotechnol. Bioeng.*, 23, 2149.
Aris, R., and Humphrey, A.E. (1977). *Biotechnol. Bioeng.*, 19, 1375.
Atkinson, B. (1973). *Pure Appl. Chem.*, 36, 279.
Atkinson, B. (1974). *Biochemical Reactors.* London: Poin.
Atkinson, B. (ed.) (1980). *Symposium on Biological Fluidized Bed Treatment of Water and Wastewater*, Water Research Centre, Stevenage Lab., Elder Way, Stevenage/Hertfordshire, Great Britain, April 1977.
Atkinson, B., and Fowler, H.W. (1974). *Adv. Biochem. Eng.*, 3, 221.
Atkinson, B., and Knights, A.J. (1975). *Biotechnol. Bioeng.*, 17, 1245.
Atkinson, B., and Kossen, N.W.F. (1978). *Proceedings of the 1st European Congress on Biotechnology*, Interlaken, Switzerland, Dechema Monograph, 82, 37
Atkinson, B., et al. (1980). *Proc. Biochem.*, May, 24.
Atkinson, B., and Mavituna, F. (1983). *Biochemical Engineering and Biotechnology Handbook.* Nature Press, Macmillan. Byfleet, Surrey England.
Atkinson, B., and Davis, I.J. (1972). *Trans. Inst. Chem. Engrs.* 50, 208.
Bahl, H., et al. (1982). *Eur. J. Appl. Microbiol. Biotechnol.*, 15, 201.
Bailey, J.E. (1973). *Chem. Eng. Commun.*, 1, 111.
Bailey, J.E., and Ollis, D.F. (1977). *Biochemical Engineering Fundamentals.* New York: McGraw-Hill, p. 642.
Bajpaj, R.K., and Reuss, M. (1982). *Canad. J. Chem. Eng.*, 60, 384.
Bajpaj, R.K., and Reuss, M. (1981). *Biotechnol. Bioeng.*, 23, 717.
Bajpaj, R.K., and Reuss, M. (1980). *J. Chem. Techn. Biotechnol.*, 30, 322.
Baker, C.G.J., et al. (1980). In Moo-Young M., et al. (eds.). *Proceedings of the 6th International Fermentation Symposium*, Vol. 1. Oxford: Pergamon Press, 635.
Barford, J.P., et al. (1982). In Bazin, M.J. (ed.). *Microbial Population Dynamics.* Boca Raton, Fla.: CRC Press, p. 55.
Baxerres, J.L., Haewsungcharern, A., and Gibert, H. (1977). *Lebensm.-Wiss. Technol.*, 10, 191.
Bazin, M.J. (ed.) (1982). *Microbial Population Dynamics.* Boca Raton, Fla.: CRC Press.
Bentley, T.L., and Kincannon, D.F. (1976). *Water Sewage Work*, R 10.
Bergter, F. (1972). *Wachstum von Mikroorganismen.* Jena: G. Fischer Verlag.
Biggs, R.D. (1963). *Canad. J. Chem. Eng.*, 60, 384.
Bischoff, K.B. (1966). *Canad. J. Chem. Eng.*, 45, 281.
Björkman, U. (1987). *Biotechnol. Bioeng.*, 29, 101 and 114.
Blenke, H. (1979). *Adv. Biochem. Eng.*, 13, 122.
de Boks, P.A., and van Eybergen, G.C. (1981). *Biotechnol. Lett.*, 3, 577.
Borzani, W., et al. (1976). *Biotechnol. Bioeng.*, 18, 623, 885.
Brown, D.E. (1981). *Instn. Chem. Eng.*, Ser. 64, N7.
Bruxelmane, M. (1983). *Tech. l'Ingen.*, 2, A5910.
Bryant, J. (1977). *Adv. Biochem. Eng.*, 5, 101.
Bryant, J., and Sadeghzadeh, N. (1979). Paper F3 presented at 3rd Eur. Conf. on Mixing, Univ. of York, Great Britain. April 4–6.
Bull, D.N., and Young, M.D. (1981). *Biotechnol. Bioeng.*, 23, 373.
Bu'Lock, J.D. (1965). *The Biosynthesis of Natural Products.* New York: McGraw-Hill, Chap. 1.
Bu'Lock, J.D., et al. (1965). *Canad. J. Microbiol.*, 11, 765.
Bungay, H.R., III, and Bungay, M.L. (1968). *Adv. Appl. Microbiol.*, 10, 269.

Bungay, H.R., III, et al. (1969). *Biotechnol. Bioeng.*, 11, 765.
Calderbank, P.H., and Moo-Young, M. (1959). *Trans. Instn. Chem. Engrs.*, 37, 26.
Cameron, I.L., and Padilla, C.M. (eds.) (1966). *Cell Synchrony*. New York: Academic Press.
Charaklis, W.G. (1981). *Biotechnol. Bioeng.*, 23, 1923.
Charaklis, W.G., et al. (1982). *Water Res.*, 0, 1.
Charles, M. (1978). *Adv. Biochem. Eng.*, 8, 1.
Charley, R.C., et al. (1983). *Biotechnol. Lett.*, 5, 169.
Chen, J.W., et al. (1965). *Ind. Eng. Chem., Proc. Des. Dev.*, 4, 421.
Chi, C.T., and Howell, J.A. (1976). *Biotechnol. Bioeng.*, 18, 63.
Chi, C.T., et al. (1974). *Chem. Eng. Sci.*, 29, 207.
Constantinides, A., et al. (1981). *Biotechnol. Bioeng.*, 23, 899.
Cooker, B., et al. (1983). *Chem. Eng., Proc. Des. Dev.*, 19, 600.
Cooney, Ch.L. (1979). *Proc. Biochem.*, 14, May, 31.
Cooney, Ch.L., and Acevedo, F. (1977). *Biotechnol. Bioeng.*, 19, 1449.
Cooney, Ch.L., and Wang, D.I.C. (1976). *Biotechnol. Bioeng.*, 18, 189.
Coughlin, R.W., et al. (1975). *Chem. Ing. Techn.*, 47, 111.
Coulman, G.A., et al. (1977). *Appl. Environ. Microbiol.*, 34, 725.
Court, J.R., and Pirt, S.J. (1976). 5th Internat. Ferment. Symp., Berlin, 6.21.
Cysewski, C.R., and Wilke, Ch.R. (1978). *Biotechnol. Bioeng.*, 20, 1421.
Daigger, G.T., and Grady, C.P.L. (1982). *Biotechnol. Bioeng.*, 24, 1427.
Dairaku, K., et al. (1981). *Biotechnol. Bioeng.*, 23, 2069.
Danckwerts, P.V. (1958). *Chem. Eng. Sci.*, 8, 93.
Dawson, P. (1972a). *J. Appl. Chem. Biotechnol.*, 22, 79.
Dawson, P. (1972b). In *Fermentation Technology Today* (Terui G., ed.) Proc. 4th Intern. Ferment. Symp. Society of Ferment. Technol. Japan Osaka, p. 121.
Dawson, P. (1980a). In *Bioconversion and Biochem. Engineering* (Ghose, T.K., ed.) Proc. 2nd Symp., Indian Institute of Technology, New Delhi Vol. 2, p. 275.
Dawson, P. (1980b). Paper at 6th Intern. Ferment. Symp., London/Ontario July 20–25, no. F-8.1.2.
Dawson, P. (1980c). Paper at 6th Intern. Ferment. Symp., London/Ontario July 20–25, no. F-8.1.10.
Dean, A.C.R., et al. (1972). *Environmental Control of Cell Synthesis and Function*. New York: Academic Press.
Deindoerfer, F.H., and Humphrey, A.E. (1959). *Ind. Eng. Chem.* 51, 809.
Demain, A.L., et al. (1979). *Proc. Symp. Soc. Gen. Microbiol.*, 29.
Dorawala, T.G., and Douglas, J.M. (1971). Amer. Inst. Chem. Eng. Journal *AIChEJ*, 17, 974.
Dostalek, M., and Häggstrom, M. (1982). *Biotechnol. Bioeng.*, 24, 2077.
Douglas, J.M. (1972). *Process Dynamics and Control*, Vol. 2. Englewood Cliffs, N.J.: Prentice Hall.
Dunn, I.J., and Mor, J.R. (1975). *Biotechnol. Bioeng.*, 17, 1805.
Eckenfelder, W.W. (1966). *Industrial Water Pollution Control*. New York: McGraw-Hill, Chap. 13.
Edwards, V.H. et al. (1970). *Biotechnol. Bioeng.*, 17, 975.
Eggers, E., and Terlouw, T. (1979). *Water Res.*, 13, 1077.
Eilers, H. (1941), *Kolloid Z.*, 97, 313.
Einsele, A., and Finn, R.K. (1980). *Ind. Eng. Chem., Proc. Des. Dev.*, 19, 600.
Einstein, A. (1906). *Ann. Physik.*, 19, 289.

Einstein, A. (1911). *Ann. Physik.*, 34, 591.
Eirich, F., et al. (1936). *Kolloid Z.*, 74, 276.
Erickson, L.E., et al. (1972). *J. Appl. Chem. Biotechnol.*, 22, 199.
Esener, A.A., et al. (1981a). *Biotechnol. Lett.*, 3, 15.
Esener, A.A., et al. (1981b). *Eur. J. Appl. Microb. Biotechnol.*, 13, 141.
Esener, A.A., et al. (1981c). *Biotechnol. Bioeng.*, 23, 1851.
Fan, L.T., Erickson, L.E., Shah, P.S., and Tsai, B.I. (1970). *Biotechnol. Bioeng.*, 12, 1019.
Fan, L.T., Tsai, B.I., and Erickson, L.E. (1971). Amer. Inst. Chem. Eng. Journal *AIChEJ*, 17, 689.
Fencl, Z. (1964). In Malek, I, et al. (eds.). *Proceedings of Symposium on Continuous Culture of Microorganisms.* Prague: Csechoslovak Academy of Sciences, p. 23.
Fencl, Z., and Novak, M. (1969). *Folia Microb.*, 14, 314.
Fencl, Z., et al. (1969). In Perlman, D., (ed.). *Fermentation Advances.* New York: Academic Press, p. 301.
Fencl, Z., et al. (1972). *J. Appl. Chem. Biotechnol.*, 22, 405.
Fencl, Z., et al. (1978). In *Proceedings of the 7th Symposium on Continuous Culture of Microorganisms.* Sikyta, B., et al. (eds.). Czechosl. Acad. of Sciences Prague: (1980) p. 49.
Fiechter, A. (1982). In Rehm, H.J., and Reed, G. (eds.). *Biotechnoloby—A Comprehensive Treatise*, Vol. 1. Deerfield Beach, Fla., and Basel, Verlag Chemie Weinheim, Chap. 7.
Finn, R.K., and Fiechter, A. (1979). *Symp. Soc. Gen. Microbiol.*, 20, 83.
Furusaki, S., and Miyauchi, T. (1977). *J. Chem. Eng.* (*Jap.*), 10 (3), 247.
Gerhardt, P., and Gallup, D.M. (1963). *J. Bacteriol.*, 86, 919.
Gerstenberg, H., et al. (1980). *Chem. Ing. Techn.*, 52, 19.
Geurts, Th.G., et al. (1980). *Biotechnol. Bioeng.*, 22, 2031.
Goma, G., et al. (1979). *Biotechnol. Lett.*, 1, 415.
Goto, S., et al. (1973). *J. Ferment. Technol.*, 51, 582.
Greenshields, R.N., and Smith, E.L. (1974). *Proc. Biochem.*, April, 11.
Grieves, R.B., et al. (1964). *J. Appl. Chem.*, 14, 478.
Gutke, R. (1980, 1982). In UNEP/UNESCO/ICRO training course, *Theoretical Basis of Kinetics of Growth, Metabolism and Product Formation of Microorganisms.* Jena: Science, Academy of East Germany, ZIMET, Vol. 1, p. 112 (1980); pps. 39 and 58 (1982).
Gutke, R., and Knorre, W.A. (1980). *Z. Allgem. Mikrobiol.*, 20 (7), 441.
Gutke, R., and Knorre, W.A. (1981). *Biotechnol. Bioeng.*, 23, 2771.
Gutke, R., and Knorre, W.A. (1982). *Biotechnol. Bioeng.*, 24, 2129.
Gutke, R., et al. (1980). *Biotechnol. Lett.*, 2, 315.
Hamer, G. (1982). *Biotechnol. Bioeng.*, 24, 511.
Han, K., and Levenspiel, O. (1987). *Biotechnol. Bioeng.*, in press.
Harremoës, P. (1978). In Mitchell, R., (ed.), J. Wiley & Sons N.Y. *Water Pollution Microbiology*, Vol. 2. Chap. 4.
Harris, N.P., and Hansford, G.S. (1976). *Water Res.*, 10, 935.
Harrison, D.E.F., and Topiwala, H.H. (1974). *Adv. Biochem. Eng.*, 3, 167.
Heckershoff, H., and Wiesman, U. (1981). *Chem. Ing. Techn.*, 53, 268.
Hegewald, E.M., et al. (1978). In Sikyta, B., et al. (eds.), Czechosl. Acad. of Sciences. *Proceedings of the 7th Symposium on Continuous Culture of Microorganisms.* Prague: (1980), p. 717.

Heijnen, J.J., et al. (1979). *Biotechnol. Bioeng.*, 21, 2175.
Heijnen, J.J., and Roels, J.A. (1981). *Biotechnol. Bioeng.*, 23, 739.
Herbert, D. (1961). In *Proceedings of Symposium on Continuous Culture of Microorganisms.* London: Elsworth R., ed. Society of Chemical Industry Monograph 12, p. 21.
Herbert, D. (1964). In Malek, I., et al. (eds.). *Continuous Cultivation of Microorganisms.* Prague: Czechoslovak Academy of Science, p. 23.
Herbert, D., et al. (1956). *J. Gen. Microbiol.*, 14, 601.
Herzog, P., et al. (1983). *Chem. Ing. Techn.*, 55, 566.
Holmes, D.B., et al. (1964). *Chem. Eng. Sci.*, 19, 201.
Howell, J.A., and Atkinson, B. (1976). *Biotechnol. Bioeng.*, 18, 15.
Hughmark, G. (1980). *Ind. Eng. Chem., Proc. Des. Dev.*, 19, 638.
Humphrey, A.E. (1978). *Am. Chem. Soc. Symp.*, Ser. 72.
Humphrey, A.E. (1980). *Adv. Biotechnol.*, 1, 203.
Imanaka, T., et al. (1973). *J. Ferment. Technol. (Jap.)*, 51, 558.
Irvine, R.L., et al. (1980). *J. Water Poll. Contr. Fed.*, 52, 1997.
Jennings, P.A., et al. (1976). *Biotechnol. Bioeng.*, 18, 1249.
Joshi, J.B. (1980). *Trans. Instn. Chem. Engrs.*, 58, 155.
Joshi, J.B., et al. (1982). *Chem. Eng. Sci.*, 37, 813.
Kan, J.K., and Shuler, M.L. (1976). *Amer. Inst. Chem. Eng. Symp. Ser.* 172, 31.
Kan, J.K., and Shuler, M.L. (1978). *Biotechnol. Bioeng.*, 20, 217.
Kargi, F., and Park, J.K. (1982). *J. Chem. Techn. Biotechnol.*, 32, 744.
Katinger, H. (1976). Paper presented at 5th Internat. Ferment. Symp., June 28–July 3 Berlin, No. 4.16.
Keller, R., and Dunn, I.J. (1978a). *J. Appl. Chem. Biotechnol.*, 28, 508.
Keller, R., and Dunn, I.J. (1978b). *J. Appl. Chem. Biotechnol.*, 28, 784.
Khang, S.J., and Levenspiel, O. (1979). *Chem. Eng. Sci.*, 31, 569.
Kincannon, D.F., and Sherrard, J.H. (1974). *Water Sewage Work*, R32.
Kipke, K.D. (1984). In *Process Variables in Biotechnology*, Bioreactor Performance working group chairman W. Crueger, Dechema Monograph, Chap. 20.
Kishimoto, M., et al. (1976). *J. Ferment. Technol.*, 54, 891.
Kitai, A., et al. (1969). *Biotechnol. Bioeng.*, 11, 911.
Kjaergaard, L., and Jørgensen, B.B. (1979). *Biotechnol. Bioeng.*, 21, 147.
Klei, H.E., et al. (1975). *J. Appl. Chem. Biotechnol.*, 25, 535.
Knorre, W.A. (1968). In *Proceedings of 4th Symposium on Continuous Culture of Microorganisms.* Prague; Malek, J., et al. (eds.) (1969): Academia, Prague p. 225.
Knorre, W.A. (1980). In Beier, W., and Rosen, R. (eds.). *Biophysikalische Grundlagen der Medizin.* Stuttgart, New York: G. Fischer, p. 132.
Koga, S., and Humphrey, A.E. (1967). *Biotechnol. Bioeng.*, 9, 375.
Kong, M.F., and Yang, P.Y. (1979). *Biotechnol. Bioeng.*, 21, 417.
Kornegay, B.H. (1969). In *Proceedings of 24th Industrial Waste Conference.* Purdue Univ., Lafayette, Ind. p. 1398.
Kornegay, B.H. (1975). *Mathematical Modelling of Water Pollution Control Bioprocess,* Keinath, T.M. and Wainielista, M. (eds.), Ann Arbor, Mich., Ann Arbor Science Publ. Inc.
Kornegay, B.H., and Andrews, J.F. (1968). *J. Water Poll. Contr. Fed.*, 40, R460.
Kosaric, N., et al. (1980). *Adv. Appl. Microbiol.*, 26, 147.
Kramers, H., et al. (1953). *Chem. Eng. Sci.*, 2, 35.
Kuhn, H.J., et al. (1980). *Eur. J. Appl. Microbiol. Biotechnol.*, 10, 303.

Küng, W., and Moser, A. (1986). *Bioprocess Engng.*, 1, 23.
Küng, W., and Moser, A. (1988). *Biotechnol. Lett.*, in press.
Lacey, R.E. (1972). *Chem. Eng.*, September 4, 56.
Laederach, H., et al. (1978). In Preprints 1st Europ. Congr. Biotechnology, Interlaken, Switzerland, Sept. 25–29, Dechema, Frankfurt.
La Motta, E.J. (1976). *Biotechnol. Bioeng.*, 18, 1359.
Lane, A.G. (1977). *J. Appl. Chem. Biotechnol.*, 27, 165.
Lee, H.H., and Yan, B.D. (1981). *Chem. Eng. Sci.*, 36, 483.
Lee, I.H., et al. (1976). *Biotechnol. Bioeng.*, 18, 513.
Lee, J.M., et al. (1983). *Biotechnol. Bioeng.*, 25, 497.
Leegwater, M.P.M., et al. (1982). *J. Chem. Technol. Biotechnol.*, 32, 92.
Levenspiel, O. (1972). *Chemical Reaction Engineering.* New York: John Wiley.
Levenspiel, O. (1979). *The Chemical Reactor Omnibook.* Corvallis, Ore.: OSU Book Stores.
Levenspiel, O. (1980). The Monod Equation; A Revisit; Biotechnology and Bioengineering, 22, 1671–1687, John Wiley and Sons, Inc.
Liepe, F., et al. (1978). In reprints, *1st European Congress on Biotechnology*, Interlaken, Switzerland, Part 1, p. 78.
Lilly, M.D., and Dunnill, P. (1971). *Proc. Biochem.*, 6 (8), 29.
Lilly, M.D., and Dunnill, P. (1972a). *Adv. Biochem. Eng.*, 3, 221.
Lilly, M.D., and Dunnill, P. (1972b). *Biotechnol. Bioeng. Symp.*, 3, 221.
Lilly, M.D. (1982). *J. Chem. Techn. Biotechnol.*, 32, 162.
Lippert, J., et al. (1983). *Biotechnol. Bioeng.*, 25, 437.
Luedeking, R., and Piret, E.L. (1959). *Biotechnol. Bioeng.*, 1, 431.
Luttman, R., et al. (1981). *Eur. J. Appl. Microbiol. Biotechnol.*, 13, 90, 145.
Maiorella, B., et al. (1981). *Adv. Biochem. Eng.*, 20, 43.
Malek, I., and Fencl, Z. (1966). *Theoretical and Methodological Basis of Continuous Culture of Microorganisms.* New York: Academic Press.
Malek, I., and Ricica, J. (1969). *Folia Microbiol.*, 14, 254.
Margaritis, A., and Wilke, Ch.R. (1972). *Dev., Ind. Microbiol.*, 13, 159.
Margaritis, A., and Wilke, Ch.R. (1978a). *Biotechnol. Bioeng.*, 20, 709.
Margaritis, A., and Wilke, Ch.R. (1978b). *Biotechnol. Bioeng.*, 20, 727.
Marsot, P., et al. (1981). *Biotechnol. Lett.*, 3, 689.
Matsumura, M., et al. (1981). *J. Ferment. Technol.*, 59, 115.
Mattiasson, B. (1983). *Trends in Biotechnology*, 1, 16.
McGrath, M.J., and Yang, R.Y.K. (1975). *Chem. Eng. J.*, 9, 187.
Melling, J. (1977). In Wiseman, A. (ed.). Topics in *Enzyme and Fermentation Biotechnology*, Vol. 1. Chichester: Ellis Horwood Ltd., p. 10.
Mersmann, A., et al. (1976). *Int. Chem. Eng.*, 16, 590.
Merz, A., and Vogg, H. (1978). *Chem. Ing. Techn.*, 50, 108.
Metcalf & Eddy Engineers. (1972, 1979). *Waste Water Engineering Treatment, Disposal and Reuse.* New York: McGraw-Hill.
Metz, B. (1981). *Biotechnol. Bioeng.*, 23, 149.
Metz, B., and Kossen, N.W.F. (1977). *Biotechnol. Bioeng.*, 19, 781.
Metz, B., et al. (1979). *Adv. Biochem. Eng.*, 11, 103.
Metzner, A.B. (1957). Americ. Inst. Chem. Eng. Journal *AIChEJ*, 3, 3.
Middleton, J.C. (1979). Paper A2 presented at 3rd Eur. Conf. on Mixing, York, Great Britain.
Molin, G., et al. (1982). *Eur. J. Appl. Microbiol. Biotechnol.*, 15, 218.

Monod, J. (1942). *Recherches sur la Croissance des Cultures Bacteriennes*. Paris: Hermann.

Monod, J. (1950). *Ann. Inst. Pasteur*, 79, 390.

Mooney, U. (1951). *J. Colloid. Sci.*, 6, 162.

Moreno, M., and Goma, G. (1979). *Biotechnol. Lett.*, 1, 483.

Moser, A. (1973). In Dellweg, H. (ed.). *Proceedings of the 3rd Symposium on Technical Microbiology*. Inst. für Gärungsgewerbe und Biotechnologie Berlin, p. 61.

Moser, A. (1977). *Chem. Ing. Techn.*, 49, 612.

Moser, A. (1980a). In Moo-Young, M., et al. (eds.). *Proceedings of the 2nd Intern. Symp. on Waste Treatment & Utilization*, Vol. 2. Oxford: Pergamon Press, p. 177.

Moser, A. (1980b). In Ghose, T.K. (ed.). *Proceedings of the 2nd International Symposium on Bioconversion and Biochemical Engineering*, Vol. 2. New Delhi: Indian Inst. Technology, p. 253.

Moser, A. (1982). *Biotechnol. Lett.*, 4, 281.

Moser, A. (1983a). *Proc. Adv. Ferment.*, 83 (*Suppl. Proc. Biochem.*), 201.

Moser, A. (1983b). In *Proceedings of the 33rd Canadian Chemical Engineering Conference*, Toronto, Canad. Soc. for Chem. Eng., Oct. 2–5. Vol. 2, p. 417.

Moser, A. (1984a). In Ghose, T.K. (ed.). *Proceedings of the 7th International Biotechnology Symposium*, Vol. 2. New Delhi: Indian Inst. Technology, p. 529.

Moser, A. (1984b). *Acta Biotechnolog.*, 4, 3.

Moser, A. (1987). In Crueger, W., et al. (eds.). *Physical Aspects of Bioreactor Performance* report of working party Bioreactor Performance of Europ. Fed. Biotechnol., Dechema, Frankfurt, Chap. 4.

Moser, A. (1985a). In Rehm, H.J., and Reed, G. (eds.). *Biotechnology—A comprehensive Treatise*, Vol. 2. Deerfield Beach, Fla., and Basel: Verlag Chemie Weinheim, Chaps. 15 and 16.

Moser, A. (1985b). In Moo-Young, M. (ed.-in-chief). *Comprehensive Biotechnology*, Vol. 1. Oxford: Pergamon Press, Part 2, Chap. 4.

Moser, A. (1985c). *Conservation & Recycling*, Vol. 8. No. 1/2, 193–210, Pergamon Press, Oxford.

Moser, A., Preselmayr, W., and Scherbaum, H. (1974). In *Proceedings of the 4th International Symposium on Yeasts*, Part I, Klaushofer, H. and U., Sleytr (eds.), Univ. of Bodenkultur, Vienna, p. 117.

Moser, F. (1977). *Verfahrenstechnik*, 11, 670.

Moser, F. (1980). In Moser, F. (ed.). *Grundlagen der Abwasserreinigung*, Vol. 2. Munich: Oldenburg, p. 431.

Moser, F., et al. (1977). *Prog. Water Tech.*, 8, 235.

Moser, F., et al. (1979). *Österr. Abwasser Rund.*, 24, 83.

Mou, D.G. (1979). Ph.D. thesis, Massachusetts Institute of Technology, Cambridge, Mass.

Mou, D.G., and Conney, Ch.L. (1983a). *Biotechnol. Bioeng.*, 25, 225.

Mou, D.G., and Conney, Ch.L. (1983b). *Biotechnol. Bioeng.*, 25, 257.

Mukataka, S. (1981). *J. Ferment. Technol.*, 59, 303.

Mukataka, S., et al. (1980). *J. Ferm. Technol.*, 58, 155.

Mulcahy, L.T., and La Motta, E.J. (1978). Rep. no. 59-78-2, Environmental Engineering Program, Dept. Civil Engng., University of Massachusetts, Amherst.

Nagai, S., et al. (1968). *J. Gen. Appl. Microbiol.*, 14, 121.

Nagel, O. et al. (1977). *Chem. Ing. Tech.*, 44, 367.

Nagel, O., Hegner, B., and Kürten, H. (1978). *Chem. Ing. Techn.*, 50, 934.

Nagel, O., Kürten, H., and Sinn, R. (1972). *Chem. Ing. Techn.*, 44, 367.
Nestaas, E., and Wang, D.I.C. (1981). *Biotechnol. Bioeng.*, 23, 2803.
Norwood, K.W., and Metzner, A.B. (1960). *Americ. Inst. Chem. Eng. Journal*, 6, 432.
Novick, A., and Szilard, L. (1950). *Proc. Nat. Acad. Sci., Wash.*, 36, 708.
Oleszkiewicz, J. (1976). *Environ. Protect. Eng.*, 2, 85.
Oleszkiewicz, J. (1977). *Prog. Water Tech.*, 9, 777.
Oosterhuis, N.M.G. (1984). Ph.D. thesis, Technical University, Delft, Netherlands.
Ottengraf, S.P.P. (1977). *Biotechnol. Bioeng.*, 19, 1411.
Paca, J., and Gregr, V. (1976). *Biotechnol. Bioeng.*, 18, 1075.
Paca, J., and Gregr, V. (1979). *Enzyme Microb. Technol.*, 1, 100.
Park, Y., et al. (1984). *Biotechnol. Bioeng.*, 26, 457, 468.
Parulekar, S.J., and Lim, H.C. (1985). *Adv. Biochem. Eng.*, 32, 207.
Pickett, A.M. (1982). In Bazin, M. (ed.). *Microbial Population Dynamics*. Boca Raton, Fla.: CRC Press, Chap. 4.
Pickett, A.M., et al. (1979a). *Proc. Biochem.*, 13, November, 10.
Pickett, A.M., et al. (1979b). *Biotechnol. Bioeng.*, 21, 1043.
Pickett, A.M., et al. (1980). *Biotechnol. Bioeng.*, 22, 1213.
Pirt, S.J. (1974). *J. Appl. Chem. Biotechnol.*, 24, 415.
Pirt, S.J. (1975). *Principles of Microbe and Cell Cultivation*. Oxford: Blackwell.
Pirt, S.J., and Righelato, R.C. (1967). *Appl. Microbiol.*, 15, 1284.
Pirt, S.J., et al. (1961). *J. Gen. Microbiol.*, 25, 119.
Pirt, S.J., et al. (1983). *J. Chem. Tech. Biotechnol.*, 33B, 35.
Powell, O., and Lowe, J.R. (1964). In Malek, I., et al. (eds.). *Continuous Cultivation of Microorganisms*. Prague: Czechoslovak Academy of Science, p. 45.
Powell, O., et al. (1967). *Microbial Physiology and Continuous Culture*. London: Her Majesty's Stationery Office.
Prochazka, J., and Landau, J. (1961). *Colln. Czech. Chem. Commun.*, 26, 2961.
Pye, E.K., and Humphrey, A.E. (1979). *Interim Report to U.S. Department of Energy*, p. 79.
Rautenbach, R., and Albrecht, R. (1982). *Chem. Ing. Techn.*, 54, 229.
Regan, D.L., et al. (1971). *Biotechnol. Bioeng.*, 13, 815.
Reisinger, C., et al. (1987). Paper at 5th International Conference on partition in Aqueous Two-Phase Systems, Oxford August 23–28.
Remson, I., et al. (1971). *Numerical Methods in Subsurface Hydrology*. New York: John Wiley.
Reuss, M., et al. (1979). *Europ. J. Appl. Microbiol. Biotechnol.*, 8, 169.
Reuss, M., and Buchholz, K. (1979). *Biotechnol. Bioeng.*, 21, 2061.
Reuss, M., et al. (1980). Paper presented at 6th Internat. Ferment. Symp., London, Ontario.
Reuss, M., et al. (1982). *Chem. Eng.*, June, 233.
Reusser, F. (1961). *Appl. Microbiol.*, 9, 361.
Ricica, J. (1969a). In Perlman, D. (ed.). *Fermentation Advances*. New York: Academic Press, p. 427.
Ricica, J. (1969b). *Folia Microbiol.*, 14, 322.
Ricica, J. (1973). *Folia Microbiol.*, 18, 418.
Ricica, J., and Necinova, S. (1967). *Mitteil. d. Versuchsstation f. Gärungsgewerbe Wien*. 11/12, 130.
Ringpfeil, M. (1980). Paper presented at 6th Internat. Ferm. Symp., London, Ontario.
Rippin, D.W.T. (1967). *Ind. Eng. Chem. Fundam.*, 6 (4), 488.

Rittmann, B.E. (1982). *Biotechnol. Bioeng.*, 24, 501 and 1341.
Rittmann, B.E., and McCarty, P.L. (1980). *Biotechnol. Bioeng.*, 22, 2349, 2359.
Roels, J.A. (1980a). *Biotechnol. Bioeng.*, 22, 23.
Roels, J.A. (1980b). *Biotechnol. Bioeng.*, 22, 2457.
Roels, J.A. (1982). *J. Chem. Techn. Biotechnol.*, 32, 59.
Roels, J.A. (1983). *Energetics and Kinetics in Biotechnology.* Amsterdam: Elsevier Biomedical.
Roels, J.A., and van Suijdam, J.C. (1980). *Biotechnol. Bioeng.*, 22, 463.
Roels, J.A., et al. (1974). *Biotechnol. Bioeng.*, 16, 181.
Rogers, P.L., et al. (1980). In Ghose, T.K. (ed.). *Proceedings of the Second Symposium on Bioconversion and Biochemical Engineering,* Vol. 2. *Indian Inst. Technology,* New Delhi p. 359.
Romanovsky, J., et al. (1974). *Kinetische Modelle in der Biophysik.* Jena: G. Fischer. (German translation by W.A. Knorre and A. Knorre).
Russel, R.M., and Tanner, R.D. (1978). *Ind. Eng. Chem. Proc. Des. Dev.*, 17, 157.
Russel, T.W.F., et al. (1974). *Biotechnol. Bioeng.*, 16, 1261.
Ryu, D.D.Y., and Lee, B.K. (1975). *Proc. Biochem.*, 10 January/February, 15.
Ryu, Y.W., et al. (1982). *Eur. J. Appl. Microbiol. Biotechnol.*, 15, 1.
Schneider, H., and Moser, A. (1987). *Bioprocess Eng.*, 2, 129.
Schügerl, K. (1977). *Chem. Ing. Techn.*, 49, 605.
Schügerl, K. (1982). *Adv. Biochem. Eng.*, 22, 94.
Schügerl, K. (1983). In Proceedings of NATO ASI, *Mass Transfer with Chemical Reactions,* Izmir Turkey (1981) Vol. 1 (72), 415.
Schultz, J.S., and Gerhardt, P. (1969). *Bact. Rev.*, 33, 1.
Seipenbusch, R., and Blenke, H. (1980). *Adv. Biochem. Eng.*, 15, 1.
Sheintuch, M. (1980). *Biotechnol. Bioeng.*, 22, 2557.
Sherrard, J.H. (1977). *J. Water Poll. Contr. Fed.*, 49, 1968.
Sherrard, J.H., and Lawrence, A.W. (1975). *J. Water Poll. Contr. Fed.*, 47, 1848.
Sherrard, J.H. (1980). *J. Chem. Techn. Biotechnol.*, 30, 447.
Sherrard, J.H., and Schroeder, E.D. (1976). *J. Water Poll. Contr. Fed.*, 48, 742.
Shieh, W.K. (1980a). *Water Res.*, 14, 695.
Shieh, W.K. (1980b). *Biotechnol. Bioeng.*, 22, 667.
Shieh, W.K., et al. (1981). *J. Water Poll. Contr. Fed.*, 53, 1574.
Shieh, W.K., et al. (1981). *J. Water Poll. Contr. Fed.*, 53, 1574.
Shimmons, B.W., et al. (1976). Biotechnol. Bioeng., 18, 1793.
Shiotani, T., and Yamane, T. (1981). *Eur. J. Appl. Microbiol. Biotechnol.*, 13, 96.
Sinclair, C.C., and Brown, D.E. (1970). *Biotechnol. Bioeng.*, 12, 1001.
Solomon, B.O., and Erickson, L.E. (1981). *Proc. Biochem.*, February/March, 44.
Sortland, L., and Wilke, Ch.R. (1969). *Biotechnol. Bioeng.*, 11, 805.
Stenberg, O. (1984). Ph.D. thesis, TU Göteborg/Sweden.
Stieber, R.W., and Gerhardt, P. (1979). *Appl. Environ. Microbiol.*, 37, 487.
Stieber, R.W., and Gerhardt, P. (1981a). *Biotechnol. Bioeng.*, 23, 523.
Stieber, R.W., and Gerhardt, P. (1981b). *Biotechnol. Bioeng.*, 23, 535.
Stieber, R.W., et al. (1977). *Appl. Environ. Microbiol.*, 34, 733.
Stiebitz, O., et al. (1987). Knorre, W. (ed.). Proceedings of the UNESCO-training course, *Modern Biotechnology: Optimization of Fermentation Processes,* Jena, East Germany, Oct. 12–31.
Stucki, J.W. (1978). *Prog. Biophys. Molec. Biol.*, 33, 99.
Sundstrom, D.W., et al. (1976). *Biotechnol. Bioeng.*, 18, 1.

Swartz, R.W. (1979). *Ann. Rep. Ferm. Proc.*, 3, 75.
Tempest, D.W. (1970). In Norris, J.R., and Ribbons, D.W. (eds.). *Methods in Microbiology*, Vol. 2. Academic Press, London, N.Y. p. 259.
Toda, K., and Dunn, I.J. (1982). *Biotechnol. Bioeng.*, 24, 651.
Topiwala, H.H. (1974). *Biotechnol. Bioeng. Symp.*, 4, 681.
Topiwala, H.H., and Hamer, G. (1971). *Biotechnol. Bioeng.*, 13, 919.
Trilli, A., et al. (1977). *J. Appl. Chem. Biotechnol.*, 27, 219.
Tsai, B.I., Erickson, L.E., and Fan, L.T. (1969). *Biotechnol. Bioeng.*, 11, 181.
Tsai, B.I., Fan, L.T., Erickson, L.E., and Chen, M.S.K. (1971). *J. Appl. Chem. Biotechnol.*, 21, 307.
Tyagi, R.D., and Ghose, T.K. (1980). *Biotechnol. Bioeng.*, 22, 1907.
Vand, W. (1948). *J. Phys. & Colloid. Chem.*, 52, 277.
van Dedem, G., and Moo-Young, M. (1973). *Biotechnol. Bioeng.*, 17, 1301.
van Suijdam, J.C. (1987). In Crueger, W., et al. (eds.). *Physical Aspects of Bioreactor Performance* report of working party Bioreactor Performance of Europ. Fed., Biotechnol. Frankfurt: Dechema. Chap. 6.
van Suijdam, J.C., and Metz, B. (1981). *Biotechnol. Bioeng.*, 23, 111.
van Suijdam, J.C., et al. (1982). *Biotechnol. Bioeng.*, 24, 177.
van Suijdam, J.C. (1986). Paper at 14th Intern. Congr. of Microbiol., Manchester Sept. 7–13.
van Suijdam, J.C., and Dusseljee, P.J.B. (1987). In Crueger, W., et al. (eds.). *Physical Aspects of Bioreactor Preformance* report working party Bioreactor Performance, Europ. Fed. Biotechnology, Chap. 6, Dechema, Frankfurt.
van de Vusse, J.G. (1964). *Chem. Eng. Sci.*, 19, 994.
Warmoeskerken, M., and Smith, J.M. (1982). Paper Gl at 4th Europ. Conference on Mixing Noordwijkerhout, Netherlands.
Wandrey, C., and Flaschel, E. (1979). *Adv. Biochem. Eng.*, 12, 148.
Wandrey, C., Flaschel, E., and Schügerl, K. (1979). *Biotechnol. Bioeng.*, 21, 1649.
Wittler, R. et al. (1983). Eur. J. Appl. Microbiol. Biotechnol. 18, 17.
Whitaker, A. (1980). *Proc. Biochem.*, May, 10.
Wolfbauer, O., Klettner, H., and Moser, F. (1978). *Chem. Eng. Sci.*, 33, 953.
Woods, J.L., and O'Callaghan, J.R. (1975). Biotechnol. Bioeng., 17, 779.
Wright, D.G., and Calam, C.T. (1968). *Chem. Ind.*, 1274.
Wu, Y.C., et al. (1980). *Biotechnol. Bioeng.*, 22, 2055.
Yamane, T., and Hirano, S. (1977). *J. Ferment. Technol.*, 55, 156.
Yamane, T., and Shimizu, S. (1982). *Biotechnol. Bioeng.*, 24, 2731.
Yamane, T., and Shimizu, S. (1984). *Adv. Biochem. Eng.*, 30, 148.
Yang, Ren der, and Humphrey, A.E. (1975). *Biotechnol. Bioeng.*, 17, 1211.
Yano, T., and Koga, S. (1973). *J. Gen. Appl. Microbiol.*, 19, 97.
Yoshida, F., et al. (1973). *Biotechnol. Bioeng.*, 15, 257.
Zeuthen, E., (ed.) (1964). *Synchrony in Cell Division and Growth.* New York: Interscience.
Ziegler, H. (1980). *Biotechnol. Bioeng.*, 22, 1613.
Ziegler, H., et al. (1977). *Biotechnol. Bioeng.*, 19, 507.
Zlokarnik, M. (1978). *Adv. Biochem. Eng.*, 8, 133.
Zlokarnik, M. (1979). *Adv. Biochem. Eng.*, 11, 157.
Zwietering, Th.N. (1959). *Chem. Eng. Sci.*, 11, 1.

APPENDIX I

Fundamentals of Stoichiometry of Complex Reaction Systems

Bioprocessing includes a large variety of metabolic reactions even on the macroscopic level (cf. Fig. 2.16) and at the same time is carried out in different modes of reactor operations. Therefore, complex reaction systems must be treated stoichiometrically, which means that not only complex reactions themselves must be considered but also complex reactor operations.

I.1 Stoichiometry of Complex Reactions

The stoichiometric equation in the case of complex reactions can be written in the form

$$\sum_{i}^{M}\sum_{j}^{N} \nu_{ij} \cdot A_j = 0 \tag{I.1}$$

while for a simple reaction the following form is valid:

$$\sum_{j}^{N} \nu_j A_j = 0 \tag{I.2}$$

where i = number of reactions ($1 \leq i \leq M$)
j = number of components ($1 \leq j \leq N$)
A_j = components of reaction mixture
ν = stoichiometric coefficients

The differential change of the number of moles n of component A_j due to the reaction i is defined as

$$(dn_j)_{r_i} = \nu_{ij} \cdot d\xi_i \tag{I.3}$$

where ξ_i is the extent of reaction, defined as the change of number of moles divided by the stoichiometric coefficient [mole].

In complex reactions, the singular reaction steps are interconnected, that is, singular components participate in different reactions, with the consequence that a part of the reactions is stoichiometrically dependent. Only the independent reactions can be determined from the change in the number of moles. The solution to the problem of stoichiometric dependence can be found with the aid of the "matrix of the stoichiometrical coefficients" of the following structure

$$\begin{array}{c|c} M \backslash N & \text{Components } 1 \leqslant j \leqslant N \\ \hline \text{Reactions } 1 \leqslant i \leqslant M & \begin{bmatrix} \nu_{11} & \cdots\cdots & \nu_{1j} \\ \nu_{21} & \cdots\cdots & \nu_{2j} \\ \vdots & & \vdots \\ \nu_{i1} & \cdots\cdots & \nu_{ij} \end{bmatrix} \end{array} \tag{I.4}$$

where the row index is the number of reactions i and the column index is the number of components j. Each row of the stoichiometric coefficient matrix expresses the stoichiometry of a reaction in terms of the number of moles of each compound converted per unit reaction rate.

The number of stoichiometrically independent reactions is given by the rank of the matrix R_β, which can be determined with e.g. the aid of the Gaussian method of elimination. As a result, the stoichiometrical coefficients of R_ν linearly independent equations for the reaction system are necessary and sufficient for, for example, calculation of the conversion of the key variables and therefore also for all other components. Thus

$$R_\nu = N - R_\beta \tag{I.5}$$

Sometimes balancing is carried out without the formulation of reaction equations. This situation, however, arises rarely when kinetics are of interest. In this case an "element–species matrix" can be written on the basis of N species (components) with k elementary balances

$$\begin{array}{c|c} k \backslash N & \text{Components } 1 \leqslant j \leqslant N \\ \hline \text{Elements } 1 \leqslant i \leqslant k & \begin{bmatrix} a_{11} & \cdots\cdots & a_{1j} \\ a_{21} & \cdots\cdots & a_{2j} \\ \vdots & & \vdots \\ a_{i1} & \cdots\cdots & a_{ij} \end{bmatrix} \end{array} \tag{I.6}$$

where row index = number of components, j
column index = number of chemical elements, i
i = number of elementary balances ($1 \leq i \leq k$)

a_{ij} = number of atoms of atomic species i present in molecule of component j ((resp. a_j, b_j, c_j, d_j, etc.)

The number of key variables R is determined again with the aid of the rank of this matrix R. Thus (Schubert and Hofmann, 1975)

$$R = N - R_\beta \tag{I.7}$$

Normally $R_\beta = k$ in the case of N species and k elements.

Hence, only $(N - k)$ net conversion rates can be chosen independently. From this reasoning it becomes clear that the number of independent kinetic equations to be postulated cannot be chosen at will—it is completely specified by the number of elementary balances k and the number of components N in the system (Roels, 1980). Which key variable or which kinetic equation is to be chosen strongly depends on the application one has in mind.

As an example for the determination of the number of independent equations, the situation of an organism with balanced growth is considered. This organism grows on one sole source of carbon and energy, a source that may contain nitrogen. One sole source of nitrogen is supplied, and this source may also contain carbon. One product is excreted; CO_2, H_2O, and O_2 are the only other components, to be exchanged with the environment.

In terms of formalism, the system of organisms will be considered to be a given quantity of mass. The organism exchanges, macroscopically speaking, an exact replica of itself with the environment; it is characterized by its gross elemental composition formula $C_{a1}H_{b1}O_{c1}N_{d1}$. The concept of the C-mole of organism, that is, the amount containing 1 mole of carbon, is adopted (see Sect. 2.2.3.2). Figure 6.53 gives a schematic representation of the system and the possible flows to and from the system. Only the elements C, H, O, and N are considered, and indeed these elements comprise in most cases about 95% of the cellular mass and the various other exchange flows. Equation 2.11 can be directly applied to this specific case, with $\nabla n_j'$ in the case of stirred tank reactors being the rate of flow of components (F_j, C-mole/m^3 hr).

The element–species matrix that represents the elemental composition of the flows of compounds can be written in this case as

k \ N	C	H	O	N
F_1 : X	1	b_1	c_1	d_1
F_2 : S_C	a_2	b_2	c_2	d_2
F_3 : P	a_3	b_3	c_3	d_3
F_4 : S_N	a_4	b_4	c_4	d_4
F_5 : O_2	0	0	2	0
F_6 : CO_2	1	0	2	0
F_7 : H_2O	0	2	1	0

(I.8)

In the present case there are seven flows, and Equ. 2.11 specifies four equations between the flows represented in the matrix of Equ. I.8. Hence, only three flows are independent variables (cf. Sect. I.2). Which kind of flows to be chosen for measurement depends on the possibilities for experimental determination. The knowledge, for example, of the respiratory quotient and the ratio of oxygen consumption to substrate consumption allows direct estimation of the biomass production rate and the product formation rate. This conclusion from the application of balancing is of the greatest importance in situations where process variables, for example, X, are very difficult to measure, which is the case in penicillin fermentation (Mou and Cooney, 1983).

I.2 Example for Determination of Number of Linearly Independent Equations ("Key Reactions")

Reaction scheme

$$\nu_A \cdot c_A + \nu_{B1} \cdot c_B \underset{2}{\overset{1}{\rightleftharpoons}} \nu_{C1} \cdot c_C \tag{I.9a}$$

$$\nu_{C2} \cdot c_C + \nu_{B2} \cdot c_B \underset{4}{\overset{3}{\rightleftharpoons}} \nu_D \cdot c_D \tag{I.9b}$$

Reaction steps

$$\nu_A \cdot c_A + \nu_{B1} \cdot c_B \xrightarrow{r_1} \nu_{C1} \cdot c_C \tag{I.9c}$$

$$\nu_{C1} \cdot c_C \xrightarrow{r_2} \nu_A \cdot c_A + \nu_{B1} \cdot c_B \tag{I.9d}$$

$$\nu_{C2} \cdot c_C + \nu_{B2} \cdot c_B \xrightarrow{r_3} \nu_D \cdot c_D \tag{I.9e}$$

$$\nu_D \cdot c_D \xrightarrow{r_4} \nu_{C2} \cdot c_C + \nu_{B2} \cdot c_B \tag{I.9f}$$

Process kinetics includes four rates of individual steps (r_1 to r_4) and the rate equations for all compounds can be written on the basis of r_1 to r_4 as follows:

$$r_A = -\nu_A \cdot r_1 + \nu_A \cdot r_2 \tag{I.10a}$$

$$r_B = -\nu_{B1} \cdot r_1 + \nu_{B1} \cdot r_2 - \nu_{B2} \cdot r_3 - \nu_{B2} \cdot r_4 \tag{I.10b}$$

$$r_C = \nu_{C1} \cdot r_1 - \nu_{C1} \cdot r_2 - \nu_{C2} \cdot r_3 + \nu_{C2} \cdot r_4 \tag{I.10c}$$

$$r_D = \nu_D \cdot r_3 - \nu_D \cdot r_4 \tag{I.10d}$$

Thus, the matrix of stoichiometric coefficients is:

$$\begin{pmatrix} -\nu_A & \nu_A & 0 & 0 \\ -\nu_{B1} & \nu_{B1} & -\nu_{B2} & \nu_{B2} \\ \nu_{C1} & -\nu_{C1} & -\nu_{C2} & \nu_{C2} \\ 0 & 0 & \nu_D & -\nu_D \end{pmatrix} \tag{I.11}$$

The rank of this matrix can be shown to be two by using the method of Gauss elimination or by finding the order of the determinant, which is not zero. Thus, for example, the second column is identical with the first when multiplied by -1; the same with the fourth and the third columns. The remaining matrix is then

$$\begin{pmatrix} 0 & \nu_A & 0 & 0 \\ 0 & \nu_{B1} & -\nu_{B2} & 0 \\ 0 & -\nu_{C1} & -\nu_{C2} & 0 \\ 0 & 0 & \nu_D & 0 \end{pmatrix} \tag{I.12}$$

With $R_\beta = 2$, the number of key reaction or key variables is two, and these can be chosen arbitrarily (e.g., A and D). Thus the remaining dependent rates can be written on the base of r_A and r_D:

$$r_B = \frac{\nu_{B1}}{\nu_A} r_A - \frac{\nu_{B2}}{\nu_D} r_D \tag{I.13a}$$

$$r_C = \frac{\nu_{C1}}{\nu_A} r_A - \frac{\nu_{C2}}{\nu_D} r_D \tag{I.13b}$$

I.3 Stoichiometry of Complex Reactor Operation

The majority of stoichiometric considerations is restricted to closed reactor operations, where $\mathrm{div}\,(c_j v) = 0$, so that according to Equ. 2.3b

$$\frac{dc_j}{dt} = \pm \nu_j \cdot r \tag{I.14}$$

A simple case appears also with continuous stirred tanks.

I.3.1 Semidiscontinuous Reactor

For semidiscontinuous processing, often used in fermentations for the production of yeast biomass or secondary metabolites, the basic balance equation is to be modified. Balancing in this case has to distinguish between components that are already present in the reactor at the beginning (number of moles $n_{j,0}$) and components that are fed later on to the reactor (flux of moles $\dot{n}_j^0$). The stoichiometric balance in this case of semidiscontinuous reactor operation is

$$n_j = n_{j0} + \dot{n}_j^0 \cdot t \pm \sum_i \nu_{ij} \cdot \xi_i \tag{I.15}$$

where n_j = number of moles of component j
n_{j0} = temporal initial value of n
n_j^0 = spatial initial value of n

This operation contains the time t as an independent process variable, in contrast to all batch reactor configurations.

I.3.2 NONSTATIONARY REACTOR OPERATION

Stoichiometry, for example, in the case of nonstationary modes of reactor operation, needs another fundamental equation. For the CSTR the balance of a component j is written as

$$\frac{dc_j}{dt} + \frac{c_j - c_j^0}{\bar{t}} = \sum_i \nu_{ij} \frac{d\xi_i}{dt} \tag{I.16}$$

which gives after integration

$$c_j = c_j^0 + (c_{j,0} - c_j^0)e^{-t/\bar{t}} + \sum_i \nu_{ij} \cdot \xi^*(t) \tag{I.17}$$

where ξ^* is the modified extent of reaction.

Characteristically this equation contains again the time t as a process variable. This term with the time t disappears and can be neglected in the case of stationarity (if $t \gg \bar{t}$) and when $c_{j,0} = c_j^0$, that is, in a batch reactor. Therefore the stoichiometric balance equation of a discontinuous system can be formally applied to a nonstationary continuous stirred tank. Similarly, the stoichiometric equations of heterogeneous reactor systems with interfacial mass transfer can be derived (Budde, Bulle, and Rückauf, 1981).

BIBLIOGRAPHY

Budde, K., Bulle, H., and Rückauf, H. (1981). *Stöchiometrie chemisch-technologischer Prozesse.* Berlin: Akademie Verlag.

Mou, D.G., and Cooney, Ch.L. (1983). *Biotechnol. Bioeng.*, 25, 225, and 257.

Roels, J.A. (1980). *Biotechnol. Bioeng.*, 22, 2457.

Schubert, E., and Hofmann, H. (1975). *Chem. Ing. Techn.*, 47, 191.

APPENDIX II
Computer Simulations*

This appendix contains a series of computer simulations that are thought to represent the most significant basic kinetic models in bioprocessing. The models are summarized in Table II.1. The simulations in the figures also contain the values of model parameters chosen for demonstration. Mainly two different kinds of plots are presented, the first showing concentration/time curves and the second the corresponding time curves of specific rates of bioprocesses. The models are as follows:

1. Simple Monod-kinetics in batch operation:

$$\text{Reaction scheme: } S \rightarrow X \tag{II.1}$$

Reactor balance equations (DCSTR):

$$r_X = \mu(s) \cdot x \tag{II.2}$$

$$-r_S = \frac{1}{Y_{X|S}} \cdot \mu(s) \cdot x \tag{II.3}$$

with kinetic equation (Monod type, cf. Equ. 5.38):

$$\mu(s) = \mu_{max} \frac{s}{K_S + s} \tag{II.4}$$

Model parameters; process variables: μ_{max}, K_S, $Y_{X|S}$; x_0, s_0.

2. Monod kinetics with lag time t_L (cf. Sect. 5.3.3.1):

$$r_X = \mu(s, t) \cdot x \tag{II.5}$$

$$-r_S = \frac{1}{Y_{X|S}} \cdot \mu(s, t) x \tag{II.3}$$

*This appendix has been added to the English edition of this book as a consequence of a critical recommendation by I.J. Dunn (Federal Technical University/ Lab of Technical Chem. Zurich). Most of the simulations were realized at Graz the University of Technology, Austria, or at ZIMET, Central Institute of Microbiology and Experimental Therapy, Jena, East Germany (Fig. II.33 and II.34).

TABLE II.1. Overview of simulated bioprocess kinetic models.

No.	*Bioreactor operation*	*Process variables*	*Kinetic model type*	*Variation*	*Equations*	*Figures*
1	DCSTR	X, S	Microbial growth: Monod	s_0	II.1–II.4	II.1
				x_0		II.2
				$Y_{X\|S}$		II.3
				μ_{max}		II.4
				K_S		II.5
2	DCSTR	X, S	Monod with t_L	t_L	II.5, II.6	II.6
3	DCSTR	X, S	Monod with k_d	k_d	II.3, II.7	II.7, II.8
4	DCSTR	X, S	Monod with m_S	m_S	II.2, II.8	II.9, II.10
5	DCSTR	X, S	Monod with k_d, m_S	k_d, m_S	II.7, II.8	II.11, II.12
6	DCSTR	X, S	Monod with S inhibition	K_{IS}	II.2, II.3, II.9	II.13, II.14
7	DCSTR	X, S	Monod with 2-S limitation	K_{12}	II.10–II.13	II.15
8	DCSTR	X, S	Monod: diauxie	s_{crit}	II.14–II.19	II.16, II.17
				$K_{12}(K_R)$		II.18, II.19
9	DCSTR	X, S, O	Monod with O_2 limitation	$k_{L1}a$	II.20–II.23	II.20, II.21
10	DCSTR	X, S, P;	multiinhibition model for microbial production (with repression and multiinhibitions)	K_{IP}, $K_{ISP}(K_R)$, K_P, K_{ISX}	II.24–II.29	II.22–II.24, II.25–II.29
11	SCSTR	X, S, P	Type 10 with S feed	F_S	II.24–II.30	II.30, II.31
12	DCSTR	S_e, S_d, S_{ads}	Biosorption	c_i/t	II.31–II.38	II.32
13	CSTR	X, S	Monod	s_0	II.39, II.40 and II.4	II.33
				x_0		II.34

Text continues p.

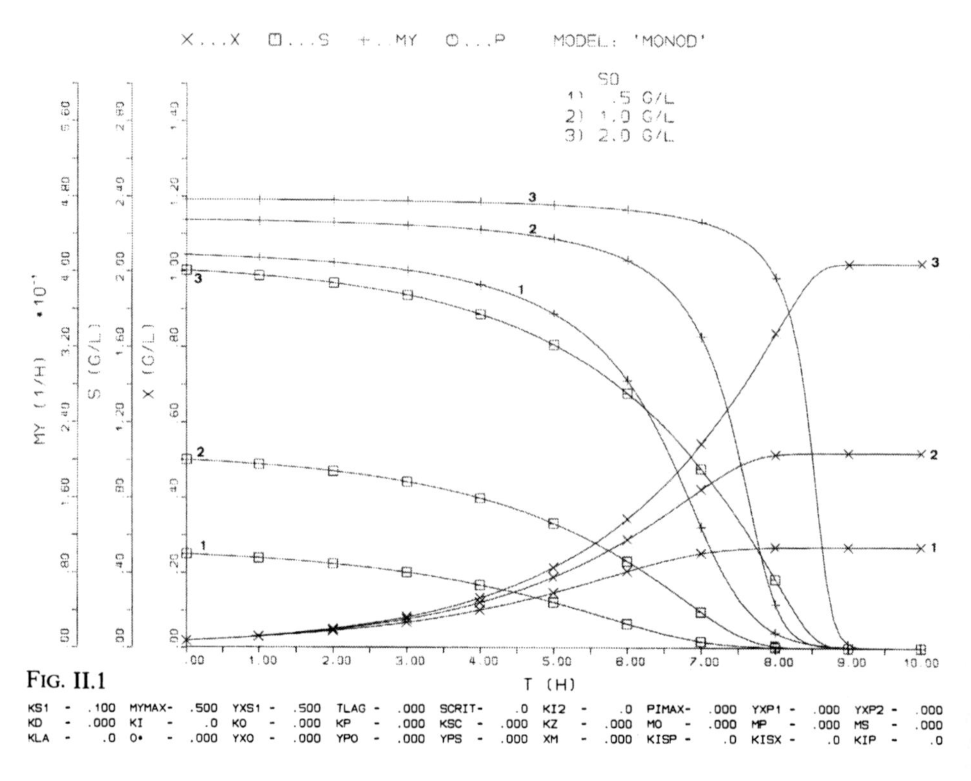

FIG. II.1

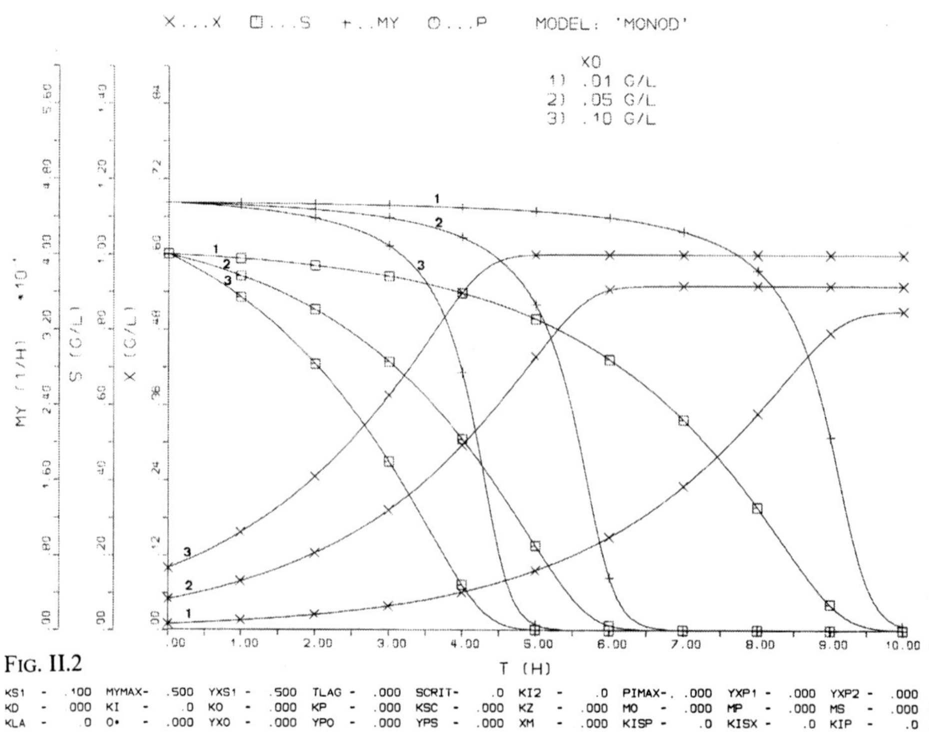

FIG. II.2

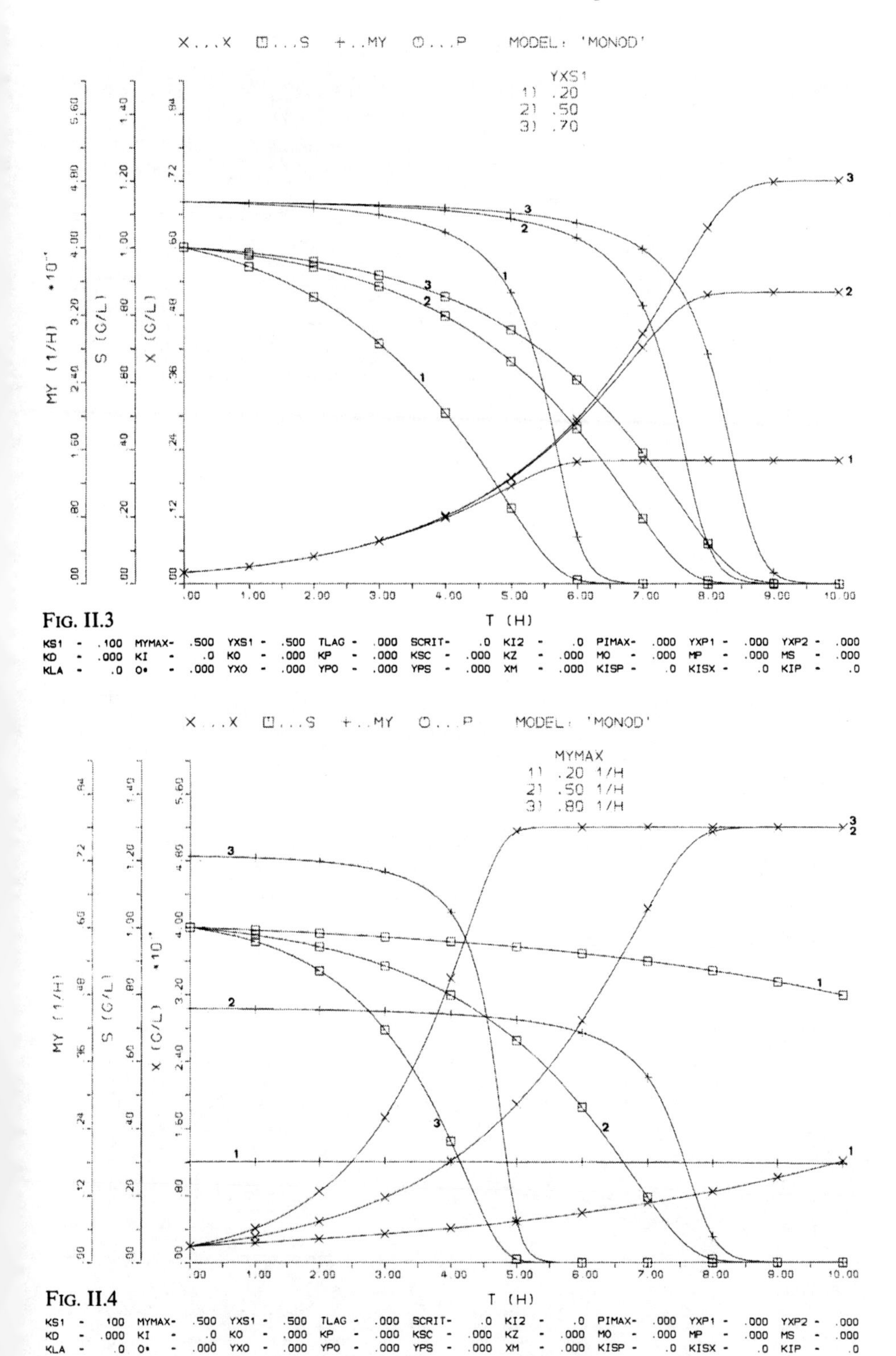

Fig. II.3

Fig. II.4

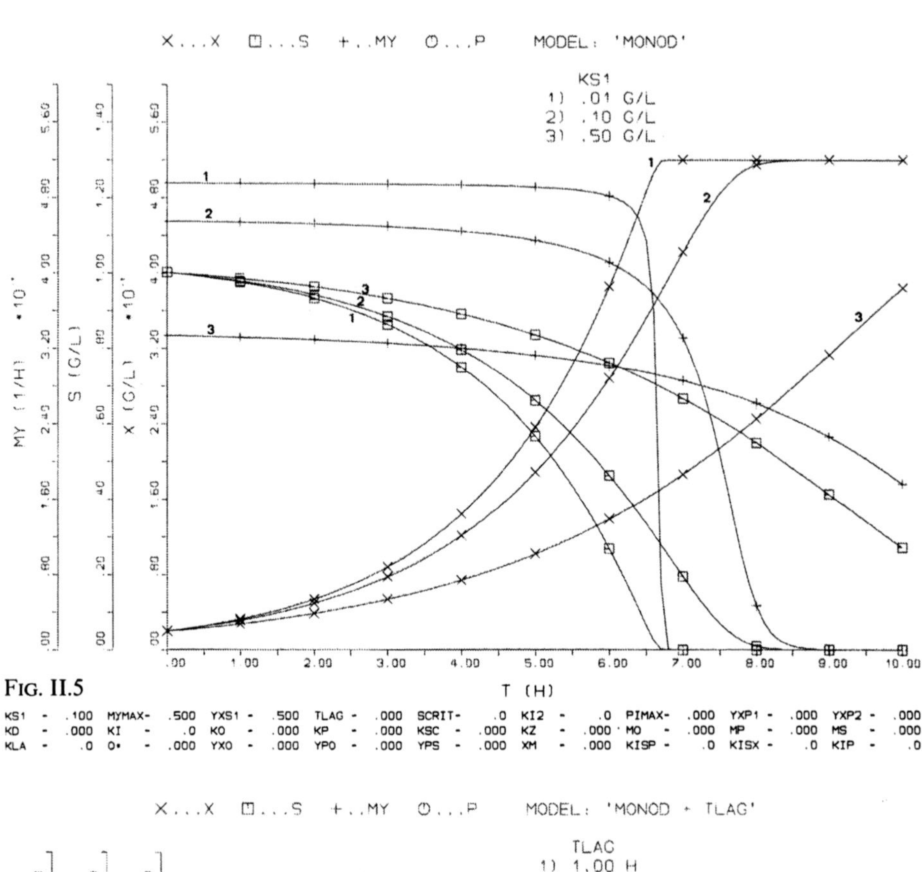

FIG. II.5

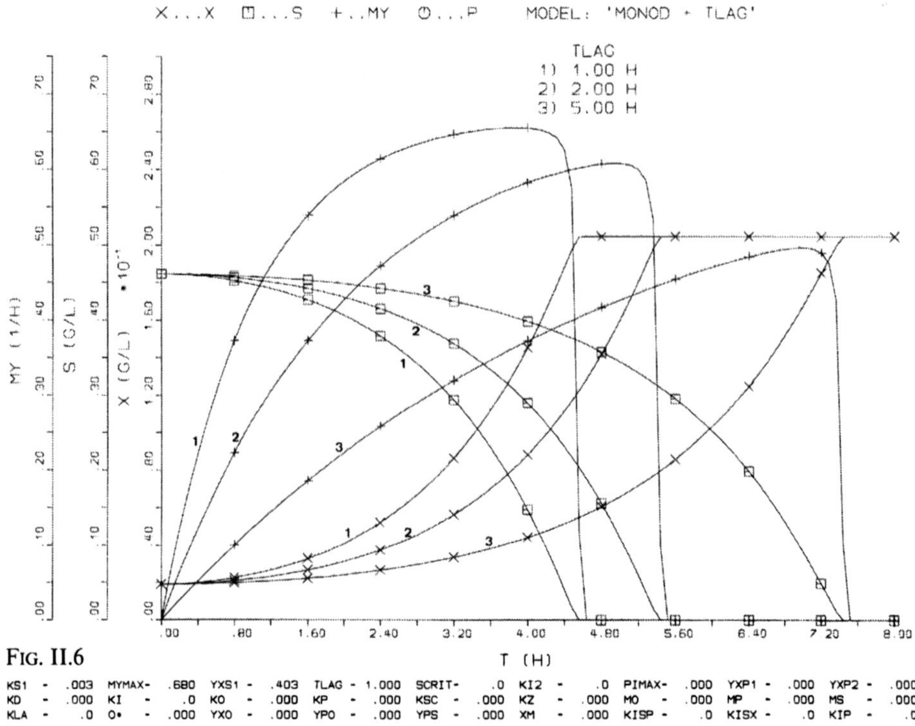

FIG. II.6

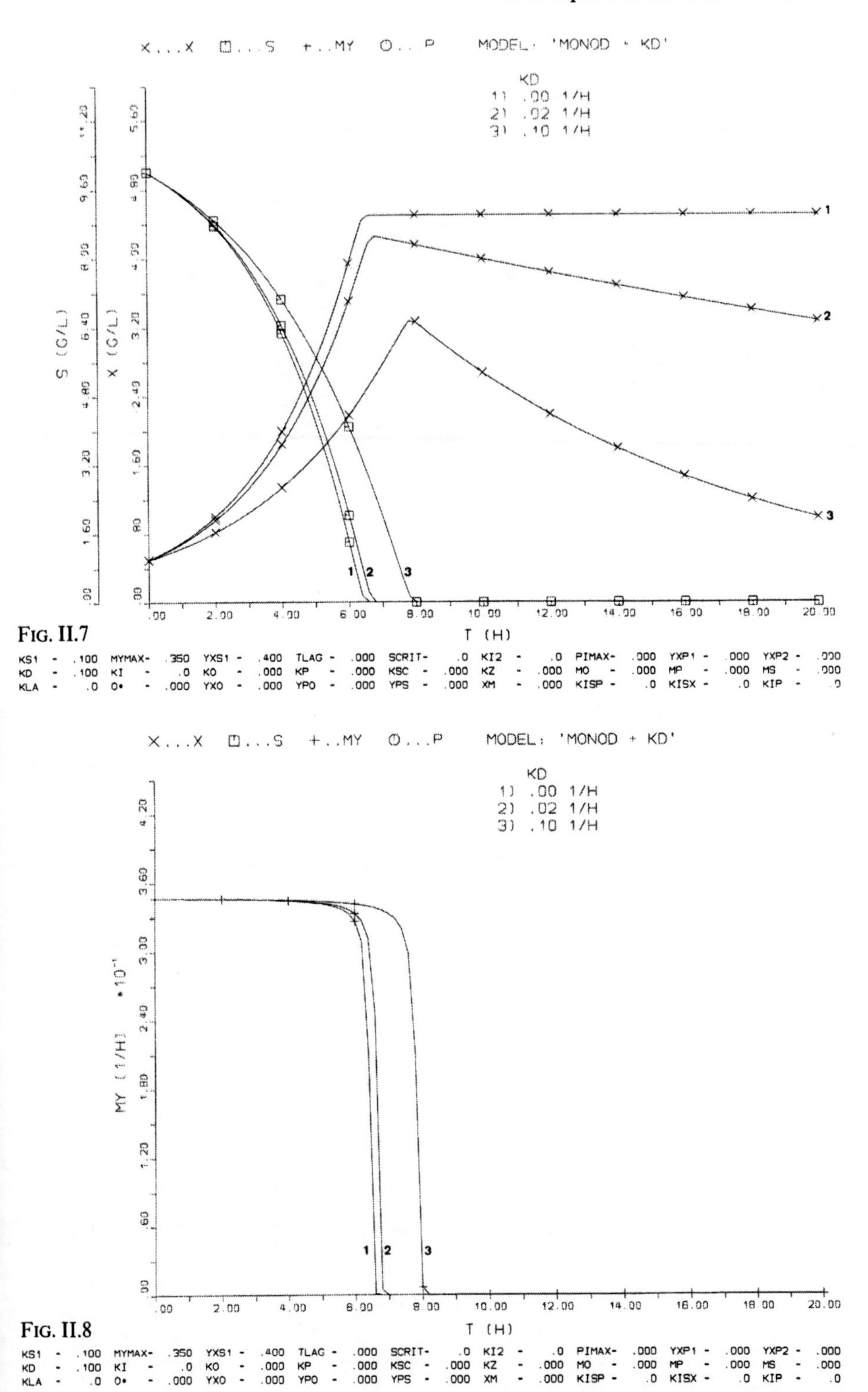

FIG. II.7

FIG. II.8

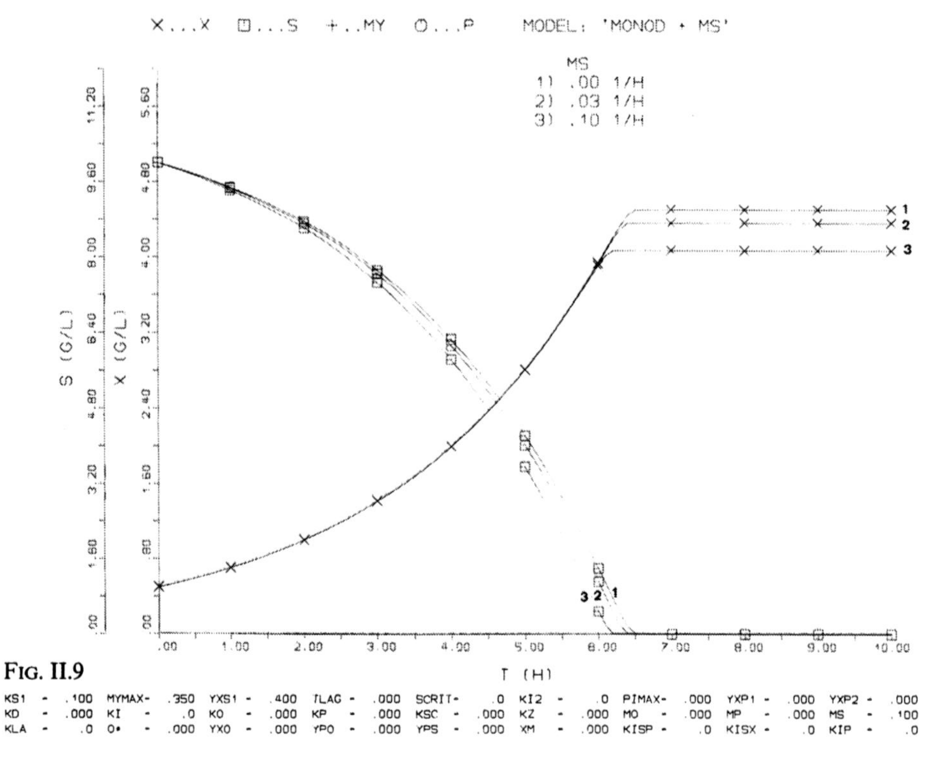

Fig. II.9

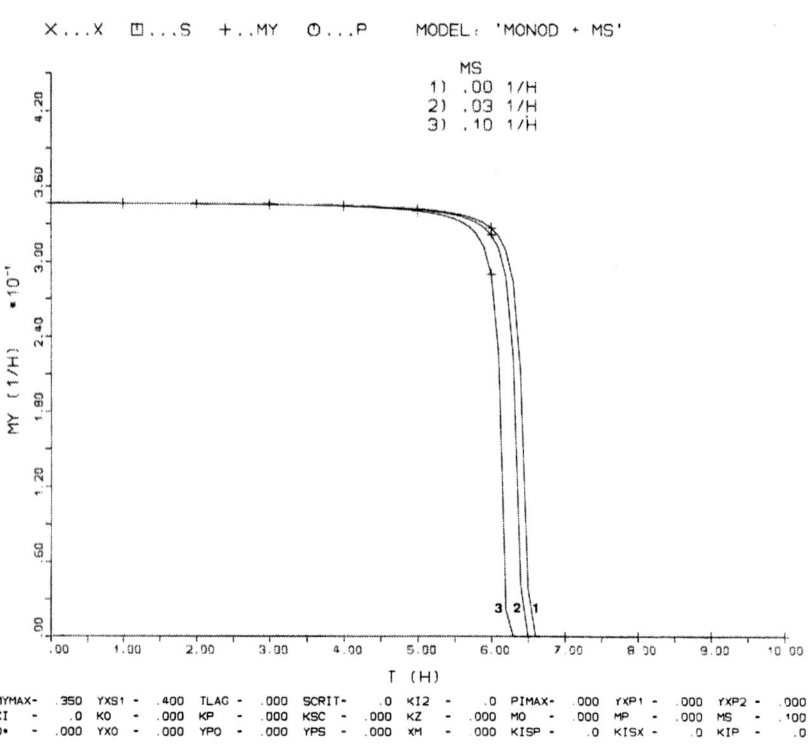

Fig. II.10

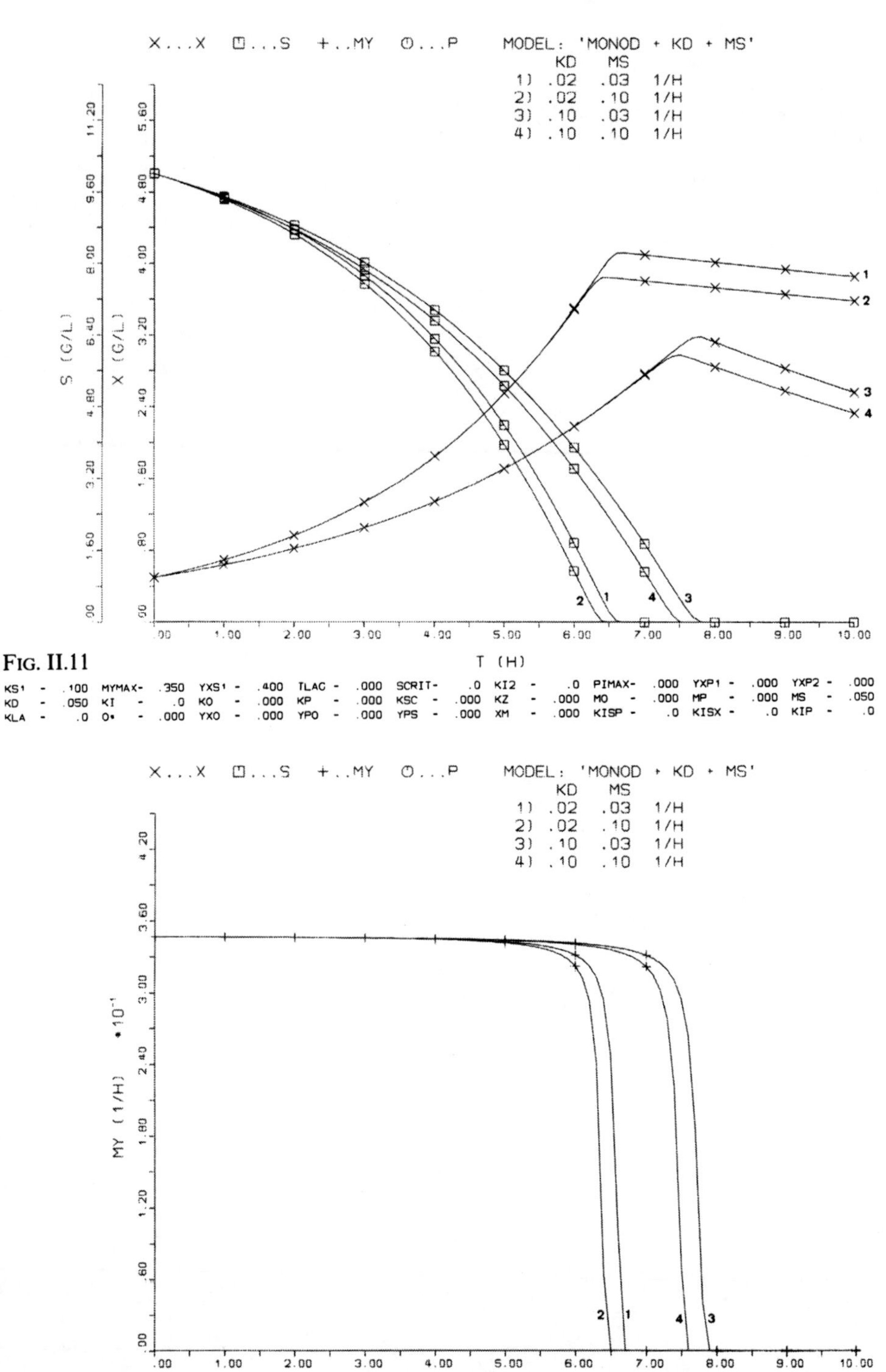

FIG. II.11

KS1 - .100 MYMAX- .350 YXS1 - .400 TLAG - .000 SCRIT- .0 KI2 - .0 PIMAX- .000 YXP1 - .000 YXP2 - .000
KD - .050 KI - .0 KO - .000 KP - .000 KSC - .000 KZ - .000 MO - .000 MP - .000 MS - .050
KLA - .0 O* - .000 YXO - .000 YPO - .000 YPS - .000 XM - .000 KISP - .0 KISX - .0 KIP - .0

FIG. II.12

KS1 - .100 MYMAX- .350 YXS1 - .400 TLAG - .000 SCRIT- .0 KI2 - .0 PIMAX- .000 YXP1 - .000 YXP2 - .000
KD - .050 KI - .0 KO - .000 KP - .000 KSC - .000 KZ - .000 MO - .000 MP - .000 MS - .050
KLA - .0 O* - .000 YXO - .000 YPO - .000 YPS - .000 XM - .000 KISP - .0 KISX - .0 KIP - .0

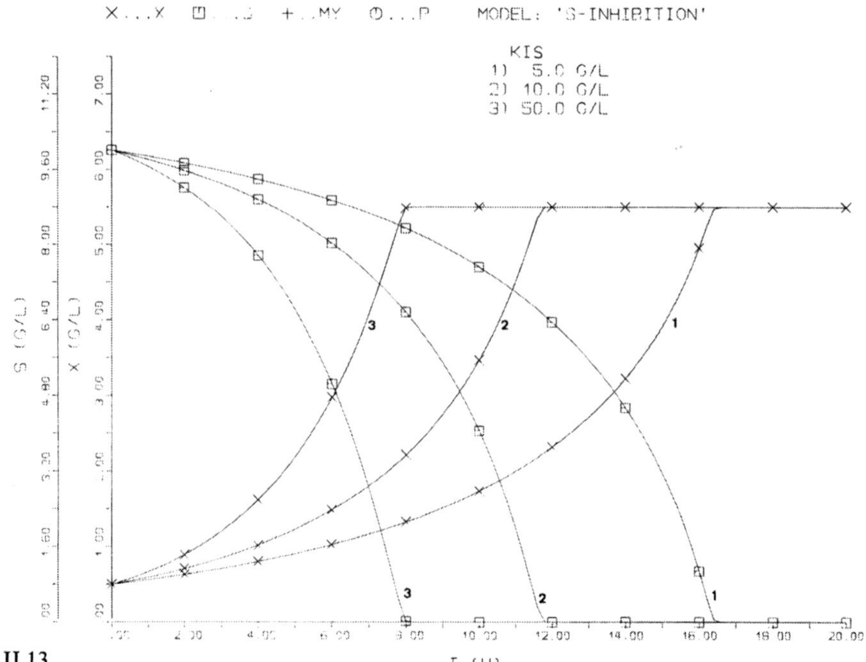

FIG. II.13

KS1 - .100 MYMAX- .350 YXS1 - .500 TLAG - .000 SCRIT- 0 KI2 - .0 PIMAX- .000 YXP1 - .000 YXP2 - .000
KD - .000 KI - 2.0 KO - .000 KP - .000 KSC - .000 KZ - .000 MO - .000 MP - .000 MS - .000
KLA - 0 O* - .000 YXO - .000 YPO - .000 YPS - .000 XM - .000 KISP - .0 KISX - .0 KIP - .0

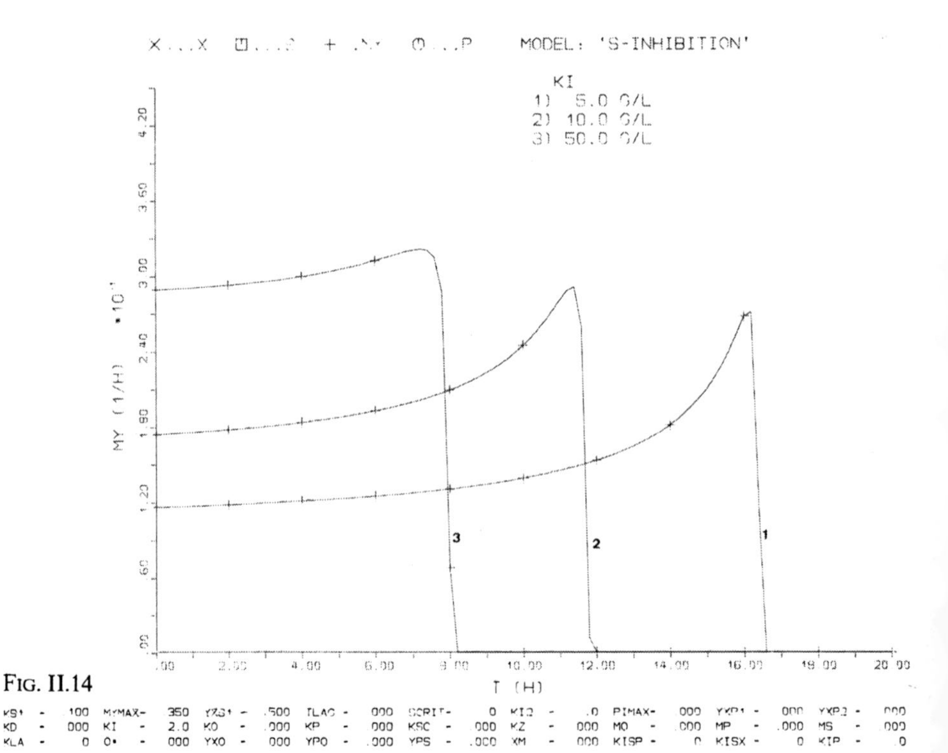

FIG. II.14

KS1 - .100 MYMAX- .350 YXS1 - .500 TLAG - .000 SCRIT- 0 KI2 - .0 PIMAX- .000 YXP1 - .000 YXP2 - .000
KD - .000 KI - 2.0 KO - .000 KP - .000 KSC - .000 KZ - .000 MO - .000 MP - .000 MS - .000
KLA - 0 O* - .000 YXO - .000 YPO - .000 YPS - .000 XM - .000 KISP - .0 KISX - .0 KIP - .0

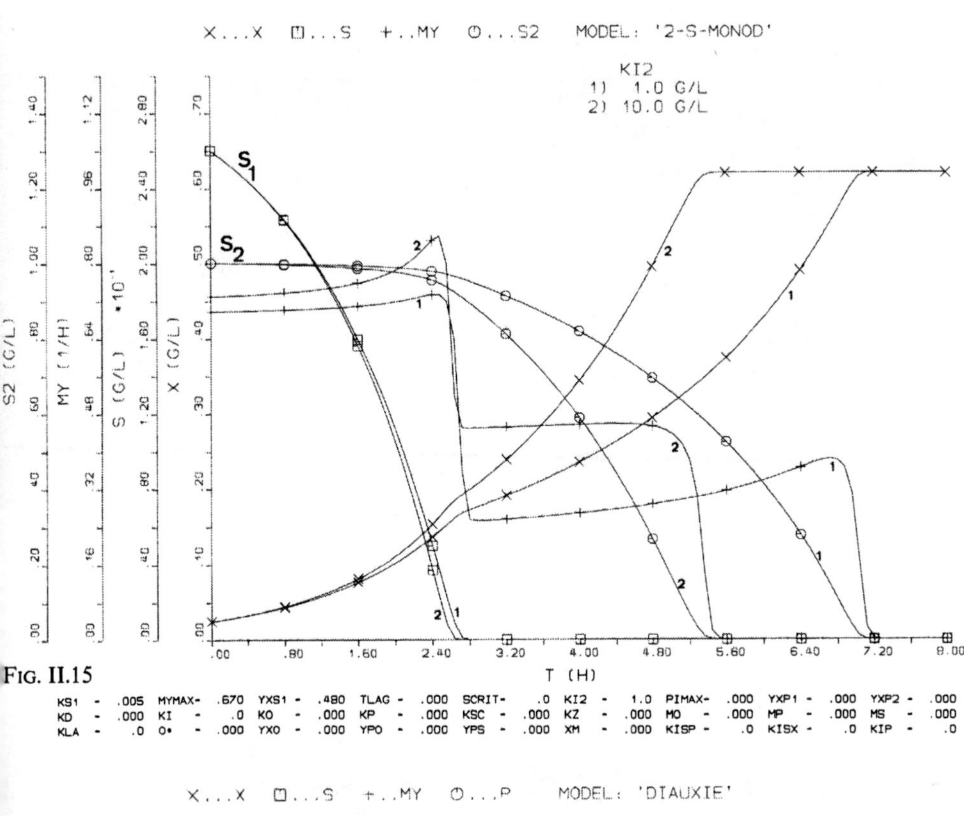

FIG. II.15

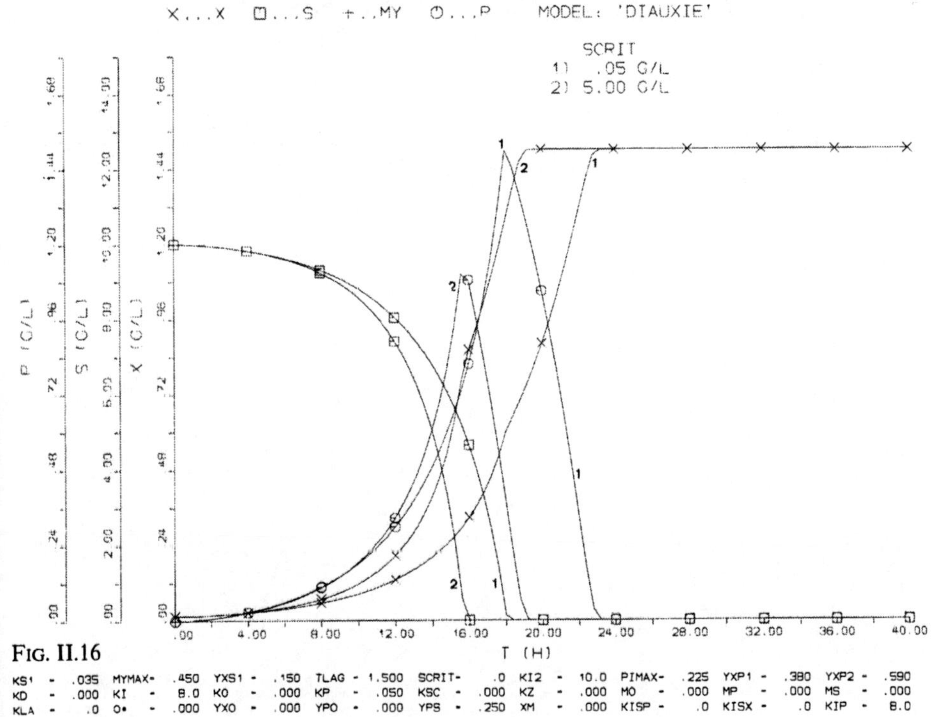

FIG. II.16

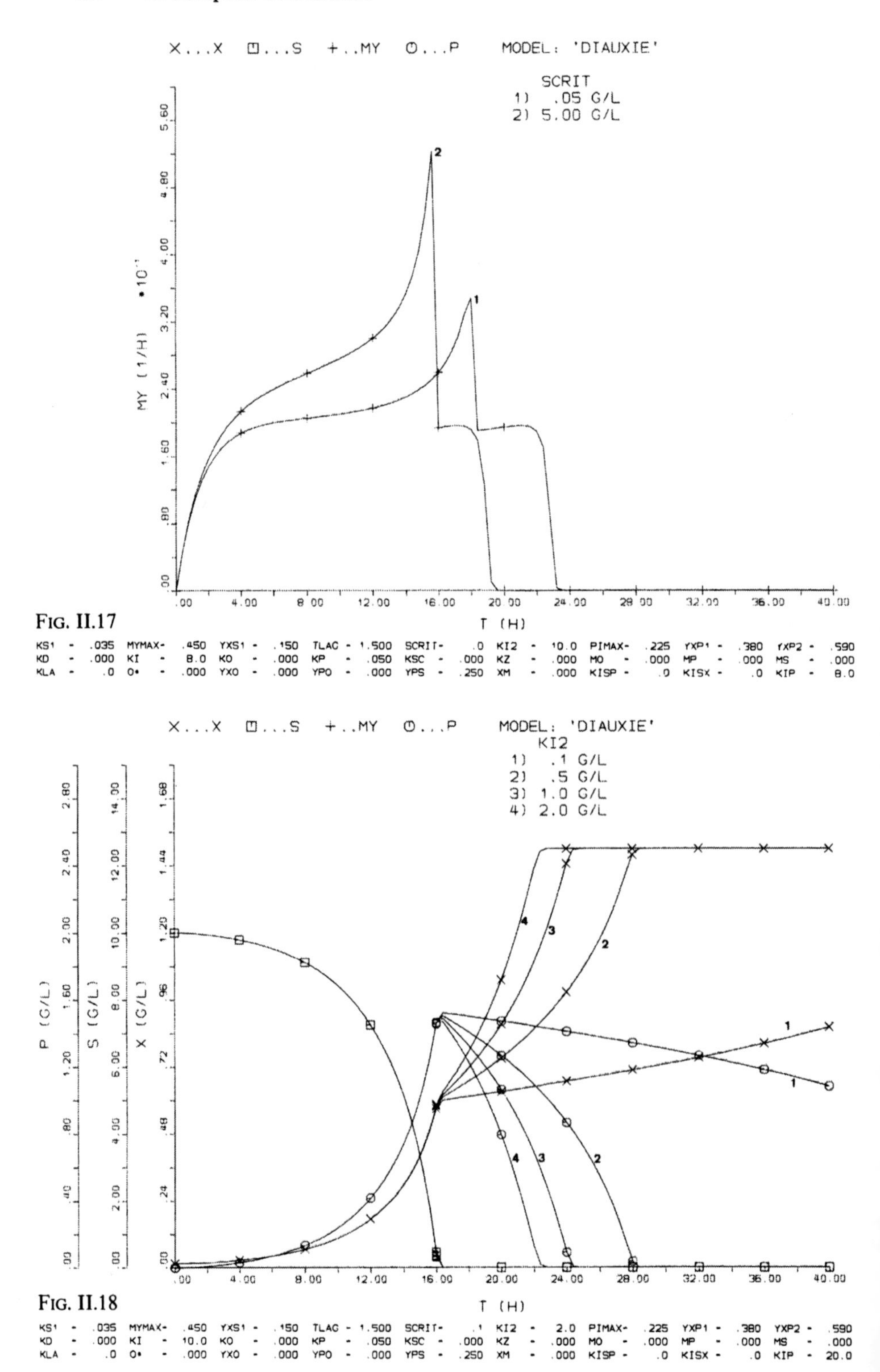

FIG. II.17

FIG. II.18

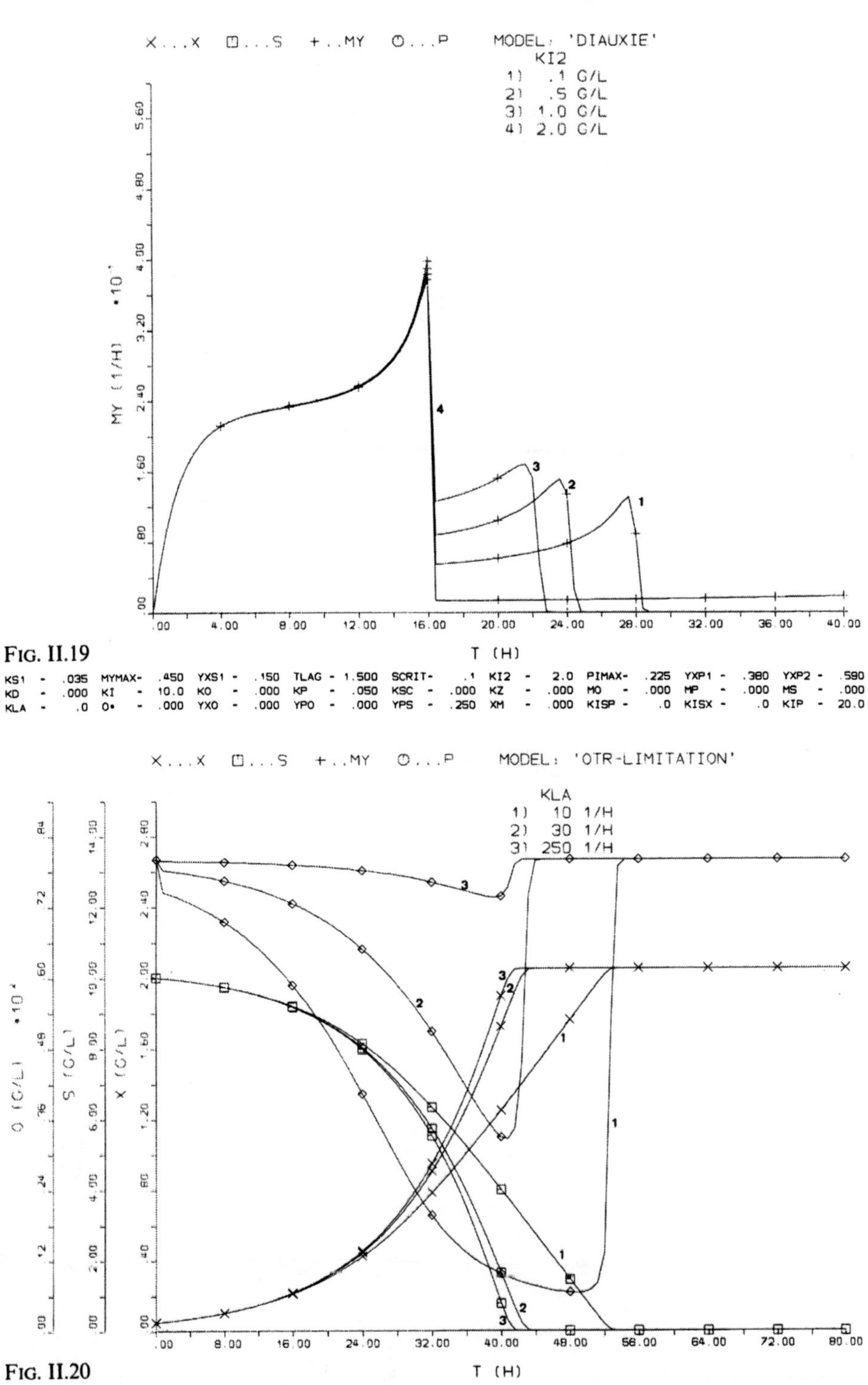

Fig. II.19

Fig. II.20

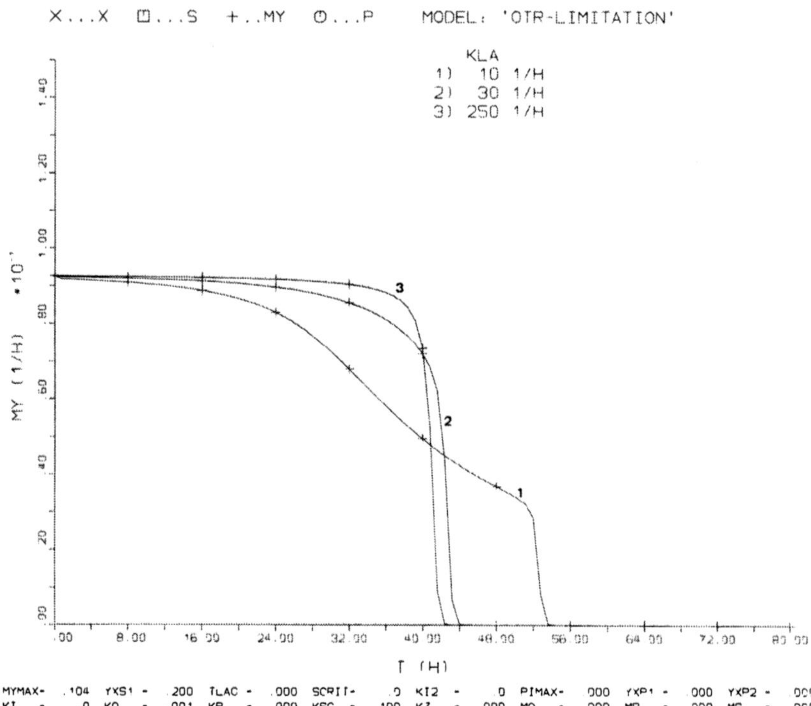

Fig. II.21

KS1	.000	MYMAX	.104	YXS1	.200	TLAC	.000	SCRIT	.0	KI2	.0	PIMAX	000	YXP1	.000	YXP2	.000
KD	000	KI	.0	KO	.001	KP	.000	KSC	100	KZ	000	MO	000	MP	000	MS	000
KLA	10.0	O•	.008	YXO	.880	YPO	.000	YPS	.000	XM	.000	KISP	.0	KISX	.0	KIP	0

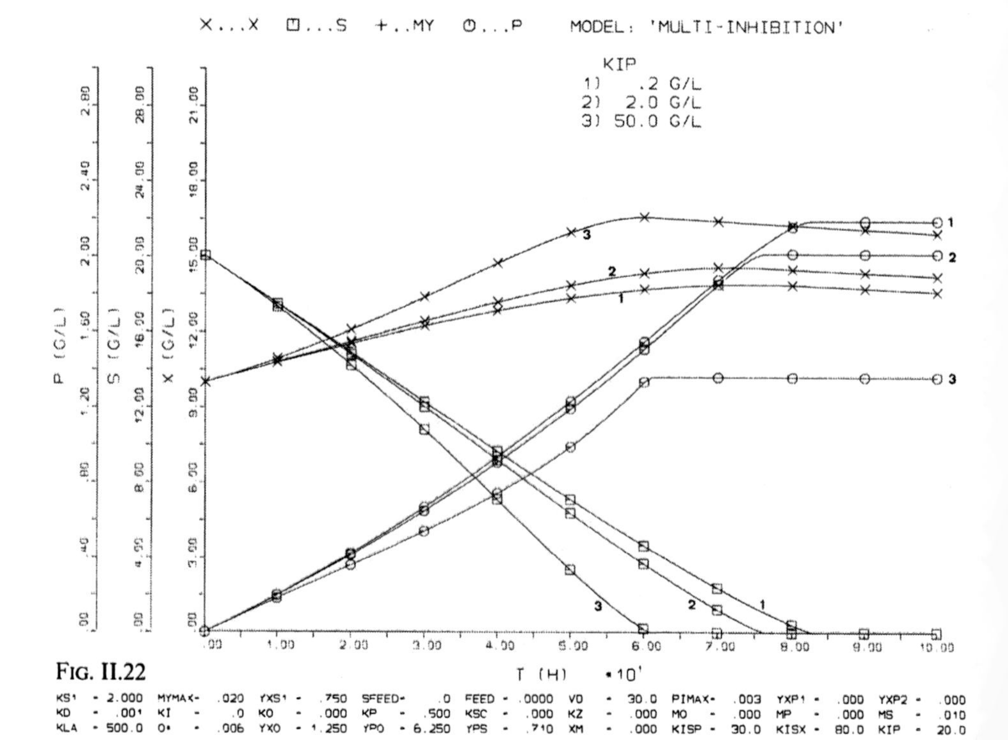

Fig. II.22

KS1	2.000	MYMAX	.020	YXS1	.750	SFEED	.0	FEED	.0000	VO	30.0	PIMAX	.003	YXP1	.000	YXP2	.000
KD	.001	KI	.0	KO	.000	KP	.500	KSC	.000	KZ	.000	MO	.000	MP	.000	MS	.010
KLA	500.0	O•	.006	YXO	1.250	YPO	6.250	YPS	.710	XM	.000	KISP	30.0	KISX	80.0	KIP	20.0

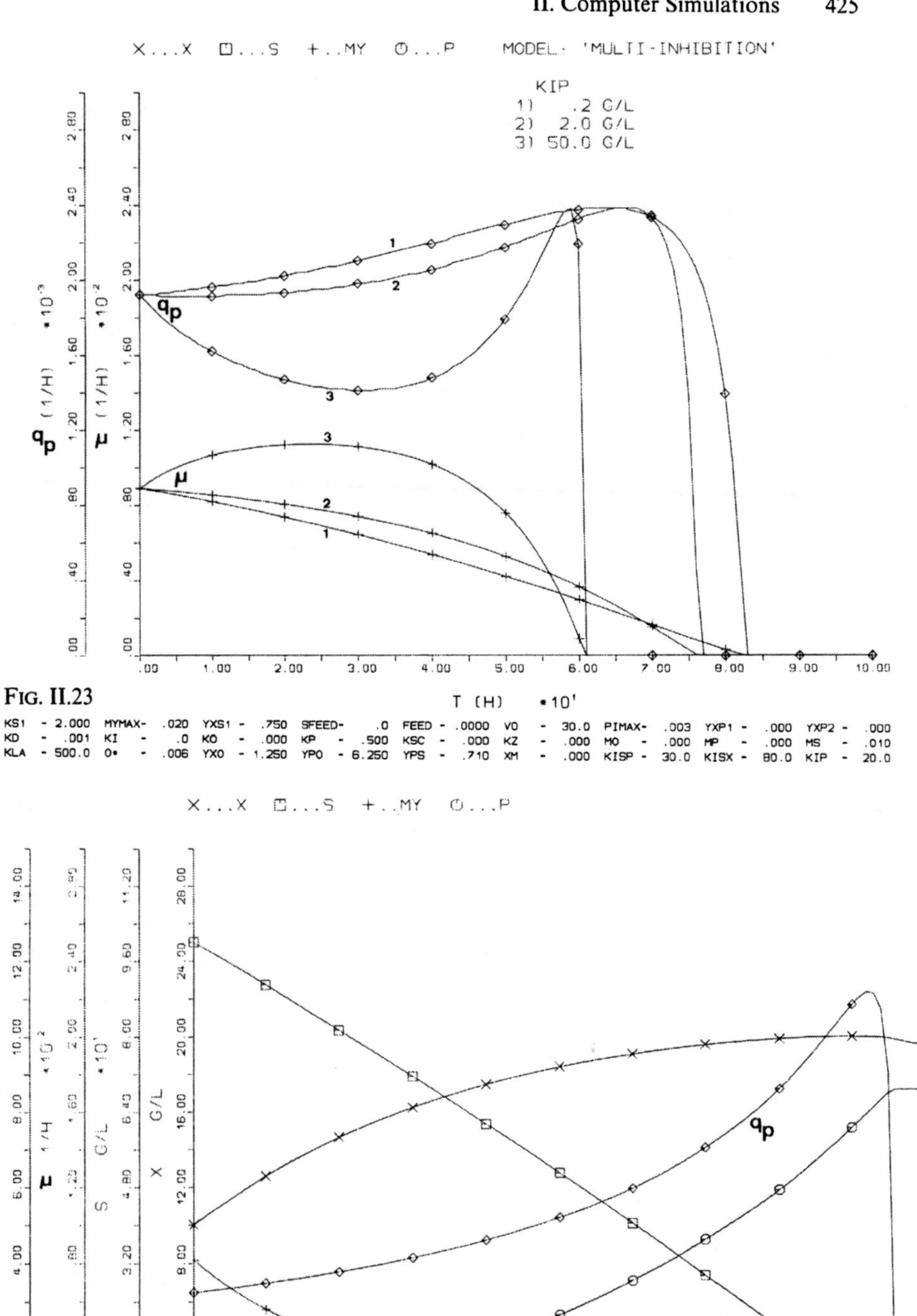

FIG. II.23

G. II.24

KS1 - 2.000 MYMAX- .020 YXS1 - .750 TLAG - .000 SCRIT- .0 KI2 - .00 PIMAX- .003 YXP1 - .000 YXP2 - .000
KD - .001 KI - .0 KO - .000 KP - .500 KSC - .000 KZ - .000 MO - .000 MP - .000 MS - .010
KLA - 500.0 O• - .006 YXO - 1.250 YPO - 6.250 YPS - .710 XM - .000 KISP - 30.0 KISX - 80.0 KIP - .6

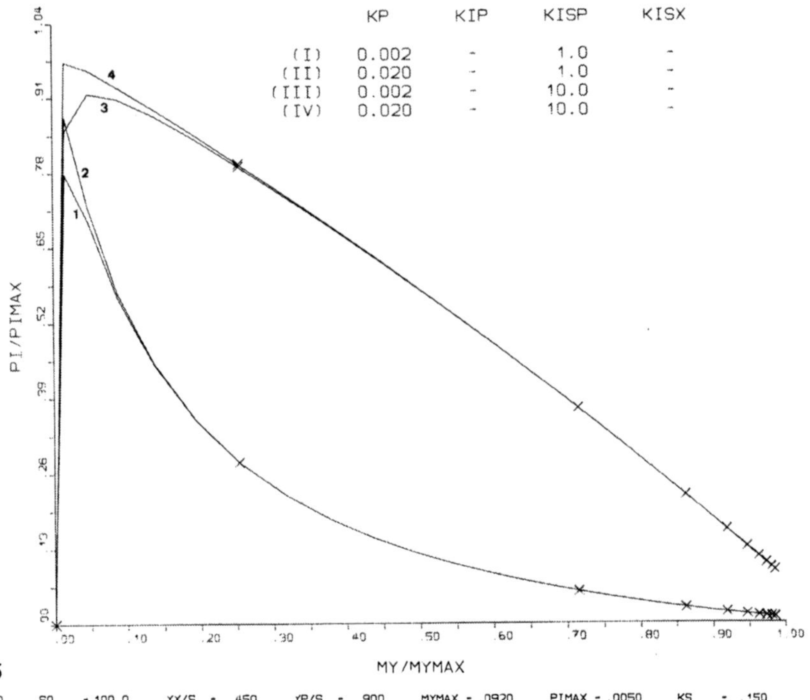

FIG. II.25

XO - 10.0 SO - 100.0 YX/S - .450 YP/S - .900 MYMAX - .0920 PIMAX - .0050 KS - .150
MS - .0140 KD - .0010

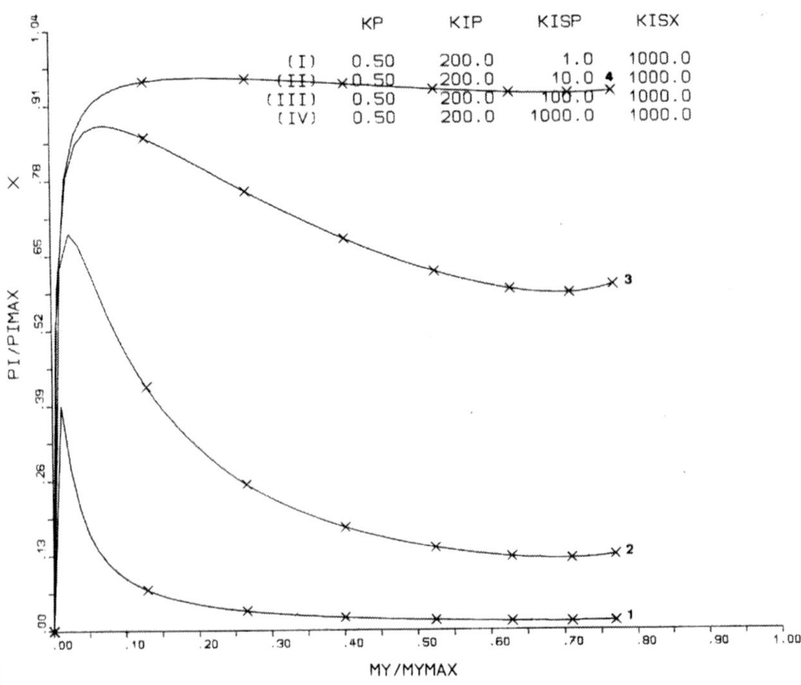

FIG. II.26

XO - 10.0 SO - 100.0 YX/S - .750 YP/S - .710 MYMAX - .0200 PIMAX - .0030 KS - 2.000
MS - .0100 KD - .0010

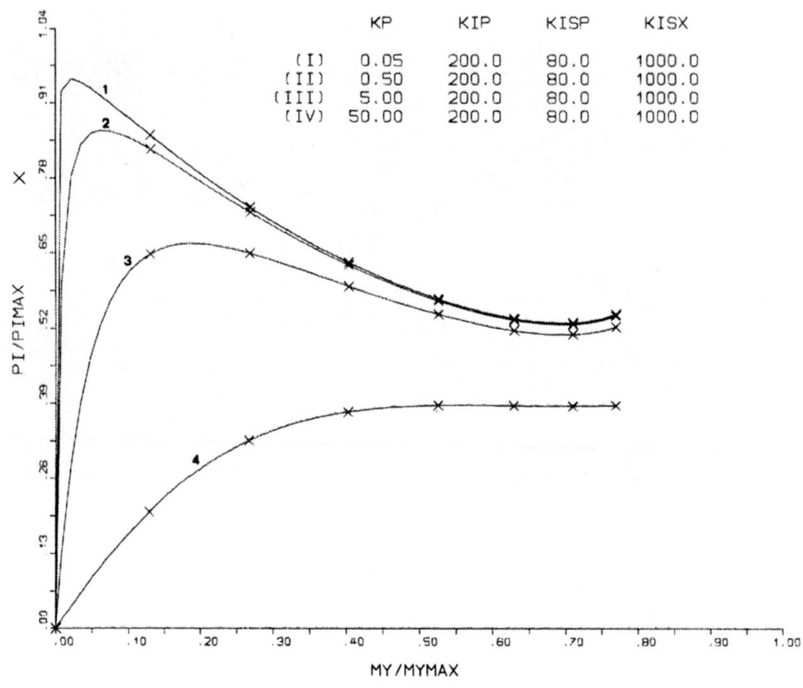

Fig. II.27

XO = 10.0 SO = 100.0 YX/S = .750 YP/S = .710 MYMAX = .0200 PIMAX = .0030 KS = 2.000
MS = .0100 KD = .0010

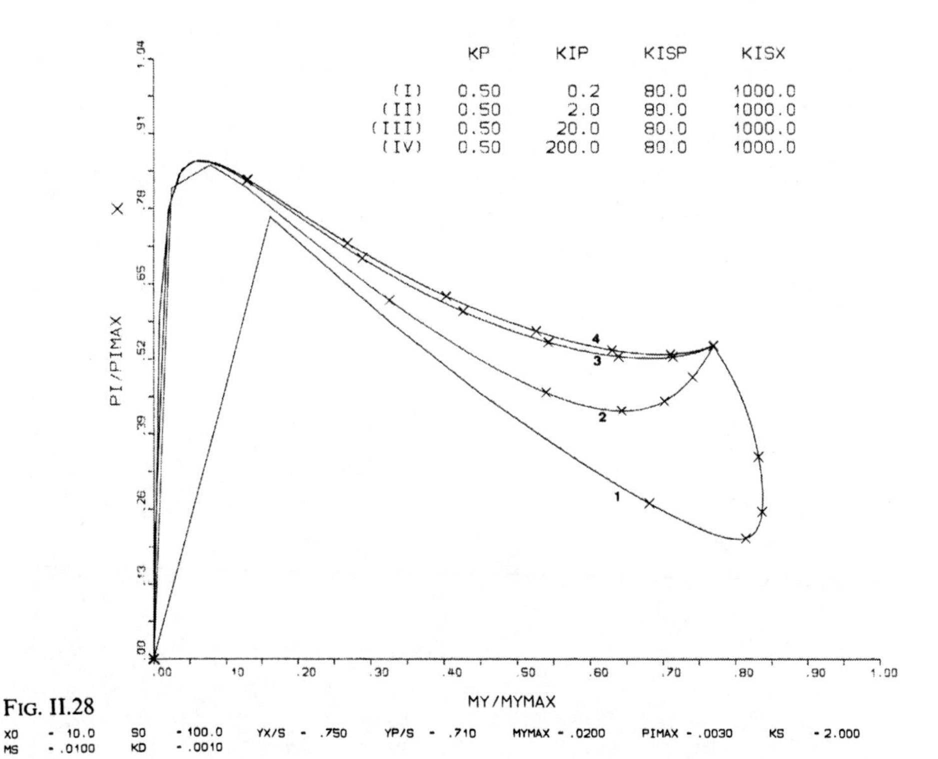

Fig. II.28

XO = 10.0 SO = 100.0 YX/S = .750 YP/S = .710 MYMAX = .0200 PIMAX = .0030 KS = 2.000
MS = .0100 KD = .0010

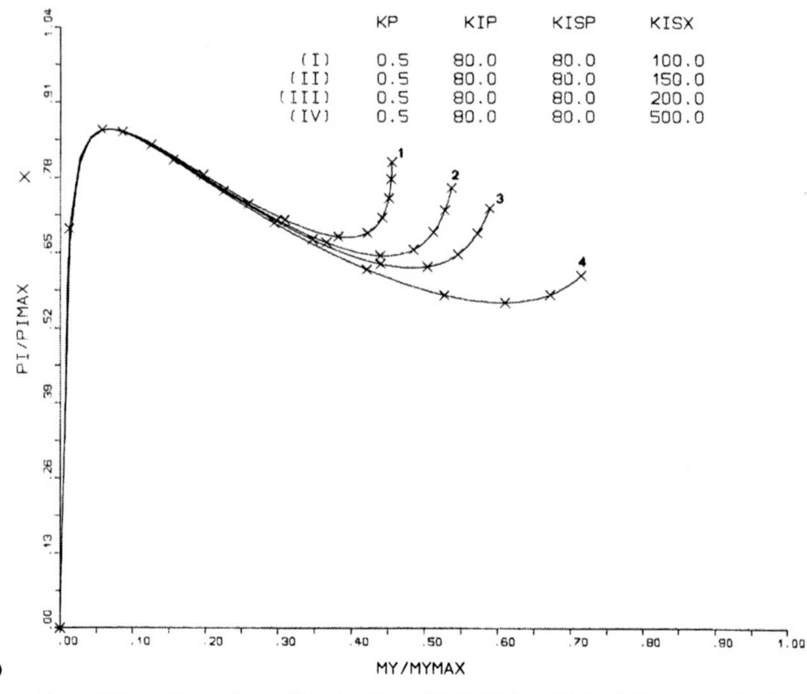

FIG. II.29

XO	- 10.0	SO	- 100.0	YX/S	- .750	YP/S	- .710	MYMAX	- .0200	PIMAX	- .0030	KS	- 2.000
MS	- .0100	KD	- .0010										

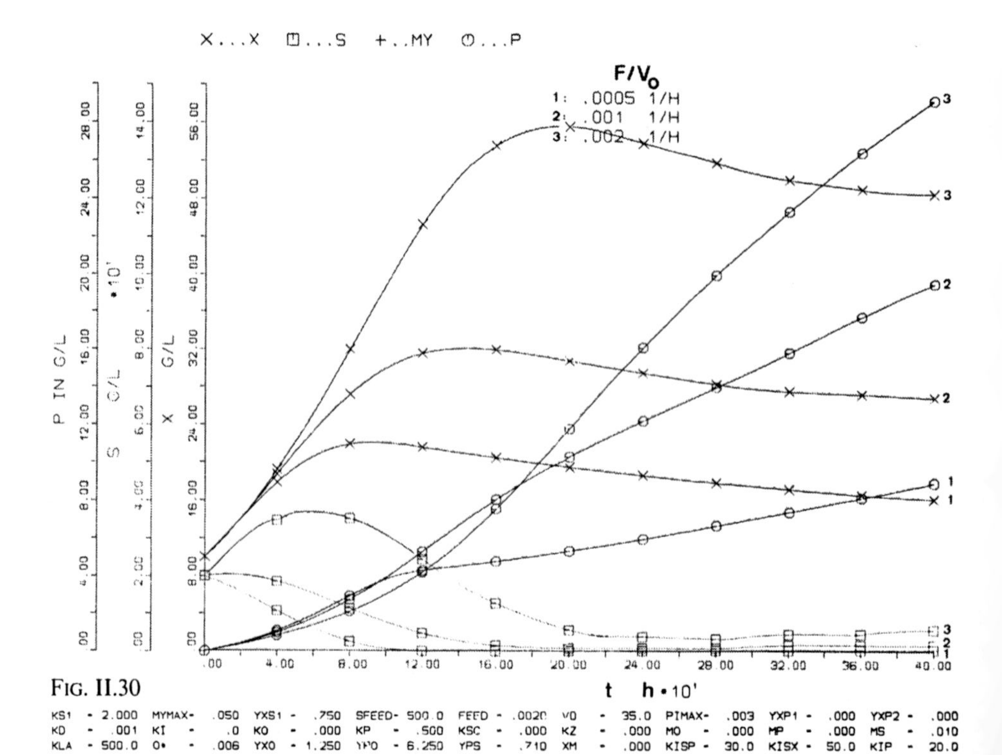

FIG. II.30

KS1	- 2.000	MYMAX-	.050	YXS1 -	.750	SFEED-	500.0	FEED -	.0020	VO	- 35.0	PIMAX-	.003	YXP1 -	.000	YXP2 -	.000
KD	- .001	KI -	.0	KO -	.000	KP -	.500	KSC -	.000	KZ	- .000	MO -	.000	MP -	.000	MS -	.010
KLA	- 500.0	O• -	.006	YXO -	1.250	YPO -	6.250	YPS -	.710	XM	- .000	KISP -	30.0	KISX -	50.0	KIP -	20.0

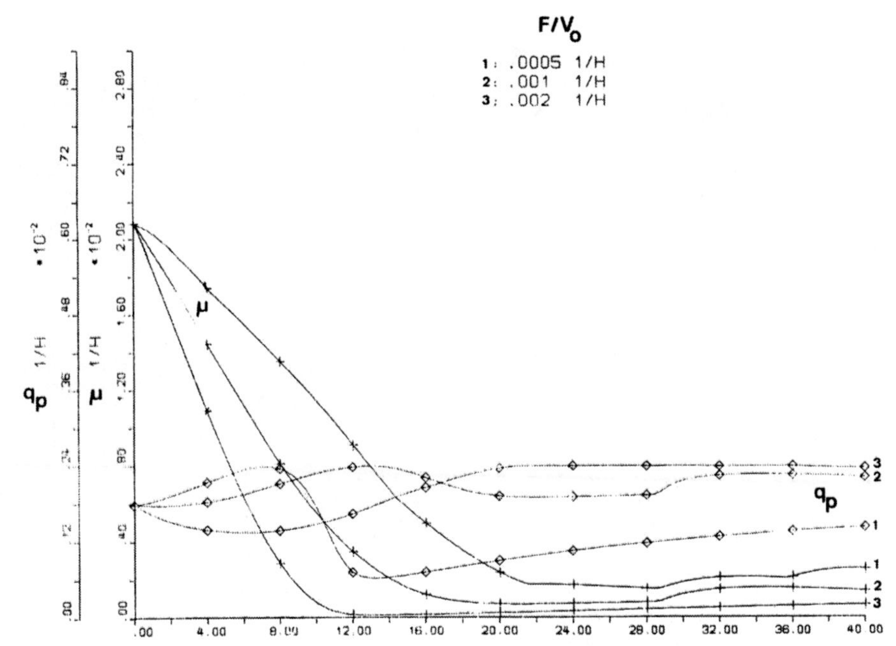

FIG. II.31

KS1	- 2.000	MYMAX-	.050	YXS1 -	.750	SFEED-	500.0	FEED -	.0020	V0	-	35.0	PIMAX-	.003	YXP1 -	.000	YXP2 -	.000
KD	- .001	KI -	.0	KO -	.000	KP -	.500	KSC -	.000	KZ	-	.000	MO -	.000	MP -	.000	MS -	.010
KLA	- 500.0	O* -	.006	YXO	- 1.250	YPO	- 6.250	YPS -	.710	XM	-	.000	KISF -	30.0	KISX -	50.0	KIP -	20.0

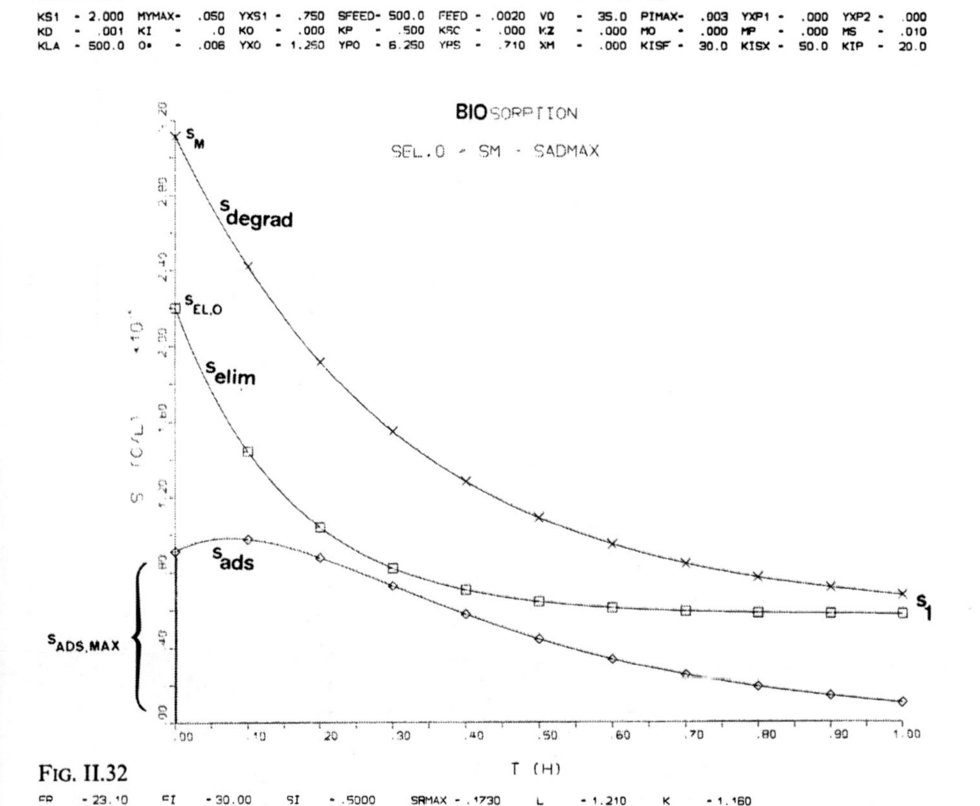

FIG. II.32

FR - 23.10 FI - 30.00 SI - .5000 SRMAX - .1730 L - 1.210 K - 1.160

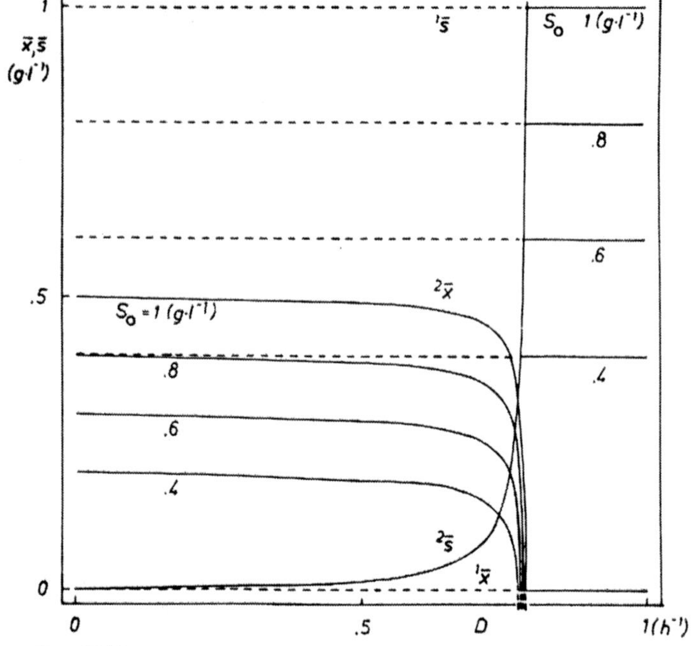

FIG. II.33

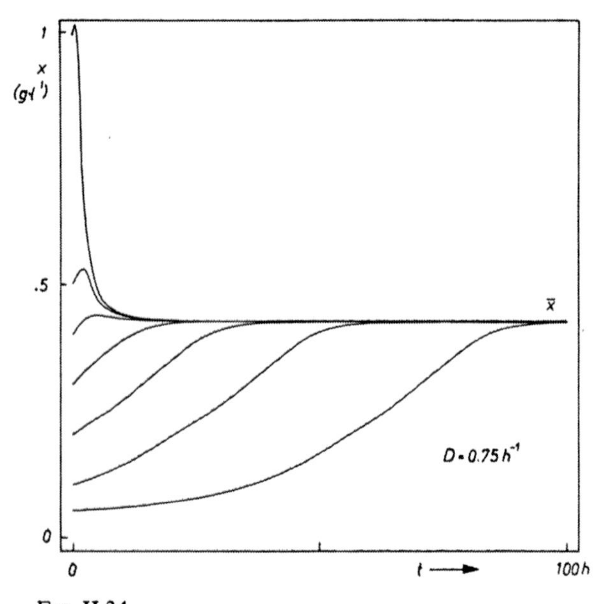

FIG. II.34

with

$$\mu(s,t) = \mu_{max}\frac{s}{K_S + s}\cdot(1 - e^{-t/t_L}) \tag{II.6}$$

Model parameters: μ_{max}, K_S, $Y_{X|S}$, t_L.

3. Monod kinetics with death rate k_d (cf. Sect. 5.3.4.1):

$$r_X = \mu(s)x - k_d \cdot x \tag{II.7}$$

$$-r_S = \frac{1}{Y_{X|S}}\cdot\mu(s)x \tag{II.3}$$

with Equ. II.4 for $\mu(s)$ and μ_{max}, K_S, $Y_{X|S}$, k_d.

4. Monod kinetics with endogeneous metabolism (m_S) (cf. Sect. 5.3.4.2):

$$r_X = \mu(s)x \tag{II.2}$$

$$-r_S = \frac{1}{Y_{X|S}}\mu(s)x - m_S \cdot x \tag{II.8}$$

with Equ. II.4 for $\mu(s)$ and μ_{max}, K_S, $Y_{X|S}$, m_S; x_0, s_0.

5. Monod kinetics with k_d and m_s as a combination of Equs. II.7 and II.8.
6. Monod kinetics with S inhibition (cf. Sect. 5.3.5.1). Thus, taking Equs. II.2 and II.3 with Equ. 5.88:

$$\mu(s) = \mu_{max}\frac{1}{1 + K_S/s + s/K_{IS}} \tag{II.9}$$

and K_{IS} being the key parameter.

7. Monod kinetics with double substrate limitation (cf. Sect. 5.5):

$$r_X = \mu(s_1,t)x + \mu(s_2)x \cdot f \tag{II.10}$$

$$-r_{S1} = \frac{1}{Y_{X|S1}}\cdot\mu(s_1,t)x \tag{II.11}$$

$$-r_{S2} = \frac{1}{Y_{X|S2}}\cdot\mu(s_2)x \cdot f \tag{II.12}$$

with Equ. II.6 for $\mu(s_1, t)$ and the following expression for the term f (cf. Equ. 5.106 or Equ. 5.152):

$$f = f_1 \cdot f_2 = \frac{1}{1 + s_1/s_{1,crit}}\cdot\frac{1}{1 + s_2/K_R} \tag{II.13}$$

with K_R being the key parameter for repression and $s_{1,crit}$ representing the substrate concentration, where both substrates are utilized simultaneously, thus losing the diauxic behavior.

8. Diauxic growth as an analogy to case 7:

scheme: $S_1 \begin{matrix} \nearrow X \\ \searrow P = S_2 \rightarrow X \end{matrix}$

$$r_X = \mu(s_1, t)x + \mu(p)x \cdot f_1 \cdot f_2 \tag{II.14}$$

$$-r_{S1} = \frac{1}{Y_{X|S1}} \cdot \mu(s_1, t)x + \frac{1}{Y_{P|S}} \cdot q_P(s) \cdot x \tag{II.15}$$

$$r_P(\equiv -r_{S2}) = \frac{1}{Y_{X|P}} \cdot \mu(s, t)x - \frac{1}{Y_{X|S2}} \mu(p)x \cdot f_1 \cdot f_2 \tag{II.16}$$

with kinetic equations

$$\mu_{S1} = \mu_{max,1} \frac{s_1}{s_1 + K_{S1} + s_1^2/K_{IS}} \cdot \frac{1}{(1 + p/K_{IP})} \cdot (1 - e^{-t/t_{L1}}) \tag{II.17}$$

$$\mu_{S2} = \mu_{max,2} \frac{s_2}{K_{S2} + s_2} \qquad (s_2 \equiv p) \tag{II.18}$$

$$q_P = \frac{1}{Y_{X|P}} \cdot \mu_{S1} \tag{II.19}$$

and Equ. II.13 for the term $f_1 \cdot f_2 = f$. This simplest model of diauxie thus contains 13 parameters: $\mu_{max,1}$, $\mu_{max,2}$, K_{S1}, $K_P(\equiv K_{S2})$, $Y_{X|S1}$, $Y_{X|P}$, $Y_{P|S1}$, $Y_{X|S2}$, t_{L1}, K_{IS}, K_{IP}, $K_R(\equiv K_{12})$, and $s_{1,crit}$.

9. Monod with O_2 limitation (cf. Sect. 5.5.3):

$$r_X = \mu(s, o)x \tag{II.20}$$

$$-r_S = \frac{1}{Y_{X|S}} \cdot \mu(s, o)x \tag{II.21}$$

$$r_O = k_{L1}a(o_L^* - o_L) - \frac{1}{Y_{X|O}} \cdot \mu(s, o)x \tag{II.22}$$

with Equ. 5.169 for $\mu(s, o)$:

$$\mu(s, o) = \mu_{max} \frac{s}{K_S + s} \cdot \frac{o}{K_O + o} \tag{II.23}$$

The parameters are μ_{max}, K_S, K_O, $Y_{X|S}$, $Y_{X|O}$, $k_{L1}a$, and o^*.

10. "Multiinhibition model" for the quantification of secondary metabolite productions (cf. Sect. 5.4):

scheme: $S + O \xrightarrow{x} X + P$

$$r_X = \mu(s, o)x - k_d \cdot x \tag{II.24}$$

$$-r_S = \frac{1}{Y_{X|S}} \cdot \mu(s, o)x + m_S \cdot x + \frac{1}{Y_{P|S}} \cdot q_P(s, o)x \tag{II.25}$$

$$r_P = q_P(s, o)x \tag{II.26}$$

$$r_O = k_{L1}a(o_L^* - o_L) + m_O \cdot x - \frac{1}{Y_{X|O}}\mu(s, o)x - \frac{1}{Y_{P|O}}q_P(s, o)x \tag{II.27}$$

with the kinetic expressions for μ and q_P

$$\mu(s, o) = \mu_{max}\frac{s}{s(1 + s/K_{ISX}) + K_S \cdot x/(1 + p/K_{IP})} \cdot \frac{o}{K_O + o} \tag{II.28}$$

$$q_P(s, o) = q_{P,max}\frac{s}{K_P + s(1 + s/K_{ISP})} \tag{II.29}$$

Combining Equs. II.4 and II.29 yielded Equ. 5.138 ($K_{ISP} \equiv K_R \equiv K_I$!) according to an approach of Bajpaj and Reuss (1981), while the full model of "multiinhibitions" according to Moser and Schneider (1988) contains four constants K_P, K_{IP}, K_{ISP} ($\equiv K_R$), and K_{ISX}. The main aim of this complex model was to achieve plots of increasing q_P at decreasing μ, as illustrated in Figs. II.23 and II.24. The corresponding successful plots of concentrations are shown in Figs. II.22 and II.24.

The behavior of the product formation model including such multi-inhibition kinetics (with repression), basically represented in Fig. 5.47, is shown in sensitivity analysis in the q_P/μ plots of Fig. II.25 through II.29, with variations in all four parameters K_P, K_{IP}, K_{ISP}, and K_{ISX}.

11. Evaluation of substrate feeding strategy for a fed-batch culture with optimal production of secondary metabolites—case study. On the basis of the mathematical model (Equs. II.24–II.29) and experimental estimation of model parameters, an optimal S feed can be found by simulations adding the expression $F_S(t)$ for S feed to Equ. II.25:

$$F_S(t) = \frac{F}{V_0} \cdot s_F \tag{II.30}$$

where s_F is the substrate concentration [$g \cdot l^{-1}$] in the feedstream and F_S is the substrate feed rate [$g/l \cdot h$], while F is the volumetric flow rate [$l \cdot h^{-1}$] and V_0 [l] is the initial volume. Figure II.30 illustrates the concentration/time curves for substrate, biomass, and product at varied S feed F/V_0 at constant S_F. The corresponding time curves of specific rates μ in respect to q_P show that choice of the right F/V_0 allows the value for μ to be maintained at a minimum while q_P is kept constant at a high level (Fig. II.31).

12. Biosorption model (cf. Sect. 5.3.9). A simple reaction scheme (see Fig. 5.40) is written as a sequence of elimination and degradation

$$S_L \xrightarrow{\text{elim}} S_S + O \xrightarrow{\text{degrad}} P \tag{II.31}$$

The balance equations in case of a DCSTR are (cf. Equs. 5.112–5.115):

$$r_{el} = \frac{d(s_{el} - s_l)}{dt} = k_{el}(s_{el} - s_l) \tag{II.32}$$

$$r_{degr} = \frac{d(s_d - s_l)}{dt} = k_{degr}(s_{degr} - s_l) \tag{II.33}$$

and the biosorption rate r_{ads} is

$$r_{ads} = r_{el} - r_{degr} \tag{II.34}$$

According to the literature (Theophilou et al., 1979; see Sect. 5.3.9), a sorption capacity ξ_{ads} is defined (see Equ. 5.115) and used in an analogy to Langmuir adsorption (see Equ. 5.114). The kinetic constants were elaborated by these authors to be load dependent:

$$k_{el} = (0.58 \cdot \sqrt{L})x \tag{II.35}$$

$$k_{degr} = (1.86 \cdot L - 0.37)x \qquad \text{for } L < 0.9 \text{ h}^{-1} \tag{II.36a}$$

$$k_{degr} = (-0.23 \cdot L + 1.53)x \qquad \text{for } L > 0.9 \tag{II.36b}$$

and

$$s_{el} = 41 \cdot \sqrt{L} + 12 \quad [\text{mg COD} \cdot \text{l}^{-1}] \tag{II.37}$$

"Sludge loading" $\hat{s}$ (= adsorbed concentration s_{ads}) was found to be L dependent

$$\hat{s} = 36.5 \cdot \text{L} + 22 \quad [\text{mg COD} \cdot \text{l}^{-1}] \tag{II.38}$$

and $s_{ads,max} \equiv \hat{s}_{max} = 173 \text{ mg} \cdot \text{l}^{-1}$ in this situation. Figure II.32 represents a typical plot of the time curves of concentrations (s_{ads}, s_{el}, and $s_d \equiv s_{degr}$).

13. Monod kinetics in CSTR (cf. Sect. 6.1.1). The following set of balance equations were used (see Equ. 6.3):

$$r_X = \mu(s)x - D \cdot x \tag{II.39}$$

$$-r_S = \frac{1}{Y_{X|S}}\mu(s)x + D(s_0 - s) \tag{II.40}$$

with Monod-type kinetics according to Equ. II.4. Whereas Fig. 6.1 showed the typical plot of "chemostat" behavior (steady-state concentrations $\bar{x}$ and $\bar{s}$ versus dilution rate D), Fig. II.33 represents a similar plot with variations in initial substrate concentration s_0. The "washout state" is indicated (${}^1\bar{x}$, ${}^1\bar{s}$) together with a second stationary state characterized by a nonvanishing biomass concentration ${}^2\bar{x}$, ${}^2\bar{s}$.

Whereas for batch cultures the final concentrations depend on the initial concentrations, in an open reactor like the CSTR "equifinality" is established, where the end value of stable concentration $\bar{x}$ is independent of initial values, as shown in Fig. II.34 in case of $\mu_{max} = 0.8 \text{ h}^{-1}$, $K_S = 10$ $\text{mg} \cdot \text{l}^{-1}$, $Y_{X|S} = 0.5$, $s_0 = \text{g} \cdot \text{l}^{-1}$ (from ZIMET, Jena).

CSTR behavior in case of more complex kinetics has been shown in Figs. 6.4 through 6.11 and 6.14 through 6.20. Alternative bioreactor operations (CPFR, NCSTR, RR, etc.) were represented in Sects. 6.4 through 6.8.

Bibliography

General:

Röpke, H., Riemann, J. (1969). Analogcomputer in Chemie und Biologie, Berlin: Springer-Verlag.

Knorre, W.A. (1971). Analogcomputer in Biologie und Medizin. Jena: VEB G. Fischer.

Romanovsky, J.M., Stepanova, N.V., and Chernavsky, D.D. (1974). Kinetische Modelle in der Biophysik. Jena: VEB G. Fischer.

Levin, S. (ed.) (1981). Modèles Mathématiques en Biologie, Berlin: Springer-Verlag.

Knorre, W.A. (1980). Kinetische Modelle in der Mikrobiologie in "Biophysikalische Grundlagen der Medizin (Beier W., Rosen R., eds.) Stuttgart: Fischer.

Spain, J.D. (1984). Basic Microcomputer Models in Biology, Addison-Wesley Publ. Comp., Reading, Massachusetts, USA.

Rogers, P.L. (1976). In Advances Biochem. Engng. 4, 125.

Special:

Bajpaj, R.K., Reuss M. (1981). *Biotechnol. Bioengng.*, 23, 717.

Moser, A., Schneider, H. (1988). *Bioprocess Engineering* 3, in press.

Theophilou, J., Wolgbauer, O., and Moser, F. (1979). Gas-Wasser-Fach-Wasser/Abwasser 120, 119.

APPENDIX III

Microkinetics: Derivation of Kinetic Rate Equations from Mechanisms

The objective of this appendix is to demonstrate the concepts of the rds (rate-determining step) and qss (quasi-steady-state). These are both of great importance in kinetic modeling, as explained in Sects. 2.4, 4.2, and 5.2. At the same time, some well-known approaches to the microkinetics of enzyme reactions are presented here; these are more fully discussed in sect. 5.2.2.1.

III.1 Simple (Chemical) Reaction

Reaction scheme:

$$A \rightarrow B \tag{III.1}$$

Reaction mechanism (collision theory):

$$2A \xrightarrow{k_1} \{A\} + A \tag{III.2a}$$

$$A + \{A\} \xrightarrow{k_2} 2A \tag{III.2b}$$

$$\{A\} \xrightarrow{k_3} B \tag{III.2c}$$

Rates of individual steps:

$$r_1 = k_1 \cdot c_A^2 \tag{III.3a}$$

$$r_2 = k_2 \cdot c_A \cdot c\{\hat{A}\} \tag{III.3b}$$

$$r_3 = k_3 \cdot c\{\hat{A}\} \tag{III.3c}$$

The qss concept (assuming that $r\{\hat{A}\} = 0$) gives

$$r_{\{A\}} = k_1 \cdot c_A^2 - k_2 \cdot c_A \cdot c_{\{A\}} - k_3 \cdot c_{\{A\}} = 0 \tag{III.4}$$

This assumption enables the derivation of an equation for the activated state $\{A\}$, which normally cannot be measured:

$$c_{\{A\}} = \frac{k_1 \cdot c_A^2}{k_2 \cdot c_A + k_3} \tag{III.5}$$

Now, the rds concept is applied, assuming that

$$r_{tot} = r_3 = k_3 \cdot c_{\{A\}} \tag{III.6}$$

resulting in an expression for the global rate r_{tot} by using Equ. III.5

$$r_{tot} = \frac{k_1 \cdot c_A^2}{(k_2/k_3)c_A + 1} \tag{III.7}$$

In case of $c_A \gg 0$, the final form of a kinetic equation is

$$r_{tot} = \left(\frac{k_1 \cdot k_3}{k_2}\right) \cdot c_A \tag{III.8}$$

which is of simple first order.

III.2 Michaelis–Menten Approach (cf. Equ. 2.53)

Reaction mechanism:

$$S + E \underset{k_{-1}}{\overset{k_{+1}}{\rightleftharpoons}} \{ES\} \xrightarrow{k_{+2}} E + P \tag{III.9}$$

rds concept:

$$r_{tot} = k_{+2} \cdot c_{\{ES\}} \tag{III.10}$$

qss concept applied for $\{ES\}$:

$$\frac{dc_{\{ES\}}}{dt} = k_{+1} \cdot s \cdot e - k_{-1}\{es\} = 0 \tag{III.11}$$

result in Equ. 2.53 with

$$K_m = \frac{k_{-1}}{k_{+1}} \tag{III.12}$$

and

$$r_{max} = k_{+2} \cdot e \tag{III.13}$$

Remember that $e = e_0 - \{es\}$.

III.3 Briggs–Haldane Approach

The same approach as in Equs. III.9 and III.10 but with a qss concept including a steady-state assumption of all k_i values gives:

$$r_{\{ES\}} = k_1 \cdot s \cdot e - k_{-1} \cdot \{es\} - k_{+2}\{es\} = 0 \tag{III.14}$$

The result is the same type of a kinetic expression (cf. Equ. 2.53) but with a different meaning of the saturation constant K_m:

$$K_m = \frac{k_{-1} + k_{+2}}{k_{+1}} \tag{III.15}$$

III.4 Langmuir–Hinshelwood Approach

Reaction mechanism:

$$E + S \underset{k_{-1}}{\overset{k_{+1}}{\rightleftharpoons}} \{ES\} \xrightarrow{k_{+2}} \{EP\} \underset{k_{-3}}{\overset{k_{+3}}{\rightleftharpoons}} E + P \tag{III.16}$$

Rates of individual components with qss:

$$\frac{ds}{dt} = -k_{+1} \cdot s \cdot e + k_{-1} \cdot \{es\} \tag{III.17a}$$

$$\frac{de}{dt} = -k_{+1} \cdot s \cdot e + k_{-1}\{es\} - k_{-3} \cdot e \cdot p + k_3\{ep\} \tag{III.17b}$$

$$\frac{d\{es\}}{dt} = k_{+1} \cdot s \cdot e - k_{-1}\{es\} - k\ _2\{es\} = 0 \tag{III.17c}$$

$$\frac{d\{ep\}}{dt} = k_{+2}\{es\} + k_{-3}\{ep\} - k_{+3}\{ep\} \tag{III.17d}$$

$$\frac{dp}{dt} = k_{+3}\{ep\} - k_{-3} \cdot e \cdot p \tag{III.17e}$$

with

$$e + \{es\} + \{ep\} = e_0 \tag{III.18}$$

and steps 1 and 3 being at equilibrium (K_S = adsorption, K_P = desorption constant):

$$\frac{\{es\}}{e} = \frac{k_{+1} \cdot s}{k_{-1}} \cdot s = K_S \cdot s \tag{III.19a}$$

$$\frac{\{ep\}}{e} = \frac{k_{-3}}{k_{+3}} \cdot p = K_P \cdot p \tag{III.19b}$$

Equation III.18, rewritten by using Equ. III.19a, b gives

$$\frac{\{es\}(1 + K_S \cdot s + K_P \cdot p)}{K_S \cdot s} = e_0 \tag{III.20}$$

which can be replaced in Equ. III.10 (rds!) resulting in the final form of a kinetic term:

$$r_{tot} = r_{max} \frac{K_S \cdot s}{1 + K_S \cdot s + K_P \cdot p} \tag{III.21}$$

This is the general form of the Langmuir–Hinshelwood equation applicable to heterogeneous chemical catalysis.

III.5 Reversible Michaelis–Menten Type

Reaction mechanism:

$$E + S \underset{k_{-1}}{\overset{k_{+1}}{\rightleftharpoons}} \{\mathrm{ES}\} \underset{k_{-2}}{\overset{k_{+2}}{\rightleftharpoons}} E + P \tag{III.22}$$

The rds concept yields

$$r_{\mathrm{tot}} = \frac{dp}{dt} = k_{+2} \cdot \{\mathrm{es}\} - k_{-2} \cdot e \cdot p \tag{III.23}$$

The qss concept gives

$$\{\mathrm{es}\} = \frac{(k_{+1} \cdot s + k_{-2}p)e_0}{k_{+1} \cdot s + k_{-1} + k_{-2}p + k_{+2}} \tag{III.24}$$

Thus

$$r_{\mathrm{tot}} = e_0 \frac{k_{+2}k_{+1} \cdot s - k_{-2} \cdot k_{-1} \cdot p}{k_{+1} \cdot s + k_{-1} + k_{-2} \cdot p + k_{+2}} \tag{III.25}$$

and in equilibrium, where

$$k_{+1} \cdot k_{+2} \cdot s = k_{-1} \cdot k_{-2} \cdot p$$

and using the definition of K_{eq} (cf. Equ. 5.22) resulted in Equ. 5.21.

Simplifications are achieved if $p = 0$ and $s = 0$ for the initial rates of forward and backward reaction, which are used in enzyme kinetic studies.

III.6 Reversible Langmuir–Hinshelwood Approach

Reaction mechanism:

$$E + S \underset{k_{-1}}{\overset{k_{+1}}{\rightleftharpoons}} \{\mathrm{ES}\} \underset{k_{-2}}{\overset{k_{+2}}{\rightleftharpoons}} \{\mathrm{EP}\} \underset{k_{-3}}{\overset{k_{+3}}{\rightleftharpoons}} E + P \tag{III.26}$$

The rds concept states that

$$r_{\mathrm{tot}} = k_{+2}\{\mathrm{es}\} - k_{-2}\{\mathrm{ep}\} \tag{III.27}$$

and qss concepts are needed for both complexes {ES} and {EP}:

$$r\{\mathrm{es}\} = k_{+1} \cdot s \cdot e - k_{-1}\{\mathrm{es}\} + k_{-2}\{\mathrm{ep}\} - k_{-2}\{\mathrm{es}\} = 0 \tag{III.28a}$$

$$r\{\mathrm{es}\} = k_{+2}\{\mathrm{es}\} - k_{-2}\{\mathrm{ep}\} + k_{-3} \cdot e \cdot p - k_{+3}\{\mathrm{ep}\} = 0 \tag{III.28b}$$

In analogy to the preceding case (cf. Equs. III.18 and III.19) expressions for both complexes are written as

$$\{\text{es}\} = \frac{e_0 \cdot s \cdot K_S}{1 + K_S \cdot s + K_P \cdot p} \tag{III.29a}$$

$$\{\text{ep}\} = \frac{e_0 \cdot p \cdot K_P}{1 + K_S \cdot s + K_P \cdot p} \tag{III.29b}$$

and Equ. III.27 can be rewritten by using Equ. III.29a and b as

$$r_{\text{tot}} = e_0 \left(\frac{k_{+2} \cdot s \cdot K_S}{1 + K_S \cdot s + K_P \cdot p} - \frac{k_{-2} \cdot p \cdot K_P}{1 - K_S \cdot s + K_P \cdot p} \right) \tag{III.30}$$

resulting in an expression already shown in Equ. 5.23. Very often, the resulting rate equation has the same form as Equ. 5.21 or Equ. 5.23 (both of them can be brought in a similar form) but with a different interpretation of the kinetic parameters. There are many mechanisms more complicated than those described here that nonetheless generate the same type of formal kinetics, for example, Equs. 2.54, 5.21, or 5.23.

Index

A
absolute deviation, 89
absolute rate, 19, 20
absorption, 221
activated state, 203, 436ff
activation, 199
activation energy, 199, 200
activity function of product formation, 276
adaptation constant, 277
adaptational parameter, 60
adaptive modeling, 50
adsorption, 240
aeration number, 106
age, 246
allosteric control, 217
allosteric inhibition, 232
allosterie, 212, 217
amensalism, 261, 265
anhydrase-method, 190
apical growth, 238
apparent morphology, 390
apparent rate, 172
apparent value, 172
apparent viscosity, 387
Arrhenius equation, 199
Arrhenius plot, 200, 201
assimilation, 220, 221
Atkinson equation, 283
ATP, 29, 30, 245, 282
attraction domains, 320
autocatalytic process, 344, 345
autocatalysis, 25
autoregulation, 205
available electrones, 31
axial dispersion, 122, 175

B
Back mix-plug flow model, 84
balance, mass, 118ff
balanced growth, 145, 157, 272
balanced system, 11, 116, 145, 272
balancing method, 53, 382ff
batch reactor *see* discontinuous reactor
bed expansion, 117
bed porosity, 369
bench scale (lab scale), 44
Bingham plastic, 385
biocoenosis, 10, 259, 260, 293
biodisc reactor, 68, 139
biofilm, 68
biofilm kinetics, 151, 283ff
biofilm operation, 69, 139
biofilm reactor operation, 69, 139, 358ff
biological inertia, 198, 212, 214
biological rate equation, 178, 283f
biological test system, 41, 87, 90, 110ff
biomass, 19
biomass holdup, 367
bioparticle terminal Re number, 369
Bioprocess design, 307ff
Bioprocess kinetics, 19, 197ff
Bioprocess technology, 5, 7, 13
bioreactor, 2, 66ff, 73, 307
bioreactor concept, 112
bioreactor dimensioning, 127
bioreactor model, 44, 45, 56, 112, 118ff, 125, 307
bioreactor operation, 112, 307
bioreactor performance, 307ff
biosociety, 1
biosorption, 238, 342
biosorption model, 238, 433

biosorption rate, 434
biotechnology, 1
Biot-number, 187
bistable system, 319
bistability, 319, 320
blads box model, 48
Blackman-kinetic, 218
bleed stream, 378
BOD, 292
BOD_5, 292
BOD_{PL}, 292
Bodenstein-number, 75, 80, 86, 347
Boltzmann-constant, 203
bottleneck, 206, 436ff, 218
boundaries (closed, open), 76
boundary conditions, 76, 169, 176
branching rate, 238
Briggs-Haldane approach, 437
Briggs-Haldane equation, 208, 437
bubble column, 67
bulk, 18, 168

C
calculation methods, 349ff
calorimetry, 104
Carbon-flow-brenching concept, 37
Carman-factor, 147
cascade, 77, 113, 333, 353
cascade of reactors with cell recycling, 353
Casson equation, 386, 389
Casson viscosity, 386
catabolite repression, 145, 251, 237
catalytic constant, 234
catalytic process, 344
cell age, 246
cell age (mean), 246
cell cycle, 379
cell model, 58, 77
cell tissue culture, 67
characteristic diameter, 105
characteristic length, 75
characteristic times, 142, 393
characteristic-time concept, 141ff
chemostat, 119
circulation time distribution, 88
circulation time, 82, 86, 89, 90
closed boundaries, 76
closed environment, 261
closed environment, 261
C-mole (of biomass), 28
COD, 292
community matrix, 261
commensalism, 261, 265
compartment 279, 144, 145
compensation, 203
competition, 261, 262f
competitive inhibition, 209
completely mixed microbial film fermenter, 128
complex reaction system, 406
computer simulation, 412ff
CO_2/NaOH-method, 190
concentration polarization, 373
concept of characteristic times, 141ff
conductivity, 81, 93
conservation of mass, 22
consistency index, 387
constant of adaptation, 277
constant growth phase, 219, 244
constant-volume continuous stirred tank reactor, 119
constant volume reactor operation, 113, 119
consumption activity coefficient, 243
continuous mode of reactor operation, 10, 113
continuous recycle reactor, 123, 351
continuous plug flow reactor (CPFR), 113, 121, 337ff
continuous stirred tank reactor (CSTR), 113, 119, 308ff, 434
Contois-kinetics, 217
controlled filmthickness, 126, 138
convection coefficient, 91
convection theory, 91
conversion, 38
costs, 40
critical diameter, 147, 150, 177
critical film thickness, 147, 150, 177
critical S-concentration, 252
critical treshold concentration, 233
critical washout point, 309
cross-inhibition, 255
crowding factor, 391
CTR control, 309
cube root equation, 289
cycle time distribution, 87, 88
cyclic operation, 116, 325

D

Damkoehler-number-1st degree, 347
Damkoehler-number-2nd degree, 148, 176, 185
damped oscillations, 270
Danckwerts boundry conditions, 169
Danckwerts reactor, 129, 191
death, microbial, 227
death, phase, 225
death rate constant, 227, 228, 230
decay rate (decomposition), 244
decimal reduction time, 202
declining growth phase, 243
deduction, 10
deductive method, 50
definition of reaction rate, 23ff, 155
degradation (substrate), 221, 239, 240
degree of Hinterland, 99
degree of (micro) mixing, 81
degree of O_2-utilization, 99
degree of reductance, 32
degree of reduction, 32
degree of segregation, 72, 83, 349, 350
depth of penetration, 179
deviation variables, 275
dialysis (membrane) reactor operation, 371
dialysis reactor, 371
diauxic growth, 237, 251, 431
diauxie, 226
differential analysis, 11, 156
differential evaluation method, 154
differential reactor, 11, 152
diffuson, 22, 75, 91
diffusion regime, control, 171, 177, 185, 189
dilution rate, 119
"direct linear" plot, 167
discontinous mode of reactor operation, 113, 119
discontinuous recycle reactor, 123
discontinuous stirred tank reactor, 119
dispersion (coeff.), 75, 122
dispersion model, 74, 122
dissipation (energy), 147, 181
distributed parameter reactor, 121, 113, 151
Dixon-plot, 211
double-substrate-limitation function, 255ff
D-value, 294
D_{10} value, 202, 294
dynamic flow equilibrium, 204
dynamic method, 90, 92, 94, 102
dynamic models, 95, 272ff
dynamic process (kinetics), 274ff
dynamics of enzyme synthesis, 213

E

Eadie-Hofstee plot, 166, 174, 180
economy (of O_2), 101
economics, 38
effective hyphal length, 238
effectiveness factor, 11, 148, 170, 182, 186, 286
effectiveness, G/L, 148
effectiveness, intraparticle, 148
effectiveness, liquid/particle, 149
effectiveness, surface based, 182
effectiveness, volume based, 170, 178
effective rate, 170, 173, 177, 179, 180, 185
effectivity, 170
efficiency general, 11, 47
efficiency of motor, 100
efficiency of O_2, 101
electrode dynamics, 97
electrode response, 97
elementary analysis, 25, 382, 406
element balance, 382
element-species matrix, 407
elimination, 221, 239, 240
elutriation, 117
empiric pragmatic approach, 11, 16
endogeneous metabolism, 198, 216, 228ff
energy dissipation, 147, 181
energy efficiency coefficient, 31
energy of activation, 199
engineering morphology, 390
enhancement factor, 170, 189
enhancement of transport, 170, 188ff
enhancing substrate, 255
enthalpy, 203
enthalpy/entropy compensation, 203
entropy, 203
enzyme kinetics, 436ff
enzyme mechanisms, 436ff
enzyme reactor, 68, 311, 338, 341, 344
enzyme regulation and control, 211

equifinality, 205, 321, 434
equilibrium constant, 57, 437ff
equivalent stage, number, 77
essential substrate, 255
exluded-volume-concept, 389
expansion index, 368
exploratory research, 44
exponential growth phase, 225, 243
extended fed-batch culture, 325
extent of reaction, 406, 411
external transport limitation, 170, 171ff, 183ff
extractive fermentation, 377

F
falling film reactor, 128
falling jet reactor, 128
fast reaction, 189, 190
fed-batch, 113
fed-batch cycle, 113
fed-batch reactor operation, 114ff, 325ff
feed rate, 114
fermentation enthalpy, 249
fermenter see bioreactor
Fick's law of diffusion, 22
filamentous growth, 237
film thickness, 91
filtration time, 391
first order kinetics, 214, 216, 239, 253
fitting parameter, 60
fixed bed reactor model, 360
flow behavior index, 385
flow rate, 114
fluidization model, 368
fluidization velocity, 117
fluidized bed, 117, 128, 366
fluidized bed biofilm reactor (FBBR), 117, 366ff
food technology, 294
formal kinetic evaluation method, 60, 242
formal kinetics, 59
formal macro-approach, 46, 59
Froude number, 108
frozen system, 11, 116, 145, 272
Fujimoto-equation, 217
fungal growth, 237
F-value, 294

G
gas analysis, 90, 98
gas/liquid reactor model, 357
gas phase dynamics, 96
gas hold-up, 91
gas-in method, 91
gas-out method, 91
Gates (MarBar) plot, 163, 164
gau β-elimination method, 407
G-compartment, 145, 279
generalization, 52, 222, 233, 236, 254, 363
glucose oxidase method, 91
glycolysis oscillations, 205, 206
Gompertz equation, 289, 290
gradientfree reactor operation, 153
gradientless reactor, 153, 157
gradient-reactor, 157
graphical methods, 156
gross stoichiometric equation, 26, 382, 406
growth, 21, 23, 25, 216
growth association, 241, 242, 244
growth-enhancing substrates, 255
growth kinetics, 216ff
growth of pellets, 288

H
Haldane-relationship, 208
half-order reaction, 179, 185, 214, 287, 369
Hatta-number, 189
heat balance, 103
heat capacity, 103
heat, general, 101
heat formation, 103
heat of combustion, 250
heat of fermentation, 103, 249, 250
heat production, 103, 247ff
heat transfer coefficient, 20, 104, 105
heat transfer rate, 20, 101ff
Henry-distribution coefficient, 92
Henry-equation, 160
heterogeneous bioprocesses, 151, 168
heterogeneous kinetics, 151, 283ff
heterogeneous reactor, 9, 70, 139
heterogeneous system, 70, 139, 168
Hill-coefficient, 212
Hill-equation, 212, 217

Hill-kinetics, 212
Hill-plot, 213
Hinshelwood's network theory, 238
hinterland, 99, 189, 190, 358
homeostasis, 261
homogeneous kinetics, 151, 216
homogeneous rate equations, 151, 216ff
homogeneous reactor, 9, 70
homogeneous system, 70
horizontal position of reactor, 126
hyperbolic type, 212
hyphal growth unit, 239

I
ideal single-stage reactor, see CSTR
idiophase, 378
inactivation of spores, 294
incomplete mixing, 313
incomplete substrate penetration, 179
induction, 145, 213
induction phase, 243
inductive method, 50
ingestion, 220
inhibition, 145, 198, 209, 233, 252, 255, 339
inhibition types, 209, 210, 233, 234
inhibitor, 209
inhomogeneity, 72, 82, 83
initial rate, 208
integral analysis, 11, 154, 156
integral evaluation method, 155
integral reactor, 11, 151, 338
integrated bioreactor system, 68
integrating strategy, 41ff, 53, 381ff, 396
interactions, 43, 61, 140, 261, 385
interactive model, 256, 257
internal transport limitation, 170, 175ff, 183ff
intrinsic rejection, 373
intrinsic viscosity, 392
ionic strength, 93, 232
invariance, 215
isokinetic temperature, 203

J
jeopardy, to put a model in, 52
joint analysis, 93, 167

K
K-compartment, 145
key reaction, 279, 409
key variable, 26, 408
kinetic model, 54ff
kinetic modeling of lag-phases, 225, 430
kinetic modeling of endogenous metabolism, 227, 431
kinetic regime, control, 171, 177, 189, 190
kinetic similarity, 45
K_La-value, 94, 106
Kolmogoroff-theory, 147, 181
Konak-equation, 222
Kono approach, 219
Kono concept, 219
Kono-kinetics, 219
Kozeny-Carmen equation, 391

L
lag-phase (growth), 145, 216, 225
lag-time, 145, 214, 216, 225, 275
Langmuir-Hinshelwood approach (reversible), 208, 439
Langmuir-Hinshelwood kinetic (irreversible), 208, 438
Langmuir-kinetics, 57, 160, 165, 166, 240
Langmuir plot, 166, 181
level of sterility, 293
limitation-substrate, 16, 18
limitation-transport, 170
limit cycle (oscillation), 269
limiting viscosity number, 392
Lineweaver-Burk plot, 166, 180, 209
linear dependence, 26, 409
linear growth, 290, 376
linearization, diagram, 159, 161ff
linearly independent stoichiometric equation, 409
lineary independent equations, 409
local isotropic turbulence, 147
logarithmic mean value, 98, 103
logistic equation (low), 226
log-mean value, 98, 103
longitudinal dispersion, 122
Lotka-Volterra analysis, 268
Lotka-Volterra model, 269

Lotka-Volterra relationship, 220, 269
lumped parameter reactor, 119, 151
lysis, 312

M
macrobalances, 25, 54
macrokinetics, 45, 56, 139
macromixing, 74, 350
macroscopic principle, 18, 46, 382
macroscopic yield, 28
Maillard reaction, 295
maintenance coefficient, 229, 230, 282, 312
Malthus type, 224
mass-action law, 143, 211
mass balance, 118ff
mass flux, 22, 373
mass loading rate per unit biomass (sludge), 275
mass transfer, 20, 169
mass transport with simultaneous reaction, 169
master reaction, 11, 206, 218, 436ff
mathematical model, 49, 168
mathematical modeling, 48, 49
matrix, 407
matrix of the stoichiometri coefficients, 407
maturation time, 246, 346
maximum mixedness, 71
mean diameter, 91, 181, 182
mechanism, 55
mechanistic kinetics, 55
mechanistic model, 55, 58
medium design (optimization), 16
membrane performance, 372
membrane permeability, 372
membrane reactor, 371
metabolic process enthalpies, 249
metabolite repression, 145, 252, 237, 389
methodology, 13, 385
Michaelis-Menten approach, 437
Michaelis-Menten equation, 56, 208, 437
Michaelis-Menten kinetics, 56, 160, 437
Michaelis-Menten kinetics (reversible), 208, 439
Michaelis-Menten mechanism, 437
Michaelis-Menten model, 437
Michaelis-Menten type (irreversible), 208
microbalances, 25
microbial death phase, 225, 227
microbial interactions, 259
microbial product formation, 240
microfiltration, 371
microkinetics, 45, 55, 204, 436
micromixing, 72, 81, 350
minimal medium, 16
mini lab reactor, 16, 44
mixedness, 71
mixed plug flow, 87
mixed population, 259
mixed population kinetics, 259ff
mixed substrate kinetics, 250ff
mixing, 70, 81
mixing decay rate constant, 88
mixing models, 84
mixing number, 105, 109
mixing time, 81, 83, 89
model, 49
model building, 50, 51, 52
model discrimination, 52
model identification, 50, 52
model parameter, 50
model reactor, 84, 128, 129, 191, 130, 395
modulus, 149, 172, 176, 182, 186, 188
mole of microorganisms, 27
momentum method, 95, 96
Monod equation (kinetics), 56, 160, 166, 216, 217, 218, 224, 412
Monod kinetics with death rate, 227, 228, 431
Monod kinetics with lag time, 225, 226, 412
Monod kinetics with S-inhibition, 216, 232, 431
Monod with O_2 limitation, 256, 432
Monod type kinetics with endogeneous metabolism, 228, 431
monostability, 309
morphology, 389
morphology factor, 389, 390, 391, 392
morphology index, 389
Moser equation (kinetics), 217
multicomponent system, 294
multiinhibition model, 432
multi-loop mixing model, 85

multiple phase bioreactor model, 124
multiple steady states, 265, 267, 321
multi-phase reactor model, 124
multi-purpose reactor, 128
multiresponse analysis, 167
multistage reactor system, 329ff
multistage system, 329, 330
multistage system with recycle, 329
multistream multistage operation, 329, 334
multistream system, 329
multisubstrate kinetics, 250ff
mutualism, 261, 266f
mycelia, 237, 389
mycelial (filamentous) growth, 237, 389

N
neutralism, 261
Newtonian fluids, 385
nomenclature, 5, 19, 21
noncompetitive inhibition, 209
non-growth-linked product formation, 242, 244, 246
noninteractive model, 258
Non-Newtonian fluids, 385
non-stationary kinetics, 272ff
non-stationary reactor operation, 411
nonvanishing biomass, 320, 434
non-wash-out state, 320
normal catalytic process, 344
number of equivalent stage, 77
number of tanks in a cascade, 77
numerical fitting, 57, 201, 237

O
observed rejection, 373
"on-off" concept, 252
open boundaries, 76
open environment, 261
optimum, biological, 198
optimum conversion, 39, 339, 341, 350
ordinary sinetics, 145
oscillation, 206, 266, 269, 270, 273
OTR control, 309
OUR control, 309
overall transfer coefficient, 186
overlapping (substrate), 252
oxygen economy, 101
oxygen efficiency, 101
oxygen saturation, 92
oxygen solubility, 92
oxygen transfer number, 106, 109, 110
oxygen transfer rate, 90, 125, 144
oxygen utilization degree, 99
oxystat, 309

P
parameter, 50
parameter estimation, 50, 151ff
parameter sensitivity analysis, 51, 414ff
parasitism, 261
partial dilution rates, 335
Pected number, 75
pellet, 193, 288, 359
penetration, 221
penetration depth, 129, 179
penetration theory, 91
perfect bioreactor, 41, 44, 126ff, 191
performance, 307
performance equations, 307ff
peripheral zone model, 289
periodic equilibrium curve, 333
periodic reactor operation, 116
period of oscillation
permeability, 372, 374
permeate, 373
phased culture techniques, 379
photo bioreactor, 238
photosynthesis, 237
pH-auxostat, 309
pH-dependence (kinetics), 237
pH-stat, 309
phyto technology, 1
pilot plant (reactor), 16, 44
pilot scale, 16, 44
Planck-constant, 203
plug flow reactor, 113, 121, 337f
plug flow reactors with recycling, 354
point, 72
polarization, 373
point, 72
Powell kinetics, 217, 287
power consumption, 100
power index, 385
power law constant, 385

power law fluid, 385
power number, 105
predation, 261
predator, 268
predator-prey interactions, 268ff
preexponential factor, 199
pressure, 101
prey, 268
principle of separation, 47
process analysis, 139ff
process design, 7, 44
process kinetics, 52, 138
process kinetic analysis, 11, 42, 44, 138ff
process parameter, 51
process variable, 19, 52, 60
product formation (kinetics), 240ff, 276f, 314f
product formation activity function, 276
product inhibition, 145, 316f
product inhibition kinetics, 216, 234ff
production of secondary metabolities, 241, 244, 246, 247, 278, 388, 432
productivity, 38, 309, 353
profit, 38
proto cooperation, 266
pseudo-differential reactor, 153
pseudohomogeneity, 11, 146
pseudohomogeneous L-phase reactor model, 169
pseudohomogeneous rates, 11, 237, 240, 285, 360
pseudo-integral reactor, 152
pseudokinetics, 139, 290f
pseudokinetic parameters, 173
pseudokinetic phenomena, 173
pumping capacity, 393f, 17

Q

Q-value, 276
q_C, 24
q_H, 24
q_O, 24, 94, 139, 223
q_S, 24
quantification, 6, 18, 47, 73, 241
qss (quasi steady state), 11, 42, 120, 157
quasi-steady state reactor operation, 120, 327

R

rank of matrix, 407
rate, 20
rate coefficient, 20
rate determining step (rds), 11, 42, 218
rate of apical growth, 238
rate of branching, 238
rate of permeation, 373
"rds concept", 11, 206, 436ff, 272, 360
reaction enthalpy of fermentation, 249
reaction order, 214
reaction rate definition, 23, 24, 25
reaction time, 92, 146
reactor with gradients, 157
real plug flow reactor, 122
realstat, 119
recycle rate, 79
recycle ratio, 79, 86
recycle reactor operation, 79, 154, 351ff
reductance, 32
reductance degree, 32
reduction, 32
reference fermentation, 112
regime analysis, 141ff
regulation of enzyme amount, 145, 206ff, 211
regulation of enzyme activity, 145, 206ff, 211
Reith-approximation, 189
relative growth rate, 23, 222
relative rate, 20
relative velocity, theory, 147
relaxation time (see characteristic time), 277
relaxed steady state, 116
repeated fed-batch, 114, 325
repression, 145, 213
repression kinetics, 237, 248, 389
residence time distribution, 74, 341
response time, 95, 142
reversed-two-environment model, 84
reverse osmosis, 371
Reynolds-number, 108
rheology, 385
ribosome, 214, 252

S

salinity, 223
saturation constant, 56

saturation-type kinetics, 56, 160ff, 198, 217ff
Sauter-diameter, 91, 182
scale-down, 45
scale-up, 17, 381
scale-up concept, 381
scale-up correlation, 107ff
Schmidt-number, 108
screening, 14
second order kinetics, 214
secondary metabolite productions, 241, 345, 432
segregated kinetic model, 49
segragation, total, 72
selectivity, 38
semibatch culture, 113, 325, 410
semicontinuous mode of reactor operation, 10, 113, 410
semicontinuous stirred tank reactor, 113, 119
sensitivity analysis, 51, 53, 414ff
separation, 47
sequential S-utilization, 251ff
shear rate, 111, 385, 387
shear stress, 385
Sherwood-number, 108, 174
shift-down, 319, 322, 323
shift-up, 319, 322, 323
sigmoidal type, 212
significant variable, 18
simplification, 46
simultaneous S-utilization, 253ff
single cell model, 58
single-stream multistage operation, 331
single-stream reactor operation, 331
sloping plane bioreactor, 126, 128
slow reaction, 171, 189
sludge adsorption, 238
sludge loading, 275
sorption capacity, 240
sorption number, 106, 109, 110
specific growth rate, 23
specific heat capacity
specific rate, 20
spores, 293, 295
spouted bed, 117
spouted fluidized bed, 117
spread of distribution, 76
stability analysis, 266, 318
stable oscillations, 269, 270
stable steady states, 309, 319, 323
standard bioreactor, 127
stationary method, 98
stationary phase of growth, 225, 226
stationary state, 119, 157, 204, 207
steady state, 119, 157, 204, 207, 320, 380
steady state growth, 157, 273, 274
sterilization process, 346
sterilization kinetics, 292ff
sterilization level, 293
sterilization process, 293
sterilizer, 68
stimulation, 198
stoichiometrical coefficients, 34
stoichiometrical dependence, 26
stoichiometric line, 256
stoichiometric matrix, 407
stoichiometry, 11, 20, 25, 34, 382ff, 406ff
stoichiometry (balancing methods), 382
stoichiometry of complex reactor operation, 406ff
strain gauge, 100
strategy, 15, 396
structured cell model, 58, 145
structured kinetics models, 145, 278ff
structured modeling of bioreactors (OTR), 144, 392
structured models, 82, 87, 144
structured reactor models, 82, 87, 144, 392ff
substrate, 145
substrate inhibition, 145
substrate inhibition kinetics, 216, 232ff, 318
substrate-utilization kinetics (sequential), 251
substrate-utilization kinetics (simultaneous), 216, 253
sugar-feeding strategies, 433
sulphite oxidation, 189ff
sulphite oxidation method, 90, 189ff
surface area, 90
surface renewal rate, 91
surface renewal theory, 91
survival of the fittest model, 262
symbiosis, 261
synergism, 261
synchronous culture operation, 378ff

synchronous/synchronized pulsed and phased culture, 379
systematic approach, 5, 15, 41

T
tanks in series model, 77
Teissier kinetics, 217, 232
temperature, 198
temperature dependence, 198
test of pseudohomogeneity, 146ff
test system of biolog. test system
theory of activated complexes, 203
thermodynamic efficiency, 31, 32, 33
Thermodynamics, 25
Thiele-modulus, 149, 172, 176, 178, 182, 187, 286
Thiele-modulus based on biofilm thickness, 188
Thiele-modulus for zero-order, 179, 186
Thiele-modulus generalized, 187
Thiele modulus f. 1st order, 178, 187
thin layer (tubular film) fermenter, 126, 128, 129
three-compartment model, 145
three-constant (parameter) equation, 218, 235
threshold concentration, 233
time lags, 97
torque, 100
TOC, 292
total hyphal length, 390
total segregation, 71f
toxic power number, 236
toxicity, 198
trajectory, 264, 321, 320, 323
transient behavior of the CSTR, 157, 321
transient operation, 157
transient reactor operation, 157, 325
transient kinetics, 145, 321
transient state, 207, 321
transient phase, 243
transition line, 257
transport coefficient, overall, 185ff
transport coefficient, volumetric, 94
transport enhancement, 140, 170, 188ff
transport limitation, 140, 170
transport rate, 19
trophophase, 378
tubular reactor, 67, 69, 238
turbidistat, 309
two-compartment model, 144, 280ff, 394
two-environment model, 84
two-film theory, 91
two-phase reactor model, 357
two-region-mixing model, 84
two-stage reactor, 330
two-zone model, 84, 192
types of kinetic models, 54, 151

U
Uhlich approximation, 103
ultrafiltration, 371
unbalanced growth, 157
uncompetitive inhibition, 209
uncontrolled film thickness, 138
unified performance, 364
unsegregated model, 49
unstable state, 319
unsteady state reactor operation, 157
unstructured models, 118, 146, 216
unstructured models (kinetics), 146, 216ff, 274ff
unstructured models (reactors), 118ff

V
variable, 52, 60
variable volume reactor operation, 119, 325
variance, 72, 89
Verhulst-equation, 224
Verhulst-Pearl's equation, 224, 225
viscosity, 385
volume-based effectiveness factor, 178, 182
volumetric heat, 18, 20, 103
volume-utilization (*see* hinterland), 99

W
Walker-diagram (plot), 161, 253
wall growth, 313
wash-out, 309
wash-out point, 309

wash-out state, 309, 320
waste water plant, 2, 66
waste water treatment, 10, 67, 260
water activity, 198
whirling bed, 117
Williams model, 145, 280
working principles, 46

Y

yield, 20, 39
yield coefficient, 20, 28, 29, 35ff, 86, 111, 223, 230, 242, 312
yield constant, 20, 28, 29, 35ff, 86, 111, 223, 230, 242, 312
yield factor, 20, 28, 29, 35ff, 86, 111, 223, 230, 242, 312
yield stress, 20, 28, 29, 35ff, 86, 111, 223, 230, 242, 312

Z

Z-value, 202, 294
zero order kinetics, 214, 216, 239
Zootechnology, 1

CPSIA information can be obtained at www.ICGtesting.com
Printed in the USA
LVOW10s1625210214

374700LV00004B/80/P